Trace Elements in Human
and Animal Nutrition—Fifth Edition
Volume 2

Trace Elements in Human and Animal Nutrition—Fifth Edition

Volume 2

Edited by

WALTER MERTZ

U. S. Department of Agriculture
Agricultural Research Service
Beltsville Human Nutrition Research Center
Beltsville, Maryland

Editions 1–4 were prepared by
the late Dr. Eric J. Underwood

1986

ACADEMIC PRESS, INC.
Harcourt Brace Jovanovich, Publishers

Orlando   San Diego   New York   Austin
London   Montreal   Sydney   Tokyo   Toronto

ACADEMIC PRESS, INC.
Orlando, Florida 32887

United Kingdom Edition published by
ACADEMIC PRESS INC. (LONDON) LTD.
24–28 Oval Road, London NW1 7DX

LIBRARY OF CONGRESS CATALOGING-IN-PUBLICATION DATA

Main entry under title:

Trace elements in human and animal nutrition.

    Rev. ed. of: Trace elements in human and animal
nutrition / Eric J. Underwood . 4th ed. c1977.
    Includes bibliographies and index.
    1. Trace elements in nutrition.   2. Trace elements
in animal nutrition.   I. Mertz, Walter, Date
II. Underwood, Eric J. (Eric John), Date
Trace elements in human and animal nutrition.   [DNLM:
1. Nutrition.   2. Trace Elements   QU 130 T75825]
QP534.T723   1986   599'.013   85-20106
ISBN 0-12-491252-4 (v. 2 : alk. paper)

86 87 88 89   9 8 7 6 5 4 3 2 1
Transferred to digital printing 2005

Contents

3. Selenium

Orville A. Levander

4. Lead

John Quarterman

5. Cadmium

Krista Kostial

6. Arsenic

Manfred Anke

7. Silicon
Edith Muriel Carlisle

8. Lithium
Walter Mertz

9. Aluminum
Allen C. Alfrey

10. Other Elements: Sb, Ba, B, Br, Cs, Ge, Rb, Ag, Sr, Sn, Ti, Zr, Be, Bi, Ga, Au, In, Nb, Sc, Te, Tl, W
Forrest H. Nielsen

11. Soil–Plant–Animal and Human Interrelationships in Trace Element Nutrition

W. H. Allaway

Contributors

Numbers in parentheses indicate the pages on which the authors' contributions begin.

ALLEN C. ALFREY[1] (399), Department of Medicine, University of Colorado, Denver, Colorado 80262

W. H. ALLAWAY (465), Department of Agronomy, Cornell University, Ithaca, New York 14853

MANFRED ANKE (347), Karl-Marx-Universität Leipzig, Sektion für Tierproduktion und Veterinärmedizin, Tierernährungschemie, 6900 Jena, German Democratic Republic

EDITH MURIEL CARLISLE (373), School of Public Health, University of California, Los Angeles, California 90024

CLARE E. CASEY (1), University of Colorado Health Sciences Center, Denver, Colorado 80262

K. MICHAEL HAMBIDGE (1), University of Colorado Health Sciences Center, Denver, Colorado 80262

BASIL S. HETZEL (139), Division of Human Nutrition, CSIRO, Adelaide, South Australia 5000, Australia

KRISTA KOSTIAL (319), Institute for Medical Research, YU-41001 Zagreb, Yugoslavia

NANCY F. KREBS (1), University of Colorado Health Sciences Center, Denver, Colorado 80262

ORVILLE A. LEVANDER (209), U.S. Department of Agriculture, Agricultural Research Service, Beltsville Human Nutrition Research Center, Beltsville, Maryland 20705

GLEN F. MABERLY (139), Department of Medicine, Endocrine Unit, Westmead Hospital, Westmead, New South Wales, Australia 2145

[1]Present address: Denver Veterans Administration Medical Center, University of Colorado Medical School, Denver, Colorado 80220.

WALTER MERTZ (391), U.S. Department of Agriculture, Agricultural Research Service, Beltsville Human Nutrition Research Center, Beltsville, Maryland 20705

FORREST H. NIELSEN (415), U.S. Department of Agriculture, Agricultural Research Service, Grand Forks Human Nutrition Research Center, Grand Forks, North Dakota 58202

JOHN QUARTERMAN (281), Department of Inorganic Biochemistry, The Rowett Research Institute, Aberdeen AB2 9SB, United Kingdom

Preface

Eric J. Underwood died from a heart ailment on 18 August 1980, in Perth, Australia. During his long, productive career he saw his field of scientific interest, trace element nutrition and physiology, expand from small, isolated activities into an area of research that has gained almost universal recognition in basic and applied sciences and transcends all barriers of disciplines. He contributed to this development more than any other person of our time, through his own research, which laid the foundation for our understanding of the physiological role of cobalt, through his numerous national and international consulting activities, and through his book *Trace Elements in Human and Animal Nutrition,* which he authored in four editions. The book has assumed an eminent place among publications in trace element nutrition, not only as a source of data but because it offered mature judgment in the interpretation of the diverse results. In his preface to the fourth edition, Underwood stated that "the overall aim of the book has remained, as before, to enable those interested in human and animal nutrition to obtain a balanced and detailed appreciation of the physiological roles of the trace elements. . . ." This goal remains the dominating guideline for the fifth edition.

The invitation of the publisher to assure the continuation of the book was discussed in detail with Dr. Erica Underwood and Dr. C. F. Mills, who had been closely associated with the previous editions. Both agreed that a fifth edition was consistent with Eric Underwood's plans and should be considered. There was little question concerning the format. In our opinion, Eric Underwood was one of the few persons, perhaps the last, whose whole scientific career had paralleled the development of modern trace element research, and who had acquired a direct, comprehensive understanding of new knowledge as it emerged. On this basis he was able to evaluate with authority the progress and problems of the field. If the fifth edition was to preserve Underwood's aim of presenting a balanced account of the field rather than a mere compilation of literature, we considered the efforts of more than one author essential. We recognize that the participation of many authors may affect the cohesiveness of the revised edition, but we believe the risk is minimized by the close past association with Underwood's work and the philosophy of many of our contributors.

The major change in the format of the fifth edition is the presentation of the book in two volumes, necessitated by the rapidly increasing knowledge of metabolism, interactions, and requirements of trace elements. Even with the expansion into two volumes, the authors of the individual chapters had to exercise judgment as to the number of individual publications that could be cited and discussed. No claim can be made for a complete presentation of all trace element research, and no value judgment is implied from citing or omitting individual publications. The guiding principle was to present the minimum of results that would serve as a logical foundation for the description of the present state of knowledge. The inclusion of part of the vast amount of new data published since 1977 was possible only by condensing the discussion of earlier results. We have tried, however, not to disrupt the description of the historical development of our field.

Recent results of research were accommodated by devoting new chapters to the subjects "Methodology of Trace Element Research" and "Quality Assurance for Trace Element Analysis" and by expanding the discussion of lithium and aluminum in separate, new chapters. The first two subjects are of outstanding importance as determinants of future progress. The concern for the quality of analytical data motivated the authors of the individual chapters to review critically and, where necessary, revise analytical data presented in the previous editions. The rapid progress of trace analytical methodology since the mid-1970s has changed what had been accepted as normal for the concentrations of many trace elements in tissues and foods. The new data reflect the present state of the art in trace element analysis, but they may be subject to future revision.

The editor thanks the contributors of this fifth edition for their willingness to devote much time, effort, and judgment to the continuation of Eric Underwood's work.

WALTER MERTZ

1

Zinc

K. MICHAEL HAMBIDGE, CLARE E. CASEY,
and NANCY F. KREBS

University of Colorado Health Sciences Center
Denver, Colorado

I. INTRODUCTION

Over a century ago, Raulin (754) showed that zinc (Zn) is essential in the nutrition of *Aspergillus niger*. Not until 1926 was the essentiality of zinc for the higher forms of plant life clearly established (861,862). Before this zinc had been shown to be a constituent of hemosycotypin, the respiratory pigment of the snail *Sycotypus* (589,590).

Following the original demonstration of the occurrence of zinc in living tissues (509), many investigations revealed the regular presence of this element in plants and animals in concentrations often comparable with those of iron and usually much greater than those of most other trace elements. Particularly high concentrations were reported in serpent venom (189), in some marine organisms, notably oysters (77,828), and in the iridescent layer of the choroid of the eye (969).

The first indications of a function for zinc in higher animals came from the work of Birckner (70) in 1919. Early attempts to demonstrate such a function using semipurified diets met with limited success (63,418,572) until 1934, when Todd and associates (909) produced the first indisputable evidence that zinc is a dietary essential for the rat. Twenty years passed before Tucker and Salmon (911) made the important discovery that zinc cures and prevents parakeratosis in pigs, although Raper and Curtin (753) had previously reported that a combined supplement of cobalt and zinc prevents "dermatitis" in pigs receiving corn–

cottonseed meal rations. Subsequently O'Dell and co-workers (677,678) showed that zinc is required for growth, feathering, and skeletal development in poultry. About this time it also became evident that zinc deficiency can occur in cattle under natural conditions in some areas (215,325,326,510). From an early stage, it was apparent that zinc is of outstanding importance for growth and development. Hurley and Swenerton (424) and Warkany and Petering (953) demonstrated that this applied to prenatal as well as to postnatal development.

The first demonstration of zinc as a constituent of an enzyme, carbonic anhydrase, was by Keilin and Mann (466) in 1939. Today more than 200 zinc proteins are known and several biological roles of zinc have been clarified, including those related to cell replication and differentiation (152,927). However, much remains to be learned about changes in zinc-dependent functions resulting from zinc deficiency and about how these relate to the manifestations of zinc deficiency.

Interest in zinc in human nutrition followed the work of Pories and Strain (724) on the relationship of zinc to wound healing, the observations of Vallee (924) on alterations in zinc metabolism in alcoholic liver disease, and especially the studies of Prasad and co-workers (726,732), which provided the first evidence for the occurrence of nutritional zinc deficiency in the human. This was followed in the 1970s by considerable progress toward clarification of the practical importance of zinc in human nutrition. Highlights included the identification of acrodermatitis enteropathica as a genetic disease of zinc metabolism by Moynahan (647), which was followed by the identification of severe acquired zinc deficiency states (460) and appreciation of the multifarious clinical consequences of severe human zinc deficiency. At the other end of the spectrum, there was recognition that mild nutritional zinc deficiency syndromes occur in North America in the free-living population (362). Observations by Golden and Golden (284) on young Jamaican children recovering from protein energy malnutrition (PEM) demonstrated the clinical significance of zinc deficiency as a component of some multinutritional deficiency disorders and have given further insight into the clinical and metabolic consequences of human zinc deficiency.

While absolute deficiency of dietary zinc is uncommon in animals or humans except under experimental conditions, there is now convincing evidence that a relative deficiency of zinc in the human diet is by no means rare. The importance of food processing (818) and food selection (695) in this context have been documented. While the details remain to be clarified, the even greater importance of variations in bioavailability due to dietary factors that inhibit absorption has been established (682,800). However, further progress toward quantifying these effects of dietary and other factors will be necessary in order to delineate nutritional requirements in different circumstances and to prevent zinc deficiency.

II. ZINC IN ANIMAL TISSUES AND FLUIDS

A. General Distribution

With the exception of some specialized tissues that may contain much higher levels, the concentrations of zinc in most mammalian tissues are in the order of 10 to 100 μg/g wet weight (30–250 μg/g dry weight), with little variation among species (128,302,408,439,540,656,864,870). Typical normal levels of zinc in the principal soft tissues of several mammalian species are given in Table I. Other orders have not been investigated extensively, but levels in tissues of birds (817) are in the same range as mammals, whereas the average zinc content of finfish is lower, ~6.5 μg/g (656).

In the late 1940s, Widdowson et al. (982) measured by direct means the chemical composition of the whole body of three adult humans. They found an average zinc concentration of 28 μg/g fat-free tissue. Calculation of total-body content by factorial means, using more recent values for tissue concentrations, yielded a very similar result of 30 μg/g body weight or ~2 g total in an adult male. Kennedy et al. (472) estimated an average total-body concentration of 34 μg/g lean body mass, using compartmental analysis of zinc-65 (^{65}Zn) turnover measurements. The whole-body concentration of zinc, on a fat-free basis, is in the range of 20 to 30 μg/g in the rat, cat, pig (870), cow (619), and sheep (302), but is higher (50 μg/g) in the rabbit and mouse (983).

Table I
Typical Zinc Concentrations in Normal Tissues[a]

Tissue	Human	Monkey	Rat	Cow	Sheep
Brain	13 (58)		13		12
Heart	23 (99)	22	17 (50)		14
Kidneys	37 (163)	29	50	(84)	25 (103)
Liver	58 (199)	19	31 (113)	(111)	40 (128)
Lungs	10 (50)	19	30	(80)	16
Muscle	42 (170)	24	13 (46)	(129)	30
Pancreas	26 (89)	48	33 (145)		15 (74)
Prostate	120 (650)		223		
Spleen	14 (61)	21	23		33
Testis	13	17	22 (152)	(82)	

[a]Compiled from References 127, 244, 278, 302, 439, 441, 455, 473, 540, 617, and 817.
[b]Values for dry weight are in parentheses.

Dietary intakes of zinc are reflected in the zinc concentrations of some tissues (e.g., blood, hair, bone, testes, liver), but others (e.g., brain, lung, muscle, heart) are insensitive to marked reductions or increases in zinc intakes (46,540,751). Zinc deficiency in the young is characterized primarily by retardation of whole-body growth rather than by lower tissue zinc content. Nonetheless, a reduction in the total-carcass concentration of zinc has been reported in calves (617) and rats (441), and in newborn lambs of ewes fed a low-zinc diet during pregnancy (561). Jackson and co-workers (441) found a marked reduction in the zinc concentration in plasma, bone, and testes, and a smaller reduction in liver zinc, in growing rats fed a low-zinc diet (4 μg/g) over 80 days. Total-carcass concentration was also reduced by 30%. When animals that had received the low-zinc diet for 45 days were subsequently repleted with 30 μg zinc per gram of diet, zinc levels in liver and testes rapidly normalized, but bone, and consequently total-body concentrations were only slowly restored. In an attempt to overcome the effect of decreased food consumption, Flanagan (244) fed growing rats a zinc-deficient diet by gastric tube and was able to maintain growth and health, comparable to that in rats fed a zinc-sufficient diet, for 8 days. In the first 2–4 days, levels of zinc in plasma, pancreas, liver, and kidney dropped rapidly to about one-half to one-third the initial levels. Thereafter levels did not change. Bone zinc concentration showed a gradual decrease over the 8 days and was only slowly repleted, compared with the complete and rapid restoration of levels in other tissues when adequate zinc was given. Miller *et al.* (617) had earlier also suggested that there was an initial rapid reduction in zinc levels in some tissues when calves were fed a zinc-deficient diet. Levels then stabilized and were only further reduced in liver, kidney, and rumen with the onset of clinical symptoms. Kirchgessner and Pallauf (482) found the zinc concentrations of the liver, bones, tail, and whole body to be significantly reduced in young rats depleted of zinc over a 35-day period. In weaned rats depleted of zinc for 10 days and then fed diets ranging in zinc content from 2 to 500 μg/g for 21 days, the levels of zinc in serum and liver increased almost linearly with increasing zinc intakes up to the optimal level (12 μg/g diet). A further abrupt rise in these tissues occurred at 500 μg of zinc per gram of diet. The heart and the testes remain relatively insensitive to very high zinc intakes, but large increases in the zinc concentrations of the plasma, liver, kidney, and spleen have been demonstrated on rats fed a normal diet plus 1000 and 2000 μg of supplementary zinc per gram for 15 days (144). Comparable large increases in tissue zinc, other than in the heart and muscles, have been observed in calves at high zinc intakes (600 μg/g), following the breakdown of zinc homeostatis (873). In pigs fed an adequate corn–soybean meal diet, zinc levels in liver, kidney, heart, muscle, pancreas, and aorta were unchanged by supplemental intakes of zinc up to 500 μg/g diet. At a supplementary level of 5000 μg/g liver, kidney, and pancreas levels increased dramatically, but the zinc concentration in aorta and muscle remained unchanged (400).

B. Zinc in Visceral Organs and Brain

The liver of the adult human contains 100–250 μg zinc per gram dry weight (30–100 μg/g wet weight) (127,439,870). In most other species, reported liver concentrations are at the lower end of this range (408,656,817,864,870,983). Liver levels are affected by the level of zinc intake (45,540,751), and species differences may in part reflect different types of diets. In liver cells the zinc is present in the nuclear, mitochondrial, and supernatant fractions, with the highest levels per unit of protein in the supernatant and microsomes (48,620).

The zinc concentration of the kidney cortex in humans is 176 μg/g dry weight, about twice the level in medulla, 81 μg/g dry weight (128). Cortex zinc has been reported to be five times greater than medulla levels in the rat kidney, with the highest concentrations occurring in the cytoplasm of the cells of the juxtaglomerular apparatus (656). The zinc concentration in the spleen is comparable in most species, about 55 to 65 μg/g dry weight (127,128,864), or 15 to 30 μg/g wet weight (408,540,982). The zinc content of the pancreas is about 25 to 40 μg/g wet weight in humans (127), monkeys (540), and dolphins (408), but the level in pigs is twofold higher (178). Zinc is found in the pancreas mainly in the cytoplasm of the α- and β-cells of the islets of Langerhans, and levels in other parts of the organ are very low (656).

Zinc concentrations do not vary widely in the different regions of the brain, and are on average 13 μg/g wet weight in most species (408,656). In the rat brain, higher concentrations of zinc are associated with the mitochondrial and synaptosomal fractions than with other cell fractions (455).

C. Zinc in Muscle

Because of its bulk, a substantial proportion of the total-body zinc in large species is present in skeletal muscle: 50–60% in adult humans (656) and lactating cows (659).

The zinc concentration of muscles varies with their color and functional activity. A threefold variation in the zinc concentration of eight bovine muscles (899) and a fourfold variation in the same eight porcine muscles (131) have been reported. The mean zinc level of the longissimus dorsi, which is light-colored and shows little activity, was 69 μg/g dry, fat-free weight, whereas that of the serratus ventralis, which is a dark, highly active muscle, was 247 μg/g (131). Such differences are minimal in newborn pigs and develop with the use of the muscles, so that marked differences between red and white muscle were apparent by 8 weeks of age (132). The higher zinc content of the red muscle is situated entirely in those subcellular fractions that are composed mainly of myofibrils and nuclei. At least some of the increase in zinc may be associated with higher levels of carbonic anhydrase III, which is found in large concentrations in highly oxidative animal skeletal muscles (449). The highest zinc concentrations are found in muscles that are highly oxidative and have a large proportion of slow-twitch fibers (551). The rat soleus muscle, in which 63% of fibers are the slow-twitch oxidative type, contains 298 μg zinc per gram dry weight (441). Conversely, the extensor digitorum longus is primarily a fast-twitch glycolytic muscle and has a threefold lower zinc content (688). Heart muscle, though highly oxidative, also has a lower zinc concentration, 50 μg/g dry weight (441), associated with its high proportion of fast-twitch type fibers.

The zinc content of muscles is not generally reduced in zinc deficiency, except for a slight decrease in the soleus (172,441,449,551,688). However, the size and number of the various types of muscle fibers may be reduced and the relative distribution altered, with a reduction of the slow-twitch oxidative and an increase in fast-twitch glycolytic type (551,688).

D. Zinc in Bones and Teeth

The concentration of zinc in bone is generally quite high, between 100 and 250 μg/g fresh or dry weight (113,127,128,439,441,751). Alhava *et al.* (15) reported a slightly lower range for human cancellous bone from the anterior iliac crest, of 50 to 120 μg/g dry fat-free tissue. The concentrations of zinc in the diaphysis of the tibia and the bony portion of the mandible in rats were higher than those in the epiphysis and condyle, respectively, but all levels were still within the overall range of 100 to 200 μg/g (62). No major differences have been reported among species for bone zinc levels. Alhava *et al.* (15) found a trend of declining levels after 40 years of age in human bone, but age-related changes were not particularly strong in the few studies in which they have been examined (126,128).

Because of the high weight of the bones and teeth and their relatively high zinc content, an appreciable portion of the total-body zinc resides in these structures: ~28% in adult humans (656) and the lactating cow (659), and 22% in the grown sheep (303).

Using autoradiography, zinc can be identified in growing compact bone at the border between the uncalcified and calcifying tissue (371). This border is also characterized by a high level of alkaline phosphatase. The zinc is progressively incorporated into preosseous tissue as mineralization occurs, and it appears to remain in the fully calcified tissue from which it is removed only by bone resorption (373). A largely intracellular distribution of zinc has been argued from the lower calcium–zinc ratio in the more cellular trabecular bone compared with cortical bone (13). However, zinc is also known to be present in the inorganic matrix and is avidly sequestered by the apatite crystal even after the deposition of bone mineral (95).

In growing animals, dietary zinc deficiency reduces the zinc concentration of bone (45,244,540,751). The decline in bone zinc is more gradual than that in other tissues and may better reflect the gradual decline in overall zinc status of the body than does plasma, for example. Jackson *et al.* (442) reported that bone (femur) zinc was a good indicator of total-body zinc in growing rats fed a diet containing 4 μg zinc per gram over 45 days. Bone zinc concentration was reduced to less than half the baseline level. When animals were repleted with zinc, bone levels increased but had not completely normalized within 35 days when zinc levels in other tissues had reached predepletion values. A similar delay in repleting bone zinc after a period of depletion has been reported for dairy cows (820). Wallwork *et al.* (946) reported that dietary zinc was reflected by femur zinc in growing rats fed zinc in diets in the amounts of 6, 12, and 18 μg/g. Bone zinc concentration was highest in the 18-μg/g diet, although maximum growth was obtained at 12 μg/g. In adult male rats, very little zinc is mobilized from bone during periods of deficiency (946).

The range of zinc concentrations in 102 samples of dental enamel collected from different parts of Finland was 122–538 μg/g, with a slight correlation with environmental levels (505). Dentine contained 100–320 μg/g, and increased with age up to 25 years. Zinc levels of >1000 μg/g have been reported in normal samples of tooth enamel (671). Zinc has been shown to be incorporated into the apatite of tooth enamel during crystal formation and later by exchange with calcium on the surface of existing crystals (95).

E. Zinc in the Integument

The mean zinc concentrations of normal human epidermis and dermis are about 80 and 12 μg/g dry weight, respectively, with an average content of about 20 μg/g dry weight in whole skin (352,439,441,630, 656). In humans and cattle about 2 to 3% of the total-body zinc is present in skin and hair (656,659), but in smaller animals covered with hair or wool a much higher proportion is present in these tissues. In the rat and hedgehog, no less than 38% of the whole-body zinc is situated in the skin and hair, or skin and bristles (870). The zinc content of epidermis is not decreased in nutritional zinc deficiency, even when taken from the site of skin lesions (441,659). Hambidge *et al.* (352) reported that the zinc concentration was reduced in the whole skin of a patient with acrodermatitis enteropathica when off zinc therapy, but returned to normal when therapy was reinstituted.

McKenzie (576) found a mean level of 118 μg/g in toenail clippings from a large group of healthy New Zealanders, but concentrations did not correlate with zinc in hair or serum. The average zinc content of nails from Tokelau islanders was found to be higher (150 μg/g) despite a dietary intake of zinc about one-half that of the New Zealanders (575). The level of zinc in fingernails of 18 normal human subjects was reported by Smith (840) to range from 93 to 292 μg/g and to average 151 μg/g. The zinc content of bovine hoof has been reported as 24 to 37 μg/g for soft horn and 50 to 70 μg/g for hard horn (550).

Head hair has been frequently analyzed in humans; concentrations in healthy adults generally range from about 100 to 300 μg/g, with a mean of ~180 μg/g (188,272,342,360,576). The average hair zinc concentrations of adult females have been reported to be ~20% higher than those of adult males (272). Hambidge and co-workers (342) obtained the following mean zinc concentrations (in micrograms per gram) for the hair of 338 apparently normal subjects living in Denver with ages ranging from 0 to 40 years: neonates ($n = 25$), 174 ± 8; 3 months to 4 years ($n = 93$), 88 ± 5; 4–17 years ($n = 132$), 153 ± 5; and 17–40 years ($n = 88$), 180 ± 4. Lower zinc levels in hair of younger children compared with those older than 5 years have also been reported by Amadar and colleagues in Cuba (20) and Erten and colleagues in Turkey (221). However, although relatively low hair zinc concentrations have been observed in young children and especially in infants by several investigators, this has not been a consistent finding and the variations are probably not directly related to age (344).

Mild zinc deficiency may cause a marked reduction in hair zinc concentrations (342), but in severe zinc deficiency states, such as acrodermatitis enteropathica, hair levels may be normal or only mildly depressed (352), because of a reduction in hair growth. The mean zinc level in the hair of normal Egyptians was 103.3 ± 4.4 μg/g and of untreated, zinc-deficient dwarfs only 54.1 ± 5.5 μg/g; oral zinc sulfate therapy increased hair zinc levels to 121.1 ± 4.8 μg/g. The hair of Iranian villagers, in a region where zinc deficiency had been suspected, averaged 127 μg/g, compared with 220 μg/g for controls whose diet was diversified and abundant (761). Cheek and colleagues (142) found that mean hair zinc in a group of 4- to 14-year-old aboriginal children in northwestern Australia was only 105 ± 2 μg/g.

The level of zinc in hair reflects dietary zinc intakes in rats (762), although Pallauf and Kirchgessner (702) found that the level only rose with successive additions of dietary zinc when the zinc intake was adequate for hair growth. With levels of 2 to 8 μg/g of dietary zinc, stepwise additions of zinc resulted in an actual reduction in the level of zinc in the hair, because of the large increase in body weight and mass of hair induced by the supplementary zinc. Hair zinc concentrations have also been reported to reflect dietary zinc intakes in pigs, cattle, goats, and monkeys (516,611,614,898). Individual variability is very high, and there is some variation with age and part of the body (611). O'Mary *et al.* (690) found that the zinc level in Hereford cattle hair did not vary with color or season. Most of their values fell within the range of 115 to 135 μg/g. Hidiroglou and Spurr (396) observed lower levels in the winter hair than in the summer hair of cattle, probably due to the lower zinc levels in the winter herbage. Hair color in humans has also been reported not to affect zinc concentrations (201).

Masters and Somers (559) reported a range of 84 to 142 µg/g for the zinc content of wool from a large series of rams and ewes in South Australia. They found that mean zinc levels in the wool from ewes were lower in the winter and spring months than in the summer. Grace (303) has reported a higher mean level of 220 µg/g in wool from young sheep grazing on ryegrass–white clover pasture in New Zealand.

F. Zinc in the Eye

The highest zinc concentrations known to exist in living tissues occur in the choroid of the eye of some species. The zinc levels in the choroids of the dog, fox, and marten have been given as 14,600, 69,000, and 91,000 µg/g dry matter, respectively (968). In the iridescent layer (tapetum lucidum) of the choroid of the dog and fox, levels of up to 8.5% (85,000 µg/g) and 13.8% (138,000 µg/g) zinc, respectively, have been reported (969). Lower levels occur in the choroid of humans (274 µg/g) and cattle (284 µg/g). Zinc concentrations also vary widely in other eye tissues according to species and may be particularly high in fishes. Eckhert (216) found that zinc concentrations in retina ranged from 3.8 µg/g dry weight in the human, up to 5000 µg/g in red snapper. The function of zinc in such high concentrations in different eyes is unknown, but may be related in part to carbonic anhydrase or retinal reductase activity (692).

G. Zinc in Male Sex Organs and Secretions

The zinc concentration of the testes of young rats was found to remain fairly constant at about 120 µg/g dry weight for the first 30–35 days; after that it increased to 200 µg/g during the second month of life, when spermatids are transformed into spermatozoa, then fell again to the adult level of 176 µg/g (705, 733). In zinc-deficient rats the testes became depleted of zinc with levels falling from 152 µg/g dry weight to 71 µg/g, but zinc content was rapidly restored when the animals were given a zinc-adequate diet (441). A reduction in testicular zinc has also been reported in severely deficient rams in which levels were 74.0 ± 5.0 µg/g dry weight, compared with 105 ± 4.4 µg/g in the healthy animals (920). Concentrations of 13 to 30 µg/g wet weight (50–125 µg/g dry weight) have been reported for zinc in the human testis (366,439,552). Jackson et al. (441) found a lower level, of 32 µg/g dry weight, in the testes of an 11-year-old boy in whom plasma and bone zinc were also below the normal range; it is not known if there are any changes in zinc in the human testes with age.

High concentrations of zinc have been reported in the prostate gland of rat, rabbit, sheep, monkey, and human (49,110,324,327,397,439). The caudal lobe of the prostate of a rhesus monkey was found to contain 650 µg zinc per gram wet weight, with 40 µg/g in the cranial lobe (49). In rhesus and cynomolgus monkeys, high levels of zinc occurred in the nuclei of the epithelial and basal cells and in secretory granules along the lateral cell membranes in both lobes of the prostate. The zinc content and subcellular distribution were similar in immature monkeys. In castrated animals, however, distribution remained the same but the total zinc content of the cranial lobe was reduced. This is opposite to the pattern in the baboon, in which castration caused a marked reduction in zinc uptake by the caudal lobe (815). The zinc content of the dorsolateral prostate of the rat, usually about 850 μg/g dry weight, is also reduced by castration. (318). The rate of zinc accumulation is greatly increased in young rats by testosterone or gonadotrophin injections (599). Chorionic gonadotrophin increases the weight and ^{65}Zn uptake in all the other accessory reproductive structures in male rats, as well as the dorsolateral prostate, whereas follicle-stimulating hormone (FSH) reduces this uptake per organ and gram of tissue (776). The normal human prostate contains 80–200 μg/g zinc per gram on a wet-weight basis (324). Habib (327) found an even wider range, 250–1200 μg/g dry weight, in prostatic tissue from patients with benign prostatic hypertrophy. Lower levels may occur in cancerous prostatic tissue (324).

Whole human semen from 456 men was found to contain (mean ± SD) 150 ± 85 μg zinc per milliliter (553). Lower concentrations have been reported in whole semen from boars (range 7.8–78 μg/ml) and rams and bulls (~10 μg/ml) (972,973). A wide range of zinc concentrations has also been reported in seminal plasma from men: 60–190 μg/ml, with no clear differences between fertile and infertile men (447,876). Levels of zinc in seminal plasma and washed spermatozoa of dogs have been reported as 1750 μg/g dry weight and 1040 μg/g dry weight, respectively (785). Zinc in rat sperm is of the same order, 890 μg/g dry weight (785), and is concentrated mainly in the tail (600). The origin of the zinc in the seminal plasma varies: in the cat, dog, and man it is contributed by the prostatic secretions, but in the boar it is secreted by the seminal vesicles (973). Human expressed prostatic fluid is rich in zinc; reported levels are usually between 300 and 500 μg/ml (553). The zinc in the prostatic secretions is initially associated with low molecular weight ligands, including citrate, from which it may readily be taken up by the spermatozoa. Addition of seminal plasma introduces higher molecular weight species, mainly proteins, to which zinc is redistributed during the coagulation and liquefaction phases (38). High molecular weight bound zinc is nondiffusible and appears to be unavailable to the spermatozoa. Uptake of zinc by the sperm after leaving the testes is necessary for viability and motility in some species (784,973). In the dog, the zinc concentration in ejaculated sperm is approximately fivefold higher than that collected from the vas (785). The zinc level in the ejaculated semen was related to both the concentration of spermatozoa and sperm motility (784). A relationship between sperm density and zinc content of seminal plasma was found in fertile but not infertile men, although concentrations of zinc are not generally different (447, 876). Rat sperm already have a high zinc content when collected from the epididymis, and prostatic fluid is not required for fertility in this species (785).

H. Zinc in Female Sex Organs and Products of Conception

In general, concentrations of zinc in the female sex organs and secretions are within the lower end of the range found in other body tissues. In the nonpregnant woman, levels of zinc in the uterus, ovaries, and cervix are between 11 and 30 µg/g wet weight (366,439,552). Human endometrial tissue contains about 12 to 25 µg/g wet weight, and cervical mucus, 0.5–2 µg/g wet weight. There were significant variations within these ranges during the menstrual cycle, with lowest levels in the late proliferative phase and highest levels just previous to the onset of blood flow (329). In the pregnant rat, uterine and placental zinc concentrations are similar to those in human tissues. Masters et al. (560) reported concentrations of 16.7 ± 0.41 µg/g wet weight (mean $\pm$ SEM) and 12.2 ± 0.31 µg/g, respectively. Placental but not uterine zinc levels were reduced by maternal zinc deficiency. Both placental zinc content and uptake of ^{65}Zn were reduced in pregnant rats treated with a diet containing 5% ethanol (275). The zinc concentration of human amniotic fluid has generally been reported to be <0.2 µg/ml (22,834). Levels do not appear to change with duration of gestation until near term; Kynast et al. (501) reported a threefold increase from 36 to 42 weeks. These workers also reported that this increase in median zinc levels was reduced in amniotic fluid in cases of toxemia and maternal diabetes and when the infant was large or small for gestational age (502). Conversely, neither Anastasiadis et al. (22) nor Shearer et al. (834) were able to find a relationship between fetal anomalies and the level of zinc in amniotic fluid at delivery, or at 13 to 19 weeks' gestation, respectively. Uterine fluid collected at the end of the first week of pregnancy, from rats fed a zinc-adequate diet, contained 7.6 µg zinc/g, which was reduced to 3.9 µg/g when a zinc-deficient diet was fed (269).

I. Zinc in the Fetus

Widdowson and Spray (981) found a total-body concentration of zinc in the human fetus of 20 µg/g fat-free tissue, with no change in concentration with age from 17 to 40 weeks' gestation. Their data were obtained by direct carcass analysis of a total of 13 fetuses. Shaw (833) later used these data to calculate the rate of accumulation of zinc in the fetal body during the third trimester, and Table II presents results for the accretion rates in fetuses growing along the tenth, fiftieth, and ninetieth percentiles. Casey (122) analyzed zinc in individual tissues

Table II

Rate of Accumulation of Zinc *in Utero* by a Human
Fetus Growing along the Tenth, Fiftieth, and
Ninetieth Percentiles[a]

Gestation (weeks)	Zinc accumulation[b] (μg/day)		
	Tenth	Fiftieth	Ninetieth
24	143	209	266
26	185	238	287
28	232	287	327
30	286	348	388
32	348	427	481
34	432	548	611
36	553	675	—

[a]From Reference 833.
[b]Equivalent to an accumulation rate of 270 μg/kg per day
(tenth), 249 μg/kg per day (fiftieth), and 211 μg/kg per day
(ninetieth) at all ages.

collected from 39 fetuses of 22 to 43 weeks' gestation and used the results to
derive an estimate of the zinc content of the whole fetal body. She calculated a
higher total-body concentration of 38 μg/g body weight at all ages.

An overall range of 100 to 300 μg/g wet weight has been reported for zinc in
livers of human fetuses from 20 weeks to term (123,585,984). There is a small
decline with gestational age, but even at term levels are usually higher than
normal adult concentrations. At term, ~20% of the total-body zinc is in the liver
(122,985), compared with only 2% in the adult (656). Concentrations of zinc in
kidney, brain, heart, lung, and bone are similar to those in adults and do not
change with gestational age (123). Levels in skeletal muscle were found to
increase with gestational age from a mean of 110 μg/g dry weight at 22 weeks to
adult levels (160 μg/g) at term (123).

For nonhuman species, very few analyses have been reported for whole-body
or tissue zinc content prior to term. In the rat, the total-body concentration (per
unit dry weight) increased by ~50% from 14 to 21 days (428). The concentration
at term is about 20 to 30 μg/g body weight (275,770,870,871). Concentrations
in the rabbit, guinea pig, and cat are similar, but the whole-body content is
somewhat lower in the pig: 10 μg/g fat-free tissue (870). In the pig, cat, rabbit,
and rat, the concentration of zinc in the liver is higher in the term fetus than the
adult animal, with levels in kitten liver being threefold elevated (983). The total-
body concentration of zinc in the sheep at term is ~20 μg/g, with an accretion
rate of 1 to 2 mg/day (989). The level in the liver is similar to that in the human,
but kidney is generally lower, 58 μg/g dry weight (561). The concentration in

the pancreas (209 μg/g dry weight) is much higher than that in the adult human, but similar to the level in the newborn rat (770).

As in the adult, much of the zinc in the fetal liver is present in the cytosol. In human fetal liver, 85% of the total zinc is in the soluble fraction (768); the proportion is slightly lower (65%) in the sheep (807), and is 40–50% in the rat (48). In both rat and human fetal liver, part of the cytosolic zinc is known to be present in a zinc-rich metallothionein, which contains 3–5 gram atoms of zinc and 1–2 gram atoms of copper per mole of protein, similar to the metallothionein found in adult liver (768).

When the zinc intake of the pregnant sheep or rat is restricted, the total-carcass content of zinc in the term fetus is reduced, as are levels in the liver (561) and pancreas (561,770).

J. Zinc in Tissues during Postnatal Growth

In general, the total-body concentration of zinc in young animals is similar to that of the adult of the species (302,983). Lambs and calves contained a range of 20 to 26 μg/g fresh weight with a slight tendency to increase with age (891). Grace (303) found that young sheep accumulate 24 μg zinc for each gram gain of fleece-free, empty body weight, whether fed fresh herbage or a semipurified diet. Spray and Widdowson (870) reported little change in total-carcass concentrations of zinc from birth to adulthood in the cat and pig. In rats and rabbits there was an increase during suckling, possibly due to increased zinc retention in growing hair, with a slight decrease thereafter to adult levels.

In the liver of the rat, rabbit, cat, and pig, levels of zinc at birth are three- to fivefold higher than in the adult and fall rapidly during the first 3–4 postnatal months (983). In humans, liver zinc concentrations at term are 100–200 μg/g wet weight and decrease slightly during the first year of life, with little change thereafter, to the adult levels of 50 to 100 μg/g wet weight (126,127,716). Levels of zinc in human kidney rise during childhood and are generally higher in the adult (127,716). Schroeder et al. (817) have reported a large decrease in the zinc concentrations of kidney, liver, aorta, heart, and spleen during the first year after birth in humans, with slight increases or little change during the subsequent growth years. Concentration measurements in this study were reported in terms of ash weight and are thus difficult to compare with other studies, as small changes in ash content with age will result in a large proportional change in zinc concentration.

K. Zinc in Blood

1. Distribution

Less than 0.5% of the total zinc content of the body is found in the blood of adult humans (656) or sheep (302). Of the total zinc in blood, 75–88% is in the red cells, 12–22% is in the plasma, and ~3% is in the leukocytes and platelets. In human serum, zinc is distributed in three pools: ~18% is tightly bound to α_2-macroglobulin, 80% is more loosely associated with albumin, and the remaining 2% is bound to other proteins such as transferrin and ceruloplasmin and to amino acids, principally histidine and cysteine, with a small proportion of free zinc (234,250,279). A similar distribution of zinc was found in plasma from sheep, rats, red deer, and cattle, but in pigs about one-third is bound to a globinlike protein of molecular weight 100,000–140,000, with a corresponding reduction in the amount bound to albumin (150). Pregnancy, stress, and disease may alter the distribution of plasma zinc (232,809,1002). Much of the zinc in the erythrocytes is accounted for as carbonic anhydrase (412), but the red cell membrane contains about 60 μg zinc per gram of protein, of which 54% is bound to the lipid phase (157).

2. Normal Levels

The concentration of zinc in whole human blood is 4–8 μg/ml (52,187,403). Packed erythrocytes contain 8–14 μg/g wet weight (52, 937), equivalent to 40 μg/g hemoglobin (980) or 10–11 μg per 10^{10} cells (403). Levels in red blood cells were not affected by short-term (up to 63 days) zinc deprivation (52). In newborn infants the zinc content of the erythrocytes is only one-quarter of the adult value, rising progressively over the first 12 years of life (61,644). This is a reflection of the low levels of carbonic anhydrase present in the red cells of the neonate (654). Erythrocyte zinc levels in other species are within the range found in humans (61,302,408,838), except in the goose and in the growing calf, for which values of 6.5 μg/g (838) and 5.9 μg/g red cells (873), respectively, have been reported.

Individual white cells contain on average 5–10 times as much zinc as individual red cells (403). Values reported for the content of zinc in human leukocytes have varied according to the cellular composition of the analyzed sample and other factors. Careful separation techniques are required to obtain relatively pure cell populations uncontaminated by zinc from other sources. Hinks *et al.* (402) found a mean (±SD) level of 76 ± 20 μg per 10^{10} cells in a mixed-leukocyte population, lower than some earlier reports that generally give a range of 100 to 200 μg per 10^{10} cells (192,669). Whitehouse *et al.* (980) reported the following zinc levels (mean ± SD) in white cells: 104 ± 12.5 μg per 10^{10} cells in neutrophils (98% pure fraction), and 115 ± 14.5 μg per 10^{10} cells in lymphocytes (50–85% pure fraction). Nishi *et al.* (669) found a slightly higher zinc content in T lymphocytes (144 μg per 10^{10} cells) compared with non-T lymphocytes (126 μg per 10^{10} cells). Monocytes contained higher levels (337 μg per 10^{10} cells) than other types of white cells (669). Purcell *et al.* (745) found a mean

(±SD) zinc concentration of 49 ± 13 μg per 10^{10} cells in neutrophils (98% pure fraction) and 109 ± 26 μg per 10^{10} cells in the mononuclear leukocyte layer (97% monocytes). Milne and co-workers (628) reported very similar levels in monocytes (103 ± 20 μg per 10^{10} cells), and a slightly higher mean in polymorphs (63.3 ± 13.0 μg per 10^{10} cells). Cheek et al. (143) reported reduced levels of zinc in lymphocytes and neutrophils in aborigine boys in conjunction with reduced zinc concentrations in plasma and hair. Neutrophil zinc was also reported to be lower in patients with sickle cell anemia who had other evidence of poor zinc status, and concentrations increased on zinc supplementation (740). Milne et al. (627) found that zinc levels were not reduced in lymphocytes from the spleen or lymph nodes of zinc-deficient rats compared with control animals, but levels were slightly reduced in blood lymphocytes. The zinc content of mixed leukocytes was not reduced in pigs fed a zinc-deficient diet (172).

Purified human blood platelets were found to contain 49.2 ± 11.0 μg/g wet weight (mean ± SD) of zinc, or about 5 fg zinc per platelet (475).

The normal concentration of zinc in serum or plasma of adult humans appears to lie between 0.7 and 1.0 μg/ml. Different laboratories report different values for their normal mean level; such variations appear to rise largely from methodological considerations (187,354,478,576,740,825,938,980,1002). In the past, the zinc content of human serum has been reported as ~16% higher than the level in plasma, probably from hemolysis and disintegration of platelets during clotting. Later studies using careful collection techniques found no differences between serum and plasma zinc levels (830). Although significant differences between the sexes have been reported (374,478,576), this is not a consistent finding (403,669). Normal plasma levels in other species include the following: 1.32 μg/ml in guinea pigs (751), 1.5 μg/ml in rats (441), 0.7–0.9 μg/ml in mice (242), 1.4 μg/ml in pigs (172,504), 0.7–1.0 μg/ml in sheep (302,559), 0.9 μg/ml in beagle dogs (465), and 1.0–1.2 μg/ml in squirrel and rhesus monkeys (291,540).

Diurnal variations have been reported in the zinc content of serum and plasma in humans, pigs, and lactating cows; much of the circadian rhythm appears to be due to ingestion of food. In fasting men, the minimum value was observed at 1500 hr with maxima at 0900 and 1800 (317). In contrast, in men who took a standard liquid diet at 4-hr intervals from 0800 hr, the serum zinc levels remained above the mean until 2400 and reached a minimum at ~0600, but levels were never >10% from the overall mean (518). When subjects are permitted to follow their ordinary meal habits, there appears to be a slight gradual decline throughout the waking day from fasting early-morning values (187,395,478,580). An immediate postprandial drop in plasma zinc has been observed by some workers (478,764) but not others (565), and appears to be related to the composition and possibly bulk of the meal (643). A lowering of plasma zinc of up to 35%

of the fasting level has been measured in fed pigs (504). Lactating cows were also found to have lower plasma zinc levels when feeding and when resting during the daylight hours compared with milking times and night (479).

Keen *et al.* (465), in California, found a variation due to season in serum zinc of beagles, with higher levels being associated with higher ambient temperatures. Masters and Somers (559), in Australia, reported that plasma zinc was also lowest in sheep in the autumn and winter months compared with levels in warmer months. However, in this study, the lower blood concentrations could be associated with lower levels and lower bioavailability of zinc in pasture during March through May, and with pregnancy during June through August, when pasture zinc levels were highest and ewes, but not rams, had low blood levels.

3. Effect of Age

In the human, levels of zinc in plasma are higher in the fetus and premature infant than later in life, and decline with gestational age (61,522,985). Sann *et al.* (803) found a strong negative correlation between gestational age at birth and serum zinc concentrations at 7 days postnatal age in 88 healthy infants born at 27 to 42 weeks. Although cord levels of zinc are higher in preterm than term infants at birth, both groups are within the same range as healthy adults, and newborn levels are significantly higher than maternal plasma levels (207,382,644,833). After birth, levels of zinc in both term and preterm infants have been reported to decline in some populations (382,522,577,687,833,916,950) but not others (202,459), and this phenomenon appears to be largely related to the type of diet and intake of zinc (169,522). Hambidge *et al.* (354) found that breast-fed infants maintained adult plasma zinc levels at 6 months of age, whereas levels in infants fed a cow's milk-based formula were significantly lower. Craig *et al.* (169) found levels were even further reduced in infants fed a soy-based formula providing a zinc intake similar to that of the cow's milk-based formulas. After 6 to 8 months of age, by which time mixed feeding has generally been instituted, plasma zinc concentrations are again similar to those in healthy adults and remain at that level throughout childhood (120,143,347,382,833,950). Plasma zinc levels have been reported to decline with age in adults (520), but this is not a consistent finding (374,940).

Plasma levels of zinc in newborn sheep (807) and monkeys (290) are similar to or slightly higher than values in the nonpregnant adult of the species and, as in the human, are significantly higher than maternal levels at term. Keen *et al.* (463) reported the serum zinc was lower in young pups than adult dogs, and levels showed a slight decline with age in older animals.

4. Effect of Zinc Intake

Most workers have observed a decline in plasma or serum zinc in zinc-deficient animals. Mills *et al.* (622) reported a decline from normal levels of 0.8

to 1.2 μg/ml to <0.4 μg/ml in the serum of severely zinc-deficient lambs and calves. In growing pigs with clinical manifestations of zinc deficiency, plasma zinc was reduced to 0.32 μg/ml compared with 1.27 μg/ml in pair-fed, zinc-adequate control animals, but leukocyte zinc levels were not reduced (172). Flanagan (244) found that in rats tube-fed a diet providing 0.4 μg zinc per gram, plasma levels fell from 2.2 μg/ml on day 1 to ~0.8 μg/ml on day 3 and thereafter remained constant through day 8. Pallauf and Kirchgessner (701) found similar degrees of decline over 4 days in rats consuming a diet containing 1.8 μg zinc per gram. In both of these studies (244,701), addition of zinc to the diet rapidly restored the serum zinc level to normal. Plasma zinc also falls in less severely zinc-deficient calves, goats (172), baby pigs (602), and squirrel monkeys (540). In an experiment lasting 45 days, Jackson *et al.* (441) found that rats fed a higher but still inadequate level of zinc intake (4 μg/g diet) also showed a fall of plasma zinc, from 1.52 to 0.8 μg/ml, a level similar to that seen in the shorter term studies. A fall in plasma zinc of guinea pigs from a normal level of 1.04 to 0.45 μg/ml was associated with an intake of zinc (9.85 μg/g diet) low enough to affect feeding behavior but not to cause other signs of zinc deficiency (750).

Filteau and Woodward (242) demonstrated that serum zinc levels of mice were also influenced by the level of dietary protein. Low serum zinc was associated with low total serum protein levels in the mice, and with reduced levels of serum albumin in pigs and monkeys (540,602), Low plasma zinc has also been reported in PEM in children, partly in conjunction with hypoalbuminemia (118, 282). However, the low zinc levels may persist after albumin values return to normal if dietary zinc is inadequate.

In humans, plasma zinc concentrations may not be depressed in mild zinc deficiency, but moderate zinc deficiency is usually associated with plasma zinc levels between 0.4 and 0.6 μg/ml (352). Concentrations <0.2 μg/ml have been reported in severe, acute zinc deficiency (460), and untreated acrodermatitis enteropathica (352). Experimental zinc depletion in human subjects (adult males) resulted in a fall in plasma zinc from 0.89 to 0.50 μg/ml over 4 to 9 weeks, when zinc intakes were reduced from 10.6 to 0.28 mg/day (52). Lukaski *et al.* (536) reported that plasma zinc levels declined gradually in healthy men during the first 60 days on a diet providing 3.6 mg zinc/day, and remained constant for the next 60 days. Repletion with an intake of 33.6 mg zinc/day caused a rapid increase in plasma levels to the starting value within the next 30 days, but plasma levels remained within the laboratory normal range at all times. Prasad *et al.* (738) found that the zinc content of plasma, erythrocytes, and leukocytes was depressed in four male volunteers receiving a semipurified diet supplying 2.7–3.5 mg zinc per day for 6 to 12 months.

Large oral doses of zinc increase whole-blood and plasma zinc concentrations in rats, rabbits, cats, pigs, sheep, and cattle (61,405,699). In pigs given supplemental zinc up to 500 μg/g diet, serum zinc was not elevated, but addition of 5000 μg/g diet resulted in a two- to threefold increase over 12 weeks, which declined back to presupplemental levels by 20 weeks (398). Cattle provided with dietary zinc intakes from 18 to 189 μg/g diet showed mean serum zinc increases from 1.5 to 2.7 μg/ml. Further massive increases in zinc intake, from 372 to 11,279 μg/g diet, caused an increase in plasma zinc levels from 3.2 to 7.5 μg/ml after 6 weeks (610, 715).

5. Effect of Female Reproductive Cycle

Plasma zinc levels of women fall during pregnancy (207,422,444,644). The decline starts very early in pregnancy: Breskin *et al.* (86) found that the mean (±SD) serum zinc concentration (0.68 ± 0.14 μg/ml) was already significantly reduced from the prepregnancy mean of 0.97 ± 0.22 μg/ml by 21 days in a group of 106 women in whom the time of conception was precisely determined from basal body temperature monitoring and chorionic gonadotrophin levels. Hambidge *et al.* (360) found that the mean level in a group of well-nourished women was significantly below that in nonpregnant controls by 2 months of gestation and continued to decline to term. These workers also found that the decline could not be prevented by a supplemental intake of zinc of 15 mg/day. The mean plasma level at term in these women was 0.57 ± 0.11 μg/ml (mean ± SEM), compared with their nonpregnant control value of 0.86 ± 0.10 μg/ml. Somewhat higher mean levels of serum zinc were reported at 7 to 8 months in low-income Mexican-American mothers, with no difference between those taking a supplement providing 21.8 mg zinc/day (mean ± SD: 0.65 ± 0.10 μg/ml) (421). However, there were significantly fewer women with levels <0.53 μg/ml in the supplemented group. Zimmerman *et al.* (1002) reported that the decline in total serum zinc was accompanied by a decrease of 26% in the amount bound to albumin and an increase of 34% in the amount bound to α_2-macroglobulin from the first to the third trimester. Giroux *et al.* (280) found that the decreased proportion of serum zinc bound to albumin in the third trimester of pregnancy was due in part not only to the lowered concentration of serum albumin but also to a decrease in the affinity of albumin for zinc binding.

Krebs *et al.* (495) reported that in lactating women, plasma zinc increased after delivery to 0.70 ± 0.11 μg/ml (mean ± SEM) at 1 month, significantly higher than the mean at 36 weeks' gestation of 0.54 ± 0.09 μg/ml. In this group, concentrations continued to rise to a maximum of 0.79 ± 0.10 μg/ml at 4 months postpartum but were still significantly lower than in nonpregnant, non-lactating control women. Moser and Reynolds (644) also found that plasma zinc had increased significantly by 1 month postpartum to 0.79 μg/ml in lactating women and 0.83 μg/ml in women who did not breast feed. Plasma levels continued to increase in the lactating women, to 0.88 μg/ml at 3 months postpartum.

A fall in plasma zinc levels during pregnancy has also been observed in the rhesus monkey, but in contrast to women, did not commence until after 45 days' gestation (291). Plasma zinc levels do not change markedly during pregnancy in well-nourished cattle and sheep (561), except that there may be a fall during and immediately after parturition.

Concentrations of zinc in plasma were found not to change significantly during the menstrual cycle in women (329).

6. Effects of Stress and Disease

Various types of stress cause a lowering of blood zinc levels. Administration of endotoxin caused a decline in serum zinc in rats (710) and pigs (152). In the latter study, endotoxin injection caused a decrease in zinc flux through the plasma in zinc-sufficient animals, but the changes were not observed in animals in which plasma zinc was already low due to nutritional zinc deficiency. Both hyperthermal and hypothermal stress have been reported to lower serum zinc levels (685,961). Conversely, maximal exercise caused a transient increase in plasma zinc levels, not entirely accounted for by volume changes in healthy men (536). The administration of large doses of corticosteroids to patients with burns and surgical stress produced a rapid and sustained depression in serum zinc (245). Wegner and co-workers (961) were unable to demonstrate consistent trends linking plasma zinc concentration directly to the adrenal response or stressful conditions in dairy cattle. Possible mechanisms for the effect of stress on plasma zinc are discussed in Section III,B.

Lower than normal plasma zinc levels have been reported in a large number of disease states, including cancers, especially squamous cell carcinomas, atherosclerosis, postalcoholic cirrhosis and other liver diseases, sickle cell anemia, chronic and acute infections, and after acute tissue injury regardless of origin (341,587,740,809,889,936). In conditions such as phenylketonuria, celiac disease, and malabsorption syndromes, decreased plasma zinc levels (7,526,658,793) may be due mainly to reduced intake or absorption of zinc rather than to the disease per se.

7. Effect of Hormones

A number of hormones have been shown to alter zinc metabolism, including adrenal cortical steroids, thyroid hormone, growth hormone, and sex steroids (58,656). Glucocorticoids cause a depression in plasma zinc associated with a concomitant uptake of zinc by the liver and other tissues (225). The lowering effect of pregnancy on plasma zinc levels is probably hormone mediated, but the mechanism is currently unknown. Earlier findings of a decrease in plasma zinc in women taking oral contraceptives (337) have not occurred in subsequent studies (345,403); differences may be due to the changes in hormonal composition of the later preparations. Sato and Henkin (805) reported that changes in plasma zinc in female rats during the estrous cycle and pregnancy were not consistently related to changes in levels of estrogen or progesterone. Plasma zinc remained unchanged after administration of estradiol-17β to ovariectomized rats, but was significantly decreased when high levels (50 mg/kg) of progesterone were administered, or by a combination of more moderate doses of both hormones together. Changes in zinc metabolism have been described in some patients with abnormalities of adrenocorticosteroid metabolism (378). Elevated serum zinc has been reported in patients with untreated adrenal cortical insufficiency and panhypopituitarism, with normalization of zinc levels after hormone replacement therapy (384). Conversely, in Cushing's syndrome serum zinc levels were depressed; treatment also resulted in a return toward normal zinc concentrations. Significant increases in serum zinc were also observed in adrenalectomized and hypophysectomized cats (384). Falchuk (232) found that an infusion of adrenocorticotrophic hormone (ACTH) over 4 hr caused an acute lowering of serum zinc in five healthy subjects, with the decrease being entirely due to changes in albumin-bound zinc. These studies indicate an inverse relationship between plasma cortisol and zinc levels. Serum zinc concentrations also show an inverse relationship to growth hormone levels: patients with acromegaly had significantly decreased serum zinc levels, whereas in those with isolated growth hormone deficiency, zinc levels were elevated. In both conditions, serum zinc levels normalized with treatment (383).

8. Hyperzincemia

Although plasma zinc levels are lowered by a large number of conditions, elevated concentrations are much less common in humans. Increases in plasma zinc levels up to ~3.5 μg/ml have been achieved in acute response to ingestion of a single oral dose of a zinc salt supplying 25–100 mg zinc on an empty stomach (683, 852), but levels rapidly return to normal within 4 hr. Chronic ingestion of zinc supplements (50–150 mg/day) usually only results in plasma levels of 1.5 to 2.0 μg/ml. However, two conditions of true hyperzincemia have been reported to occur in humans on normal dietary zinc intakes. Smith *et al.* (843) described a familial hyperzincemia in members of a family in the eastern United States in which plasma levels of the affected individuals ranged from 2.50 to 4.05 μg/ml. These blood levels were not related to any pathological changes, and zinc concentrations in red cells, hair, and bone were normal. An extraordinarily elevated level of plasma zinc has also been reported in a case of pyoderma gangrenosum, in which a young adult male was found on repeated testing to have a plasma concentration of ~10 μg/ml (363).

L. Zinc in Milk

The zinc content of milk varies with species and stage of lactation. A very wide range of zinc concentrations (3–23 μg/ml) has been found in human colostrum (day 1) (61,129), but the average level in most women is between 7 and 12 μg/ml (591,655,704). Levels fall rapidly to an average of about 3 to 5 μg/ml at the end of the first postpartum week. By 1 month postpartum the concentration is about 2.5 to 3.0 μg/ml, and it continues to decline in an exponential fashion to ~0.6 μg/ml by 1 year (236,489,495,591,636,644, 655,704,718,941). Variation among individuals may be high, but there appears to be no consistent effect of factors such as time of day or duration of feed on measured zinc content (496,664,718). Data on the variation of zinc content of milk are less complete for other species, particularly with respect to longitudinal changes throughout the duration of lactation. Table III gives typical values for the zinc concentration of colostrum and mature milk in some nonhuman species.

Regional differences have not been reported to occur in the zinc content of cow's milk, nor are geographical variations apparent in concentrations in human milk (174,179,495,655,941). Zinc levels in milk up to 6 months postpartum did not correlate with dietary intakes in women consuming their habitual diets providing 8–12 mg zinc per day (644,941). Krebs *et al.* (495) examined the effect on milk zinc of giving lactating women a small daily supplement throughout lactation that increased their intake from 12 mg/day up to 25 mg/day, the current U.S. recommended dietary allowance (RDA) for lactating women (657). The rate of decline of milk zinc with duration of lactation was significantly lower

Table III
Zinc Concentrations in Nonhuman Milks[a]

Species	Zinc concentration (μg/ml)	
	Colostrum	Mature milk
Cow	7.2	3.0–5.0
Rat	16.0–18.0	5.0
Dog	9.6	8.7
Goat	13.0	5.0–6.0
Sheep	14.0	1.0–2.0
Pig	13.0–16.0	4.0–6.0
Dolphin	—	11.0
Horse	2.0	1.0–2.0
Cat	6.0–7.0	5.0
Mouse	—	12.0
Monkey	5.2	1.8

[a]Compiled from References 399, 408, 464, 528, 529, 532, 644, 653, 719, 882, and 917.

in the zinc-supplemented group. Feeding a zinc-deficient diet to rats and dairy cows resulted in a reduction in the zinc concentration of milk in both species and in the total milk production in rats (653,819).

Very large amounts of additional zinc provided to cows and pigs receiving a zinc-adequate diet resulted in an increased output of zinc in milk (610,839,882). Miller *et al.* (610) found that an additional 5 g zinc daily provided to dairy cows normally consuming 0.5 g zinc/day caused an increase in milk zinc concentration from 4.2 up to 6.7 μg/ml. Higher levels of intake resulted in further increases but of decreasing magnitude. Zinc concentrations in sow milk did not reflect additional zinc supplements of 100 μg/g diet at the colostral stage, but by 35 days milk zinc was 40% higher than the level in milk from unsupplemented animals (882). Milk zinc concentrations were not increased in lactating women, with normal serum zinc levels, when they took a supplement of 50 to 150 mg zinc daily for 1 week, although serum zinc was significantly elevated (634).

Lowered levels of zinc have been found in the inflamed quarter of the udder of mastitic cows (935).

The aqueous fraction contains 60–70% of the total zinc in human milk, with 18% occurring in the fat layer and the remainder in the pellet when fresh milk is separated by ultracentrifugation (75,258). Considerably less (1%) of the zinc in bovine milk was found in the fat, with 15% in the aqueous layer and up to 75% precipitating with the caseins (258). Other workers have also reported that zinc in cow's milk binds strongly to caseins with little bound to whey proteins (370). In several other species, the proportion of zinc associated with the casein is similar to that in cow's milk (528). Of the total zinc in the fat fraction, 60–70% in both human and cow's milk is present in the outer fat globule membrane, probably in the form of zinc metalloenzymes such as alkaline phosphatase (259). The distribution of zinc in the aqueous fraction is affected by types and affinities of zinc-binding ligands, the total zinc concentration, and the concentrations of other competing divalent cations, and may readily be disturbed by the methods of investigation. In the ultrafiltered portion of human milk, zinc appears to be bound to whey proteins such as lactoferrin and albumin, and is also associated with low molecular weight molecules such as citrate (12,75,527,554,563).

M.   Zinc in Other Body Fluids

The zinc content of human parotid saliva collected from healthy children is ~60 ng/ml (355). Greger and Sickels (312) found that whole mixed-saliva levels did not correlate with serum and hair zinc levels, and did not reflect dietary intakes of zinc in adolescent girls. Parotid saliva obtained from adults contains between 50 and 100 ng zinc per milliliter, with some correlation with flow rate (955). Concentrations were not related to normal dietary intakes over a period of 1 to 2 months (955), and did not fall in young men consuming <1 mg zinc per day for 1 to 2 months (52). However, Freeland-Graves *et al.* (266) reported that the zinc content of saliva sediment, mostly epithelial cells, was significantly reduced (from 126 ± 28 μg/g to 93 ± 14 μg/g) after a period of low zinc intake in healthy young women. Hambidge *et al.* (360) reported that parotid saliva zinc was higher at term (73 ng/ml) in a group of pregnant women who received a supplement of 15 mg zinc per day, compared with unsupplemented pregnant women (55 ng/ml). Henkin *et al.* (386) found very low levels of zinc (10 ng/ml) in parotid saliva from patients with hypogeusia (diminished sense of taste). Grazing sheep may lose 0.4 mg, or 2% of the dietary intake, of zinc daily in parotid saliva (301).

Human pancreatic secretions contain 0.5–5.0 μg/ml of zinc, both in the fasting state and when stimulated by cholecystokinin or secretin. There was no difference in zinc content of secretions from treated acrodermatitis enteropathica patients compared with healthy individuals (888, C. E. Casey *et al.*, unpublished observations). Sullivan *et al.* (888) found that zinc levels in pancreatic secretions and bile were normal in patients with pancreatitis and chronic cirrhosis, in the fasting state, but were much lower than normal after stimulation with secretin. The concentration of zinc in pancreatic secretions of rats is similar to that in humans, 3–4 μg/ml. Levels in bile collected by drainage are of the same order in rats and cats, and generally <1 μg/ml (814). Sheep fed a pellet-type diet with varying levels of zinc were found to secrete about 3 mg zinc daily in mixed bile and pancreatic juice, with zinc intakes of up to 400 mg/day. When intakes were increased to 700 mg/day, secretion in bile and pancreatic juice increased to 7 mg zinc (301). In zinc-deficient pigs, the concentration of zinc was reduced in pancreatic secretions but not in bile (890). Human cerebrospinal fluid contains about 5 to 20 ng/ml of zinc (703).

N. Zinc in the Avian Egg

The total zinc content of eggs from hens consuming a good laying ration is generally between 0.5 and 1.0 mg (70,195). Sandrock *et al.* (787) reported a total of 476 ± 18.2 ($\pm$SEM) μg zinc in fertilized eggs weighing 55–60 g, with a slightly higher level of 589 ± 61.6 μg in unfertilized eggs. Over 99% of the zinc was present in the yolk, where it is associated with the lipoprotein, lipovitelin (912). At the time of hatching of the fertilized eggs, ~86% of the zinc was found to have been assimilated into the chick, with ~66 μg zinc remaining in the residual yolk (787).

III. ZINC METABOLISM

A. Absorption

The mechanisms and control of zinc absorption are still not fully understood. Apparently conflicting results of experimental studies both *in vitro* and *in vivo* may arise in part from differences in experimental conditions, as zinc absorption appears to be readily influenced by a wide variety of host and environmental factors (856).

In rats zinc is absorbed mainly from the duodenum, jejunum, and ileum, with very little being absorbed from the stomach (31,185,930). Zinc can be absorbed from the colon, but this route is of minimal importance in the intact animal (186, 274). Davies (185) found that in fasted rats, absorption (transfer of ^{65}Zn from lumen to carcass) was greatest from the duodenum, which contributed 60% of the total absorption of ^{65}Zn, with 30% from the ileum and 10% from the jejunum. However, Antonson *et al.* (31) showed that in nonfasted rats, zinc absorption was greatest from the ileum (60%), with ~20% from both the duodenum and jejunum. In all segments a proportion of the zinc removed from the lumen was recovered in the intestinal mucosa; this was ~10% of the total absorbed in the duodenum and ileum but 30% in the jejunum. These workers also found that ligation of the bile and pancreatic ducts considerably enhanced zinc absorption from the duodenum to 32% of the total, but the amount retained in the mucosa did not differ; the difference was not due to a decrease in the amount of zinc available, as the bile and pancreatic juice contributed very little zinc to the amount in the perfusate. This is in contrast to the situation in humans, in whom consumption of food resulted in a large increase, to 144 to 300% of that ingested, in the amount of zinc in the lumen contents leaving the duodenum and indicated that net absorption occurs mainly below this level (562). In cattle about one-third of an oral dose of ^{65}Zn was apparently absorbed from the abomasum, with further absorption occurring throughout the small intestine (606). Zinc absorption also occurs throughout the small intestine in calves, with the amount absorbed per unit of length being as great in the distal as in the proximal ends (367). Substantial zinc absorption from the proventriculus, as well as from the small intestine, occurs in chicks (607).

Cousins (167) has proposed four phases in the intestinal absorption of zinc: uptake by the intestinal cell, movement through the mucosal cell, transfer to the portal circulation, and secretion of endogenous zinc back into the intestinal cell. The first step in zinc absorption involves the transfer of zinc from the lumen of the intestine into the mucosal cell. The details of this mechanism have not been defined as yet, but transport of zinc across the brush border appears to be a carrier-mediated process that probably involves interaction with the metal in a chelated form (856). A large number of low molecular weight binding ligands have been shown to enhance mucosal uptake and absorption of zinc under experimental conditions, including citrate, picolinate, ethylenediaminetetraacetic acid (EDTA) and amino acids such as histidine and glutamate (684,824,952). Unequivocal evidence of a specific physiological role of one or more of these has not yet been presented, however, and the availability of zinc for uptake by the brush border may be determined by factors such as pH and the relative distribution of zinc among small ligands and larger molecules within the intestinal lumen.

Zinc uptake by the brush border exhibits saturation kinetics at physiological lumen zinc concentrations (185,365,848). Davies (185) found that mucosal uptake in isolated, vascularly perfused segments of rat duodenum was saturable up to a lumen zinc level of 0.8 mM; with higher lumen concentrations uptake was linear, suggesting nonspecific binding to the cell surface. Steel and Cousins (881) used a vascular perfusion preparation in rats that included most of the intestine, from 1 cm distal to the bile duct to the ileocecal valve, to study the mechanisms of zinc absorption in animals previously fed a zinc-adequate or a zinc-deficient diet. They found that zinc absorption (transfer from lumen to portal perfusate) was saturable in both groups, but in the zinc-adequate animals, saturation occurred at a lower lumen zinc concentration (100 μM) than that reported by Davies (185). In the zinc-adequate rats, the maximal rate of zinc absorption was 7.4 nmol/min. These workers also confirmed that uptake of zinc was increased by zinc deficiency: the zinc-depleted animals showed a more rapid rate of zinc absorption at all lumen zinc concentrations. In the depleted rats, zinc absorption appeared to include both mediated and nonmediated components, with the mediated component accounting for more of the total absorption rate at lower (<100 μM) luminal zinc concentrations. The intestines from zinc-depleted animals had higher cytosolic zinc concentrations after perfusion, suggesting that enhanced uptake rather than increased basolateral transport may account for some of the difference in absorption rates. Thus it appears that zinc uptake at the brush border is a process that can be regulated in the control of zinc homeostasis.

The zinc in the mucosal cell includes both that entering from the lumen of the intestinal tract and that resecreted into the cell from the serosal surface. A substantial amount of the cytosol zinc is associated with high molecular weight proteins, and a variable amount, according to zinc intake and status, with metallothionein (878,887,975). In small intestine from zinc-sufficient, postabsorptive rats, about 20 to 30% of cytosol zinc was found in the metallothionein fraction (673). Binding to metallothionein impedes the movement of zinc from the cell, thus contributing to the regulation of zinc flux between the intestinal lumen and portal circulation (167). The amount of dietary zinc bound to metallothionein fluctuates with the zinc supply. Menard et al. (588) demonstrated that metallothionein mRNA activity in the intestine is regulated by the dietary zinc content, and when the rate of metallothionein synthesis is at a maximum, net zinc absorption is least. Under conditions of zinc loading, the reduction in the absorption of a [65]Zn dose correlated with the increase in metallothionein-bound stable zinc (878). Induction of metallothionein by factors other than a high zinc supply, such as dexamethasone or stress, results in increased zinc uptake and retention by the mucosal cell (79,878).

The transfer of [65]Zn from the mucosal cell to the portal circulation is slower than the uptake and accumulation within the cell, and appears to be the rate-limiting step in absorption in rats (185,848) and ruminants (618). At low lumen concentrations of zinc levels, much of the available supply is transported to the plasma, but as the lumen zinc content increases, less is taken up into the body. Davies (185) described two phases in the transfer of zinc from the lumen of the rat duodenum to the carcass: (1) a rapid transfer over the initial 30 min, (2) followed from 30 min to 6 hr postdosing by a much slower transfer, probably of zinc that is bound within the mucosal cell and given up gradually to the portal circulation. Transfer of zinc across the serosal surface into the blood appears to have an obligatory requirement for albumin (847), and newly absorbed zinc is carried in the circulation largely bound to albumin.

Under highly controlled experimental conditions in which zinc is administered alone, it can be shown that zinc supply and status are major factors in the homeostatic control of zinc absorption, operating at the level of the mucosal cell (186,856,878,962). Weigand and Kirchgessner (962) reported that secretion of endogenous zinc increased in rats as the dietary supply increased. Smith *et al.* (846) found that when rats were preloaded with zinc intravenously, three times more of a subsequent intravenous dose of ^{65}Zn accumulated in the mucosal cytosol of the intestine compared with control animals. Most of the extra tracer was bound to metallothionein. Induction of intestinal metallothionein by high circulating levels of zinc allows the mucosal cell to accumulate both dietary and endogenous zinc for secretion into the lumen, thus contributing to the overall control of zinc homeostasis, although the extent to which this occurs under free-living conditions is uncertain (856).

In rats, physiological factors such as age (274), pregnancy, and lactation (182) influence the level of absorption, but increased zinc absorption has not been measured in pregnant women under practical conditions (896). At very low dietary intakes of zinc, endogenous fecal excretion is reduced in rats, humans, and cows (52,614,962), and absorption is increased in rats and cows (484, 588). Absorption has not been measured in humans under conditions of dietary zinc deficiency.

Numerous dietary factors affect zinc absorption and are undoubtedly of greater significance than physiological factors under practical conditions (125,856). These substances act within the lumen of the intestine to make zinc more or less available for uptake by the mucosal brush border. The quantitative effect of such dietary factors on the amount of zinc absorbed is discussed in Section V.

B. Intermediary Metabolism

Zinc absorbed in the intestine is carried to the liver in the portal plasma bound to albumin (847). About 30 to 40% of the zinc entering in the hepatic venous supply is extracted by the liver, from which it is subsequently released back into the blood (1). Circulating zinc is incorporated at differing rates into various extrahepatic tissues, which have different rates of zinc turnover (1,573,766,865). Zinc uptake by the central nervous system and the bones is relatively slow, and this zinc remains firmly bound for long periods; bone zinc is not normally readily available for metabolic use. The zinc entering the hair also becomes unavailable to the tissues and is lost as the hair is shed. The most rapid accumulation and turnover of retained zinc occurs in the pancreas, liver, kidney, and spleen (573). Uptake and exchange of zinc in red cells and muscle is considerably slower. Aamodt *et al.* (1) showed that uptake of [69m]Zn by thigh, reflecting muscle and bone, increased only gradually after injection of the tracer and by 5 days was <5% of injected activity compared with 45% in the liver.

Absorbed zinc from orally ingested tracer, and intravenously administered [65]Zn, are eliminated from the body with kinetics best fitted by a two-component model. The initial rapid phase has a half-life in humans of 12.5 days, and the slower pool turns over with a half-life of ~300 days. This latter pool appears to include most of the skeletal muscle zinc (39,766,865). Elimination of radiozinc from the body of dogs and sheep followed a time course similar to that in humans, but the initial elimination was much greater in rats and mice (766). Zinc entering the plasma is rapidly cleared within hours, largely through entry into the blood cells and soft tissues rather than by excretion from the body (1,631).

The distribution and retention of zinc in the whole body and in individual organs and tissues is altered by a number of conditions. Turnover of [65]Zn in mice was accelerated when the animals were given an oral zinc load or high-zinc diet, but distribution within the body was not affected. Intraperitoneal injection of a zinc load also increased loss of [65]Zn and altered the distribution of the tracer among the body zinc compartments (164,165). The specificity of the zinc pathway through the body was disturbed when a large dose of stable zinc was injected and normal absorption bypassed. When dietary zinc was restricted, the retention of [65]Zn in the body was enhanced; uptake of [65]Zn into soft tissues and organs, but not bone, was enhanced and turnover decreased, although there was no measurable decline in the total zinc concentrations of the tissues (617,618,661). The low-zinc diet may have altered the small labile zinc fraction in many tissues, and this in turn caused changes in the metabolism of [65]Zn (661). A redistribution of zinc from bone to skeletal muscle has also been reported in rats fed a deficient diet (709).

Accelerated uptake and increased retention of zinc are observed at wound sites. Trauma and stress also cause changes in zinc distribution remote from the site of injury. Serum levels of zinc are decreased, uptake by the liver is increased, and uptake by bone and muscle is decreased (932).

Entry of zinc into cells occurs in two phases: an early, rapid uptake that is saturable and probably carrier mediated, followed by a slower phase that is apparently passive (872,879). Uptake of zinc by isolated cells in culture is temperature and energy dependent, and maximum entry occurs at the normal plasma zinc concentration. Further uptake can be stimulated by glucocorticoids, and the increased zinc accumulated is bound to newly synthesized metallothionein (228,229).

Intracellular zinc is largely (60–80%) found in the cytosol, with about 10 to 20% in the crude nuclear fraction and smaller amounts in the microsomal and mitochondrial fractions (48,807,905). Some zinc is also found in the cell membrane. Zinc in the cytosol is mostly bound to proteins. Beckèr and Hoekstra (57) found a pool of firmly bound zinc in rat liver cells, together with three additional pools of cellular zinc bound with differing intensities. Other workers have also reported that cytosol zinc from a number of different tissues is distributed in three or four pools, mainly larger molecular weight proteins, which would include many of the zinc metalloenzymes (673,686). Under normal dietary conditions only small amounts of zinc are bound to metallothioneins (901). Although increases in dietary zinc intake result in increased binding of zinc to all cell fractions, most of the additional cytosol zinc is found in the metallothionein fraction of liver, pancreas, kidney, and muscle but not heart and testes (686,975).

The major organ involved in zinc metabolism is the liver. The liver cytosol of ruminants (557,975), rats (673), chicks (686), and humans (872) contains zinc-binding components of differing molecular weight and lability, the amounts and proportions of which vary with the zinc status and age of the animal (557,768). When liver zinc content is increased above normal levels (30 $\mu g/g$ wet weight), the additional zinc is mainly associated with metallothionein. Similarly short-term fluctuations in metallothionein-bound zinc in rat liver could be directly related to daily variations in zinc intake (166). Such experiments suggest that metallothionein has an integral role in hepatic zinc metabolism (199, 976). McCormick *et al.* (570) found that radioactive zinc included in the diet was recovered in the liver in association with newly formed metallothionein. Fluctuations in total liver ^{65}Zn were accompanied by concomitant changes in metallothionein-bound ^{65}Zn and changes in the rate of metallothionein synthesis. The regulation of liver metallothionein is under the control of both dietary and hormonal signals. The primary inducer of metallothionein mRNA appears to be zinc per se (166,831,975), but synthesis of metallothionein is stimulated indirectly by glucocorticoids (226) and catecholamines (83), possibly through an effect on zinc transport into the hepatocyte. Synthesis of metallothionein is also induced by a number of physiological stimuli, including stress, acute infection, and shock (200,685). Induced hepatic metallothionein has a half-life of 18 to 20 hr; removal of the metal occurs at the same time as proteolysis, and the zinc becomes available for cellular efflux. The degradation rate may be increased by low-zinc status (199).

Small amounts of the metallothionein MT-I are detectable by radioimmunoassay in the plasma of normal rats (804). In dietary zinc deficiency, both zinc and MT-I levels in plasma were reduced, in contrast to endotoxin-treated animals in which plasma zinc was lowered but plasma MT-I was elevated. Endotoxin stress also caused a decrease in plasma zinc levels in zinc-sufficient pigs, and zinc flux through the plasma was significantly decreased. These changes were not seen in zinc-deficient pigs (151). Pekarek and Evans (711) have demonstrated that the decline in plasma zinc due to stress and acute infection is associated with a marked flux of zinc into the liver, induced by leukocyte endogenous mediator. The increased entry of zinc into the liver coincides with increases in hepatic metallothionein mRNA (200). Regulation of the hepatic metallothionein gene may thus account for part of the alterations in zinc metabolism observed in stress states.

C. Excretion

In normal dietary circumstances, the feces are the major route of zinc excretion (103,300,874,964). Thus in the healthy human adult with an intake of 10 to 15 mg zinc per day, and whose zinc balance is in equilibrium, $\sim$90% of this amount will be excreted in the feces (102,895). When a tracer dose of ^{65}Zn or enriched zinc-stable isotopes are given orally to a healthy adult, 2–10% is recovered in the urine, with most of the remainder eventually appearing in the feces (1,853). Most of an intravenous dose of zinc isotope will also appear in the feces over a prolonged period (865).

Fecal excretion is not limited to excretion of unabsorbed dietary zinc, which normally accounts for a large proportion of the intake, but includes excretion of endogenous zinc. The excretion of endogenous zinc in the feces varies according to the balance between true absorption and metabolic needs, and this variation is one of the primary mechanisms for maintaining zinc homeostasis. Weigand and Kirchgessner (964) have elegantly demonstrated in rats on a zinc depletion diet that inevitable endogenous fecal zinc excretion is so low that it is exceeded by urinary zinc losses, which are not affected by dietary zinc in the rat. Figure 1 demonstrates the increasing gap between true absorption and apparent absorption as the dietary zinc intake is increased (487); this is attributable to increased excretion in feces of endogenous zinc. The amount of endogenous fecal excretion at high dietary zinc concentrations far exceeded the total dietary zinc intake at the lower dietary zinc levels in this study, which illustrates the magnitude of the changes that are possible to maintain zinc homeostasis (964). Also shown in Fig. 1 is the decrease in percentage true absorption as dietary zinc intake is increased.

An increase in endogenous fecal excretion of zinc when dietary zinc intake is increased was first demonstrated in mice by Cotzias (164). Similar variations in endogenous zinc excretion in the feces according to dietary intake have been

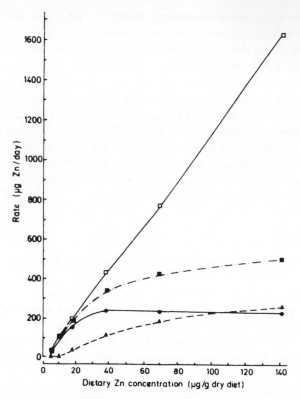

Fig. 1. True (■) and apparent (●) absorption of zinc and fecal excretion of endogenous zinc (▲) in relation to dietary intake (□) in young rats. From Kirchgessner and Weigand (487).

demonstrated in the human (442). The actual amount of endogenous zinc lost from the body in the feces thus depends on zinc intake and status and, in the human, may range from <1 mg to several milligrams daily (562). Both total and endogenous fecal excretion of a tracer dose of ^{65}Zn and of stable dietary zinc were significantly reduced in calves and goats on low-zinc diets (618). Conversely, in sheep, as zinc intake was increased, endogenous fecal losses appeared to remain constant at ~0.11 μg/g live weight per day, and homeostasis was maintained by reducing absorption (892).

Endogenous fecal losses of zinc arise from the zinc that is secreted into the gut from the body and is not subsequently reabsorbed. The enteral circulation of zinc is not well understood, but may be much larger in quantitative terms than the amount of zinc of endogenous origin finally appearing in the feces. Mateseshe *et al.* (562) used a marker perfusion technique to measure the intraluminal quantities of zinc in the upper gastrointestinal tract of healthy adults consuming several different test meals. Over the 5 hr following the meals, between 144 and 300% of the amount of zinc ingested was calculated to be recoverable at the first aspiration site, at the ligament of Treitz. This represented an entry into the gut of an additional 2.2–4.8 mg zinc. Much of this zinc must eventually be reabsorbed to avoid negative zinc balance.

The pancreatic secretions are a major contributor to endogenous zinc secretion in a number of species (301,594,633,888). Montgomery *et al.* (633) showed that 10% of a intravenous dose of ^{65}Zn was secreted in pancreatic juice, with <1% appearing in the bile of the dog; ligation of the pancreatic and bile ducts reduced recovery of radiozinc from the intestinal lumen (71). In contrast, Pekas (713) found that although ^{65}Zn was secreted in the pancreatic juice and other duodenal secretions of the pig, ligation of the ducts did not reduce fecal excretion of the tracer. Miller *et al* (601) also calculated that the amount of a dose of ^{65}Zn appearing in pancreatic, biliary, and gastroduodenal secretions in human subjects was not enough to account for the total excreted in the feces, suggesting another route of endogenous zinc loss. This other major pathway of zinc excretion is provided by the intestinal mucosal cells. The zinc in the mucosal cells may include both unabsorbed zinc taken up from the intestinal lumen and zinc reentering the cell from the plasma under conditions of net efflux from the serosal surface to the lumen (922). Mucosal zinc is associated with high molecular weight proteins and with a metallothionein, synthesis of which is sensitive to zinc status, thus enhancing the value of this route of excretion to the homeostatic control of zinc (588).

The quantity of zinc excreted in the urine of healthy adults is about 300 to 500 μg/day (102,360,895). Ritchey and colleagues (769) summarized data from several studies for preadolescent children with an overall mean of 380 μg zinc excreted in the urine daily. Young children aged 33–90 months have been reported to excrete a mean (±SD) of 187 ± 97 μg per 24 hr (355); Cheek (143) found similar excretion rates for young children and observed that rates correlated better with creatinine excretion than with age. Relatively high urine zinc excretion rates have been reported in both term (137) and premature (130) neonates, but by 2 months of age, rates were down to 10 to 20 μg zinc per kilogram body weight per day.

The amount of zinc excreted in the urine varies with the level of intake: in adult men consuming a severely zinc-deficient experimental diet providing zinc at 0.28 mg/day, urinary zinc was decreased threefold to 140 μg/day (52). Catabolic states such as follow severe burns, major surgery or other trauma, and total starvation cause clinically significant increases in urinary zinc losses (238,460,793,868). Urinary excretion of zinc is also well above normal in hepatic porphyria and postalcoholic cirrhosis (731), and may be elevated by administration of chelating agents such as EDTA (867).

Urinary excretion of zinc by sheep and calves is generally <1 mg/day, or ~0.25 μg/ml urine in dairy cows, with little effect due to zinc supply in the diet (617,819,892). Excretion in the urine of radiozinc given either orally or intravenously to steers was 0.3 and 0.25% of the dose, respectively (235). These data were similar to those obtained for calves and lactating cows (874).

Urinary zinc arises largely from the ultrafilterable portion of plasma zinc (6). Net reabsorption, of up to 95% of filtered zinc under basal conditions, occurs in the distal parts of the renal tubule (6,939). The amount of zinc excreted correlates highly with the rate of urine (volume) production and creatinine excretion (119,507); the modest increases in urinary zinc output under conditions of volume expansion and pregnancy may thus be due to increased tubular flow rates (958). The diuretic, natriuetic agent chlorothiazide further increases zincuria by altering renal tubular handling of zinc, causing net secretion of zinc into the far distal tubule and inhibiting reabsorption in the distal tubule (939,958). Hyperzincuria frequently occurs in patients receiving total parenteral nutrition (130, 319), and short-term losses as high as 23 mg/day have been measured (461). Experimental studies with proprietary preparations and infusions of single amino acids have confirmed that infused amino acids, particularly cysteine and histidine, are potent stimulators of renal zinc excretion (6,931,997). The causes of the increased zincuria may include both an increase in the ultrafilterable portion of plasma zinc and prevention of resorption of zinc by binding to amino acids secreted into the renal tubule (6).

Healthy adults living in a temperate North American climate have been reported to lose from 0.4 to 2.8 mg zinc daily in the sweat (52,393,626,816). Dermal and sweat losses are particularly difficult to measure accurately, however, and this wide range may reflect technical problems rather than true differences; lower figures obtained in later studies are probably more reliable estimates. Milne and co-workers (626) found that whole-body sweat losses were related to dietary intake: when intakes of zinc were 8.3 mg/day, young men lost an average of 0.5 mg zinc in sweat daily; losses were reduced to 0.24 mg/day on a marginal intake of zinc, and increased to 0.62 mg/day when subjects were repleted with a zinc intake of 34 mg/day. This reduction of sweat losses during deficiency may represent a homeostatic mechanism for conservation of zinc. Prasad *et al.* (730) reported that normal sweat contains 1.15 ± 0.03 μg zinc per milliliter, ~75% of which was in the cell-free portion and the remainder associated with desquamated epithelial cells. In zinc-deficient Egyptian males, the zinc content of sweat was reduced to 0.6 ± 0.27 μg/ml.

Menstrual losses of zinc are small; average losses of between 0.1 and 0.5 mg zinc per period have been reported (393,918), which represents a loss of about 5 to 15 μg/day over a typical 30-day cycle, a negligible fraction of the normal daily zinc intake.

IV. ZINC DEFICIENCY AND FUNCTIONS

A. Molecular Biology

Zinc is a IIB. element with a completed d subshell and two additional s electrons, which chemically combines in the $+2$ oxidation state. Zinc shares with the transition metals an ability to form stable complexes with the side chains of proteins, a property that is relevant to its specific biological functions. The involvement of zinc in enzyme functions and structures is the best known function of this metal. Vallee (927) has conjectured that the biological abundance of zinc, its ability to interact with widely varied coordination geometries, and its resistance to oxidation–reduction have generated selective evolutionary pressure for it to serve in biological catalysis. Carbonic anhydrase was the first zinc metalloenzyme to be isolated and purified in 1940 by Keilen and Mann (467). This was followed by the identification of bovine pancreatic carboxypeptidase A as a zinc metalloenzyme in 1954 (923). Subsequently, identification of additional zinc enzymes and proteins has proceeded rapidly, and there are now over 200 zinc enzymes or other proteins that have been identified from various sources encompassing all phyla (927) (Table IV). At least one zinc enzyme can be found in each of the six major categories of enzymes designated by the IUB Commission on Enzyme Nomenclature.

Zinc has several recognized functions in metalloenzymes including catalytic, structural, and regulatory roles (927). Vallee and Williams (925) postulated that binding of the metal to the enzyme protein at the site of catalysis in metalloenzymes results in a distorted and partial coordination sphere around the metal ion. The tension associated with this distorted geometry results in an entatic state that, it is speculated, results in zinc being poised for its catalytic function. The energy resulting from this distortion is then released or transferred to the substrate when the latter is bound. This theory fits with observations on the structure of carbonic anhydrase, carboxypeptidase, and some other zinc metalloenzymes. Zinc is thought to act catalytically, either by combining directly with the substrate or through the metal-bound water molecule, resulting in a zinc hydroxide. A combination of these two is most likely (927).

In its structural role, zinc usually stabilizes the quaternary structure of the enzyme protein as in superoxide dismutase. Alcohol dehydrogenase from horse liver contains 4 gram atoms of zinc per mole; 2 zinc ions are essential for catalytic activity, while the other 2 have a structural role. Fructose biphosphatase is an example of a zinc metalloenzyme in which zinc ion may have a physiologically important function as a regulatory inhibitor (708). Chesters (152) has emphasized, however, that assessment of the significance of these regulatory roles *in vivo* is difficult because of uncertainties about the intracellular concentrations of readily exchangeable zinc ion.

Table IV

Zinc Metalloenzymes 1982[a]

Name	Number	Source	Role[b]
Class I, oxidoreductases			
Alcohol dehydrogenase	9	Vertebrates, plants	A, D
Alcohol dehydrogenase	1	Yeast	A
D-Lactate dehydrogenase	1	Barnacle	?
D-Lactate cytochrome reductase	1	Yeast	?
Superoxide dismutase	12	Vertebrates, plants, fungi, bacteria	(A)[b], D
Class II, transferases			
Aspartate transcarbamylase	1	*Escherichia coli*	B
Transcarboxylase	1	*Penicillium shermanii*	?
Phosphoglucomutase	1	Yeast	?
RNA polymerase	10	Wheat germ, bacteria, viruses	A
DNA polymerase	2	Sea urchin, T_4 phage	A
Reverse transcriptase	3	Oncogenic viruses	A
Terminal dNT transferase	1	Calf thymus	A
Nuclear poly(A) polymerase	2	Rat liver, virus	A
Mercaptopyruvate sulfur transferase	1	*E. coli*	?
Class III, hydrolases			
Alkaline phosphatase	8	Mammals, bacteria	A (C)[c], D
Fructose 1,6-bisphosphatase	2	Mammals	C
Phosphodiesterase (exonuclease)	1	Snake venom	A
Phospholipase C	1	*Bacillus cereus*	A
Nuclease P_1	1	*Penicillium cirtrinum*	?
α-Amylase	1	*Bacillus subtilis*	B
α-D-Mannosidase	1	Jack bean	?
Aminopeptidase	10	Mammals, fungi, bacteria	A, C
Aminotripeptidase	1	Rabbit intestine	A
DD-Carboxypeptidase	1	*S. albus*	A
Procarboxypeptidase A	2	Pancreas	A
Procarboxypeptidase B	1	Pancreas	A
Carboxypeptidase A	4	Vertebrates, crustacea	A
Carboxypeptidase B	4	Mammals, crustacea	A
Carboxypeptidase (other)	5	Mammals, crustacea, bacteria	A
Dipeptidase	3	Mammals, bacteria	A
Angiotensin-converting enzyme	3	Mammals	A
Neutral protease	16	Vertebrates, fungi, bacteria	A, (B)[c]
Collagenase	4	Mammals, bacteria	A
Elastase	1	*Pseudomonas aeruginosa*	?
Aminocyclase	1	Pig kidney	?
β-Lactamase II	1	*B. cereus*	A
Creatininase	1	*P. putida*	?
Dihydropyrimidine aminohydrolase	1	Bovine liver	?
AMP deaminase	1	Rabbit muscle	?
Nucleotide pyrophosphatase	1	Yeast	A

Table IV (*Continued*)

Name	Number	Source	Role[b]
Class IV, lyases			
Fructose 1,6-bisphosphate aldolase	4	Yeast, bacteria	A
L-Rhamnulose 1-phosphate aldolase	1	*E. coli*	A
Carbonic anhydrase	22	Animals, plants	A
δ-Aminolevulinic acid dehydratase	2	Mammalian liver, erythrocytes	A
Glyoxalase I	4	Mammals, yeast	A
Class V, isomerases			
Phosphomannose isomerase	1	Yeast	?
Class VI, ligases			
tRNA synthetase	3	*E. coli, B. stearothermophilus*	A
Pyruvate carboxylase	2	Yeast, bacteria	?
Total	162[d]		

[a]From Reference 927, pp. 5–6.

[b]A, Catalytic; B, structural; C, regulatory; D, undefined. ?, Available information is insufficient to make an assignment. See the text for further details.

[c]Letters in parentheses refer to roles fulfilled by metals other than zinc.

[d]Since this chapter was written, the total number of zinc enzymes has increased to beyond 200.

Zinc possibly has important structural roles at other sites. Substantial quantities of firmly bound zinc stabilize the structures of RNA (943), DNA (835), and ribosomes (741). Zinc may also play a critical physiological role in the structure and function of biomembranes, and in plasma membranes. Bettger and O'Dell (69) have hypothesized that some of the features of zinc deficiency are mediated through a decrease in the zinc content of biomembranes. They showed that erythrocytes of zinc-deficient rats lose membrane zinc but not intracellular zinc (66), and suggested that loss of cell membrane zinc was an early event in zinc deficiency states. Zinc deficiency in the rat increases the fragility of red cells subjected to hypotonic stress (64). They concluded that zinc deficiency may cause oxidative damage to membranes, structural strains, altered function of specific receptors and nutrient absorption sites, altered activity of membrane-bound enzymes, altered function of permeability channels, and altered function of carrier and transport proteins in the membrane. Other specific examples of membrane effects include the inhibition of platelet aggregation and serotonin release by supraphysiological quantities of zinc that appear to stabilize platelet membranes (154). Zinc also prevents histamine release from mast cells (462), perhaps through masking of specific receptor sites for histamine-releasing agents. Membrane sodium transport is affected adversely by zinc deficiency (706). Chapil (155) has proposed that membrane-bound zinc alters the fluidity and stabilization of membranes and that the effects of zinc in inhibiting macrophage and neutrophil function result from membrane changes. Pharmacological quantities of zinc increase the filterability of sickle cells (87) and decrease the number of irreversibly sickled cells, apparently through effects on the eythrocyte membrane. Zinc also appears to play a role in regulating the function of the calcium protein calmodulin, especially as it affects the cell microtubules and microfilaments. Zinc ions at micromolar concentrations inhibit effects of calcium calmodulin (88) on activating enzymes such as Ca^{2+}-ATPase and beef brain phosphodiesterase. A second calcium–zinc protein has been identified in the brain that may have a role in the tubulin–microtubule system (51). Hesketh (392) has found that the polymerization of brain tubulin is decreased in brain extracts from zinc-deficient pigs and rats, and concluded that zinc may have some function in microtubule polymerization *in vivo*.

B. Enzyme Activities in the Tissues

The levels and activities of zinc metalloenzymes and zinc-dependent enzymes in the tissues of zinc-deficient animals have been studied extensively. Histochemical determinations carried out by Prasad and co-workers (734) disclosed reduced activities of several enzymes, accompanied by reduced zinc levels in the testes, bones, esophagus, and kidneys of zinc-deficient rats, compared with those of restricted-fed controls. Reductions of the following occurred: in the testes, lactic dehydrogenase (LDH), malic dehydrogenase (MDH), alcohol dehydrogenase (ADH), and reduced nicotinamide adenine dinucleotide (NADH) diaphorase; in the bones, LDH, MDH, ADH, and alkaline phosphatase; in the esophagus, MDH, ADH, and NADH diaphorase; and in the kidneys, MDH and alkaline phosphatase. In a zinc-repleted group of rats the activities of these enzymes and the level of zinc increased in all the tissues examined. Similar results were obtained with baby pigs (735), although liver ADH and liver glutamate dehydrogenase (GDH) activities are unaffected in the zinc-deficient baby pig (602). The activities of LDH and ADH remained unaltered in the muscle tissue of zinc-deficient young rats, but MDH lost much of its activity in this tissue when an extreme state of zinc depletion was reached (779).

Huber and Gershoff (419) studied the effects of alterations in dietary zinc on the tissue levels of four zinc metalloenzymes (carbonic anhydrase, LDH, GDH, and alkaline phosphatase) in 4-week-old rats over a 2- to 4-week period. Zinc deficiency per se was found to significantly depress (1) carbonic anhydrase activity in the stomach and pituitary, (2) LDH in heart, kidney, and gastrocnemius muscle, and (3) alkaline phosphatase in the duodenum, stomach, and serum. Mitochondrial GDH was not affected by high or low dietary zinc intakes in any tissues examined. In a further study with rats with gestational–lactational zinc deprivation the concentration of brain GDH was unchanged, but a significant reduction in brain 2′,3′-cyclic nucleotide 3′-phosphohydrolase (CNP) was found, implying a delay in myelination. Zinc restriction only in the lactational period resulted in smaller brain changes with no difference in the levels of the cerebellar enzyme superoxide dismutase, GDH, or CNP between the deficient pups and those from dams pair-fed a zinc-adequate diet (744).

Alkaline phosphatase is notable for its rapid loss of activity following the induction of zinc deficiency even before food consumption decreases (486). Roth and Kirchgessner (778) found that the activity of alkaline phosphatase in rat serum decreased by 25% after 2 days, and by 50% after 4 days of dietary zinc depletion, with restoration of normal activity after 3 days on a zinc-adequate diet. Preincubation of serum with zinc *in vitro* did not normalize activity, and it was concluded that there was not a normal quantity of apoenzyme present. Partial correction of the impaired growth of zinc-deficient rats has been achieved with injections of alkaline phosphatase (780). Serum alkaline phosphatase is abnormally low in severe human zinc deficiency (782), including untreated acrodermatitis enteropathica (663). In milder zinc deficiency states activity may be low-normal and increase following zinc supplementation (738). A rapid and substantial decline also occurs in the alkaline phosphatase activity of the bones of young rats fed a zinc-deficient diet, with restoration to nearly normal levels after 8 days of zinc repletion (779). Pancreatic carboxypeptidase A and B are also sensitive to zinc depletion, but erythrocyte carbonic anhydrase activity is less sensitive (486).

Among other recognized zinc metalloenzymes in mammalian tissues the activity of which is reduced by zinc depletion is δ-aminolevulinic acid dehydrogenase. Activity of this enzyme is reduced in zinc-deficient cultures of rat hepatocytes (323). Although retinol dehydrogenase in the retina is thought to be very similar to alcohol dehydrogenase in other tissues, it appears that further confirmation of its status as a zinc metalloenzyme is required (844). The activity of retinol dehydrogenase has been reported to be diminished in zinc-deficient animals (420), although studies including control of growth and food intake are needed (844). This enzyme catalyzes the conversion of retinol to retinaldehyde, a necessary step for normal dark adaptation. Dark adaptation has been reported to improve with zinc supplementation in alcoholic cirrhotic patients who have failed to improve with vitamin A therapy alone (568). Folate metabolism may also be affected by human zinc deficiency, which appears to result in a decreased hydrolysis of pteroylpolyglutamate, indicating that intestinal conjugase is a zinc-dependent enzyme (900).

Differences in the sensitivity of zinc-dependent enzymes to dietary zinc depletion result from both differences in affinity for zinc and differences in turnover rates. Tissue differences exist for the activity of the same enzymes. Early changes in activity of some enzymes with dietary zinc depletion, before total tissue zinc concentrations decrease, indicate that there is a rapid turnover or that zinc is very freely exchangeable at these particular sites.

C. Cell Replication and Differentiation

Zinc is involved extensively in nucleic acid and protein metabolism and hence in the fundamental processes of cell differentiation and, especially, replication. Zinc is detectable histochemically in the nucleus, nucleolus, and chromosomes. Firmly bound zinc stabilizes the structures of RNA (943), DNA (835), and ribosomes (741). Key enzymes required for nucleic acid synthesis and degradation are zinc dependent, including many, from various species, that are zinc metalloenzymes. These zinc-dependent enzymes include the potentially rate-limiting enzymes involved in DNA synthesis, that is, DNA polymerases (927); aspartate transcarbamylase (212), and thymidine kinase (205,211), which is required for optimal rates of thymidine nucleotide synthesis in rapidly growing (including embryonic) tissues. Thymidine kinase activity is reduced by zinc deficiency in fetal rats (211), in primary cultures of kidney cells (517), in subcutaneous proliferations of human and rat connective tissue (736), and in ascites cells (717). Other zinc-dependent enzymes involved in nucleic acid metabolism are RNA synthetase, nucleoside phosphorylase (29), deoxynucleotidyl transferases, and viral reverse transcriptase. In contrast to these zinc-dependent enzymes, ribonuclease is inhibited by zinc and degradation of RNA is increased in zinc-deficient animals (860).

The activity of DNA and RNA polymerases has been found to be decreased by zinc deficiency in a variety of tissues and cell cultures (904). Vallee and associates (233) have found that the three normal DNA-dependent RNA polymerases that are typical of eukaryotes are absent in zinc-deficient *Euglena gracilis* and are replaced by only a single RNA polymerase as normally found in prokaryotes. In the same organism the composition of mRNA is altered (173); although it retains its functional capacity and many normal proteins are synthesized, several normal proteins and polypeptides are not synthesized while unusual ones appear. The latter include several arginine-rich polypeptides.

While the potential importance of reduced activity of DNA polymerase and thymidine kinase in zinc-deficient tissues cannot be dismissed, there is evidence to suggest that impaired activity of these enzymes is not the primary event leading to disturbances of cell replication and differentiation. Both Chesters (152) and Vallee (928) have concluded that the major effects of zinc deficiency on cell replication and differentiation can be best explained by interference with normal chromatin restructuring and gene expression. The extent to which this interference affects different phases of the cell cycle depends on the type of individual cell or tissue. In rapidly dividing cells, the major effects are often at the S phase, resulting in reduced DNA synthesis. For example, impaired DNA synthesis in the liver (99,268,823) and brain (792) of zinc-deficient rats has been demonstrated in several studies. Studies of the time course of DNA, RNA, and protein synthesis in phytohemagglutinin (PHA)-stimulated lymphocytes deprived of zinc, indicate a primary involvement of zinc in DNA synthesis (147). Zinc was found to be of prime importance in reversing the inhibition by EDTA of the expression of the genetic potential of these cells to synthesize the enzymes required for DNA synthesis and cell division. Evidence has also been obtained of a zinc requirement for DNA synthesis in cultured chick embryo and mammalian cells (783).

In some cells, such as *Euglena gracilis* (231), zinc deficiency arrests the cell cycle at the G_2–M transition; hence the DNA content of the cells is twice normal. A more complete inhibition of mitosis than of DNA synthesis has also been observed in zinc-deficient cultures of lymphocytes (147). Since the G_2–M transition requires alteration in chromatin structure involving chromosome condensation, the same mechanisms that may be responsible for interruption of gene activation by zinc deficiency, which are discussed below, could also explain this arrest.

Although, in general, DNA synthesis and cell replication may appear to be impaired by zinc deficiency to a greater extent than protein synthesis (987), zinc is also clearly necessary for cell hypertrophy and differentiation. For example, impaired synthesis of collagen in skin associated with immature fibroblasts is found in zinc-deficient rats (569). Amino acid utilization in the synthesis of protein is impaired (414,796). The total protein and RNA contents of the testes of zinc-deficient rats are reduced (542), and it is only with more severe zinc deficiency that the DNA content is reduced and the ribonuclease activity is increased (860). Failure of cell differentiation has also been reported in the epiphyses of zinc-deficient chicks and rats, in those regions of bone that are most remote from the capillary network (973). The period of maximal fetal sensitivity to maternal zinc deficiency in rats coincides with the period of maximum tissue differentiation (426). The esophagus and buccal mucosa of some zinc-deficient animals become parakeratotic (237,597), with an unusually high mitotic index, in contrast to the reduced DNA synthesis that is characteristic of most other tissues. This is associated with a decreased ability of the continually dividing germinal layer to differentiate (19,149,198). Although zinc deficiency facilitates the induction of esophageal carcinoma in rats, the growth of malignant cells is in general inhibited by zinc deficiency.

Possible mechanisms for the failure of normal chromatin restructuring and gene expression have been proposed that would account for the disturbances of the cell cycle at various phases (152,564,875,926,927). Histones have a major role in the chromatin restructuring that is required for DNA replication, mitosis, and transcription. Phosphorylation of this histone appears to be related to DNA synthesis (457), and displacement of histone H1 has been implicated in gene activation (837). *In vitro* histone H1 is only phosphorylated when the system is activated by zinc (47). Histones from zinc-deficient systems migrate differently in an electric field from when adequate zinc is present (448). The sum of these observations is highly suggestive of a role for altered histone metabolism in mediating the effects of zinc deficiency on gene expression. Zinc deficiency may also affect histones by interfering with methylation and/or by the interaction of arginine-rich peptides with histones (927). Arginyl residues synthesized in zinc-deficient systems may also bind directly to phosphate groups of DNA to activate or repress genes. RNA polymerases also have a key role in transcription and gene expression, and changes in RNA polymerase activity or structure could adversely affect gene expression during differentiation and in the S phase of DNA synthesis. Abnormal RNA polymerases have been shown to be associated with abnormal mRNA and partially abnormal protein synthesis in zinc-deficient *Euglena gracilis*. Polyadenylate polymerase requires zinc for activity but has relatively low affinity for this metal. Hence zinc deficiency could influence the survival and translation of new mRNA.

In vitro experiments that have demonstrated the effects of zinc on histones have used free Zn^{2+} ions. Chesters (152) has hypothesized that the effects of zinc on gene expression *in vivo* are dependent on a small intracellular pool of readily exchangeable zinc and that this pool is rapidly depleted when dietary zinc is restricted. This could explain the profound and rapid effects of zinc restriction on normal growth and development when there is no measurable decrease in total tissue zinc concentrations (152).

D. Glucose and Lipid Metabolism

In 1937, Hove, Elvehjem, and Hart (410) first observed small differences in oral glucose tolerance curves of zinc-deficient and control rats. Impaired oral glucose tolerance has also been observed in association with human zinc deficiency (788,851). However, these apparent effects of zinc deficiency on oral glucose tolerance have been inconsistent (486) in animals and in the human. In 1966, Quarterman and colleagues (747) reported that glucose given intraperitoneally was poorly tolerated by zinc-deficient rats. In the same year, Macapinlac *et al.* (539) were unable to demonstrate an effect of zinc deficiency on tolerance for intraperitoneally injected glucose, but support for Quarterman's findings came from three other groups of investigators (81,376,419,781). However, in 1972 Quarterman and Florence (748) suggested that the differences observed between zinc-deficient and pair-fed control animals were attributable to differences in feeding patterns. This suggestion has since been refuted, at least in part, by Reeves and O'Dell (759), who fed zinc-deficient and control animals in an identical manner and measured the incorporation of [^{14}C]glucose into adipose tissue fatty acids and glycogen. In the zinc-deficient animals [^{14}C]glucose incorporation into fatty acids was reduced by 75% in comparison to the control animals, confirming that zinc deficiency reduces parenteral glucose utilization in experimental animals. Pharmacological quantities of zinc have been reported to have a hyperglycemic effect in rats (227).

In 1934 Scott (821) found that crystalline insulin contained considerable quantities of zinc. Zinc appears to affect the solubility of insulin and could therefore play a role in the mechanism of insulin release. It has also been postulated that insulin degradation is increased in zinc-deficient rats possibly due to increased activity of glutathione-insulin transhydrogenase (486). Though not a consistent finding, decreased circulating insulin levels have been reported to result from zinc deficiency in some studies (486). Kirchgessner and colleagues (485) found that insulin levels were significantly decreased in zinc-deficient lactating cows despite no decrease in food consumption, thus ruling out depressed food intake as the cause of the lowered circulating insulin levels. In addition, there may be an increase in peripheral resistance to the action of insulin (760,851).

A reduction in glucose utilization appears to lead to an increase in lipid catabolism in zinc-deficient animals. In rats, fasting plasma fatty acid levels are at least twice as high as those of control animals (748). Severely zinc-deficient rats rapidly lose visible adipose tissue. In the human, a drop in the fasting respiratory quotient suggests that relatively more lipid is being oxidized in zinc deficiency (851). There are a number of other separate interrelationships between zinc and various aspects of lipid and fatty acid metabolism. Both Atkinson and colleagues (41) and Chandra (140) have observed low plasma zinc concentrations in obese humans. Hooper and associates (409) have shown that high-dose zinc therapy administered to normal adult men results in a significant depression in circulating high-density lipoprotein cholesterol levels. However, somewhat lower doses in healthy young women have not been associated with any significant changes in total cholesterol or high-density lipoprotein cholesterol concentrations (265).

There is evidence of substantial interactions between zinc and essential fatty acids, but the picture remains confused. Some of the clinical features of zinc deficiency may be ameliorated by essential fatty acid supplementation (417), although this requires further confirmation. Zinc appears to have specific effects on the elongation–desaturation pathway of essential fatty acid metabolism (158). For example, Cunnane and Wahle (175) have reported that Δ^6-desaturation of linoleic acid was 3.4 times greater in the microsomes of mammary tissue from zinc-deficient rats compared with controls. Bettger and colleagues (65) have shown that skin lesions and growth failure in zinc-deficient rats are accentuated by essential fatty acid deficiency. Zinc deficiency increases the proportion of arachidonic acid in foot skin, especially in the essential fatty acid-deficient animal. In contrast, in zinc-deficient chicks (67), low dietary polyunsaturated fatty acids resulted in a significantly higher growth rate and decreased the dermatitis. This effect is opposite to that which the same investigators observed in the rat, but in both species a higher than normal proportion of arachidonate is found in the fatty acids of zinc-deficient skin.

Meydani and Dupont (595) found that zinc deficiency did not decrease prostaglandin synthesis below that of pair-fed controls. Only in the gut lumen were prostaglandin levels decreased by zinc deficiency, and this appeared to be due to an impairment of an active process of prostaglandin secretion involving zinc. These investigators also found that under physiological conditions inhibitors of prostaglandin synthesis did not alter a variety of zinc-dependent matabolic functions (596). In 1977, O'Dell and colleagues (680) showed that the effects of aspirin (a prostaglandin synthesis inhibitor) toxicity in the pregnant rat were similar to those of zinc deficiency under the same experimental conditions, suggesting a role for zinc in prostaglandin metabolism. It was subsequently found that while short-term zinc deprivation decreased the platelet response to minimal levels of aggregating agents, production of arachidonate metabolites was unimpaired (296). Hence it was postulated that the platelet response to products of arachidonate metabolism is reduced. They further showed that prostaglandin synthesis is not inhibited in either male zinc-deficient rats (94) or pregnant female zinc-deficient rats (681). Rather, the results of their studies suggested a failure of prostaglandin function. In the pregnant rat the prostaglandin levels were actually increased rather than decreased. Impaired platelet aggregation in zinc-deficient guinea pigs was also demonstrated (297), and it was concluded that this is a general sign of zinc deficiency in mammals and that the function of physiological eicosanoids is impaired. These workers postulated that their findings reflect a decrease in both number and function of prostaglandin receptors as a result of membrane damage. However, Hwang and co-workers (435) have been unable to confirm an effect of zinc deficiency on platelet aggregation in relatively short-term studies.

E. Zinc and Hormones

Zinc has many recognized and biologically significant interactions with hormones. Zinc plays a role in the production, storage, and secretion of individual hormones as well as in the effectiveness of receptor sites and end-organ responsiveness. Among the most notable effects of zinc deficiency on hormone production and secretion are those related to testosterone and adrenal corticosteroids. Changes in the hormonal milieu can also have profound effects on zinc metabolism. These effects are not covered in this section, but changes in circulating levels of zinc secondary to hormonal changes are considered in Section II,K. The interrelationships of zinc with some hormones, such as insulin (Section IV,D), are considered elsewhere; other selected hormones are included in this section.

1. Growth Hormone

Pituitary tissue levels of growth hormone (GH) are reduced in zinc-deficient rats compared with ad libitum-fed controls but not with pair-fed animals (775). Root and colleagues (775) found that circulating GH levels were significantly lower than those of pair-fed controls in both immature and mature male rats. Kirchgessner (488) showed that while GH levels of severely restricted young rats (<4 μg zinc per gram of diet) are not reduced compared with pair-fed controls, GH levels of those with mild zinc deficiency (6 μg/g) or with a diet in a marginal range (8–12 μg/g), when feed intake is only marginally reduced, are decreased significantly compared with pair-fed animals. The growth retardation of zinc-deficient rats (539,691,734) or pigs (498) did not respond to bovine GH. Lack of a normal rise of GH in response to insulin-induced hypoglycemia has been observed in association with human zinc deficiency, but a failure to respond was also noted in some control children from the same rural area in Egypt (160). Circulating GH has been reported (273) to increase in some GH-deficient children after zinc therapy, but this is not a typical finding. Although some GH-deficient children have also been reported to grow more rapidly when zinc therapy is administered together with GH (273), such findings have not been confirmed by others (765). The growth spurt resulting from GH treatment will increase zinc requirements, and it is quite possible that the zinc supply may then become growth limiting in some children. In these circumstances, a response to zinc supplements would be expected. The balance of available evidence indicates that the growth-limiting effects of zinc deficiency are unlikely to be mediated to a significant extent by changes in GH secretion.

Somatomedin (SM) is presumed to mediate the skeletal growth-promoting effects of GH. Serum SM levels are decreased in zinc-deficient rats, compared to those of ad libitum or pair-fed controls (691). As bovine GH administration only minimally increased these low SM levels, Oner and co-workers (691) concluded that zinc deficiency may decrease the effectiveness of GH in stimulation of SM production. However, they also concluded that zinc deficiency directly decreases the biological effectiveness of SM in stimulating cartilage growth.

2. Gonadotrophins and Sex Hormones

A substantial number of investigations of male sex hormones and spermatogenesis have been reported, but relatively little is known about the effects of zinc deficiency on sex hormones in the nonpregnant female. The major abnormality in the male is testicular hypofunction affecting both spermatogenesis (Section IV,K) and the production of testosterone by the Leydig cells. The cumulative evidence points strongly to a primary defect in Leydig cell function with secondary effects on the pituitary–gonadal axis.

Salem and co-workers (786) found that levels of testosterone and dihydrotestosterone were reduced in the testes of young zinc-deficient animals compared with control animals that were fed the same high-protein diet but with adequate zinc ad libitum. The difference was apparent at 35 and 45 days of age, but had disappeared by day 65. At 49 days, testosterone concentrations of control and zinc-deficient testes were 24 ± 2.4 and 6 ± 1.1 ng/g, respectively. Neither Lei (511) nor Root (775) and their colleagues found decreased circulating testosterone levels in zinc-deficient rats compared with pair-fed controls, but there was an abnormally low gonadal response to luteinizing hormone-releasing hormone (LHRH), indicating deficient reserves of testosterone within the Leydig cells. Prasad and associates have found decreased circulating testosterone levels in zinc-deficient humans, including experimentally zinc-deficient subjects (4), sickle cell disease patients (740), and hemodialysis patients (547). In each instance, testosterone levels increased significantly after zinc therapy. The increases in circulating testosterone in response to LHRH was lower than normal before treatment and increased significantly with zinc therapy. Castro-Magana *et al.* (133) have observed a linear relationship between serum zinc and circulating testosterone levels in boys with constitutional growth delay and familial short stature. Not all studies have been so positive. Two double-blind studies, including that of Brook *et al.* (92), have failed to confirm the findings of Antoniou *et al.* (30) that administration of zinc in the dialysis bath to patients on hemodialysis altered their endocrine status. There were no significant increases in basal testosterone or in testosterone after LHRH administration. Administration of zinc did not normalize testosterone levels or correct impotence and hypogonadism in patients with hepatic cirrhosis (289).

Tissue and circulating levels of hypothalamic–pituitary hormones are consistent with a primary failure of Leydig cell function. No changes in hypothalamic levels of immunoreactive LHRH were found by Gombe and co-workers (292) or Root and associates (775) in zinc-deficient rats. Pituitary LH was normal in zinc-deficient female rats (292), and both LH and FSH were normal in the pituitary of zinc-deficient mature male rats (775). In the pituitaries of immature male rats levels of these gonadotrophins were increased. Circulating basal FSH levels in zinc-deficient animals and humans have been found to be either normal (485,775) or, more frequently, increased (4,511,775,786). Corresponding circulating basal levels of LH have been within the normal range or occasionally (786) elevated. In zinc-deficient hemodialysis patients (547), basal circulating levels of FSH and LH decreased after zinc supplementation. The secretion of LH and FSH after injection of LHRH is significantly higher in zinc-deficient rats than in pair-fed controls (775). Similar enhanced responses have been observed in humans with sickle cell disease, and this abnormal response has been modified with zinc therapy (740).

These changes in male sex hormones in zinc-deficient animals and humans are in contrast to those resulting from protein deficiency, which is associated with decreased levels of FSH and LH and a normal or depressed response to LHRH. When protein deficiency and zinc deficiency coexist, the pattern of change resembles that of protein deficiency rather than zinc deficiency (786). Thus, in retrospect, the sex hormonal changes found in young men in rural Egypt with hypogonadism and short stature (160) suggest that protein deficiency was a significant factor.

Zinc may affect testicular function by activation of adenylate cyclase, thus stimulating testicular steroidogenesis (670). Alternative or additional roles include enhanced biological activity of the gonadotrophin receptor molecules (468, 469).

3. Prolactin

Prolactin concentrations in the pituitary or in the serum were not changed by zinc deficiency in mature or immature male rats (775). Judd and co-workers (453) found that perfusion of dispersed female rat pituitary cells *in vitro* with medium containing 50 μmol zinc acetate caused an acute, sustained, and rapidly reversible inhibition of prolactin secretion. This was a specific effect on prolactin with no change in the release of other anterior pituitary hormones. Reduced prolactin secretion, and to some extent synthesis, have also been observed by Login and co-workers (523), with the use of physiological quantities of zinc *in vitro*. They concluded that this trace element may have a role in the *in vivo* release of prolactin.

4. Thyroid Hormones

In zinc-deficient rats, hypothalamic thyrotrophin-releasing hormone (TRH) and circulating thyroid-stimulating hormone (TSH) did not differ from values for pair-fed control animals (637,775). Morley and co-workers (637) found that thyroid weight, thyroxine (T_4), and ^{125}I thyroidal uptake were also similar for zinc-deficient and pair-fed animals. However, although both groups had serum triiodothyronine (T_3) levels that were significantly lower than that of ad libitum-fed controls, the mean for the zinc-deficient group was also significantly lower than that of the pair-fed controls (637). They postulated that the extrathyroidal conversion of T_4 to T_3 may be impaired by zinc deficiency. Morley (638) later found that T_3 concentrations were low in patients with chronic hepatic and gastrointestinal disorders who also had hypozincemia, but T_3 levels did not increase with zinc supplementation. King's group (944) have reported different findings with mild experimental human zinc deficiency. Specifically, serum TSH, total T_4, and free T_4 declined significantly ($\sim$20%) after 1 month on a low-zinc diet. This was followed 1 month later by a significant reduction in basal metabolic rate from 1.00 to 0.9 kcal/kg per hour.

5. Corticosteroids

Using isolated adrenal glands *in vitro*, Flynn and co-workers (246) found that ACTH did not stimulate corticosterone synthesis when a zinc-chelating agent was added to the medium. The activity of ACTH was, however, restored if zinc was also added to the medium in excess of the chelator. These findings suggested that ACTH is functionally dependent on zinc.

Several investigators have reported adrenal hypertrophy, particularly of the zona fasciculata (514), with zinc deficiency in rats, mice, and pigs (33,193, 194,602,680,750). There have been conflicting results, however, regarding the effects of zinc deprivation on adrenal function. Reeves *et al.* (757) found no significant effects on either resting serum corticosterone concentrations or the ACTH-induced corticosterone response, although there was a tendency for an increase in the latter in the zinc deficient animals. Quarterman and Humphries (750) reported an increased resting adrenal cholesterol concentration as well as a larger decrease in adrenal cholesterol following ACTH administration in zinc-deficient rats compared to controls. These data were interpreted to be incompatible with an increase in ACTH secretion but were suggested to indicate an increased adrenal sensitivity to ACTH in the zinc-deprived animals. These investigators also observed increased plasma hydroxysteroid levels in the zinc-deficient animals, although this effect was more consistent with longer periods of restriction. DePasquale-Jardieu and Fraker (193,194) have documented significantly increased levels of plasma corticosterone in zinc-deficient mice compared to control mice. The differences in corticosterone levels in these studies were also dependent on the duration of the zinc deprivation, and this factor may partially explain seemingly conflicting observations among various studies.

F.   Zinc and Growth

Growth retardation was observed in the original demonstration of zinc deficiency in rats (909) and has been a feature of this deficiency in all subsequent investigations of the young of all species studied.

When the zinc content of the diet of weanling rats is severely restricted (<1 μg/g), growth arrest occurs after 4 to 5 days, as shown by a number of investigators including Williams and Mills in 1970 (986). Lambs and calves fed a severely zinc-deficient diet (1.2 μg zinc per gram) cease weight gain abruptly within 2 weeks (622). There is a graded response to increasingly severe levels of dietary zinc restriction in rats (Fig. 2). When the restriction is only mild (e.g., 9 μg/g in diet under experimental conditions where 12 μg/g allows maximal growth rates), a slight decrease in weight gain is detectable, that is, a 20% reduction in comparison with ad libitum-fed control animals (986). Cyclical changes in food consumption and weight are conspicuous features of severely zinc-restricted rats, but are not apparent with mild dietary zinc restriction. These cyclical changes in weight and the overall reduction in weight gain especially affect skeletal muscle (281). In pregnant rats, maternal dietary zinc deficiency severely impairs fetal growth (424), while such a diet fed during lactation impairs growth in the

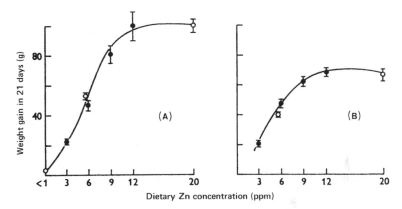

Fig. 2. Influence of dietary zinc concentration on the weight gain of (A) male and (B) female rats during a 21-day period from weaning. From Williams and Mills (986).

suckling pups (54). In the zinc-deficient suckling pup, liver growth is notably depressed compared with overall growth. In the weanling rat, on the other hand, there is a specific skeletal muscle growth depression. However, the zinc content of the skeletal muscle is almost identical to that of control animals (281).

Although most reports of growth in zinc-deficient animals have focused on weight gain, length is also decreased (54). Skeletal growth as measured by tibial–epiphyseal width has been shown to be diminished. There was a high correlation between tibial–epiphyseal widths and serum or femur zinc concentrations (691).

In the human, zinc deficiency has been reported to cause growth retardation in premature infants, infants delivered at term, preadolescent children, and adolescents. Weight loss has been observed in adults. Cessation of weight gain is abrupt in infants with severe zinc deficiency and may occur concurrently with the onset of the acro-orificial skin rash (1001). However, in infants fed intravenously, weight gain may continue after onset of the rash (782). The syndrome of adolescent nutritional dwarfism has been observed in many areas of the world (340). This syndrome may be a composite of several nutrient deficiencies, including protein, energy, and iron, and improvement may occur without zinc supplementation (159). However, since the original hypothesis of Prasad (726), substantial evidence was accumulated from studies in Egypt and Iran during the 1960s and early 1970s to indicate that zinc deficiency can be a major factor contributing to the severe chronic growth retardation that characterizes this syndrome (340,727,773,788). Similar degrees of growth retardation that appear to respond to zinc therapy have been noted in association with certain disease states including Crohn's disease (791) and sickle cell disease (737). However, the data related to Crohn's disease are not definitive, and a growth response to zinc supplements was not observed in another study of sickle cell disease patients (796).

In Jamaica, Golden and Golden (284) found very young children recovering from severe malnutrition on a milk-based diet had low plasma zinc concentrations, but children recovering on a soy protein-based diet had much lower plasma zinc concentrations, lower rates of weight gain, and higher energy costs of tissue deposition (Fig. 3). The latter children had a positive correlation between plasma zinc and rates of weight gain, while the children receiving the milk-based formula had a negative correlation, suggesting that the zinc requirements for rapid growth were responsible for lowering plasma zinc in mild zinc depletion while the rate of growth was dependent on zinc status with more pronounced zinc deficiency. Sixteen children fed the soy formula were then given variable quantities of a zinc supplement (288), which was followed by an immediate increase in weight gain in 14 of the 16 children, indicating that zinc deficiency had been growth limiting during recovery from malnutrition. Unusually high energy costs

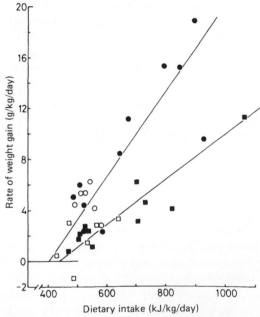

Fig. 3. The relationship between the rates of weight gain (*RWG*) and the dietary energy intakes (*E*) in children before and after zinc supplementation. ■, Soy formula fed children before supplementation; □, cow's milk formula fed children before supplementation; ●, soy formula fed children after supplementation; ○, cow's milk formula fed children after supplementation. Before supplementation: $E = 58RWG + 435, r = .88, p < .001$. After supplementation: $E = 30RWG + 402, r = .84, p < .001$. From Golden and Golden (288).

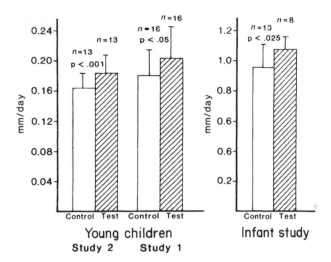

Fig. 4. Linear growth increments of male infants and young children in three studies in Denver, Colorado. Test children received a daily zinc supplement and control children a placebo for 6 months in study 1 (351) and 12 months in study 2 (951). Test infants received a zinc-supplemented cow's milk formula from 0 to 6 months of age, and control infants received the same formula without a zinc supplement in the infant study (948).

of tissue deposition were noted prior to zinc supplementation. The energy cost of tissue deposition fell with zinc supplementation, indicating improved synthesis of lean body tissue. These investigators suggested that chronic zinc deficiency during recovery from severe malnutrition could predispose to obesity.

Several studies in Western countries (106,109,342,347) have indicated an association between biochemical indices of zinc status and physical growth rates of infants and children. These studies alone have been inconclusive with respect to a growth-limiting role of suboptimal zinc nutrition. However, the results of a series of double-blind controlled studies of dietary zinc supplementation by Hambidge and colleagues (351,948,951) have provided strong evidence that a growth-limiting chronic mild zinc deficiency state is a specific single nutrient deficiency problem in some otherwise healthy infants and children in North America. One of these studies was directed to formula-fed infants from birth to 6 months (948). Two have included young children in Denver, Colorado, with low height-for-age percentiles (351,951). In both of these studies the quantities of zinc supplement were very small. In each a significant effect of the zinc supplement on linear growth increments was demonstrated (Fig. 4). As there is no evidence that zinc can have a pharmacological effect on growth rates, it was concluded that these children had a mild, growth-limiting zinc deficiency state. A significant difference in linear growth rates between zinc-supplemented and placebo-treated children was consistently observed for combined sexes and males alone but not females. The reason for this sex difference is unclear, although several other sex-related differences in zinc status have been reported (315,897,986). The mean height increment for the study boys was 12–14% greater than for the pair-matched control boys.

The growth inhibition of zinc deficiency results partly from impaired appetite (i.e., reduced food consumption) and partly from impaired food utilization (602,612,733,859). Decreased appetite is discussed in Section IV, G, and biochemical abnormalities of zinc deficiency that may account for growth inhibition are covered in Section IV,C. Direct local effects of zinc deficiency on the epiphysis are also likely to diminish longitudinal growth (691), and some of the effects on growth could conceivably be mediated through changes in the production or secretion of hormones, including GH and SM.

G.   Food Consumption and Food Efficiency

The adverse effects of zinc deficiency on food intake have been known since the original descriptions of zinc deficiency in animals. Loss of appetite is one of the first signs of zinc deficiency; it occurs rapidly following the introduction of a zinc-restricted diet and, with poor growth, may be the only overt sign of mild zinc deficiency (146,986). Voluntary food intake of severely zinc-restricted rats has been reported to fall as low as one-third that of ad libitum-fed controls (224), and there is a cyclical pattern of food consumption (146,986). Day-to-day variation in food intake varies markedly, and there is increased wastage. Fluid intake is also significantly reduced (224). Force-feeding of zinc-depleted rats with 140% of their voluntary intake, rapidly causes signs of ill health (146). Food intake is less severely affected with mild zinc deficiency, and a cyclical pattern is not evident. The zinc-deficient rat is able to discriminate between diets containing 1 and 6 µg/g of dietary zinc. Food intake increases within 1 to 2 hr of giving a zinc supplement to a zinc-deficient rat (146). A curious observation is that mild stress tends to reverse the anorectic effects of zinc deficiency (224).

In a double-blind study of young children with putative mild zinc deficiency, zinc supplementation was associated with a significant increase in food consumption, compared with placebo-treated controls (494) (Fig. 5). Numerous clinical observations have linked anorexia to severe human zinc deficiency states (348). Anorexia has also been reported as a feature of acute experimental human zinc deficiency (381), although another similar study did not cause zinc deficiency or anorexia (810). Anorexia in patients with renal failure has been found to improve with zinc therapy (42). Pica has also been associated with low zinc levels (178) and has been reported to improve with zinc supplementation (343, 446). Geophgia has been suggested as a contributory cause of zinc deficiency in

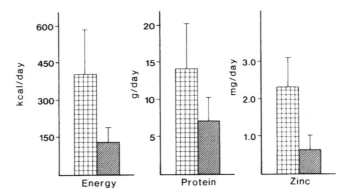

Fig. 5.   Increases in dietary intakes of energy, protein, and zinc between beginning and end of 1-year study period. Children aged 2–6 years, combined sexes. Blocked bars indicate zinc-supplemented children; slashed bars indicate control children. Means ± SEM indicated. Differences between test and control were significant ($p < .05$). From Krebs *et al.* (494).

Iran and in Turkey (135,336), but it is also possible that clay provides a source of zinc for zinc-deficient subjects (841).

Food efficiency ratios are markedly decreased in zinc-deficient rats compared with weight-matched, pair-fed or ad libitum-fed controls (224). Energy and protein utilization are extremely inefficient (758). Neither an increase in basal metabolic rate nor decreased absorption has been demonstrated. The apparent digestibility of food is unaffected, at least in ruminants (615,859). The fecal excretion of nitrogen and sulfur is also unaffected, but the urinary excretion of these elements is greatly elevated in zinc-deficient lambs compared with pair-fed controls (859). Increased urinary excretion of total nitrogen, urea, and uric acid, but not of creatinine has been documented in zinc-deficient rats (416). These rats also had increased activity of liver tryptophan pyrrolase and arginase, indicating increased protein catabolism. The energy cost of growth was markedly increased as a result of zinc deficiency in Jamaican infants with PEM, indicating that synthesis of fat was favored over increase in lean body mass (284).

The pathophysiology of the anorexia of zinc deficiency remains uncertain. Zinc-deficient rats will eat more but not grow more if the protein content of the diet is reduced, and will select a lower protein diet if given a choice (148,758). The cyclical pattern of intake will also disappear with a 5% protein diet. Thus it is clear that there are interactions between zinc and protein in the regulation of food intake. Increased ratios of tyrosine or tryptophan to other neutral amino acids have been reported in the serum of zinc-deficient rats, but abnormal circulating levels of amino acids have not been a consistent finding and did not correlate with the feeding habits of zinc-deficient rats (946,947). In the young child who is zinc deficient, there is no anorexia if severe protein deficiency is also present (284). Zinc may also have direct effects on the central nervous system in the regulation of appetite. Three modulators of feeding responses— norepinephrine, the γ-aminobutyric acid (GABA) agonist muscimol, and the dopamine agonist bromerogocryptine were relatively ineffective in zinc-deficient rats (222). These findings were compatible with the concept that zinc deficiency produces a generalized decrease in receptor responsiveness. Zinc deficiency also depresses the effects of endogenous opiate peptides, which are potent inducers of spontaneous feeding and are thought to have a key role in appetite regulation in the rat (223). Some of the depressant effects of zinc deficiency on appetite could also be mediated through impairment of taste sensation.

H. Zinc and Taste Perception

A physiological role for trace metals in normal taste sensation was first demonstrated by Henkin and associates (377,379). Subsequently, this group worked primarily with zinc because of its ease of administration and low toxicity. They produced experimental human zinc deficiency and taste dysfunction with oral administration of histidine, which depleted total-body zinc and caused, sequentially, anorexia, hypogeusia, and dysgeusia (disordered taste) (385). Higher doses of L-histidine produced cerebellar dysfunction and acute psychosis. Treatment with zinc, even while continuing histidine administration, rapidly restored normal taste function and corrected the other clinical abnormalities. A subsequent single-blind study of zinc therapy in adults with hypogeusia or dysgeusia demonstrated a significant effect of zinc in improving taste function (808). These patients included postinfluenza, idiopathic, post-head trauma, and post-allergic rhinitis cases. These positive results were, however, not confirmed with a double-blind study (387). The failure of this study to confirm a beneficial effect of zinc may have been related to case selection. Plasma zinc concentrations were not correlated with taste perception and appeared to be of no value in selecting subjects for zinc therapy. Parotid saliva samples may be more useful (832), but currently it appears that the measurement of subtle abnormalities in the intestinal absorption of this metal may be required to identify "responders" (389).

Improvement in taste perception following zinc supplementation has been reported in two double-blind studies of adult patients with chronic renal failure receiving hemodialysis (42,546). Conflicting results have been reported for similar studies with children (217,959). Several investigators have found no correlation between plasma zinc and taste thresholds in patients with chronic renal failure (27). In another double-blind study, Greger and Geissler (311) failed to demonstrate any beneficial effect of zinc supplementation on impaired taste perception found in the elderly. Patients with regional enteritis (Crohn's disease) and evidence of zinc deficiency have had demonstrated hypogeusia (567), and taste dysfunction in association with changes in zinc metabolism has been observed in acute viral hepatitis (380). Hypogeusia has also been noted in preadolescent children who were otherwise normal apart from relatively poor growth and evidence suggestive of chronic mild zinc deficiency (342). Taste sensation normalized after very small zinc supplements in these children in an uncontrolled study. Taste perception for moderately salty solutions decreased significantly when mild experimental zinc deficiency was produced in healthy young men (994). The repletion data were very limited but suggested that the changes in taste perception induced by zinc depletion are reversible. In the rat, nutritional zinc deficiency has been associated with altered preference for salt, sweet, sour, and bitter tastants (134). Altered preference was normalized with correction of the zinc deficiency.

Although the evidence points to a physiological role for zinc in taste sensation, the precise mechanism remains uncertain. It appears that zinc is most likely to be of importance in preneural events. A zinc-containing protein in parotid saliva, gustin, is present in unusually low quantities in some cases of idiopathic hypogeusia. It has been hypothesized that gustin influences the function of taste buds (388). Alternative or additional possibilities have been considered (508). In zinc-deficient rats the impairment of taste perception is apparently not due to any alteration in the taste buds (443).

I. Zinc and Immunocompetence

Zinc is essential to the integrity of the immune system. Although the predominant influence of zinc deficiency is on various T-cell functions, the apparent diversity of its effects is illustrated by a partial list of functions reported to be influenced by zinc deprivation: thymic hormone production (438) and activity (43,180), lymphocyte function (147,177), natural killer function (18,139,239,241), antibody-dependent cell-mediated cytotoxicity (139), immunological ontogeny (53,55), neutrophil function (105,121,174), and lymphokine production (60).

Thymic atrophy has been associated with selective zinc deficiency in a variety of animals (255,602,749,750,829) including Friesian cattle with hereditary thymic hypoplasia (lethal trait A46), which arises from a failure of zinc absorption (96). Similar observations have been made in humans with acrodermatitis enteropathica (771) and in infants and children with PEM (850,960). A zinc supplementation trial in Jamaican children who had recently recovered from PEM resulted in an increase in thymic size, supportive of the hypothesis that a persisting zinc deficiency was the cause of the thymic involution (286).

Studies with experimental animals have clarified the nature of the thymic atrophy. While growth retardation or weight loss is characteristic of zinc deficiency (see Section IV,F), the associated loss of lymphoid tissue in general, and the thymus in particular, is of greater magnitude than that observed in any other organ or in total-body weight (139,255). In neonatal mice body weight remained static while thymic weight decreased to a nearly athymic state after 11 days on a severely zinc-deficient diet (1004). Pair-fed adult mice had mean thymic weights somewhat less than that of the zinc-adequate controls, but significantly greater than the mean for the deficient animals (535). Repletion of zinc has been shown to result in complete regrowth of the thymus whether the deficiency was imposed in adult (256) or neonatal (1004) animals, observations consistent with thymic restoration following zinc supplementation of A46 cattle (97) and children recovering from PEM (286).

Although a persistent severe zinc deficiency state will eventually lead to atrophy of the entire organ, there is in the early stages of involution a preferential loss in the cortex, the region populated primarily by immature corticosteroid-sensitive thymocytes (194,256). This led to the question of whether the thymic atrophy of zinc deficiency was due to an increase in circulating glucocorticoids. In support of such a hypothesis, Quarterman and Humphries (750) found increased adrenal gland weight, increased plasma 11-hydroxysteroid concentrations, and decreased thymic size in intact but not in adrenalectomized, zinc-deficient weanling rats. These results and others (294) were interpreted to suggest that the thymic involution and the immunodeficiency associated with it are caused by zinc deficiency-induced adrenal changes. Subsequent work (194), however, demonstrated that in both adrenalectomized and intact mice ~50% of the functional loss in T cells with zinc deficiency occurred before any elevation in corticosterone levels. As glucocorticoid levels rose with continuation of the zinc deficiency in the intact animals, there was a further (20%) reduction in T helper cell function. However, the adrenalectomized zinc-deficient animals eventually exhibited a similar second reduction in immune responsiveness despite significantly lower steroid levels and without significant thymic atrophy.

Although these experiments suggest a critical role for zinc in lymphocyte integrity and function, the specific mechanism underlying the thymic atrophy associated with zinc deficiency is as yet unclear. Good and associates (294) suggested that decreased activity of terminal deoxyribonucleotidyltransferase, a zinc-containing DNA polymerase that is nearly exclusive to the thymus and immature thymocytes (566), may account for the unique sensitivity of this organ to zinc deficiency. With the depression of the DNA content of the thymus observed in zinc deficiency (497), an arrest in the growth, division, and differentiation of precursor cells into immunocompetent thymocytes might be expected. Bach (43) proposed that zinc deficiency may alter thymic epithelial function and impair thymic hormone production, which in turn affects T-cell maturation in the thymus and in the periphery. Another hypothesis is that of zinc acting to inhibit glucocorticoid-induced DNA cleavage by endonucleases, thus protecting the thymocyte (161).

Using moderately zinc-deficient weanling rats in which growth and lymphoid organ weights were not significantly lower than those of pair-fed controls, Gross *et al.* (316) demonstrated significantly reduced blastogenic responses to various mitogens, including PHA, pokeweed mitogen (PWM), and concanavalin A (Con A), all of which are T-cell mitogens or require intact T-helper function. Differences in responses of thymic lymphocytes and splenic and peripheral lymphocytes of the zinc-deficient animals to the Con A mitogen was suggested to indicate possible variations in zinc sensitivity among subsets of thymic cells. These findings have been corroborated in studies of human zinc deficiency syndromes (17,72,140,689,712), in which *in vitro* lymphocyte responses to PHA stimulation have been significantly reduced relative to control values prior to zinc therapy and have normalized to control levels following zinc supplementation.

In general, investigations of the effects of zinc deficiency on antibody-mediated responses (139,239,255,256,535,1004) suggest that the ability of zinc-deficient animals to mount a primary immune response (mainly T cell-independent IgM antibodies) to antigens such as sheep erythrocytes may be dependent on the duration of the zinc-deficient diet and the age of the animals. Prolonged deprivation in adult animals or imposing the deficiency on young animals (1004) will result in impairment of B-cell response, demonstrated by depression of both IgM production and response to the B-cell mitogen dextran. These results suggest that B-cell function is not independent of zinc status but is less sensitive to zinc deficiency compared to the T-cell system. The secondary response (mainly IgG antibodies) is, however, consistently reduced and to a greater extent than the primary response in zinc-deprived animals. The dependence of the secondary response on T-cell helper function to promote the IgM-to-IgG "switch" and its susceptibility to zinc deficiency were demonstrated by a recovery of the secondary responses from 10 to 61% of control values when zinc-deficient mice were injected with viable thymocytes prior to immunization (255). Similar results for primary and secondary antibody-mediated responses were found with severely zinc-deficient mice and with pair-fed controls receiving adequate zinc, but the latter required an extended period of time on the restricted diet to observe significant reductions in both thymic weight and in immune responsiveness (535). These and other (239) observations suggest that while the food intake restriction associated with the zinc deficiency eventually contributes, possibly synergistically, to the immune depression, the zinc deficiency per se has a specific effect on immune function that precedes the severe inanition and growth failure.

Further evidence for a critical role of zinc in normal T-cell function is provided by the effects of zinc deficiency on delayed-type hypersensitivity (DTH) responses, which involve a complex interaction between macrophages and a subpopulation of effector T cells called T_{dth} cells. Adult mice fed zinc-deficient diets and subsequently percutaneously sensitized with dinitrofluorobenzene gave significantly lower DTH responses compared to both pair-fed and ad libitum control mice (257), suggesting that zinc deficiency per se, not just inanition, has a negative effect on DTH sensitivity. Repletion of the zinc-deficient mice in this experiment resulted in completely normalized DTH responses. This is consistent with a number of observations in the human clinical literature that have reported depressed DTH responses in patients with apparently isolated zinc deficiency, typically in the moderate to severe range, that responded positively to zinc supplementation. These include such diverse clinical conditions as acrodermatitis enteropathica (140,771), prolonged hyperalimentation with inadequate zinc (17,689,712), obesity (140), and Down's syndrome (72). Infants and children with PEM showed significantly larger DTH reactions to intradermal injection of *Candida* antigen on the arm that had received a topical ointment containing zinc sulfate compared to the arm that received only a placebo ointment (287). Although these children had multiple nutrient deficiencies, the positive responses to the provision of zinc alone suggests that zinc deficiency was specifically responsible for the diminished responses on the placebo arms. Exactly where the impact of zinc deprivation is in the complex DTH response has not been elucidated. It is also not clear whether the positive effects observed in the cases using systemic zinc supplementation were mediated by the same mechanisms as those of the topical zinc application.

The effects of zinc deprivation on humoral immunity are much less clear. Cunningham-Rundles and associates (177) have reported that zinc is an effective B-cell activator as well as being capable of acting synergistically with B-cell mitogens. Other investigators (998) found B-cell mitogenesis unaffected by zinc deficiency *in vitro*. Reports of disturbed immunoglobulin profiles associated with human zinc deficiency syndromes have also been inconsistent (18,89,105,176,454,689,990). Studies in experimental animals by Beach *et al.* (53,55) suggest that zinc deprivation during gestation and in early postnatal life may profoundly affect immune system ontogeny as reflected by altered immunoglobulin profiles.

Work by Chvapil and associates with animal models (155,156) has suggested that the actual intracellular quantity of zinc may be critical for functional neutrophil activity. In the presence of 1.2 mM magnesium, a wide physiological range of zinc concentrations had no effect on oxygen consumption of resting neutrophils. However, a marked effect of zinc was observed in cells that had been stimulated to become phagocytic. Oxygen uptake in the stimulated cells was progressively inhibited by increasing concentrations of zinc. Bactericidal activity was also inhibited by 36 and 60 μM zinc. Experiments in which cells were incubated with varying concentrations of zinc in the media demonstrated that the inhibitory effects of zinc reflected the intracellular content rather than the zinc concentration of the media. Moreover, the prompt reversibility of the inhibition of increased intracellular zinc concentrations suggested that the intracellular zinc remained in an inorganic, readily diffusible form. Comparable studies in humans are limited, but observations by Lennard *et al.* (512) on neutrophils from patients with severe burns also suggested that maximum phagocytic activity occurs in a relatively narrow range of *in vitro* zinc concentrations. Optimal zinc concentrations for neutrophilic phagocytosis were below those of normal human plasma but were equivalent to the low plasma zinc levels typically found during bacterial infections and other disease processes associated with a prominent inflammatory response.

In vitro studies on cells from patients with zinc deficiency have helped to elucidate the role of zinc in neutrophil responsiveness. Weston *et al.* (974) found defective chemotaxis in the neutrophils from peripheral blood of patients with acrodermatitis enteropathica. Improved chemotactic responses were associated both with *in vitro* incubation of cells with zinc and with oral zinc therapy. Briggs *et al.* (89) also found reduced chemotaxis in cells from hemodialysis patients who were not receiving zinc supplements. These investigators reported a significant correlation between neutrophil zinc concentration and "chemokinetic" response.

Neutrophils may regulate plasma zinc concentrations by their secretion (following activation) of hormonelike substances that lead to hepatic sequestration of zinc. The lowered plasma zinc concentrations are proposed to enhance neutrophil phagocytic and bactericidal functions (59).

J. Zinc and Keratogenesis

Alopecia and gross skin lesions were observed in the early investigations of zinc deficiency in rats and mice. Histological studies disclosed a condition of parakeratosis, that is a thickening or hyperkeratinization, with failure of complete nuclear degeneration, of the epithelial cells of the skin and esophagus (248). In more severe zinc deficiency, scaling and cracking of the paws with deep fissures develop, in addition to loss of hair and dermatitis (252). In pigs, parakeratosis mostly occurs around the eyes and mouth and on the scrotum and lower parts of the legs. A similar distribution occurs in zinc-deficient ruminants, with the legs becoming tender, easily injured, and often raw and bleeding (74, 920). Severe skin lesions occur in the lethal trait A46 in Friesian cattle, which results from an inherited defect in the intestinal absorption of zinc (966). In the zinc-deficient squirrel monkey (*Saimiri sciureus*), parakeratosis of the tongue has been observed in addition to alopecia (542). The healing response of the parakeratotic lesions to supplemental zinc is rapid and dramatic. Experimental zinc deficiency in the dog (802) and the cat (456) also causes characteristic disorders of keratogenesis.

The influence of zinc deficiency on keratogenesis is particularly evident in sheep through changes in the wool and horns. In horned lambs the normal ring structure disappears from new horn growth, and the horns are ultimately shed, leaving soft, spongy outgrowths that continually hemorrhage (622,920). In breeds that show no horn growth, prolonged zinc depletion leads to the development of keratinous outgrowths or "buds" in the positions occupied by horns in other breeds (622). Changes in the structure of the hooves can also occur, and the effects of zinc deficiency on wool growth in lambs are striking. The wool fibers lose their crimp and become thin and loose, and are readily shed. Sometimes the whole fleece is shed and no further wool growth occurs until additional zinc is supplied, which results in immediate regrowth (920). Posthitis and vulvitis have also been observed in experimentally induced zinc deficiency in lambs, associated with enlargement of the sebaceous glands followed by an increased secretion of sebum (190).

In the zinc-deficient chick (677,678), poult (493), and pheasant (822), feathering is poor and abnormal, and dermatitis is usual. In the Japanese quail, feathering is severely affected (253) owing to degeneration of the feather follicles resulting from hyperkeratosis (678). Involvement of zinc in keratogenesis is further evident from the gross disturbances of the integument, observed in chick embryos from the eggs of severely zinc-deficient hens (476).

Facial eczema, a disease of sheep and cattle occurring in New Zealand on pastures infected with the fungus *Pithomyces chartarum* and producing the toxin sporidesmin, shows some clinical response to supplemental zinc at high doses (767). The interdigital skin lesions of infectious pododermatitis in male cattle have also been shown to respond to zinc therapy (82,191), following a suggestion by Bonomi (80) that a high incidence of infectious pododermatitis (*zoppina*) in cattle in some parts of Italy might be related to a low-zinc status of the animals. Rapid and complete repair of severe interdigital lesions of infectious pododermatitis was obtained by Dermertzis and Mills (191) in 11 of 12 young bulls given zinc supplements.

Severe human zinc deficiency is also characterized by disorders of keratogenesis including cutaneous lesions with a characteristic acro-orificial distribution, alopecia, and nail dystrophies (359,647). The dermitis quite commonly also involves flexures and friction areas and may become more generalized. Eczematoid, psoriaform, vesiculobullous, and pustular lesions may coexist. The earliest skin lesions are bright reddish, nonscaly macules and patches. Acute lesions may peel off, leaving moist erosive areas. Vesicles, pustules, and bullae are seen frequently but not in all cases. After 1 to 3 weeks the skin lesions turn into either scaly brownish patches or psoriaform plaques. Mucous membranes are characteristically involved with stomatitis, cheilitis, glossitis, and conjunctivitis (662).

Although the histology of the cutaneous manifestations of acrodermatitis enteropathica and other severe zinc deficiency states has generally been regarded as nonspecific, Gonzalez and colleagues (293) have attempted a more detailed classification of the histopathological findings according to their time of presentation. Early lesions are characterized by loss of the granular layer of the epidermis, replacement of this layer by clear cells, and focal parakeratosis. If untreated, the epidermis becomes increasingly psoriaform, the parakeratosis becomes more and more confluent, and the pallor of the upper part of the epidermis becomes more prominent (Figs. 6 and 7). Still later the pallor disappears but the psoriaform hyperplasia persists. As the lesions involute, only parakeratosis remains. While the lesions are primarily located in the epidermis, dilation and tortuosity of vessels is seen in the dermis. These investigators concluded that pallor of the upper part of the epidermis, attributable to the formation of balloon cells with pyknotic nuclei, is the single most important histological finding. Abnormal accumulation of cytoplasmic lipid droplets and intracellular edema in the spinous layer are the electron microscope findings that correspond to the epidermal pallor under the light microscope. Other frequent epidermal changes, depending on the stage of the disease, are confluent parakeratosis and epidermal hyperplasia. Spongiosis, vesiculation, crusts, cleft formation, epidermal necrosis, dyskeratosis, acantholysis, and exocytosis may also

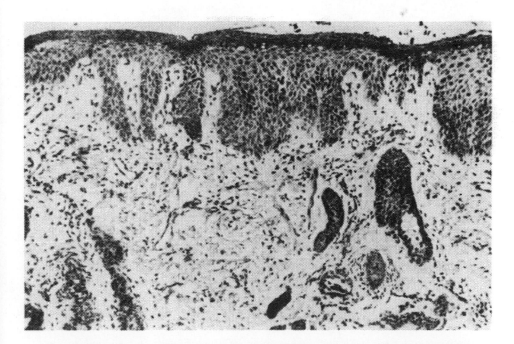

Fig. 6. Histological appearance of a psoriaform plaque of acrodermatitis enteropathica. This photomicrograph demonstrates the marked epidermal hyperplasia and confluent parakeratosis like those of psoriasis. Pallor in the epidermis has been lost, and only a few vacuolated cells are present in the granular layer (×10). From Gonzalez et al. (293).

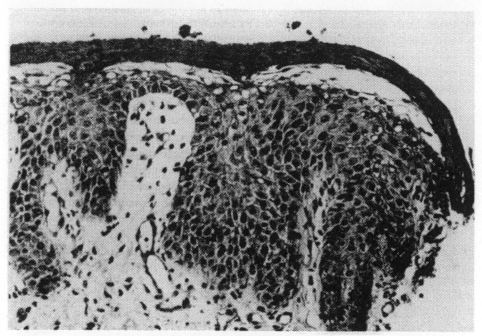

Fig. 7. High-power view of a psoriaform plaque of acrodermatitis enteropathica. Note the confluent parakeratosis overlying epidermal pallor, which in turn overlies an area of psoriasiform hyperplasia (×20). From Gonzalez *et al.* (293).

be seen. Montgomery (632) has reported subcorneal or intraepidermal serum-filled vesicles and bullae associated with inflammatory-cell infiltrate in the dermis.

Zinc-related changes in the metabolism of skin acid mucopolysaccharides (AMPs) occur in pigs (907). No differences in total or sulfated AMPs in the skin from zinc-deficient and pair-fed pigs were observed, but deficiency caused a significant increase in hyaluronic acid. Hsu and Anthony (415) similarly found no alteration in sulfated AMPs of zinc-deficient rat skin by ^{35}S incorporation. However, reduced collagen synthesis and apparent alterations in collagen cross-linking in zinc-deficient rat skin have been reported (569). A decrease in collagen production in swine skin has also been reported.

K. Zinc and Reproduction

Spermatogenesis and the development of the primary and secondary sex organs in the male and all phases of the reproductive process in the female from estrus to parturition and lactation can be adversely affected by zinc deficiency.

Atrophic seminiferous tubules were observed in the earliest study of the histopathology of zinc deficiency in the rat (248). Millar *et al.* (598) found that the degree of reversal of testicular damage depended on the degree of degeneration. Barney *et al.* (46) fed a zinc-deficient diet to weanling rats for 44 days and found immaturity of the germinal epithelium with a decreased number of spermatids. Diamond *et al.* (197) showed that lesions present after 4 weeks on a zinc-deficient diet were reversible after 15 days of zinc repletion. Mason *et al.* (556) also found lesions to be completely reversible in weanling rats that received a severely zinc-deficient diet for 4 weeks, but when similar rats were fed a marginally deficient diet for 13 weeks degenerative changes were not completely reversed. They also found that if the zinc-deficient diet was fed after puberty, degenerative changes were not reversible. Zinc may also be important for normal head–tail attachment of ejaculated human sperm (73), and an antibacterial substance secreted by the prostate gland has been reported to be a zinc complex (230).

Impaired development and functioning of the male glands are apparent in other species. Hypogonadism occurs in zinc-deficient bull calves (614,720), kids (609), and ram lambs (920). In lambs (920), testicular growth was greatly impaired and spermatogenesis ceased within 20 to 24 weeks on a diet containing 2.4 µg zinc per gram, whereas no such effects were observed in comparable pair-fed lambs on a zinc intake of 32.4 µg/g. The body growth and food consumption of the two groups of lambs receiving unrestricted diets containing 17.4 and 32.4 µg zinc per gram were similar, but testicular growth and sperm production were significantly greater in the rams receiving the larger zinc supplement. These differences were further reflected in the histological ratings given to the testes of the animals from the different treatments. Remission of all signs of zinc deficiency and recovery of testicular size, structure, and sperm production was achieved during a zinc repletion period lasting 20 weeks, indicating that the tissue had not been permanently damaged by the severe zinc deficiency imposed. Complete recovery was similarly obtained by Pitts and co-workers (720) with repletion of zinc-deficient calves.

Hypogonadism, with retarded development of the secondary sexual characteristics, is a conspicuous feature of the zinc deficiency observed in young men in Iran and Egypt (726,727,773,774), but there are indications that this cannot be attributed entirely to zinc deficiency (Section IV,E). Zinc deficiency has also been reported to be a reversible cause of gonadal dysfunction in some cases of sickle cell disease (3) and uremia (545). Zinc supplements may restore normal sexual development and raise testosterone levels and sperm count in these circumstances. Reversible oligospermia has been reported by Abbasi *et al.* (4) in mild human experimental zinc deficiency. The specific adverse effects of zinc deficiency on the male reproductive system are the result of local effects in the gonads, rather than at the level of the hypothalamus or pituitary.

The importance of adequate maternal zinc nutrition for reproduction was first established in 1959 by Turk *et al.* (913), who showed that chicks hatched from hens that were fed a zinc-deficient diet were weak and died within 4 days. Gross congenital malformations affecting 50% of the embryos of more severely zinc-deficient hens were reported by Blamberg and co-workers (76) the following year. These findings were confirmed and extended by Kienholz *et al.* (476). The first report of congenital malformations resulting from zinc deficiency in mammals was published by Hurley and Swenerton in 1966 (424). The effects of zinc deficiency in the female rat were found to depend on the severity, timing, and duration of the deficiency. When Hurley and Swenerton (424) fed a nearly zinc-free diet to female rats from weaning to maturity, the animals failed to grow and displayed severe disruption of the estrous cycles. In most cases no mating took place and the animals were infertile. Similar observations have been made in bonnet monkeys (898). Watanabe *et al.* (957) found reduced ovulation in zinc-deficient hamsters and failure of normal maturation of oocytes in zinc-deficient mice. Even marginal zinc deficiency was found to have a possible adverse effect on the maturation process of oocytes. Hurley and Schrader (430) found that zinc was required for normal development of the preembryonic conceptus; zinc deficiency during the first days of pregnancy resulted in abnormal cleavage and blastulation in preimplantation eggs as early as day 3 postfertilization.

The period of greatest apparent vulnerability to maternal zinc deficiency occurs during embryogenesis (208). When Hurley (431) fed a marginally deficient diet (9 μg zinc per gram) from weaning until mating and then a severely zinc-deficient diet (<1 μg/g) during pregnancy, 4% of the implantation sites were resorbed and 99% of the total sites were affected. Of the young living at term, 98% exhibited gross congenital malformations involving numerous organ systems (Table V, Fig. 8). Similar results were obtained even when rats had received a zinc-adequate diet prior to breeding. Malformations of the central nervous and skeletal systems were most prominent. Zinc restriction during early neurogenesis affected the development of many derivatives of the basal, alar, and roof plates of the primitive neural tube, and defects included agenesis and dysmorphogenesis of the brain, spinal cord, eyes, and olfactory tract. Forty-seven percent had brain malformations, 42% had eye malformations, and 3% had spina bifida. Developmental defects of the anterior brain appeared to involve primarily the closure of the aqueduct, resulting in hydrocephalus. Anencephaly and exencephaly, the closest equivalent of human anencephaly, also occurred quite frequently. The commonest lesions of the eye were microphthalmia or anophthalmia. Warkany and Petering (953,954) found similar central nervous system lesions and noted that they occurred with only 3 days of maternal dietary zinc restriction. Moreover, microscopic lesions were apparent even in brains that were grossly normal. Mills *et al.* (623) found hydrocephalus and microphthalmia in the offspring of female rats that received a diet containing 1 μg zinc per gram.

Table V
Types of Fetal Abnormalities
Associated with *in Utero* Zinc
Deficiency in Rats[a,b]

Curly tail
Syndactyly or missing digits
Hydrocephalus   ,
Exencephalus
Microphthalmia or anophthalmia
Cleft palate
Spinal curvature
Clubbed foot
Herniations
Heart abnormalities
Lung abnormalities
Urogenital abnormalities

[a]Data from References 204, 424, 623,
and 953; summarized by I. E. Dreosti in
Reference 208.
[b]Abnormalities are arranged in approximate order of frequency.

Defects in the closure of the neural tube and other problems in differentiation of the nervous system have also been observed in early chick embryo explants cultivated in a zinc-deficient medium (436). A large number of skeletal defects were seen, including 83% with tail malformations, 80% with fused or missing digits, 47% with scoliosis or kyphosis, 34% with clubbed forefeet, 34% with cleft palate, and 38% with short or absent mandible. Long bones were often short and in some cases were missing. Doming of the skull and poor ossification of the cranial bones were noted. Even with milder zinc deficiency (1–3 μg/g of dietary zinc) the frequency of skeletal malformations is high, with severe anomalies of long bones, vertebrae, and ribs. Soft tissue abnormalities were also very common including malformations of the heart and urogenital system. Abnormal lungs were found in 54% of fetuses. Biochemical development of the pancreas was found to be impaired (770), as evidenced by decreased activity of pancreatic chymotrypsin and carboxypeptidase A, and decreased glucagon concentration.

Dreosti (208) has also studied the teratogenic effects of maternal zinc deficiency and has defined the teratogenic zone of diet zinc concentration in the rat (0.25–6.5 μg/g). Transitory dietary zinc restriction during embryogenesis also produced congenital malformations (428). When the zinc-deficient diet was imposed from day 6 to day 14 of pregnancy almost half the young showed gross congenital malformations. Transitory periods of zinc depletion alter the pattern of anomalies produced according to the embryological events occurring at the time of the dietary deficit; thus organs developing early in gestation such as the

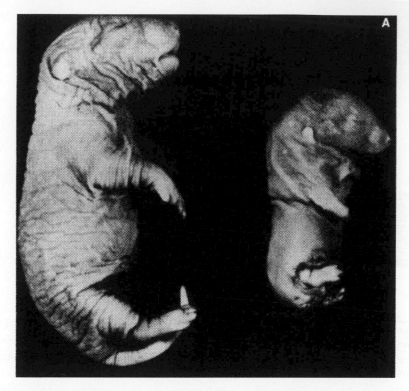

Fig. 8. (A) Typical appearance of full-term fetus from zinc-deficient rat (right) compared with control (left). Note small size, abnormal shape of head and body, short limbs (micromelia), fused or missing digits (syndactyly), short lower jaw (micrognathia), and absence of tail in zinc-deficient

brain and the eye were affected by zinc deficiency early in pregnancy. Later events such as closure of the palate were seldom affected by deficiency before day 12 of pregnancy. Hurley concluded that the rapid and severe effects of even short periods of zinc deficiency on fetal development suggested that the pregnant rat cannot mobilize zinc from body stores in amounts sufficient to supply the needs of developing embryos.

Record and colleagues (756) have investigated the effects of cyclical feeding resulting from maternal zinc deficiency on maternal plasma zinc concentrations and on embryogenesis. Zinc-deficient rats exhibited individual feeding–fasting cycles with a period of ~4 days. Maternal serum zinc levels were inversely correlated with dietary intake during the previous night. Comparison of the feeding cycles with the known times of development of individual organ systems suggested that a period of high food intake and hence low serum zinc levels was related to the production of specific organ malformations. When the cyclical

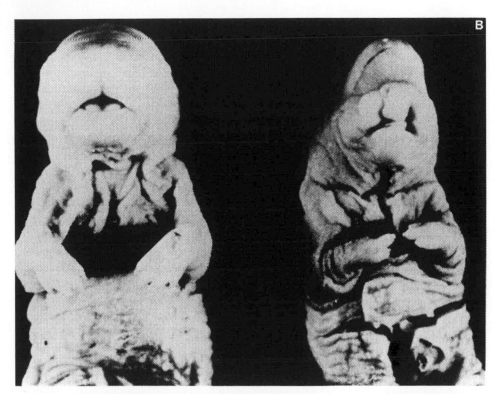

fetus. (B) Cleft lip and domed skull (due to hydrocephalus) in full-term rat fetus from zinc-deficient female (right) compared with control (left). Note also syndactyly in zinc-deficient fetus. From Hurley (431).

pattern of feeding of the zinc-deficient diet was deliberately controlled with low intake on days 6 and 7 followed by high intake on days 8 and 9, examination of the embryos on day 11 revealed 70% with major malformations especially of the central nervous system. This incidence was reduced to <5% in a separate group consuming large amounts of a diet on days 6 and 7 followed by low intakes on days 8 and 9. These observations provided an explanation for the wide variation in incidence of congenital malformations between litters.

There is suggestive but as yet inconclusive evidence that maternal zinc deficiency may be a cause of abnormal human fetal development. Sever and Emanuel (827) noted that areas of the world where zinc deficiency had been identified (e.g., Egypt and Iran) were also areas where there are high incidences of anencephaly. The limited data available on the obstetrical histories of humans with acrodermatitis enteropathica who had never received zinc therapy are also suggestive (346). Of three women recorded in the literature, two had offspring with lethal congenital malformations, one each of the nervous and skeletal systems. In Sweden, Jameson (444) noted that 5 of 8 mothers whose offspring had congenital malformations had the lowest plasma zinc concentrations in a series of 234 pregnant women. These malformations were a ventricular septal defect, hypospadias, undescended testes, dislocated hip, and multiple skeletal malformations. Serum zinc concentrations during the second trimester of pregnancy have been reported to be lower in women whose offspring have neural tube defects than in control pregnancies (98). Turkish mothers of anencephalic offspring have been reported to have significantly lower plasma zinc concentrations at delivery than mothers who had had normal offspring (136). In Ireland, maternal serum zinc at delivery was lower than normal when offspring had malformations (857). In Germany (360), women whose offspring had anencephaly were reported to have higher hair concentrations of zinc than normal, but the opposite has been reported in Ireland (665).

Fetal growth retardation may be the only apparent physical defect in the offspring of zinc-depleted mothers. Reduced birth weights have been reported in rats (429,988), grazing sheep (558), penned sheep (560) and the male offspring of zinc-depleted rhesus monkeys (432). In rats and monkeys, reduced birth weight compared with that of offspring of pair-fed controls, indicated that intrauterine growth retardation cannot be attributed to reduced maternal food intake. However, zinc-deficient pregnant sheep, rats, and monkeys exhibit decreased food intake, and there is a failure of normal weight gain during pregnancy. Apgar (35) found that zinc restriction starting after day 120 of gestation in the pregnant ewe was associated with fetal growth retardation. Zinc depletion during the post-embryonic period has also been associated with behavioral abnormalities in the offspring (Section IV,N).

Human serum or plasma zinc concentrations in the second trimester have shown a negative rather than a positive correlation with birth weight (145,207,579,593,742). Mothers of offspring with evidence of intrauterine growth retardation have been reported to have abnormally low leukocyte zinc concentrations in midpregnancy (582,706); the nutritional significance of these observations is unclear.

Zinc deficiency in late gestation in the rat causes severe problems in parturition. Apgar (32) found that rats fed a zinc-deficient diet throughout gestation delivered their litters with extreme difficulty, suffered excessive bleeding, and failed to consume afterbirths or to prepare a nest site (Fig. 9). Institution of a low-zinc diet as late as day 15 of pregnancy can also result in difficult parturition. Administration of zinc by day 19 of pregnancy can protect a rat that has become zinc deficient during pregnancy from severe stress at parturition. Administration as late as day 21 may protect some though not all of the females. O'Dell and co-workers (680) noted, in addition, that the zinc-deficient dam had significant decreases in body temperature, hypotension at or immediately following

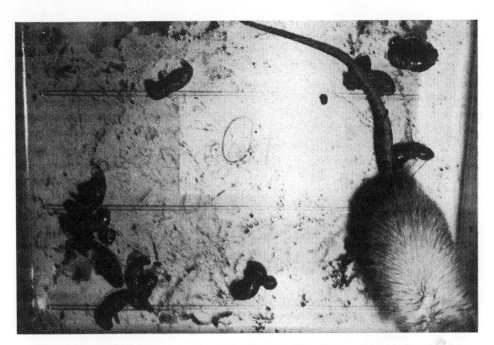

Fig. 9. Cage of zinc-deficient female rat 2 hr after parturition. Note excessive bleeding, failure to consume afterbirths, and lack of a nest site. Photo courtesy of Dr. J. Apgar.

parturition, and delayed onset of parturition; perinatal mortality was increased in both dams and fetuses. The principal inhibitory agent to parturition is progesterone, which in the rat is produced by the corpora lutea. Bunce and co-workers (101) showed that the preparturient decline in plasma progesterone in zinc-deficient rats began at the same time as in control groups, but the progesterone did not decline to the same extent in the zinc-deficient rats. Induction of ovarian 20-α-hydroxysteroid dehydrogenase activity was delayed for about 8 hr by zinc deficiency. This is an enzyme that converts progesterone into the biologically inactive C-20 hydrogenated metabolite. This delay was not observed if prostaglandin $F_{2\alpha}$ was injected previously. They concluded that there was a defect in a zinc-dependent step in uterine synthesis and/or release of prostanoids. Apgar (35) has not found evidence of delayed and prolonged parturition in moderately zinc-deficient ewes, although the zinc deficiency was sufficient to produce fetal growth retardation. Abnormal parturition was not specifically reported in rhesus monkeys maintained on a zinc-deficient diet (4 μg/g) throughout pregnancy that was sufficiently severe to produce dermatitis during the third trimester (291).

Jameson (444) reported that women with abnormal deliveries, especially inefficient labor and atonic bleeding, as well as those who gave birth to immature infants, showed lower serum zinc concentrations during early pregnancy than women with normal deliveries and normal infants. Initially low (446) and subsequently decreasing serum zinc levels were associated with a high incidence of complications. Jameson (445) reported that women with relatively low serum zinc at 14 weeks' gestation had a lower incidence of obstetrical complications if given zinc therapy. Low maternal plasma zinc has been associated with increased blood pressure (145), and zinc supplementation was associated with a lower incidence of pregnancy-induced hypertension (423). In a third study in the United States, however, there was no relationship between plasma zinc and blood pressure (649). In the latter study, low maternal plasma zinc in early and midpregnancy was associated with a higher incidence of fetal distress and an increased total incidence of obstetrical complications. McMichael *et al.* (578) found that initially low and decreasing serum zinc levels are associated with a moderate increase in complications of delivery and neonatal functioning but not with abnormalities of fetal and placental development. A possible association between low urine zinc and complications of delivery has also been noted (360). Lower activities of selected zinc metalloenzymes in non-zinc-supplemented women compared with those receiving a dietary zinc supplement have been reported (360,423).

The poor survival of rat young born to females fed a zinc-deficient diet for a short period during pregnancy could not be due to poor postnatal zinc nutrition, because the concentrations of zinc in both the milk and plasma of mothers after birth were unaffected, as were the plasma zinc levels of the offspring (428). The presence of congenital abnormalities and failure to suckle due to weakness at birth were thought to be the factors involved. Mutch and Hurley (653) subsequently showed that zinc deficiency imposed on rats during lactation rapidly reduced plasma zinc levels and caused an impairment in milk production that was specifically due to the lack of zinc rather than to inanition. The zinc level in the milk was also reduced, so that the pups received only half the amount of zinc that the control pups received on the basis of body weight. These pups became zinc deficient, as evidenced by reduced plasma zinc levels, impaired growth, and increased mortality. The zinc level in the milk of mildly zinc-depleted dairy cows (619) and probably in the mildly zinc-deficient human (495), is reduced. The effects of this reduction on the offspring have not been determined.

L.   Zinc and Wound Healing

At a molecular and cellular level, there are clear indications that zinc requirements must be relatively high at the site of wound healing. Zinc is necessary for rapid cell replication and differentiation as tissue is repaired and epithelialization occurs, and for achievement of normal tensile strength of the collagen matrix

(789). Both DNA and collagen content of early wounds are reduced by zinc deficiency, and increased with correction of the zinc deficiency (218,240). The effects of zinc deficiency and repletion on collagen synthesis appear to be part of a generalized effect of zinc on protein synthesis. In zinc-deficient animals the evidence seems unequivocal that the rate of wound healing is impaired and that this can be accelerated by supplemental zinc (933). Strain and co-workers (884) observed in 1953 that zinc supplements resulted in accelerated wound healing in rats. Retrospectively, it has been determined that the chow fed these rats was fortuitously deficient in zinc (725). Subsequently, the need for zinc for optimal rates of wound healing has been demonstrated in a variety of other animals including hamsters (592), guinea pigs (50), and calves (613). Different types of wounds, including ulcers and incised and excised wounds, in several different tissues have been examined. Although skin wounds have been studied most extensively, zinc requirements for healing have also been observed in gingival mucosa, in alveolar bone after tooth extraction, and in fractured long bone (50). Uptake of ^{65}Zn into wounds is very rapid (806). Zinc deficiency affects both the epithelial and fibroblastic components of wound healing. Impaired epithelialization may be apparent as soon as 3 to 4 days (592), but differences in the tensile strength of the collagen matrix cannot be detected until 10 to 15 days (789). Zinc supplements given before the wound occurrence may be particularly advantageous in counteracting the effects of zinc deficiency. Even if given only 30 min preoperatively, zinc supplements result in greater hydroxyproline levels in the wound by the fourth day, compared to zinc given on the day after injury (902, 903).

Although there have been conflicting reports on any pharmacological effect of zinc therapy in accelerating wound healing in animals whose zinc status is normal, the strong consensus of opinion is that the effects of zinc on wound healing are limited to correcting impairment of healing secondary to an underlying zinc deficiency state (933). The normal zinc status of the animals used in some experiments accounts for some of the reports of failure of an effect of zinc supplements on the rate of wound healing. In other studies this failure may have resulted from the delay period between the critical events of cellular proliferation and protein synthesis and their subsequent effect on tensile strength. If the latter is measured too early, the effects of zinc deficiency are likely to be missed.

Numerous studies of the effects of zinc on wound healing in the human have produced conflicting results the reasons for which are not completely understood. The first indications of a role for zinc in human wound healing came from the studies of Pories and associates (723), who demonstrated a significantly increased rate of healing, compared with untreated controls, when zinc sulfate at the rate of 50 mg zinc three times daily was administered to young men following surgery for pilonidal sinus. Van Rij and Pories (932) have summarized subsequent observations that have demonstrated beneficial effects of zinc therapy.

These studies include controlled trials on healing of venous leg ulcers (433), graft acceptance and epithelialization in children with major burns (506), healing of gastric ulcers (267) and of indolent ulcers, and improved healing in surgical patients with abnormally slow healing wounds (391), and in the wounds of patients on prolonged corticosteroid therapy (247). Although apparent responses have occurred with or without the presence of low serum zinc levels, slow wound healing has been associated with low serum zinc levels in several studies. Several groups of workers however have failed to demonstrate beneficial effects of zinc therapy on wound healing, results that van Rij and Pories (932) have attributed primarily to the absence of a zinc deficiency state. The quantities of zinc characteristically administered in both human and animal studies have been much greater than those likely to be required to treat a zinc deficiency state, even accounting for the abnormalities of zinc metabolism in the postsurgical patient that may aggravate any preexisting deficiency. Studies using much smaller quantities of zinc could help to differentiate further between a physiological and pharmacological effect of zinc in accelerating wound healing.

M. Zinc and Skeletal Development

Skeletal abnormalities are a prominent feature of zinc deficiency in growing birds. This has been demonstrated in chicks, poults, pheasants, and quail (253,493,677,678,822,996). The long bones are shortened and thickened in proportion to the degree of zinc deficiency (1000). Changes and disproportions occur in bones, giving rise to a perosis histologically similar to that of manganese deficiency (677,678,996). Leg defects do not develop in zinc-deficient chicks on all types of diets, even when other signs of the deficiency are apparent (668). These "perosislike" or "arthritislike" abnormalities can be alleviated by histamine, histidine, and various antiarthritic agents without affecting the other manifestations of deficiency (666,667). Supplementary zinc prevents both the leg disorder and the other signs of zinc deficiency. The nature of the interaction between zinc and the various agents used remains obscure, although it is known that they do not act by increasing the availability of dietary zinc.

Zinc-deficient calves exhibited bowing of the hind legs and stiffness of the joints that were corrected with zinc repletion (604,605), but the skeletal system is not affected adversely in cattle to the extent that it is in chicks and swine. A reduction in the size and strength of the femur in zinc-deficient baby pigs has been reported (602), but comparisons with pair-fed controls indicated that these changes were due to the reduced food intake. Bone growth was affected in direct proportion to the effect of zinc deficiency on body growth and maximum strength of the femur to bone size. In a later study with weanling pigs, the reduced skeletal growth of zinc deficiency was most apparent in low activity of the epiphyseal growth plate and at other points of osteoblastic predominance (672).

The mechanism of action of zinc in bone formation is still not fully understood, but zinc deficiency adversely affects chondrogenesis as well as retarding
the normal calcification process (56). In zinc-deficient chicks the epiphyseal
plate cartilage cells were found to be disorganized and the epiphyseal–diaphyseal
junction was abnormally narrow. There was increased total collagen, and the
extracellular matrix was increased in avascular areas. Bone collagen synthesis
and turnover were markedly reduced (877). The activity of tibial collagenase, a
zinc metalloenzyme, was reduced. Osteoblastic activity was reduced in the bony
collar of the long bones. In rats, high concentrations of zinc have been shown
histochemically at the site of calcification in developing osteons (372). Less
force was required to displace the epiphyses of zinc-deficient male weanling rats
than was required for controls (894). In chicks the chondrocytes that are near
blood vessels appeared normal, while those cells remote from blood vessels
showed cellular changes, were shaped differently, were surrounded by more
extracellular matrix, and did not stain normally for alkaline phosphatase activity.
As Westmoreland (973) has stated, these changes may affect the normal maturation and degeneration of these cells, which may in turn have an effect on
calcification. Depressed skeletal development may occur in human zinc deficiency, as evidenced by delayed radiographic bone age (339,772,773) and by the
increase in linear growth that occurs with the treatment of mild human zinc
deficiency (364).

In the young growing animal, bone zinc concentrations change markedly with
variations in zinc intake and zinc status. In young growing rats and Japanese
quail, there was a net loss of total bone zinc during feeding of a zinc-deficient
diet, but insufficient zinc was mobilized to prevent the development of deficiency symptoms (114,368). Young Japanese quail fed high-zinc diets appeared to be
able to store some of the excess zinc in bone to provide protection against a
subsequent depletion period (368). In general, the extent to which zinc can be
mobilized from bone appears to depend on the extent of bone remodeling that is
in progress, although the mobilization of zinc appears to be independent from
calcium and phosphorus mobilization (26). Murray and Messer (652) examined
the turnover of bone (humerus) zinc in relation to the rate of remodeling, in both
weanling and adult female rats, fed diets with and without adequate zinc and
calcium. The largest increase in bone zinc was in rats of both age groups fed a
diet sufficient in both calcium and zinc, with total deposition being much greater
in the younger animals with the more rapid bone growth. In the weanling animals, both zinc and calcium deficiency impaired deposition of zinc, but in the
older animals, only zinc deficiency reduced zinc deposition, although the net loss
of zinc from bone was greatest in the calcium-deficient, zinc-deficient state.
There was a substantial redeposition of resorbed bone zinc even in severe zinc
deficiency: zinc deficiency alone did not promote removal of zinc from bone,
and there was not a substantial loss of zinc from the skeleton unless bone

resorption was stimulated by calcium deficiency. These workers suggested that in younger animals zinc loss or deposition in bone was from the cellular matrix, whereas in the older rats, only inorganically sequestered zinc was involved (652). Hurley and Swenerton (425) found that only 1% of total zinc was mobilized from maternal rat bones when provided with a zinc-deficient diet during pregnancy, and this was insufficient to prevent the teratogenic effects of the zinc deficiency. However, a concomitant calcium deficiency in the dams allowed mobilization of up to 20% of bone zinc, sufficient to prevent effects of deficiency on the fetuses (427).

N. Brain Development and Behavior

Severe maternal zinc deficiency during the period of embryogenesis in the rat has severe teratogenic consequences for the central nervous system. When maternal zinc deficiency occurs during the postembryonic fetal period and/or during the critical period of brain growth in early postnatal life, brain function can be permanently affected. If the deficiency is imposed throughout the latter part of pregnancy, brain size is decreased, brain cell number is decreased, and the cytoplasmic–nuclear ratio is increased, implying an impairment of cell division in the brain during the critical period of macroneural proliferation (574). When nursing dams are fed a zinc-deficient diet, synthesis of brain DNA in pups, as measured by the *in vivo* incorporation of thymidine, is significantly retarded compared with that of pair-fed controls (210), as is the synthesis of brain histone proteins (790). Levels of DNA, RNA, and protein are decreased in the cerebrum and cerebellum (100). However, Dreosti *et al.* (206) found that DNA synthesis was not seriously affected but that the activities of two zinc-dependent enzymes, the myelin marker enzyme $2',3'$-cyclic nucleotide $3'$-acid dehydrogenase and L-glutamic acid dehydrogenase, were reduced.

The hippocampus is formed largely in the first 3–4 weeks postnatally in the rat, and at this stage zinc accumulates in the intrahippocampal mossy-fiber pathway, in the large terminal boutons of the granular cell axons (170). The hippocampus is involved in memory, cognitive function, and the integration of emotion, and there is reason for conjecture that some of the outstanding behavioral defects of zinc deficiency are attributable to zinc depletion in the mossy-fiber boutons. Behavioral changes in zinc-deficient animals may also be linked in part to higher levels of brain norepinephrine and dopamine (945) and/or to increased circulating levels of glucocorticoids (394). Zinc ions may also be a physiologically important modulator of opiate receptor function in the hippocampal mossy-fiber system (880).

Histological abnormalities of the zinc-deficient cerebellum have been reported. Granular cells are decreased in number and persist near to the surface of the cerebellum (214). The dendritic arborization of Purkinje, stellate, and basket cells is retarded (213).

In 1970 (111) and subsequently (112), Caldwell and co-workers reported a significantly inferior learning ability, as measured by water maze and platform avoidance conditioning tests, in the surviving offspring of relatively mildly zinc-deficient rat dams compared with similar rats from pair-fed control dams. This work was followed by numerous studies especially by Halas, Sandstead, and co-workers (333,792). Most of the studies by this group and others have used rats that have been investigated during the latter third of pregnancy and/or in lactation, while occasionally the design has involved experimental zinc depletion in more mature male rats. Prenatal and postnatal zinc deficiency caused rats to be less active than normal animals. Rats who had severe perinatal zinc deficiency were subsequently less able to cope with stress. Female zinc-deficient rats, but not males, had a significant increase in shock-induced aggression compared with ad libitum- and pair-fed controls (330). Perinatal zinc deficiency was shown to impair long-term or reference memory (331) and probably also short-term or working memory. Halas (330) concluded that many of the apparent learning defects exhibited by rats that were zinc deficient in the perinatal period could be attributed either to emotional effects or to decreased motivation for food. Although the methodological problems have been important, there is strongly suggestive cumulative evidence that perinatal zinc deficiency does have specific effects on subsequent learning ability, and this has been supported by more recent studies by the same research group (332). More difficult to explain are the apparent effects of dietary zinc deprivation commencing after brain maturation. Moderate zinc deficiency induced after the age of 45 days in male rats has been reported to impair cognitive performance (295).

The adverse effects of zinc depletion during the critical period of brain development have also been studied in pregnant rhesus monkeys (794) that were fed a severely zinc-deficient diet through most of the third trimester of pregnancy and were then nutritionally rehabilitated throughout the remainder of pregnancy and lactation. Infants of the zinc-deprived dams played and explored less than the infants of pair-fed and ad libitum-fed control monkeys. They were less active and associated with their mothers a greater percentage of the time. Testing of ability to solve complex learning sets late in the first and in the second years of postnatal life led to the conclusion that the formerly zinc-deficient monkeys had a maturational delay resulting in impaired ability to solve complex learning sets (885).

Of particular interest is the finding (330) that mild zinc deficiency (10 μg/g in dam's diet) from the day of conception until rat pups are weaned at 23 days postpartum results in identical learning impairment and short-term memory deficits to those evidenced in the offspring of dams whose diet was severely zinc restricted during lactation alone. Mild zinc deficiency during gestation and lactation also results in detectable injury to the hippocampus, which is smaller, has fewer neurons, and has greater cellular density.

Formal studies of brain function in the zinc-deficient human have been almost nonexistent, but abnormalities of behavior are a notable feature of severe human zinc deficiency states (949). Thus irritability, lethargy, and depression are notable features of severe acquired zinc deficiency states and of acrodermatitis enteropathica when untreated or even when in relatively mild relapse. Improvement in hedonic tone is rapid and dramatic when zinc therapy is commenced. Cerebellar dysfunction and acute psychosis have been reported to result from experimental acute human zinc deficiency induced by administration of large quantities of histidine (387).

O. Zinc and Vision

The retina–choroid complex contains the highest quantities of zinc in the eye and has high levels of several zinc metalloenzymes including carbonic anhydrase and retinol dehydrogenase. The latter is responsible for the oxidation of retinol to retinal (i.e., retinaldehyde), the photochemically active form of vitamin A in the rods. Night vision is dependent on an adequate supply of retinal to the rods in order to form the visual pigment, rhodopsin. Zinc deficiency appears to contribute to some cases of diminished dark adaptation in the human. McClain and colleagues (568) and Morrison and colleagues (642) have demonstrated the need for zinc therapy as well as vitamin A to achieve normal dark-adaption thresholds in some alcoholic cirrhotics. Abnormalities of cone function may be more pronounced than the abnormalities of the rods. In children with acrodermatitis enteropathica, a characteristic gaze aversion to eye-to-eye contact has been reported (648). Central cone vision appears to be distressing, and these patients rely more heavily on peripheral vision, which is predominantly rod dependent (458). Photophobia, which is also commonly seen in this condition and which may be very severe (349), may be associated with this gaze aversion phenomenon. As is the case with the other clinical features of acrodermatitis enteropathica, these abnormalities disappear once zinc treatment has been instituted. Zinc deficiency was thought to be a factor in a case of reversible macular degeneration in a young man with Crohn's disease who had a serum zinc of 0.3 μg/ml (995).

The tapetum lucidum is a modified choroidal structure that reflects light off its shiny surface back through the retina, thus enhancing the effect of low-level illumination. It is found in many animals but is not present in most primates, including the human. In the dog, zinc accounts for 7% of the dried weight of the tapetum lucidum. The tapetal cells contain a zinc–cystine complex that exhibits photoelectric properties amplifying the physiological response to low-level illumination. Administration of the chelator ethambutol, which causes a substantial drop in serum and ocular zinc levels, resulted in irreversible tapetal degeneration (108). Administration of the zinc chelator diphenylthiocarbazone to rats results in inclusion bodies in the retinal pigment epithelium, which has been termed the blood–retina barrier (28).

Corneal changes in acrodermatitis enteropathica were first reported by Wirshing (991) and consisted of radial fanlike stripes and 1-mm-wide bands concentric with the superior limbus. The opacities were both intraepithelial and subepithelial (Fig. 10). Histologically the epithelium varied in thickness; there was fragmentation of the basement membrane, focal scarring of Bowman's layer, and loss of normal lamellar architecture of the stroma. Corneal edema has been described in an infant with acrodermatitis enteropathica (533). These lesions appear to be part of the primary disease process, and respond to either diiodohydroxyquin or zinc therapy. A fairly extended period of chronic severe zinc deficiency is apparently required to produce the corneal changes. Cataracts may

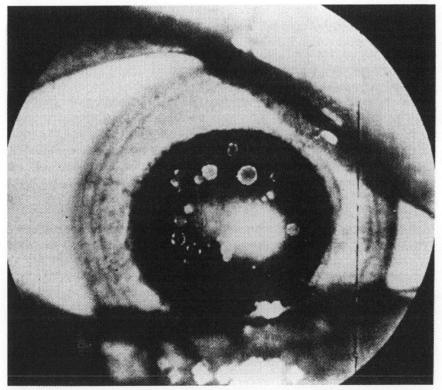

Fig. 10. Multiple rounded opacities in the lens and the absence of cilia in a 3-year-old child with acrodermatitis enteropathica. From Karcioglu (458).

also complicate acrodermatitis enteropathica, and zinc deficiency causes cataract formation in fish lenses (474). There are no indications that milder degrees of zinc deficiency can cause cataracts, and the mechanism by which zinc deficiency may contribute to cataract formation is unclear.

Other eye problems seen in acrodermatitis enteropathica are blepharitis and conjunctivitis. Hyperkeratinization of the lid margins and stenosis of the lacrimal punctum have been reported. Optic atrophy has been reported in association with chronic zinc deficiency and in some cases of acrodermatitis enteropathica that were treated with diiodohydroxyquin. It is unclear whether the latter were attributable directly to a toxic effect of the diiodohydroxyquin or resulted from zinc deficiency, but the latter appears to be more likely (886).

P.   Zinc and the Gastrointestinal System

Zinc deficiency may cause a number of abnormalities in the oral cavity. Excessive salivation occurs in zinc-deficient sheep, calves, and goats (660). Zinc-deficient calves have an increase in the number of bacteria in the mouth (622). Abnormalities of taste bud function may result from zinc deficiency (377). Among other miscellaneous abnormalities of the oral cavity, parakeratosis of the tongue and esophagus occurs in the zinc-deficient squirrel monkey (541,542). Parakeratosis of the esophagus has also been reported in the zinc-deficient chick, rat, and swine, but similar lesions have not been found in cattle (621).

In the small intestine of the zinc-deficient rat, Otto and Weitz found Paneth cell lesions that consisted of pleomorphic secretory granules, some of which contained crystalloid structures (693). Koo and Turk (491) found that enterocytes of zinc-deficient rats had supranuclear cytoplasmic vacuolation, an increased number of lysosomes, reduced granular endoplasmic reticulum, degenerative Golgi complex, and pyknotic nuclei. Membrane-bound autophagic vacuoles were frequent. Elmes and Jones (219) found that acute severe zinc deficiency in young rats resulted in ultrastructural degenerative changes in the enterocytes and damage to the microvilli that was not confined to the physiological changes that occur at the villous tip. Increased numbers of apoptotic bodies resulting from cell death were found. These changes differed from those preceding normal desquamation of enterocytes. No significant changes in Paneth cell ultrastructure were observed, but the investigators postulated that the period of experimental zinc deficiency may have been inadequate to produce Paneth cell lesions.

Microscopic lesions of the intestinal mucosa have been reported in association with severe human zinc deficiency in acrodermatitis enteropathica. The mucosa has been reported to be thin and flat with a decrease in crypt numbers and some crypt necrosis (21). Loss of villous architecture with flattening of villi and an inflammatory infiltration of the lamina propria have also been reported (470). Zinc therapy caused reversion to a normal mucosa. Ultrastructural abnormalities of the Paneth cells in acrodermatitis enteropathica consisting of pleomorphic secretory granules and abnormal fibrillar inclusion bodies, have been reported by Lombeck *et al.* (524), Braun *et al.* (84), and Bohane *et al.* (78). These ultrastructural abnormalities have been reported to persist after treatment with zinc (721), as well as with diiodohydroxyquin, in contrast to the resolution of other clinical and laboratory abnormalities. This led to speculation that these inclusion bodies may be related to the primary defect in acrodermatitis enteropathica. However, persistence of this abnormality on zinc therapy has not been confirmed by Bohane *et al* (78). In 1983, Jones *et al.* (452) reported abnormal lysosomal inclusion bodies in the intestinal epithelial cells of two patients with acrodermatitis enteropathica when in relapse, but fewer smaller ones after zinc therapy. A third patient who was in remission on zinc therapy had no inclusion bodies. The smaller inclusion bodies were similar to those found in celiac disease. These investigators suggested that some of the large abnormal lysosomal inclusion bodies were derived from phagocytosis of damaged goblet cells by adjacent epithelial cells, and that the goblet cell abnormalities were due to zinc deficiency. They also postulated that zinc is essential for normal secretory granule formation, stabilization, and release in Paneth cells, goblet cells, and pancreatic acinar cells.

In severe zinc deficiency states, diarrhea is typically a prominent feature. Golden and Golden (283) have concluded that diarrhea also occurs as a direct result of zinc deficiency in more moderate zinc deficiency states. The diarrhea is associated with particularly high fecal sodium losses (285). Changes in the intestinal bacterial flora, secondary to deficits in immunocompetence, may play a significant role in the diarrhea of zinc deficiency.

Q. Laboratory Changes in Zinc Deficiency

This section includes miscellaneous laboratory findings in zinc deficiency that have not been mentioned in previous sections.

The hematocrit values or erythrocyte numbers are above normal in the zinc-deficient rat (483,777), baby pig (602), and Japanese quail (253). In the zinc-deficient baby pig there is a relative and absolute reduction in lymphocytes (602). In male weanling rats fed a low-zinc diet for 14 days the percentage of polymorphonuclear leukocytes (P) markedly increased and that of lymphocytes (L) decreased (203). Similar changes in the P:L ratio occurred in adult female rats fed a low-zinc diet during pregnancy but not in nonpregnant rats fed a similar diet for 3 weeks (203). The physiological significance of these white cell changes and their relationship to zinc deficiency remain obscure.

A decline in plasma protein levels has been observed in zinc deficiency in some investigations (411,752) but not in others (253,602). The electrophoretic patterns of these proteins are greatly altered in the zinc-deficient pig (602) and, after fasting, in quail (253,254). The plasma levels of the specific retinol-binding proteins are significantly reduced in the zinc-deficient rat (845). It remains unclear whether the reduction in plasma vitamin A concentration that occurs in zinc-deficient rats (842,844) is entirely a secondary effect from the reduced food intake or whether it results in part from a primary effect on vitamin A metabolism including reduced hepatic synthesis of retinol-binding protein. Zinc supplementation of young children with PEM who were not vitamin A deficient, resulted in a significant increase in plasma vitamin A and retinol-binding protein (836). Low levels of serum albumin, transferrin, and, especially, thyroxine-binding pre-albumin have been reported in the serum of severely zinc-deficient patients receiving total parenteral nutrition. Zinc supplementation was associated with significant improvement in these levels. Hypoproteinemia and edema during the second month of postnatal life in premature infants have been attributed to zinc deficiency (500).

Elevated blood ammonia levels have been observed in experimental human zinc deficiency, with normalization following dietary zinc supplementation (738). Correction of zinc deficiency in patients receiving total parenteral nutrition has been associated with improved nitrogen balance, increases in circulating insulin, and concurrent decreases in blood glucose to more normal levels (992). Abnormalities of glucose tolerance and insulin release have been reported in acrodermatitis enteropathica (662). Low serum cholesterol and triglyceride concentrations occur inconsistently in untreated acrodermatitis enteropathica and normalize following zinc therapy (349). Abnormalities of circulating essential fatty acids and of prostaglandin metabolism have also been observed in severe human zinc deficiency (357).

R. Genetic Disorders of Zinc Metabolism

Two genetic disorders of zinc metabolism leading to severe zinc deficiency states have been recognized in animals (153). These are the lethal trait known as A46 or Adema disease found in Friesian cattle and the lethal milk mutation occurring in the C57BL/6J (B6) strain of mice. In the human, recognized genetic disorders of zinc metabolism are acrodermatitis enteropathica and familial hyperzincemia, which is described in Section II,K.

Adema disease (Fig. 11) has been described from Scotland (581), Italy (271), the Netherlands (929), Federal Republic of Germany (883), and Denmark (24). Affected animals are normal at birth but produce excessive frothy saliva somewhere between 2 and 6 weeks of age. This is followed by hair loss around the mouth, eyes, and anus, erosions within the mouth, and a characteristic distribution of skin lesions on the lower region of the legs, abdomen, and thorax. The lesions develop initially as areas of erythema and pruritus that then become tense, painful, and wet from serum exudate. The latter dries to form crusts that

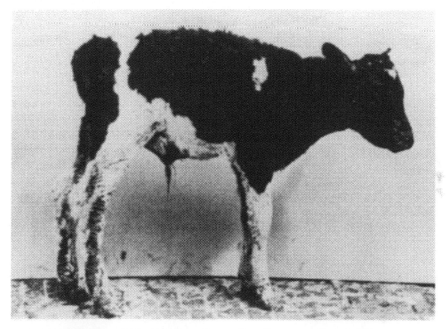

Fig. 11. Calf with Adema disease. From Brummerstedt *et al.* (97).

are shed with loss of associated hair. Animals may suffer from conjuctivitis, rhinitis, bronchopneumonia, and diarrhea. Postmortem examination reveals marked hypoplasia of the thymus (Fig. 12A). In one study (97) the average weight of the thymus was found to be 18 g compared with the normal weight of 250 to 400 g (Fig. 12A,B). Histologically the thymus is characterized by a severe depletion of small lymphocytes, particularly in the cortical region. Hypoplasia of the spleen, some regional lymph nodes, Peyer's patches, and the lymphoid tissue around the gut also occurs. Humoral immune response to tetanus toxoid and cell-mediated immune reaction to *Mycobacterium tuberculosis* and dinitrochloroben-zene are impaired. If left untreated the calves die within 4 to 8 weeks after clinical onset of the disease. Provided animals are treated before the terminal stages of the disease, supplementation with zinc can result in complete remis-sion. Discontinuation of zinc treatment causes a relapse after 2 weeks. The key defect in this syndrome appears to be in the intestinal absorption of zinc (243). Inheritance is probably autosomal recessive, and the lethal factor has complete penetrance in homozygous individuals.

Acrodermatitis enteropathica in humans is a rare autosomal recessively inher-ited disorder that corresponds closely to lethal trait A46 in cattle. The precise molecular defect has not been determined, but it is known to cause a partial block

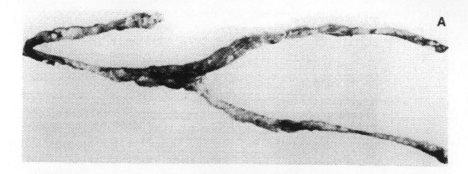

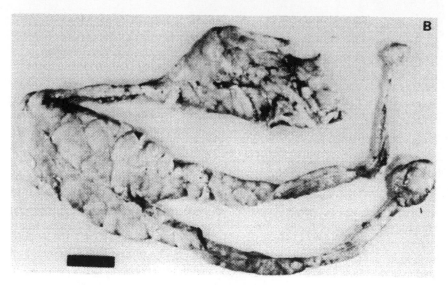

Fig. 12. (A) Hypoplastic thymus from calf with Adema disease. (B) Normal-size thymus from a zinc-treated calf. From Brummerstedt *et al.* (97).

in the intestinal absorption of zinc (525). *In vitro* studies suggest that defective uptake of zinc by enterocytes is the primary abnormality responsible for the zinc deficiency state (40). The fundamental importance of the disturbance of zinc homeostasis in acrodermatitis enteropathica was first recognized by Moynahan and Barnes (646). It quickly became appreciated that the phenotypic expression of this disorder could be attributed entirely to a severe zinc deficiency state (647,663). The most dramatic clinical feature is the skin rash, which has a characteristic acral and orificial distribution (Fig. 13). The skin lesions are vesiculobullous, pustular, and/or eczematoid. In acute stages their appearance is markedly erythematous, and the rash may become generalized. Secondary infec-

tion is common, especially with staphylococci or *Candida*. In chronic stages
hyperkeratotic plaques may be present, and the nails may become severely
dystrophic (359) (Fig. 14). Early mucosal lesions include gingivitis, stomatitis,
and glossitis.

The rash occurs first in early infancy, but onset is typically delayed until after
weaning in breast-fed infants. Diarrhea is present in 90% of cases and may be
severe. Alopecia including loss of eyebrows is another notable feature, but this
may not be readily apparent until later in infancy. The infant ceases to gain
weight concurrently with the appearance of the skin lesions, and subsequently
the clinical course is marked by severe failure to thrive. Anorexia varies in
severity but can be a major feature of the disease. Psychological and behavioral
abnormalities are notable, with irritability, lethargy, and depression occurring
even during relatively mild stages of the disease. Improvement in mental func-

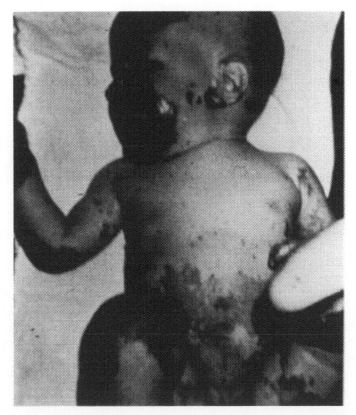

Fig. 13.   Acrodermatitis enteropathica: distribution of acute skin lesions. From Hambidge *et al.*
(348).

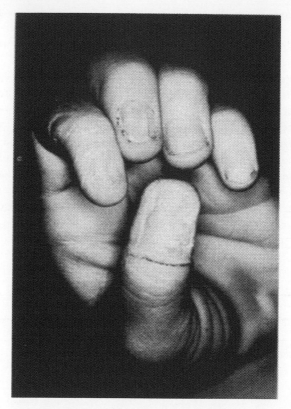

Fig. 14. Acrodermatitis enteropathica: nail dystrophy. From Hambidge (357).

tion, especially in hedonic tone, is quite dramatic and very rapid once zinc therapy is commenced. Other features that occur in some cases include hoarseness, blepharitis, conjunctivitis, photophobia, and corneal opacities. In untreated or poorly treated cases, prior to the advent of zinc therapy, there was usually a fluctuating but overall progressively downhill course, complicated by recurring infections and frequently with a fatal outcome in later infancy or early childhood. Prolonged survival in the years before zinc therapy was available was associated with frequent relapses, poor growth, hypogonadism in adolescents, and an apparent high risk of lethal congenital malformations in the offspring (346). Atrophy of the thymus and lymphoid tissue and impaired cell-mediated immunity have been reported (141), although abnormalities of T-cell function do not appear to be a consistent feature (359). Hypozincemia is characteristically severe. Oral zinc therapy in moderate quantities (i.e., 30–50 mg/day) is effective in correcting the severe zinc deficiency state and leading to a complete clinical remission. This remission can be maintained indefinitely provided zinc therapy is continued.

The lethal milk mutation in mice is recessively inherited, and the phenotypic expression is related to an abnormality in the dams' milk. Pups of this strain nursed by normal dams grow normally, and normal pups nursed by homozygous lethal milk dams die. The signs of the lethal milk syndrome in the pups are similar to those of experimental zinc deficiency, and zinc in the milk of these dams has been found to be 34% lower than normal. The primary defect appears to be a failure to concentrate zinc adequately in milk (719). A human counterpart to the lethal milk mutation has been postulated but not proven (1001). Thus a failure of the mother's mammary gland to secrete normal quantities of zinc into her milk, despite an otherwise apparently normal maternal zinc status, is responsible for some cases of severe acquired zinc deficiency syndromes in infants born prematurely. The premature infant appears to be peculiarly vulnerable to severe zinc deficiency states, and infants born at term can develop normally if breast fed by one of these mothers (1001).

S. Overview of Zinc Deficiency

This section provides a summary of the features of zinc deficiency and of the circumstances in which a deficiency of this nutrient may occur. Emphasis is given to current knowledge of human zinc deficiency.

Experimental zinc deficiency has been studied in a variety of mammals and birds including rats, monkeys, guinea pigs, hamsters, mice, cattle, pigs, sheep, dogs, cats, chickens, ducks, geese, Japanese quail, and humans. Nonexperimental zinc deficiency has been documented in pigs (911), cattle (215,325,510), sheep (549,921), and humans. Reduced growth rate and feed intake are the first effects of naturally occurring and experimental zinc deficiency in the growing animal. The next obvious feature to develop is parakeratosis and hair loss. Other features of zinc deficiency in animals will depend on a number of factors including the species, age, sex, and the severity and duration of the zinc depletion. These features may include excessive salivation, increased bacteria in the oral cavity, skeletal defects, stiffness of the joints, abnormalities of horns and hooves, swelling of the feet, lethargy, increased susceptibility to infections, impaired reproductive performance, and increased mortality.

Human zinc deficiency can be considered conveniently under subdivisions according to the severity of the effects (357). However, these divisions are quite arbitrary and not always clearly distinguishable. The total-body zinc deficit may not be very great even in clinically severe deficiency states (440). The features of severe acute acquired human zinc deficiency (357,460) are similar to those of acrodermatitis enteropathica. The pathognomonic clinical feature is the acro-orificial skin rash, although the lesions can occur elsewhere and the rash may become generalized. The skin lesions are described in Section IV,J. In premature infants a characteristic change in the anterior neckfold may occur at an early stage (37); there is poorly marginated erythema at the depth of the fold, which becomes well demarcated and scaling within 5 days. Weight gain typically ceases abruptly concurrently with the appearance of the skin rash (1001), although this effect of zinc deficiency may not occur immediately in the intravenously fed patient (782). Anorexia can be severe. If untreated, failure to thrive will become progressively more severe. Severe zinc deficiency in the adult is also accompanied by weight loss. Diarrhea is a frequent manifestation and may also be severe. Depressed mood is a consistent and notable feature. Premature infants with severe zinc deficiency exhibit excessive crying and difficulty in consolation. Mucosal lesions, including stomatitis and glossitis, may be evident at an early stage. Susceptibility to candidal and bacterial infections is increased and has been attributed to abnormalities of the immune system. Other features of chronic severe zinc deficiency that have been observed in poorly treated acrodermatitis enteropathica are nail dystrophies, several different eye lesions (458), hypogonadism in the male, and poor reproductive performance in the female (346). Severe zinc deficiency in the infant and young child may terminate fatally if untreated.

In more moderate zinc deficiency states, skin lesions are not evident, although the skin may be more vulnerable to the effects of zinc deficiency when there is concurrent PEM (282). Other features attributed to zinc deficiency in this same population of malnourished young Jamaican children included atrophy of the thymus (286), impairment of delayed cutaneous hypersensitivity (287), stunting and wasting (282), diarrhea (283), and an increased energy cost of growth (288). Plasma zinc concentrations fell markedly during recovery from malnutrition, and there was a negative correlation between rate of weight gain and plasma zinc level. These two parameters were lower in children rehabilitated on a soy protein-based diet than in children rehabilitated on a cow's milk formula. Zinc supplementation of the soy protein formula was associated with more rapid weight gain (288).

More moderate zinc deficiency has been cited as a major etiological factor in the syndrome of adolescent nutritional dwarfism in Egypt and Iran (340,728,729), the cardinal features of which are severe delay of sexual maturation and dwarfism. Hypozincemia and other laboratory abnormalities compatible with zinc deficiency have been documented, and favorable responses to zinc therapy have been reported from most but not all studies. A similar syndrome may exist among the aboriginal population in Australia (142), although confirmation is needed. The high serum copper and ceruloplasmin levels in the latter communities have not been explained satisfactorily and serve as a reminder that it is premature to accept the modest hypozincemia typically observed in this population as a definite indication of zinc deficiency. Syndromes similar to those described by Prasad and colleagues in the Middle East have also been reported in association with certain chronic diseases in North America, especially regional enteritis (567) and sickle cell disease (737). Further confirmation of these reports is necessary. Other features of more moderate zinc deficiency states may include impaired wound healing (932), impaired dark adaptation (642), and abnormalities of fetal development and of delivery (444).

The feature of mild human zinc deficiency states that has received most attention is impairment of physical growth velocity. Growth-limiting zinc deficiency in some otherwise apparently normal infants and children in North America has been confirmed with controlled studies of dietary zinc supplementation using extremely small quantities of zinc (350,948,951). Zinc supplementation has been associated with increased linear growth increments. In the latest of these studies, a significant effect of zinc supplementation on food intake (494) was also demonstrated. In other, less definitive studies in children, low zinc concentrations in plasma or in hair have been found to be associated with relatively poor growth, poor appetite, and hypogeusia (109,342).

A wide range of factors causing or contributing to the occurrence of human zinc deficiency have been documented, including genetic and acquired disorders. During the past decade there have been numerous reports of severe zinc deficiency states developing in patients who have to be maintained with intravenous nutrition (37,460). The most important etiological factor is failure to add zinc supplements to the intravenous infusates. Contributory factors include excessive urine losses of zinc, and in some cases, extraordinary gastrointestinal losses attributable to the gastrointestinal pathology for which intravenous nutrition had to be administered (992). There have been several reports of severe zinc deficiency states developing in both intravenously and orally fed premature infants. Some of the latter have been fed with their own mother's milk, and the zinc deficiency has been attributed in part to an inability of the mother's mammary gland to secrete normal quantities of zinc into her milk, despite an otherwise apparently normal maternal zinc status (1001). Positive zinc balance and positive net absorption are both difficult to achieve in the very low birth weight infant during the first few months of postnatal life (181). Similar negative balances have not been observed in equally small infants who were born at term and had severe intrauterine growth retardation. These findings suggested that the negative balance in very premature infants was attributable to immaturity of gastrointestinal absorption mechanisms for zinc.

Dietary zinc intake and, probably of even greater practical importance, other dietary factors that affect zinc absorption are the major determinants of zinc status in the general population. The quantity of zinc ingested each day can vary widely depending on the food eaten. Some diets that are adequate in protein, calories, and other nutrients can have a surprisingly low zinc content (695). In some instances, this low zinc content is attributable primarily to food processing

(818). This is of special potential concern in the formula-fed infant (169,948) because of his or her dependence on one food staple as the primary source of nutrients. Variations in absorption of zinc from different milks and formulas also appear to have a major impact on the zinc status of the infant (354), with especially favorable absorption of zinc from human milk (124,800). Infants fed soy protein formulas (169) and perhaps some synthetic formulas (124) are at special risk of zinc deficiency. Zinc deficiency associated with the use of soy formulas is probably due at least in part to interference with zinc absorption by dietary phytate (800). The high phytate content of rural diets in Iran is probably the major cause of the zinc deficiency that appears to be endemic in that country and to be a major etiological factor in the syndrome of adolescent nutritional dwarfism (763). The high fiber content of these diets has also been considered to be a factor, but only some types of fiber appear to affect zinc absorption adversely. Excessive supplemental inorganic iron is among other factors that may have an adverse effect on zinc absorption both in adults and infants (852).

The increased physiological requirements for zinc at times of rapid growth or tissue repair may make a crucial practical difference to the adequacy of the dietary zinc supply that may otherwise be sufficient. Butrimovitz and Purdy (106) have demonstrated an inverse relationship between growth rates and plasma zinc in the general population, with lowest plasma zinc concentrations in early childhood and adolescence. These data have been interpreted to indicate a marginal situation with respect to the adequacy of the zinc intake at times of increased requirement. In some subjects this may progress to a mild zinc deficiency state. Hambidge and colleagues (364) have documented the occurrence of growth-limiting mild chronic zinc deficiency in young free-living otherwise normal children in Denver, Colorado. These particular studies involved children from low-income families, but there are indications that this syndrome is not limited to any one socioeconomic or ethnic group or to one age group or geographical area (109,342).

An increase in zinc requirements during pregnancy may increase the risk of nutritional zinc deficiency (791), but the incidence of this circumstance is unknown. This uncertainty stems partly from conflicting interpretation of plasma zinc data during gestation. The balance of evidence indicates that there is a very substantial physiological decline in plasma zinc concentrations during the first 2 months of gestation (86,360), followed by a further slow decline until the ninth month of gestation (353,360). Hence, interpretation of plasma zinc data in abnormal pregnancies requires comparison with control data matched for the same duration of gestation. Mild zinc deficiency may also occur during prolonged lactation and cause depression of milk zinc concentrations (495). An increase in physiological requirements may also result from increases in sweat losses with activity and in hot climates. A variety of abnormal circumstances can cause an increase in requirements due to impaired absorption or excessive losses of this element (793).

The combination of clinical features, especially the skin lesions, and severe hypozincemia makes the detection of severe human zinc deficiency relatively simple. Plasma zinc concentrations are usually <0.4 $\mu g/ml$ and are quite frequently <0.2 $\mu g/ml$ (normal $0.65-1.10$ $\mu g/ml$). In more moderate zinc deficiency states, plasma zinc concentrations will most commonly be between 0.4 and 0.6 $\mu g/ml$. Diagnosis of zinc deficiency in these circumstances is more difficult, because the features are nonspecific and factors other than zinc deficiency, especially stress and pregnancy, can be responsible for this degree of hypozincemia. Detection of mild zinc deficiency states is especially difficult; plasma zinc concentrations may be within the normal range, and the features are nonspecific. Confirmation depends primarily on the response to zinc supplementation.

Other laboratory indices have not proved particularly useful. Hair zinc concentrations may be very low in mild zinc deficiency (356), but interpretation is difficult (358) and attempts to develop this parameter as a useful diagnostic tool have been disappointing. Although the zinc content of red cell membranes appears to be sensitive to zinc deficiency (66), total erythrocyte zinc does not provide a sensitive index of zinc status. Measurement of neutrophil zinc has been suggested as a useful aid to confirmation of zinc deficiency (143,980), but experimental data have not been supportive of the use of mixed-leukocyte zinc levels (172). Urine zinc excretion rates are decreased in more severe zinc deficiency states, but again this does not provide a sensitive index. Serum alkaline phosphatase activity is depressed in severe human zinc deficiency, and in more moderate zinc deficiency states there may be an increase in activity with zinc supplementation despite initial values within the normal range. However, despite clearly documented evidence of the sensitivity of this assay to experimental zinc depletion in other species (486), the plasma activity of this enzyme is not sufficiently sensitive to mild degrees of human zinc deficiency to provide the kind of laboratory test that is needed. There is some expectation that serum metallothionein levels will prove more useful (85). Meanwhile, plasma or serum zinc concentrations remain the best single available laboratory test for zinc deficiency.

V. ZINC REQUIREMENTS

Minimum zinc requirements vary with the age and physiological state and with the composition of the diet, particularly the amounts and proportions of the many factors, organic and inorganic, that effect zinc absorption and utilization. Zinc requirements are also influenced by unusually large losses, as in profuse sweating or parasitic infestation with its attendant blood losses. Data for the human and many other species are far from complete at this time.

A.   Animals

1.   Laboratory Species

The minimum zinc requirement for growing rats has been given as 12 µg/g on diets with casein or egg whites as the protein source (251). This level has been confirmed by others (986). Luecke *et al.* (534) found that the minimum zinc requirement of the rat for growth on an egg white diet was close to 11 µg/g, although higher serum zinc levels occurred at an intake of 13 µg/g. Kirchgessner and Pallauf (482,700) obtained maximum growth at 8 µg/g of dietary zinc and reported that at least 12 µg/g diet was necessary for optimal zinc content in serum and liver of weanling rats on casein diets. Forbes and Yohe (251) reported in 1960 that zinc was less available to rats from soybean protein and that this was related to its high phytate content (679). When dietary phytate levels are deliberately raised further, growth of young rats has been shown to be depressed even at a dietary zinc concentration of 48 µg/g (184). In this study, phytate–zinc molar ratios > 15 : 1 and 10 : 1 were associated with impaired growth and hypozincemia, respectively, at various absolute levels of dietary zinc and phytate. The significance of similar ratios has been documented by others (521) using different techniques. Excess calcium was found to depress growth rates in the presence of phytic acid but not in its absence (519,674). The very substantial influence of dietary calcium level has been confirmed and quantitated by others (639). The phytate calcium–zinc molar ratio at which weight gain in rats is depressed has been calculated to be 3.5 : 1 (186).

The minimum requirements for optimal reproductive performance in the male and female are higher than those just given for growth. In the male rat, Swenerton and Hurley (897) have produced evidence that 60 µg zinc per gram is inadequate to prevent long-term testicular changes. They maintained that rat diets containing soybean protein should have at least 100 µg zinc per gram, if extraneous sources of the element are minimal.

The zinc requirements for maximal growth in young guinea pigs have been given as 12 µg/g added to casein diets containing 2 µg zinc per gram, and 20 µg/g added to soybean protein diets containing 2 µg zinc per gram (14). Monkeys on a casein diet require at least 15 µg/g of dietary zinc to prevent signs of deficiency (34). Requirements for other laboratory species are less well documented.

2.   Pigs

On purified soybean protein diets, 45 µg/g of dietary zinc is adequate for growth in female baby pigs but not in males (603). Weanling pigs fed a soybean protein diet containing 0.66% calcium and 16 µg zinc per gram required zinc supplementation to give a total of 41 µg/g to achieve freedom from parakeratosis. A further increase to 46 μg/g improved the growth rate (849). Essentially similar results were obtained by Miller *et al.* (603), with no differences between male and female weanling pigs. Diets containing protein from animal sources, such as fish meal or meat meal, and therefore lower phytate, would reduce zinc requirements slightly below the 45–50 μg/g just indicated. Where the diets contain calcium levels twice normal or higher, these requirements would be increased, as judged by the increased incidence and severity of parakeratosis on low-zinc diets and the signs of zinc deficiency induced on diets otherwise marginal or adequate in zinc (404,515,911). High dietary intakes of copper also significantly increase zinc requirements above those supplied by most normal rations.

The reproductive performance of breeding sows was found to be satisfactory on a corn–soybean meal diet containing 35 μg zinc per gram and 1.4% calcium. No improvement was observed when this diet was supplemented with 50 μg zinc per gram (722). Hennig (390) also observed no effect on the reproductive performance of sows from zinc supplementation of barley–fish meal rations containing 36–44 μg zinc per gram, even when given excess calcium up to 1.5–2.2% of the diet. On the other hand, the addition of zinc at 100 μg/g to a corn–soybean diet containing 30–34 μg zinc per gram and 1.6% calcium has been shown in two trials to increase significantly the number of live pigs per litter without affecting birth or weaning weights (406). On this evidence, corn–soybean meal rations high in calcium must be considered marginal in zinc for reproduction in swine.

3. Poultry

The minimum zinc requirement of chicks for growth and health is given as 35 μg/g when fed on soybean protein diets containing 1.6% calcium and 0.7% phosphorus (677,678). Lowering the calcium to 1.1% slightly decreased this requirement, but raising it to 2.1% had no effect. This estimate has been confirmed, but where casein or egg white is the protein source total zinc requirements are lower (629,714,996). The minimum dietary level for zinc for growth in chicks can therefore be given as 35 to 40 μg/g for soybean protein-type diets and 25 to 30 μg/g for diets in which the protein comes mainly from animal sources. Furthermore, the chick is less vulnerable than the pig to excess calcium (714). There is ample evidence that the zinc intakes just given are also adequate to meet the requirements of egg production and hatchability.

4. Sheep and Cattle

Ott and co-workers (697) reported that 18 μg zinc per gram did not support maximal growth in lambs consuming a diet in which egg white was the nitrogen source and suggested a requirement between 18 and 30 μg/g. Mills *et al.* (622)

found 7 μg zinc per gram to be adequate for growth in lambs on a diet in which 60% of the nitrogen came from urea, while 15 μg/g were necessary to maintain normal plasma zinc levels. Underwood and Somers (920), also employing a diet in which 60–70% of the nitrogen came from urea, observed that ram lambs grew just as well when the diet supplied 17 μg zinc per gram as when it supplied 32 μg/g. However, testicular growth and spermatogenesis were markedly improved at the higher zinc intake. It was concluded that 17 μg/g is an adequate zinc intake for body growth and appetite on such diets, but is quite inadequate for normal testicular growth and function.

A dietary zinc concentration of 8 to 9 μg/g is adequate for the growth of calves (608,622), and 10–14 μg/g is necessary to maintain normal plasma zinc levels (622). These estimates of zinc requirements are based on semipurified diets and are lower than would be suggested from field observations. For example, Perry *et al.* (715) obtained increases in daily weight gain from supplementary zinc in two of four experiments with cattle fed practical fattening rations containing 18 and 29 μg zinc per gram. Raum and co-workers (755) similarly reported a small improvement in the growth of steers from zinc (and cobalt) supplementation of barley rations containing 29–33 μg zinc per gram, and Demertzis and Mills (191) observed lesions of infectious pododermatitis in young bulls, responsive to supplementary zinc, on rations containing 30–56 μg zinc per gram. Signs of zinc deficiency responsive to zinc have been observed in cattle where pastures or fodder contain 18–42 μg zinc per gram of dry matter (510), 19–83 μg/g (215), and estimates of 19 to 28 μg/g (325). Since herbage zinc levels of this magnitude are also commonly found in areas where clinical zinc deficiency of cattle has not been reported, it seems that factors must be present in some types of feed that reduce zinc absorption or impair its utilization by the animal. Limited evidence with ruminants suggests that neither phytic acid nor calcium can be incriminated in this respect (622,696).

The factorial approach has been used to derive estimates of the zinc requirements of cattle and lambs (10). Weigand and Kirchgessner (965) evaluated the zinc requirements of lactating dairy cows on the basis of a model concept developed in rats (963), using quantitative experimental data. They calculated that the daily net zinc requirement for maintenance is 53 μg/kg body weight, on the basis of minimal fecal, urinary, and body surface losses. These were calculated to be 27.1 mg (44 μg/kg per day), 3.75 mg (6 μg/kg per day) and 1.4 mg (31 μg/kg per day), respectively. Requirements for milk production were calculated to be 5 mg zinc per kilogram of milk per day. Dietary zinc utilization was estimated as low as 25%. Hence dietary requirements for a 600-kg dairy cow, producing 30 kg milk per day, would be 728 mg zinc or ~40 μg/g dry matter in the total ration. However, these investigators emphasized the inadequacies of current data and recommended a safety margin that brings the final gross estimated zinc requirement to 50 μg/g dry matter. This figure was a little higher than the 45 μg/g estimated by the Agricultural Research Council (11). Net zinc requirements for growth in calves and first-lactation cows have been estimated to be 24 µg/g fresh carcass weight (482,659,891). Suttle (891) estimated the maintenance requirement of a 50-kg calf to be 2.65 mg zinc per day, which is only a small percentage of the growth requirement. He recommended that requirements be computed in absolute amounts rather than as a concentration in the feed. The Agricultural Research Council (11) estimated the net maintenance requirements for sheep to be 76 µg zinc per kilogram of live weight per day. The zinc concentration in the lamb carcass, and hence requirement for growth, is similar to that in the calf (302,891).

B. Humans

1. Concepts

Estimates of human zinc requirements have been based most frequently on data derived from traditional metabolic-balance studies, in which intake has been compared with the sum of urine plus fecal excretion (8). In some investigations, additional information has been derived from performing balance studies with varying quantities of zinc intake and then undertaking regression analysis of calculated retention on dietary intake; percentage absorption has been calculated from the slope, and estimates of the dietary zinc intake required for equilibrium or zero balance have been derived from the intercept of the regression line with the x axis (307). Although the difficulties and limitations of such regression analysis and other limitations of the metabolic balance technique are well recognized, alternative techniques have not been readily available. At best, balance studies usually do no more than provide some of the data necessary for a complete factorial approach to determining requirements. The major components of the factorial approach are as follows:

1. Fractional or true absorption
2. Endogenous zinc losses in the feces. Minimal obligatory endogenous zinc losses in the feces and urine by healthy individuals with normal zinc status have been calculated by measuring excretion sequentially in normal volunteers after the introduction of a virtually zinc-free experimental diet. By extrapolating the regression line of excretion on time since introduction of the diet back to zero time, it was possible to calculate minimal endogenous losses before such calculations were complicated by the development of zinc deficiency (44). Metabolic-balance studies and some isotope techniques provide measurement of net or apparent absorption rather than separate measurements of fractional absorption and endogenous losses of zinc in the feces
3. Urinary zinc excretion
4. Losses in sweat, which have been especially difficult to measure accurately and in some instances have probably been grossly overestimated
5. Other losses including those in shed hair and skin, and in semen

6. Secretion in milk
7. Retention required for growth, including growth of the conceptus during pregnancy

One of the outstanding factors governing dietary zinc requirements is the wide variation in bioavailability of zinc from different food sources. Fractional absorption from a variety of meals appears to average ~20% (798,799), and this figure is used in estimates of requirements in this section. The development of stable zinc isotope methodology should lead to a very substantial increase in the potential for quantitating absorption, including studies in infants. Finally, information of value in defining minimal requirements has been derived from the identification of mild nutritional zinc deficiency states in the human, coupled with calculations of dietary intakes of the zinc-deficient subjects, and measurements of zinc intake by the breast-fed infant.

2. Healthy Adults

Numerous zinc balance studies have been performed on adult subjects under a variety of different experimental conditions with variable results. Typically, excretion measurements have been limited to feces and urine. For example, Spencer and colleagues (694,869) studied zinc balance in adults receiving 11–15 mg zinc per day and found it was in equilibrium or slightly positive. Results of other balance studies in which zinc intake ranged from 9 to 14 mg/day varied from −4.0 to +8.8 mg zinc per day (34). White and Gynne (978) studied young adult women receiving an average of 11.5 mg zinc per day. Four of their subjects were in positive balance, but apparent absorption and hence balance was negative in another four. Sandstead (795) performed prolonged balance studies on normal adult men living in a controlled environment who were fed a mixed diet containing ~16% of dietary energy as protein including 70% animal protein. The apparent zinc requirement for zero balance was calculated by regression analysis to be 12.8 mg zinc per day. This same research group, on the basis of the results of very careful measurements, calculated that dermal losses averaged 0.5 mg zinc per day in normal adult men (625). Assuming an average 20% absorption, an additional 2.5 mg dietary zinc per day would be required to compensate for these dermal losses and to maintain equilibrium, that is, a total of 15.3 mg zinc per day.

Baer and King (44) fed an experimental zinc-free (0.28 mg zinc per day) diet to normal young adult male volunteers and noted a progressive decline in urine and fecal zinc. Linear-regression analysis of excretion (urine + fecal zinc) on time was performed, and extrapolation back to day 0 of the experimental diet (to exclude the effects of the progressive zinc deficiency state) gave a mean calculated daily loss of zinc in feces and urine combined of 1.36 ± 0.24 mg. This figure represented minimal endogenous losses of zinc via the feces and urine.

Individual variation was considerable, with a maximum individual value of 1.98 mg zinc per day. Measured integumental zinc losses in the same study averaged 0.81 mg/day. Hair and nail zinc losses have been calculated (817) at 0.03 mg/day, bringing the total endogenous losses to 2.2 mg zinc. Assuming 20% absorption, these data indicate an average requirement of 11 mg zinc per day for normal young adult men. However some individuals would require an additional 3 mg/day dietary zinc to meet endogenous losses, again assuming 20% absorption. An additional 0.63 mg zinc was calculated to be lost in semen per ejaculum. Similar measurements on normal young women by this same research group (393) yielded a calculated figure of 0.72 mg zinc per day minimal endogenous losses in urine and feces, 0.67 mg zinc per day dermal losses, and an average of only 5 μg zinc per day attributable to menstrual losses. Total minimal endogenous zinc losses per day were calculated to average 1.63 mg. It is apparent from these studies that minimum endogenous losses can decline further in response to a deficiency state, and conversely, losses of endogenous zinc in the feces can increase rapidly and substantially as dietary zinc intakes increase above minimum requirements (442).

3. Elderly

Turnland et al. (914) carried out two extended balance studies in six elderly male subjects confined to a metabolic unit. Mean intake of zinc was 15.5 ± 0.3 mg. During the final stages of the study, when protein intake, though adequate, was quite low, retention ranged from a mean of −1.1 ± 1.3, to a mean of +0.1 ± 0.5 mg/day. Retention for four of these subjects on a higher protein intake, averaged +0.6 ± 1.0 mg/day. Only fecal and urine losses were measured. It was concluded that these data suggested that 15 mg dietary zinc per day is adequate for elderly adults. Bunker et al. (102) reported slightly positive (+0.2 mg zinc per day) average zinc balance in elderly subjects fed an average of 8.8 mg zinc per day, which it was concluded was an adequate intake, although no allowance was made for dermal losses.

4. Pregnancy and Lactation

Swanson et al. determined zinc balance during the third trimester of pregnancy (895,896). For 10 subjects at an average of 29 weeks' gestation, mean balance was zero on diets providing 16 mg zinc per day (896). Balance was not influenced by whether dietary protein was primarily from animal or plant sources. These subjects had a daily average intake of 26 mg zinc for 10 days prior to the balance period. Balance did not differ from that of 5 nonpregnant control women included in the same study. In an earlier study (895), 8 pregnant women averaging 31 ± 4 weeks' gestation were fed 20 mg zinc per day in a semipurified diet throughout the 21-day confined metabolic study. Mean calculated retention was

1.9 ± 0.6 mg zinc per day, which was significantly greater than that for 10 nonpregnant controls (+0.9 ± 0.5 mg zinc per day). Some individual results were negative, and there was no correlation between calculated zinc and nitrogen retentions.

Sandstead (791) has calculated that a daily retention of 750 μg zinc is required during the last 20 weeks of gestation for the fetus plus placenta. Concentrations in the fetus may in fact be higher (122) than the figures on which Sandstead based his calculations. Hence retention required for the conceptus may exceed 1 mg/day or ~5 mg additional dietary zinc per day, assuming 20% absorption. It is not known if zinc absorption increases in the human as a direct result of pregnancy, as has been reported in the rat (182).

No published data are available on zinc balance and zinc requirements during lactation. However, Krebs *et al.* (495) found evidence that an average zinc intake of 10.7 ± 4.1 mg/day throughout lactation was suboptimal. Casey *et al.* (129) found that the peak intake of zinc by the fully breast-fed infant was 2.7 mg/day, and the intake was >2 mg/day for several weeks; at 20% absorption, this would require an additional 10 mg dietary zinc per day for equilibrium.

5. Adolescents and Preadolescent Children

Greger and associates (308) found that the majority of adolescent girls did not achieve positive zinc balance when fed 7.4 mg zinc per day. Zinc retention was greater, but not always positive, when the same girls were fed a diet containing 13.4 mg zinc per day. Subsequently (307) a similar group of girls was fed a diet containing zinc in quantities either approaching the RDA (14.7 mg/day) or similar to that consumed by adolescent females surveyed previously (11.5 mg) (309). Regression analysis applied to the balance data from both studies combined (i.e., with four different levels of dietary zinc) indicated that a zinc intake of 11.0 ± 2.2 (mean ± SEM) mg/day was necessary for equilibrium. These investigators allowed 0.7 mg zinc for dermal losses and 0.6 mg zinc for peak growth, and calculated that absorption was 25%; therefore, an additional 5.2 mg zinc would have been required to achieved desired retention at the peak growth periods, or a total of 16 mg zinc per day. The intake required for equilibrium in these studies appeared relatively high in relation to the corresponding figure calculated from studies for young adult women (393) and for preadolescent girls (220).

Two early studies of zinc balance in preadolescent children (544) indicated remarkably high zinc retention on intakes ranging from 6 to 18 mg zinc per day. Since 1966 (220) a series of balance studies in preadolescent girls has been performed at Virginia Polytechnic Institute (586,743,769). Balances were generally near equilibrium or positive with zinc intakes ranging from 4.5 to 14.6 mg/day. Ritchey *et al.* (769) concluded that a zinc intake of 7 to 8 mg/day was adequate to meet requirements of preadolescent girls including retention required for growth. Earlier, Engel *et al.* (220) had estimated that 2.75 mg dietary zinc was required for zero balance, excluding dermal losses. Assuming dermal losses of 0.5 mg, maximal growth requirements of 0.25 mg and 20% absorption, an additional 3.5 mg/day would be required for desired retention, that is a total of 6.5 mg/day. Chronic mild, growth-limiting, zinc deficiency states have been identified in boys aged 3–6 years whose calculated average zinc intake was between 5 and 6 mg/day (351,951).

6. *Infants*

Zinc balance data are limited for normal infants delivered at term. Cavell and Widdowson (137) reported negative zinc balance in 9 of 10 breast-fed neonates at the age of 1 week. Casey (122) had more positive results during the first week in infants fed diluted cow's milk and found uniformly positive results by the second week. Ziegler and colleagues (999) provided a brief report on the results of 423 balance studies of 72 hr duration in 55 normal healthy infants and young children. Details of age and results in relation to age were not given. Formula or whole cow's milk provided most or all of the dietary intake. Of 20 different formulas, 10 were based on cow's milk, 6 were based on soy protein isolate, and 4 contained whey protein and casein. Equilibrium between zinc intake and fecal plus urine zinc excretion occurred on average at zinc intake of 0.21 mg/kg body weight per day. However, the intercept of the lower limit of the 95% confidence interval of regression of zinc retention on zinc intake with zero retention was at 0.82 mg/kg body weight per day, and it was concluded that consistently positive balance occurred only with intakes at or above this level. A factorial approach to the estimate of zinc requirements is particularly difficult during infancy because of the lack of necessary data, especially on endogenous zinc losses via the feces and dermal zinc losses. Although data on some factors are missing, the most important individual item in determining requirements in the infant is that related to new tissue deposition. Assuming 30 μg zinc per gram of fat-free tissue (993), the amount of zinc required for growth between months 1 and 3 would average ~150 μg/kg body weight per day, or a total of 725 μg/day. Corresponding figures between 3 and 6 months are 67 μg/kg and a total of 450 μg/day; between 6 and 9 months, 40 μg/kg or 330 μg/day; between 9 and 12 months, 25 μg/kg or 250 μg/day. If it is assumed that, as in the adult, dermal losses are of the same order as those in urine (i.e., 10–20 μg/kg per day), then the net amount of zinc that must be absorbed to meet urine and dermal losses plus growth requirements would average ~180 μg/kg per day between 1 and 3 months of age, or about 800 to 900 μg/day. The dietary intake necessary to meet these requirements will, of course, depend on percentage absorption from different infant foods, which is discussed below.

One unique source of data for nutrient requirements of the infant is provided by the nutrient intake of the fully breast-fed infant. Zinc intakes have been found to peak at an average of 2.7 mg/day at 1 week of lactation and then to decline progressively (129). At 1 month, means are still >2 mg/day. Comparable data for later lactation have not been reported, but calculations based on an assumed volume of 750 ml/day give an intake of ~500 µg/day for the fully breast-fed infant between 6 and 9 months of age (495). These intakes would meet requirements calculated from a factorial approach, assuming favorable bioavailability of zinc from human milk. However, consumption of cow's milk-based infant formula containing 1.8 mg zinc per liter has been associated with growth-limiting zinc deficiency (948). In the orally fed very low birth weight preterm infant, optimal zinc retention appears difficult to achieve whatever the level of dietary zinc (183,361), and oral zinc requirements have not been defined in this group.

7. Effects of Dietary Factors

The most outstanding factor affecting zinc requirements is the variation in percentage absorption of zinc from different dietary sources. When zinc is administered as a simple zinc salt in the postprandial state without food, absorption averages ~65% with a range from 40 to 90% (2,525,631,766,967). With the probable exception of human milk, percentage zinc absorption is much lower when ingested as part of or with a meal. The bioavailability of zinc also varies extensively according to the type of meal consumed. For example, Sandström and colleagues (798,799), employing radiozinc techniques in normal adult volunteers, found that the percentage absorption varied from 8 to 38% with a mean of 18% absorption. Only two meals, white bread and chicken, were associated with ^{65}Zn absorption >30%, and the zinc content of these two meals was very low. These and other data (631,854) suggest that a wide variety of foods and composite meals have a marked inhibitory effect on zinc absorption. The studies of Sandström and colleagues also demonstrated the importance of the absolute quantity of zinc in the diet with respect to total quantities of zinc absorbed in contrast to percentage absorption.

Phytate, myoinositol hexaphosphate, which is found in all plant seeds and in many roots and tubers, is the individual dietary item that has attracted most interest with regard to interference with the intestinal absorption of zinc. The phytate-rich flat bread *tanok*, the major staple of the rural diet in central and southern Iran, has been suspected of playing an important role in the etiology of the zinc deficiency in children, adolescents, and young adults, which has been reported in that country (763). Despite the publicity that phytate has received, and the hypothesis that phytate–zinc ratios in the human diet are of outstanding importance in determining human zinc status (676), experimental data are confusing if not contradictory. For example, the extensive studies of Morris and co-workers (640,641,646), employing careful and prolonged metabolic balance techniques have failed to demonstrate a significant inhibitory effect of phytate on apparent zinc absorption with phytate–zinc molar ratios as high as 27.5 : 1. Other investigators (797) have reported an inhibitory effect of a high-phytate diet on apparent zinc absorption with metabolic-balance studies. Using stable isotope techniques, Solomons *et al.* (853) found little difference between zinc absorption from soy formula or cow's milk formula, although the mean absorption of ^{70}Zn with soy protein was significantly lower than the corresponding mean with beef protein. On the other hand, Turnlund and colleagues (915) found that addition of phytate to a basal diet fed to normal young men reduced apparent zinc absorption from 34.0 ± 6.2% to 17.5 ± 2.5%. The phytate–zinc molar ratio was 15 : 1. The decrease in zinc absorption was accompanied by decreased urine zinc and increased fecal zinc. Sandström and co-workers (789,799) found that absorption of ^{65}Zn with a meal containing soybean protein (phytate–zinc molar ratio 24 : 1) did not differ significantly from that with a meal containing chicken and beef (phytate–zinc molar ratio 3 : 1). The zinc content of these meals was similar. However, this group (80) has since observed that addition of soy protein to bread and partial or total replacement of meat protein by soy protein did impair the absorption of zinc. This inhibition applied more or less equally with the use of soy flour, soy concentrate, or soy isolate. For example, zinc absorption from white bread and milk was 37 ± 11%. When 50% of the protein in bread was replaced by soy protein absorption averaged ~24%.

The effects of dietary calcium on zinc absorption from meals containing phytate have been much less clear in the human than in the experimental rat. Spencer and co-workers found that the intestinal absorption of ^{65}Zn was similar during a low calcium intake of 200 mg and a high calcium intake of 2000 mg/day (866). Price *et al.* (743) found that variations in dietary calcium had no effect on zinc balance in children. Wester reported that high calcium intake changed zinc balance in a position direction (971). Sandstead (797) has reported that variations in dietary phosphate have a substantial effect on calculated human zinc requirements.

The effects of dietary fiber on zinc absorption in humans are even less well defined than those of phytate. The type of fiber and interactions with other dietary constituents may be important in determining the effects of fiber on zinc absorption. Reinhold's group (437) noted that the inhibitory effect of the Iranian rural flatbreads on zinc absorption could not be explained entirely by the phytate content of these unleavened breads, and he hypothesized that the dietary fiber in these 100% extraction whole-meal breads also exerted an inhibitory effect on zinc. In balance studies limited to three individuals, he demonstrated that feeding 10 g cellulose per day reduced the apparent absorption of zinc. Subsequently Drews *et al.* (209) found that addition of 14 g hemicellulose, but not cellulose or pectin, to a basic U.S. diet containing 7 g of intrinsic fiber had a negative effect on apparent absorption. Guthrie and Robinson (322) found no consistent effect of adding 14 g of bran to the diet of four female subjects undergoing balance studies in New Zealand. Similarly, Turnlund *et al.* (915) found no effect of 0.5 g cellulose per kilogram body weight on absorption of ^{70}Zn in normal young adults. Sandstead (797) found a small, but statistically significant effect of dietary fiber on estimated zinc requirements in healthy adult volunteers; wheat bran and carrot powder significantly impaired zinc retention when given in 26-g/day quantities. Wheat germ, dry milled corn bran, soybean hulls, and apple powder did not have a significant effect. Kelsay (471) and colleagues found that the addition of fiber to the diet had an effect only in the presence of oxalic acid. Kies and colleagues (477) observed that the addition of 14.7 g of fiber to the diet depressed zinc balance more in omnivores than in ovolactovegetarians. Hemicellulose and cellulose but not pectin were found to have depressing effects on apparent zinc absorption. Ascorbic acid also depressed zinc absorption.

Despite the indications that phytate and some types of fiber may impair zinc absorption in some circumstances, studies of the zinc status of various groups of vegetarians have revealed relatively minor differences compared with omnivores (5,23,262,477,480). This may be partly explained by the observation that increasing the zinc content of a high-phytate, high-fiber food such as whole-meal bread by the addition of milk and cheese results in an increase of percentage absorption to the same level as that observed with white bread (798). This suggests that the effects of phytate in a composite meal are different from those when the source of phytate is given alone.

There have been conflicting reports on the relationship between total protein intake and zinc retention. High protein intakes have been associated with increased fecal excretion of ^{65}Zn and with increased dietary zinc requirements (797). Some studies have shown no effects of variations in dietary protein (162, 548). In contrast, higher protein intakes have been associated with improved zinc retention. For example, Greger and Snedeker (313) found that adult male subjects retained more zinc when they consumed a diet containing 80 g nitrogen per day rather than 24 g/day. Sandström *et al.* (798) found a positive correlation between ^{65}Zn absorption and the protein content of meals based on whole-meal bread with various combinations of milk, eggs, cheese, and beef.

Using change in plasma zinc concentration as an index of zinc absorption, Solomons and Jacob (852) observed a significant reduction in apparent zinc absorption when the iron–zinc ratio was increased to 2 : 1 or more. These observations were made with ferrous iron. Subsequently (855), ferric iron was found to have a lesser effect. Heme iron did not manifest any intestinal interaction with zinc. These findings have essentially been confirmed by Aggett *et al.* (9) and Meadows *et al.* (584) using similar investigative techniques. The iron preparation used by Meadows *et al.* also contained folate, which has also been reported to diminish zinc absorption (626). Hambidge *et al.* (360) found an inverse relationship between the plasma zinc concentration of pregnant women in the third trimester and the amount of iron intake from iron supplements. Solomons and Cousins (856) have concluded that the effects of iron on zinc absorption could be important when the iron is given as a supplement or in liquid formula diets, but is unlikely to have a major effect in composite meals. This conclusion has received support from studies by Sandström and colleagues (801). The mechanisms of the iron–zinc interactions are not known, but a combination of intraluminal and intracellular interactions has been suggested (855).

Variations in zinc absorption from different milks and formulas employed in infant feeding are of particular practical concern, because they are unlikely to be modified by the effects of other food items, particularly during the first 6 months of postnatal life. There is considerable evidence to suggest that the bioavailability of zinc from or with human milk is especially favorable (124,354,451,1001). The superior bioavailability of zinc from human milk has been confirmed with ^{65}Zn studies in adults in whom absorption with mature human milk averaged 57% compared with 32% for cow's milk (800). Hence, zinc requirements of infants fed with cow's milk-based infant formula appear to be very substantially higher than those of the breast-fed infant. The mean plasma zinc concentration of infants fed with an iron-fortified cow's milk formula was significantly lower than that of comparable infants fed a non-iron-fortified cow's milk formula (169). It has been suggested in retrospect (852) that iron fortification of an infant formula that had not been supplemented with zinc may have contributed to a growth-limiting zinc deficiency state (948). Zinc absorption from soy-based infant formulas is especially poor (531), and infants fed with soy protein formulas have notable depression of plasma zinc concentrations (169). Golden and Golden (288) found a lower rate of weight gain and plasma zinc response in infants recovering from malnutrition who were fed a soy formula as compared to cow's milk formula. The poor absorption of zinc from soy formulas has been found to be attributable to the phytate present in these formulas (531). It appears that the absorption of zinc from semisynthetic formulas may be especially low (124).

8. Recommended Dietary Allowances

The most widely publicized recommendations for zinc are the RDA of the Food and Nutrition Board, National Academy of Sciences, United States (657). Specific recommendations are 3 mg zinc for infants <6 months of age, 5 mg for older infants, 10 mg for preadolescent children, 15 mg for adolescents and adults, 20 mg during pregnancy, and 25 mg during lactation. In light of the current incomplete data pertaining to requirements that have been discussed above, it is, in general, difficult to improve on these recommendations at this time. In order to cover the requirements of formula-fed infants, recommendations for the first 6 months of life should be increased, while those of nonpregnant, nonlactating adult women may be overgenerous.

VI. SOURCES OF ZINC

A. Human Foods and Dietaries

Extensive data on the zinc content of foods have now become available (36,163,260,264,298,320,328,650,707,863). Results of zinc determinations from different laboratories and with different methods of analysis have been in satisfactory agreement for similar foods (163). Analyses have now been reported on a large number of food items, including mixed dishes, convenience items, and vegetarian foods, as well as commonly used prepared foods and staples. Table VI gives the zinc content of a variety of selected foods. The major dietary sources of zinc are animal products, with muscle and organ meats and some seafoods being the richest sources. However, as indicated in the table, the zinc content of these and other animal foods varies widely, ranging from 0.02 mg per 100 g for egg white and 1.00 mg per 100 g for light chicken meat to 75 mg per 100 g for Atlantic oysters. Fruits and fats are generally the poorest sources of zinc. Whole cereal grains are relatively rich in total zinc, with small differences among species. Most of the zinc is contained in the bran and germ portions and hence, a considerable proportion, nearly 80% for wheat, of the total zinc is lost in the milling process (818). For example, Zook *et al.* (1003), in a study of North American wheats and their products, reported the following mean values in micrograms per gram on a dry basis: common hard wheat, 24.0 ± 4.5; common soft wheat, 21.6 ± 7.0; baker's patent flour, 6.3 ± 1.0; soft patent flour, 3.8 ± 0.8; and white bread, conventional dough, 8.9 ± 0.5. Nuts and legumes are also relatively good plant sources of zinc, ranging from approximately 1 to 5 mg zinc per 100 g, and are less subject to processing losses. These foods, however, are also high in phytate (675), which may affect the bioavailability of the mineral constituents.

Variation in zinc content may be high within types of foods, as well as among the different classes of foods, because of the effects of soil types and fertilizer treatment. This is well illustrated by the data obtained by Warren and Delavault (956) for vegetables obtained from a wide range of locations. Underwood (919) found the zinc concentration in wheat grown with the aid of zinc-containing fertilizers to be about twice that of wheat grown on the same soils without zinc applications. Welch *et al.* (970) obtained marked increases in the zinc content of pea seeds from plants grown in solution culture when zinc sulfate was added to the culture. They suggested that the nutritional value of legume seeds with respect to zinc content could be increased by applying zinc fertilizers "possibly in excess of requirements for optimal plant yields." Other investigators (513, 746) have reported that treatment of soil with municipal compost or sewage sludge resulted in higher levels of available zinc in soils for up to 6 years beyond the last year of application. The increased availability of zinc was reflected in increases of zinc that were greater in the leaves than in the roots of plants. Contamination with industrial sources of zinc provides a further source of variation. Such factors as age of the animal or plant and seasonal variations have not been shown to be of major influence (650). In contrast, genetic factors are sometimes important considerations, such as with different species of oysters and different cultivars of wheat. The practical importance of these variations may not be significant in countries where the food supply is relatively homogeneous due to the wide distribution of food staples. This is supported by reports from several different investigators (93,310,645,979) of close agreement between analyzed and calculated zinc intake values, using similar data bases but with analyses of locally purchased foods. Additionally, comparison of food tables from different countries (163,320,707,863) shows, in general, similar zinc concentrations for comparable foods.

The variation in the zinc content of processed and mixed foods is due primarily to differences in composition (i.e., differences in the proportions of zinc-rich ingredients) rather than losses due to cooking, which are minimal. These differences are reflected in the zinc content of the final prepared product. An additional source of variation in the zinc content of processed foods as purchased in some countries is that of fortification. Although zinc lost in processing is not routinely added back in the standard enrichment process, individual manufacturers may fortify their products to levels higher than those found in the comparable basic product. This applies mainly to cereal products (163) and infant formulas. Zinc concentrations in milk-based formulas may be diluted in conjunction with processing aimed at dilution of the protein fraction. Such treatment results in lower zinc levels than in whole cow's milk, while fortification may yield higher concentrations. Thus, Lönnerdal et al. (530) reported zinc concentrations in infant formulas from eight countries to range from 0.21 to 13.48 μg/ml. For comparison, zinc concentrations in cow's· milk are ~4 μg/ml, and in human milk are 2.5–3.0 μg/ml at 1 month, and 0.6 μg/ml at 12 months postpartum (495).

Total dietary intakes are greatly influenced by the choice of foods consumed. Zinc intakes are most strongly correlated with protein intakes, but even this relationship is strongly dependent on the choice of protein source. This is strikingly illustrated in a comparison of different weight-reducing diets made by Kramer et al. (492). Diets with approximately the same caloric content (1000 kcal) and protein contents differing by only 8 g, differed in their zinc content by a factor of 1.8. In seven diets (providing approximately 600–1200 kcal/day), zinc (mg) to protein (g) ratios ranged from 0.08 and 0.19, a range similar to those reported in other studies (495,695). These reports have emphasized the large variations in the zinc–protein density observed in diets, depending on whether protein sources are primarily foods such as eggs, milk, poultry, and fish, all relatively low in zinc, or beef and other red meats, which are relatively rich in

Table VI
Zinc Content of Selected Foods

Food	Zinc (mg per 100 g)	Zinc (mg per portion[a])	Zinc–protein ratio (mg/g)
Meat, poultry, and seafood			
Beef			
Ground (77% lean), cooked	4.40	3.74	0.18
Separable lean, cooked	5.80	4.93	0.20
Separable fat, raw	0.50	—	—
Liver, cooked	5.09	2.90	0.19
Chicken			
Breast, meat only, cooked	1.00	0.85	0.04
Drumstick, meat only, cooked	3.18	1.40	0.11
Eggs, fresh			
Whites	0.02	0.01	0.00
Yolks	3.38	0.57	0.21
Whole	1.44	0.72	0.12
Frankfurter, all meat	1.60	0.72	0.12
Pork			
Loin, separable lean, cooked	2.59	2.20	0.11
Ham, medium fat, cooked	4.00	3.40	0.13
Seafood			
Crabs, blue, steamed	4.30	3.66	0.25
Fish, white varieties, cooked	1.00	0.85	0.04
Oysters, Atlantic, raw	75.00	42.00	8.89
Oysters, Pacific, raw	9.00	5.04	0.85
Tunafish, canned, drained solids	1.10	0.94	0.04
Dairy products			
Cheese			
Cheddar	3.11	0.87	0.12
Cottage, creamed	0.37	0.42	0.03
Cream	0.54	0.15	0.07
Parmesan	3.19	0.16	0.08
Processed spread, American	2.99	0.84	0.15
Milk, fluid			
Whole (3.3% fat)	0.38	0.93	0.12
Nonfat	0.40	0.98	0.12
Yogurt, low fat, plain	0.89	2.02	0.17
Legumes, nuts, and seeds			
Beans, common mature, cooked	1.00	0.92	0.13
Nuts			
Almonds	2.56	0.72	0.14
Brazil nuts	5.06	1.42	0.35
Walnuts, English	2.26	0.63	0.15

Table VI (*Continued*)

Food	Zinc (mg per 100 g)	Zinc (mg per portion[a])	Zinc–protein ratio (mg/g)
Peas, green, mature, cooked	1.10	2.20	0.14
Peanuts			
Butter	2.9	0.46	0.12
Roasted	3.0	0.84	0.11
Seeds			
Sesame	10.25	0.82	0.39
Sunflower	4.58	0.37	0.19
Breads and cereals			
Breads			
White	0.60	0.15	0.07
Whole wheat	2.00	0.56	0.22
Rye	1.60	0.40	0.18
Corn flakes	0.28	0.08	0.04
Macaroni, cooked	0.50	0.35	0.15
Oats			
Rolled, dry	3.40	0.95	0.24
Oatmeal, cooked	0.50	1.20	0.25
Rice			
Brown, cooked	0.60	0.51	0.24
White, cooked	0.40	0.34	0.19
Wheat cereals, ready to eat			
Germ, toasted	15.39	0.92	0.51
Shredded	2.79	0.78	0.28
Vegetables			
Beans, green, cooked	0.30	0.30	0.19
Broccoli, cooked	0.15	0.14	0.05
Cabbage, raw	0.40	0.18	0.31
Carrots, cooked	0.30	0.23	0.33
Peas, green, immature, cooked	0.69	0.55	0.13
Potato			
Boiled	0.30	0.23	0.16
Chips	0.80	0.22	0.15
French fried	0.30	0.20	0.07
Squash, winter varieties, cooked	0.30	0.30	0.27
Tomatoes, raw	0.20	0.25	0.18
Fruits (raw)			
Apple	0.04	0.07	0.20
Banana	0.16	0.19	0.15
Orange	0.20	0.26	0.20
Peach	0.20	0.20	0.33

[a]Edible part of common household units.

zinc for similar amounts of protein. The zinc–protein ratios for individual foods are also listed in Table VI. Vegetarian diets have also been observed to exhibit a wide range of total zinc content as well as zinc–protein density. Freeland-Graves *et al.* (263) reported mean zinc intakes for lactoovo- and lactovegetarians of 11.2 ± 7 and 11.3 ± 8 mg/day, respectively, and 7.9 ± 8 mg/day for vegans. These authors observed that subjects in all categories with higher intakes tended to eat liberal quantities of legumes, whole grains, nuts, and cheeses, foods with relatively high zinc–protein density, while subjects with lower intakes tended to subsist primarily on fruits and vegetables, which are poor sources of zinc. Mean zinc intakes of adult lactoovovegetarians in other reports (23,369,480) have ranged from 6.4 to 9.2 mg/day.

The mean zinc content of adult self-selected mixed diets in the United States has been reported to range from 8.6 to 14 mg/day (249,407,910,977). Mean estimated intakes by English men and women (858), West German women (811), Japanese men and women (480), New Zealand women (321), Scottish men and women (538), and Canadian women (276) have also generally ranged from 9 to 14 mg/day. Zinc intakes of elderly populations in the United States and England have been reported to average from 7 to 10 mg/day, with free-living populations tending to have lower mean intakes (102,434) than those living in institutions (305,826) or participating in feeding programs (306). For pregnant and lactating women in the United States (107,360,421,480,495,645,934), Finland (942), Sweden (445), and Iran (270), intakes have been reported to fall within a range similar to that of adult mixed diets. Mean zinc intakes of 10.7 to 11.3 mg/day have been reported for adolescent girls in the United States (309,571). The daily zinc intakes of children 8–13 years of age have been estimated to be 7.3–9.7 mg in Scotland (538), 10.2 ± 1.4 mg in Norway (812), and 9.8 mg in the United States (571). Intakes of younger children in the United States (355, 494), Norway (812), and Canada (277) have been reported in the range of 5.2 to 8.2 mg/day. Reports of zinc intakes of infants are quite limited. MacDonald *et al.* (543) reported mean monthly zinc intakes for Canadian infants aged 1–6 months. For the breast-fed infants, mean zinc intakes were 1.9 ± 0.2 mg/day at 1 month, and 2.7 ± 0.5 mg/day at 6 months, at which time the intake of nonmilk foods supplied 14% of total dietary zinc. Bottle-fed infants' comparable intakes were 3.6 ± 0.6 mg/day at 1 month, and 4.6 ± 0.7 mg/day at 6 months, with 18% of the total dietary zinc being supplied by nonmilk sources at the latter point.

B. Animal Feeds and Fodders

Typical zinc concentrations in pasture herbage remote from industrial areas have been given by Mills and Dalgarn (624) as 25 to 35 µg/g. Levels from 5 to 50 times higher were obtained for such herbage from agricultural land exposed to contamination from industrial sources. Grace (299) obtained values ranging from

23 to 70 (mean 38) μg/g from 10 improved mixed pastures in the North Island of New Zealand, 17 to 27 (mean 27) μg/g for such pastures in the South Island of that country, and the very wide range of 8 to 48 μg/g for tussock grassland from hill country. Pastures in Western Australia were reported to be lowest in zinc in autumn (20.1 μg/g) and highest in winter (36.9 μg/g) (559). In northern Ontario, winter grazing contained 20–30 μg/g and summer pasture 40–60 μg/g (396). The zinc concentration in plants usually falls with advancing maturity (481,503), and leguminous plants invariably carry higher zinc levels than grasses grown and sampled under the same conditions (906). Heavy dressings with lime and to a lesser extent with superphosphate can greatly reduce pasture zinc levels (25).

The cereal grains used as the basis of pig and poultry rations typically contain 20–30 μg zinc per gram, with appreciably higher levels in most materials used as protein supplements. Typical values for soybean, peanut, and linseed meals may be given as 50 to 70 μg/g (919). The zinc contents of fish meal, whale meal, and meat meal are normally much higher than that of soybean meal. Levels of 90 to 100 μg/g or more are common (537,919).

VII. TOXICITY

Zinc is relatively nontoxic to birds and mammals. Rats, pigs, poultry, sheep, cattle, and humans exhibit considerable tolerance to high intakes of zinc; nonetheless zinc toxicity has been reported to occur under nonexperimental conditions in a number of species, from oral or intravenous administration of excessive supplemental zinc and from environmental sources such as galvanized iron containers (16,91,115,314,635). In an early study with rats, dietary zinc intakes of 0.25% or 2500 μg/g induced no discernible effects whether ingested as the metal, chloride, or carbonate. At intakes of 5000 μg/g, growth was severely depressed and mortality was high in young animals when ingested as the chloride, with little mortality and only slight growth depression as the oxide (375). Subsequently Sutton and Nelson (893) confirmed that 5000 or 10,000 μg zinc per gram as the carbonate induced subnormal growth and anorexia, and the higher rate caused heavy mortality in young rats. They also observed a severe anemia. Adult female rats fed a diet containing 2000 μg zinc per gram as the oxide maintained normal pregnancies with no malformations in the fetuses, but at 4700 μg/g variable degrees of resorption and death of the fetuses were observed (813). The zinc-poisoned rats developed a microcytic, hypochromic anemia accompanied by high levels of zinc and subnormal levels of iron, copper, cytochrome oxidase, and catalase (168,304). The anemia was typical of copper deficiency, despite a normally adequate copper intake, and both anemia and biochemical changes were reversed by supplemental copper and iron. Excessive intakes of zinc thus interfere with absorption and utilization of iron and copper.

Campbell and Mills (116) found that in weanling rats maintained on diets low or marginal in copper, additional dietary zinc as low as 300 μg/g diet reduced plasma ceruloplasmin activity; more zinc, to 1000 μg/g diet, caused growth depression, hair depigmentation, and depressed liver copper levels.

Pigs appear to have a high tolerance to zinc. Intakes of up to 500 μg/g diet had little effect on growing animals or reproductive performance of sows, as determined by size and viability of litters through two parities (398,400). Addition of 5000 μg zinc per gram to the diet of sows increased zinc levels and depressed copper concentrations in plasma and liver (401). This level of maternal zinc intake also resulted in a decrease in the number of piglets weaned and their weight at weaning, although the number and weight of the litter at birth was not reduced (400). Abnormal articular cartilage was observed at the articulating surfaces of several long bones, especially the distal humerus of animals receiving 500 μg/g diet (400). Cartilaginous surfaces were fractured and showed areas of cartilage proliferation; excessive synovial fluid occurred in some joints. In weanling pigs, dietary zinc supplements of 1000 μg/g diet had no ill effects (516). Above this level, Brink *et al.* (90) reported that zinc intakes of 2000, 4000, or 8000 μg/g diet caused severe signs of toxicity including reductions in weight gain, feed intake, and feed efficiency, as well as arthritis, hemorrhaging, and gastritis. Raising the dietary calcium level from 0.7 to 1.1% had a protective effect against the toxic effects of a zinc intake of 4000 μg/g diet (413). Older (30 kg) but still growing gilts tolerated 8000 μg/g of dietary zinc well, with only a slight depression in weight and elevation in serum zinc and alkaline phosphatase over 20 weeks (398).

Broilers and layer hens exhibit a tolerance to high intakes of zinc similar to that seen in pigs. When growing chicks were given extra zinc oxide to levels of 2000, 4000, or 6000 μg/g diet from 2 to 6 weeks, they grew poorly and exhibited lesions of the gizzard and exocrine pancreas (196). Histologically, gizzard lesions included desquamation of epithelial cells and erosion of koilin, glands, and pits. In the pancreas, dilation of the acinar lumina, necrosis of the exocrine cells, and interparenchymal fibrosis were observed. High intakes of zinc can induce a pause in egg production in laying hens, accompanied by a decrease in feed intake and weight loss (171). Intakes on the order of 10,000 to 20,000 μg/g diet caused lesions of the gizzard and pancreas similar to those seen in the growing chick (196).

Ruminants, particularly young and pregnant animals, are more susceptible to zinc toxicity than rats, pigs, and poultry. In "naturally" occurring cases of chronic zinc poisoning in sheep and calves, animals were in poor condition and anemic (16). The sheep had subcutaneous edema, ascites, and proteinuria, whereas calves were dehydrated. In both species, the abomasal mucosa was mottled and ulcerated; the pancreas and kidneys of the sheep also showed distinctive gross and microscopic abnormalities. Zinc concentrations in liver, kidney, and pancreas were elevated; iron levels were also increased, but copper concentrations were not affected. The tissue zinc levels suggested that the zinc intakes of these animals, which were not directly measured, were of the order of 1000 μg/g diet by the sheep and 500–900 μg/g diet by the calves. Allen et al. (16) reproduced these findings experimentally in sheep given either acute doses of 2 g zinc daily for 13 days or 0.8 g zinc daily (equivalent to 2000 μg/g diet) for 12 days, followed by 1.2 g zinc daily for up to 10 weeks. Clinical manifestations of zinc toxicity, including loss of appetite and condition, profound weakness, and jaundice, in addition to tissue abnormalities, were very similar to the "natural" cases. These workers also noticed a breed difference in that Southdown rams were considerably more tolerant of high zinc intakes than were Merinos and Poll Dorsets. Weaned lambs given diets containing 1500 μg zinc per gram exhibited reduced food consumption and weight gains, and decreased feed conversion efficiency, but no renal damage (698). However, in younger, suckling lambs, Davies et al. (183) produced extensive renal damage, and poor growth and appetite, by feeding a diet containing 134 μg/g of dietary zinc. Zinc intakes of 750 μg/g diet by pregnant sheep caused a high incidence of abortions and stillbirths, and feed consumption, weight gain, and feed efficiency were all reduced compared with ewes receiving 30 or 150 μg/g diet (117). The animals had copper deficiency manifested by reduced plasma levels of copper, ceruloplasmin, and amine oxidase. Nonviable lambs had elevated zinc and severely reduced copper levels in the liver and arrested bone growth. Provision of supplemental copper to the ewes prevented the development of copper deficiency but did not alleviate the effects of excess zinc on maternal feed intake or fetal viability. Steers and heifers have been shown to be unaffected by dietary zinc levels of 500 μg/g or less, but 900 μg/g caused reduced weight gains and lowered feed efficiency, and 1700 μg/g induced, in addition, a depraved appetite characterized by excessive salt consumption and wood chewing (698). Dairy cows, however, appear to tolerate zinc intakes of 1300 μg/g diet with no ill effects, possibly because of the additional route of excretion in milk (610). The tissue mineral changes induced by high zinc intakes in ruminants differ somewhat from those reported in rats: liver zinc was generally increased and copper levels were depressed in liver but elevated in other tissues; iron concentrations were generally considerably elevated (16,699). It has been suggested (656) that the greater toxicity of zinc to ruminants compared with some other species may be due to adverse effects on rumen microorganisms. At high levels of zinc intake in lambs there was a reduction in the volatile fatty acid concentration and acetic acid–propionic acid ratio in the rumen (699). Cellulose digestion by rumen bacteria in vitro is reduced by zinc concentrations in the medium of 10 to 20 μg/ml (555).

Because of the effective homeostatic control mechanisms and the low inherent toxicity of zinc among the divalent cations, orally ingested zinc is generally regarded as being relatively nonhazardous to humans. Nonetheless, both acute and chronic zinc toxicity syndromes have been reported in humans, with one of the principle features being a direct effect of zinc on the gastrointestinal tract resulting in epigastric pain, diarrhea, nausea, and vomiting (656). As well as these gastrointestinal disorders, acute oral ingestion of large amounts of zinc, especially in liquids, may cause irritability, headache, and lethargy. A 16-year-old boy who took 12 g elemental zinc over 2 days exhibited light-headedness and lethargy, and had difficulty writing, along with an elevated blood zinc level and increased serum amylase and lipase activities, suggestive of an effect of the zinc on the pancreas (651). Daily intake of 150 mg zinc as the sulfate for 16 to 26 weeks was not associated with evidence of toxicity in patients with venous leg ulcers (copper was not investigated) (335); however, a case of bleeding gastric erosion was reported in a 15-year-old girl, otherwise healthy, who took 100 mg zinc daily (698). A fatal outcome occurred in a woman who accidentally received 1.5 g zinc intravenously over 60 hr (91). She developed hypotension, pulmonary edema, vomiting, jaundice, and oliguria, with kidney damage apparent on necropsy. Metal fume fever has been reported to occur after inhalation of zinc oxide fumes, as well as other types of metals (656). Symptoms develop within 4 to 8 hr after exposure and include hyperpnea, shivering, profuse sweating, and general weakness. The attack is short in duration (24–48 hr) and accompanied by leukocytosis.

Prolonged use of oral zinc supplements, even at relatively modest levels of zinc intake, may not be without undesirable consequences. Copper deficiency, evidenced by microcytic anemia, neutropenia, and decreased plasma levels of copper and ceruloplasmin, was documented in a young man with sickle cell anemia who had been taking 150 mg zinc per day for 2 years (739). Reduced ceruloplasmin levels were also found in other sickle cell patients who had been receiving zinc therapy for 4 to 24 weeks (739). Hematological and radiological evidence of copper deficiency was observed in an infant with acrodermatitis enteropathica who was receiving 30 mg supplemental zinc daily (350).

The observation that high intakes of zinc caused an increase in serum cholesterol levels in rats (490) caused some concern that zinc therapy may be atherogenic in humans. Hooper *et al.* (409) investigated blood lipids in 20 healthy, nonobese young men before and after zinc supplementation with 160 mg daily for 5 weeks. Total cholesterol, triglyceride, and low-density lipoprotein cholesterol levels remained unchanged, but high-density lipoprotein cholesterol concentrations in the plasma were significantly reduced and returned to normal by 7 weeks postdosing. However, zinc supplementation with 100 mg daily for 8 weeks did not affect blood lipids in healthy young women (265).

ACKNOWLEDGMENTS

This work was supported by the National Institute of Arthritis, Metabolic, Digestive, and Kidney Diseases, Grant No. 5 R22 AM12432, and by Grant No. RR 69 from the General Clinical Research Centers Program of the Division of Research Resources, National Institutes of Health.

REFERENCES

1. Aamodt, R. L., Rumble, W. F., Johnston, G. S., Foster, D., and Henkin, R. I. (1979). *Am. J. Clin. Nutr.* **32,** 559.
2. Aamodt, R. L., and Rumble, W. F. (1983). *In* "Nutritional Bioavailability of Zinc (G. E. Inglet, ed.), p. 61. Am. Chem. Soc., Washington, D.C.
3. Abbasi, A. A., Prasad, A. S., Ortega, J., Congco, E., and Oberleas, D. (1976). *Ann. Intern. Med.* **85,** 601.
4. Abbasi, A. A., Prasad, A. S., Rabbani, P., and DuMouchelle, E. (1980). *J. Lab. Clin. Med.* **96,** 544.
5. Abu-Assal, M. J., and Craig, W. J. (1984). *Nutr. Rep. Int.* **29,** 485.
6. Abu-Hamdan, D. K., Migdal, S. D., Whitehouse, R., Rabbani, P., Prasad, A. S., and McDonald, F. D. (1981). *Am. J. Physiol.* **241,** F487.
7. Acosta, P. B., Fernhoff, P. M., Warshaw, H. S., Elsas, L. J., Hambidge, K. M., Ernest, A., and McCabe, E. R. B. (1982). *J. Inherited Metab. Dis.* **5,** 107.
8. Aggett, P. J., and Davies, N. T. (1980). *Proc. Nutr. Soc.* **39,** 241.
9. Aggett, P. J., Crofton, R. W., Khin, C., Gvozdanovic, S., and Gvozdanovic, D. (1983). *In* "Zinc Deficiency in Human Subjects" (A. S. Prasad, A. O. Cavdar, G. J. Brewer, and P. J. Aggett, eds.), p. 117. Alan R. Liss, Inc., New York.
10. Agricultural Research Council (1965). *The Nutrient Requirements of Farm Livestock,"* Ruminants Tech. Rev. No. 2. ARC, London.
11. Agricultural Research Council (1980). "The Nutrient Requirements of Ruminant Livestock." Slough Commonw. Agric. Bur., ARC, London.
12. Ainscough, E. W., Brodie, A. M., and Plowman, J. E. (1979). *Inorg. Chim. Acta* **33,** 149.
13. Aitken, J. M. (1976). *Calcif. Tissue Res.* **20,** 23.
14. Alberts, J. C., Lang, J. A., and Briggs, S. M. (1975). *Fed. Proc., Fed. Am. Soc. Exp. Biol.* **34,** 906 (abstr.).
15. Alhava, E. M., Olkkonen, H., Puittinen, J., and Nokso-Koivisto, V.-M. (1977). *Acta Orthop. Scand.* **48,** 1.
16. Allen, J. G., Masters, H. G., Peet, R. L., Mullins, K. R., Lewis, R. D., Skirrow, S. Z., and Fry, J. (1983). *J. Comp. Pathol.* **93,** 363.
17. Allen, J. I., Kay, N. E., and McClain, C. J. (1981). *Ann. Intern. Med.* **95,** 154.
18. Allen, J. I., Perri, R. T., McClain, C. J., and Kay, N. E. (1983). *J. Lab. Clin. Med.* **102,** 577.
19. Alvares, O. F., and Meyer, J. (1968). *Arch. Dermatol.* **98,** 191.
20. Amadar, M., Gonzalez, A., and Hermelo, M. (1973). *Rev. Cubana Pediatr.* **45,** 315.
21. Ament, M. E., and Broviac, J. (1973). *Gastroenterology* **64,** A9/692.
22. Anastasiadis, P., Atassi, S., and Rimpler, M. (1981). *J. Perinat. Med.* **9,** 228.
23. Anderson, B. M., Gibson, R. S., and Sabry, J. H. (1981). *Am. J. Clin. Nutr.* **34,** 1042.
24. Andreson, E., Flagstad, T., Basse, A., and Brummerstedt, E. (1970). *Nord. Veterinaer Med.* **22,** 473.
25. Anonymous (1971). *N. Z. Agric.* **20,** 5.
26. Anonymous (1978). *Nutr. Rev.* **36,** 152.
27. Anonymous (1981). *Nutr. Rev.* **39,** 207.
28. Anonymous (1982). *Nutr. Rev.* **40,** 218.
29. Anonymous (1984). *Nutr. Rev.* **42,** 279.
30. Antoniou, L. D., Sudhaker, T., Shalboub, R. J., and Smith, J. C. (1977). *Lancet* **2,** 895.
31. Antonson, D. L., Barak, A. J., and Vanderhood, J. A. (1979). *J. Nutr.* **109,** 142.
32. Apgar, J. (1968). *Am. J. Physiol.* **215,** 160.
33. Apgar, J. (1972). *J. Nutr.* **102,** 343.
34. Apgar, J. (1978). *CRC Handb. Ser. Nutr. Food* **2,** 315.

35. Apgar, J., Figueroa, J. P., and Nathanielsz, P. N. (1985). *In* "Trace Element Metabolism in Man and Animals-5" (C. F. Mills, P. J. Aggett, I. Bremner, and J. K. Chesters, eds.). CAB Publications, Farnham Royal, U.K. (in press).
36. Appledorf, H., and Kelly, L. S. (1979). *J. Am. Diet. Assoc.* **75,** 35.
37. Arlette, J. P., and Johnston, M. M. (1981). *J. Am. Acad. Dermatol.* **5,** 37.
38. Arvar, S., and Eliasson, R. (1982). *Acta Physiol. Scand.* **115,** 217.
39. Arvidsson, B., Cederblad, Å., Björn-Rasmussen, E., and Sandström, B. (1978). *Int. J. Nucl. Med. Biol.* **5,** 104.
40. Atherton, D. J., Muller, D. P. R., Aggett, P. J., and Harries, J. T. (1979). *Clin. Sci.* **56,** 505.
41. Atkinson, R. L., Dahms, W. T., Bray, G. A., Jacob, R., and Sandstead, H. H. (1978). *Ann. Intern. Med.* **89,** 491.
42. Atkin-Thor, E., Goddard, B. W., O'Nion, J., Stephen, R. L., and Kolff, W. J. (1978). *Am. J. Clin. Nutr.* **31,** 1948.
43. Bach, J. F. (1981). *Immunol. Today* **2,** 225.
44. Baer, M. T., and King, J. C. (1984). *Am. J. Clin. Nutr.* **39,** 556.
45. Barge, M. T., and Mazzocco, P. (1982). *Z. Tierphysiol., Tierernaehr. Futtermittelkd.* **48,** 36.
46. Barney, G. H., Orgebin-Crist, M. C., and Macapinlac, M. P. (1968). *J. Nutr.* **95,** 526.
47. Barrett, T. (1976). *Nature (London)* **260,** 576.
48. Bartholemew, M. E., Tupper, R., and Wormall, A. (1959). *Biochem. J.* **73,** 256.
49. Battersby, S., Chandler, J. A., Harper, M. E., and Blacklock, N. J. (1983). *J. Urol.* **129,** 653.
50. Battistone, G. C., Rubin, M. I., Cutright, D. E., Miller, R. A., and Harmuth-Hoene, A. E. (1972). *Oral Surg., Oral Med. Oral Pathol.* **34,** 542.
51. Baudier, J., Haglid, K., Haiech, J., and Gerard, D. (1983). *Biochem. Biophys. Res. Commun.* **114,** 1138.
52. Bauer, M. T., and King, J. C. (1984). *Am. J. Clin. Nutr.* **39,** 556.
53. Beach, R. S., Gershwin, M. E., Makishima, R. K., and Hurley, L. S. (1980). *J. Nutr.* **110,** 805.
54. Beach, R. S., Gershwin, M. E., and Hurley, L. S. (1982). *J. Nutr.* **112,** 1169.
55. Beach, R. S., Gershwin, M. E., and Hurley, L. S. (1983). *Am. J. Clin. Nutr.* **38,** 579.
56. Becker, W. M., and Hoekstra, W. G. (1966). *J. Nutr.* **90,** 301.
57. Becker, W. M., and Hoekstra, W. G. (1968). *J. Nutr.* **94,** 455.
58. Beisel, W. R., Pekarek, R. S., and Wannemacher, R. W. (1976). *In* "Trace Elements in Human Health and Disease" (A. S. Prasad and D. Oberleas, eds.), Vol. 1, p. 87. Academic Press, New York.
59. Beisel, W. R. (1982). *In* "Clinical, Biochemical, and Nutritional Aspects of Trace Elements" (A. S. Prasad, ed.), p. 203. Alan R. Liss, Inc., New York.
60. Bendtzen, K. (1980). *Scand. J. Immunol.* **12,** 489.
61. Berfenstam, R. (1952). *Acta Paediatr. Scand.* **41,** Suppl. **87.**
62. Bergman, B. (1970). *Acta Odontol. Scand.* **28,** 425.
63. Bertrand, G., and Benson, R. (1922). *C. R. Hebd. Seances Acad. Sci.* **175,** 289.
64. Bettger, W. J., Fish, T. J., and O'Dell, B. L. (1978). *Proc. Soc. Exp. Biol. Med.* **158,** 279–282.
65. Bettger, W. J., Reeves, P. G., Moscatelli, E. A., Reynolds, G., and O'Dell, B. L. (1979). *J. Nutr.* **109,** 480–488.
66. Bettger, W. J., Fernandez, M. F., and O'Dell, B. L. (1980). *Fed. Proc., Fed. Am. Soc. Exp. Biol.* **39,** 896.

67. Bettger, W. J., Reeves, P. G., Moscatelli, E. Z., Savage, J. E., and O'Dell, B. L. (1980). *J. Nutr.* **110,** 50–58.

69. Bettger, W. J., and O'Dell, B. L. (1981). *Life Sci.* **28,** 1425.

70. Birckner, V. (1919). *J. Biol. Chem.* **38,** 191.

71. Birnstingl, M., Stone, B., and Richards, V. (1956). *Am. J. Physiol.* **186,** 377.

72. Bjorksten, B., Back, O., Gustavson, K. H., Hallmans, G., Gagglof, B., and Tarnvir, A. (1980). *Acta Paediatr. Scand.* **69,** 183.

73. Bjorndahl, L., and Kvist, U. (1982). *Acta Physiol. Scand.* **116,** 51.

74. Blackmon, D. M., Miller, W. J., and Morton, J. D. (1960). VM/SAC, *Vet. Med. Small Anim. Clin.* **62,** 265.

75. Blakeborough, P., Salter, D. N., and Gurr, M. I. (1983). *Biochem. J.* **209,** 505.

76. Blamberg, D. L., Blackwood, W. B., Supplee, W. C., and Combs, C. F. (1960). *Proc. Soc. Exp. Biol. Med.* **104,** 217.

77. Bodansky, H. (1920). *J. Biol. Chem.* **44,** 399.

78. Bohane, T. D., Cutz, E., Hamilton, J. R., and Gall, D. G. (1977). *Gastroenterology* **73,** 587.

79. Bonewitz, R. F., Foulkes, E. C., O'Flaherty, E. J., and Hertzberg, V. S. (1983). *Am. J. Physiol.* **244,** G314.

80. Bonomi, A. (1964). *Conv. Soc. Ital. Sci. Vet. Prescara, 18th.* [cited by Demertzis and Mills (191)].

81. Boquist, L., and Lernmark, A. (1969). *Acta Pathol. Microbiol. Scand.* **76,** 215.

82. Bosticco, A., and Bonomi, A. (1965). *Prog. Vet. Anno.,* p. 1 [cited by Demertzis and Mills (191)].

83. Brady, F. O., and Helvig, B. S. (1985). *In* "Trace Element Metabolism in Man and Animals-5" (C. F. Mills, P. J. Aggett, I. Bremner, and J. K. Chesters, eds.). CAB Publications, Farnham Royal, U.K. (in press).

84. Braun, O. H., Heilmann, K., Rossner, J. A., Pauli, W., and Bergmann, K. E. (1977). *J. Pediatr.* **125,** 153.

85. Bremner, I., Mehra, R. K., and Sato, M. (1985). *In* "Trace Element Metabolism in Man and Animals-5" (C. F. Mills, P. J. Aggett, I. Bremner, and J. K. Chesters, ed.). CAB Publications, Farnham Royal, U.K. (in press).

86. Breskin, M. W., Worthington-Roberts, B. S., Knopp, R. H., Brown, Z., Plovie, B., Mottet, N. K., and Mills, J. L. (1983). *Am. J. Clin. Nutr.* **38,** 943.

87. Brewer, G. J., and Oelshlegel, F. J. (1974). *Biochem. Biophys. Res. Commun.* **58,** 854.

88. Brewer, G. J., Aster, J. C., Knutsen, C. A., and Kruckeberg, W. C. (1979). *Am. J. Hematol.* **7,** 53.

89. Briggs, W. A., Pedersen, M. M., Mahajan, S. K., Sillix, D. H., Prasad, A. S., and McDonald, F. D. (1982). *Kidney Int.* **21,** 827.

90. Brink, M. F., Becker, D. E., Terrill, S. W., and Jensen, A. H. (1959). *J. Anim. Sci.* **18,** 836.

91. Brocks, A., Reid, H., and Glazer, G. (1977). *Br. Med. J.* **2,** 1390.

92. Brook, A. C., Ward, M. K., Cook, D. B., Johnston, D. G., Watson, M. J., and Kerr, D. N. S. (1980). *Lancet* **2,** 618.

93. Brown, E. D., McGuckin, M. A., Wilson, M., and Smith, J. C. (1976). *J. Am. Diet. Assoc.* **69,** (6), 632.

94. Browning, J. D., Reeves, P. G., and O'Dell, B. L. (1983). *J. Nutr.* **113,** 755.

95. Brudevold, F., Steadman, L. T., Spinelli, M. A., Amdur, B. H., and Gron, P. (1963). *Arch. Oral Biol.* **8,** 135.

96. Brummerstedt, E., Flagstad, T., Basse, A., and Andresen, E. (1971). *Acta Pathol. Microbiol. Scand.* **79,** 686.

97. Brummerstedt, E., Basse, A., Flagstad, T., and Andresen, E. (1977). *Am. J. Pathol.* **87,** 725.

98. Buamah, P. K., Russell, M., Bates, G., Ward, A. M., and Skillen, A. W. (1984). *Br. J. Obstet. Gynaecol.* **91,** 788.

99. Buchanan, P. J., and Hsu, J. M. (1968). *Fed. Proc., Fed. Am. Soc. Exp. Biol.* **27,** 483.

100. Buell, S. J., Fosmire, G. J., Ollerich, D. A., and Sandstead, H. H. (1977). *Exp. Neurol.* **55,** 199.

101. Bunce, G. E., Wilson, G. R., Mills, C. F., and Klopper, A. (1983). *Biochem. J.* **210,** 761.

102. Bunker, V. W., Lawson, M. S., Delves, H. T., and Clayton, B. E. (1982). *Hum. Nutr.: Clin. Nutr.* **36C,** 213.

103. Bunker, V. W., Hinks, L. T., Lawson, M. S., and Clayton, B. E. (1984). *Am. J. Clin. Nutr.* **40,** 1096.

105. Businco, L., Menghi, A. M., Rossi, P., D'Amelio, R., and Galli, E. (1980). *Arch. Dis. Child.* **55,** 966.

106. Butrimovitz, G. P., and Purdy, W. C. (1978). *Am. J. Clin. Nutr.* **31,** 1409.

107. Butte, N. F., Calloway, D. H., and van Duzen, J. L. (1981). *Am. J. Clin. Nutr.* **34,** 2216.

108. Buyske, D. A., Sterling, W., and Peets, E. (1966). *Ann. N.Y. Acad. Sci.* **135,** 711.

109. Buzina, R., Jusic, M., Sapunar, J., and Milanovic, N. (1980). *Am. J. Clin. Nutr.* **33,** 2262.

110. Byar, P. D. (1974). *In* "Male Accessory Sex Organs" (D. Brandes, ed.), p. 164. Academic Press, New York.

111. Caldwell, D. F., Oberleas, D., Clancy, J. J., and Prasad, A. S. (1970). *Proc. Soc. Exp. Biol. Med.* **133,** 1417.

112. Caldwell, D. F., Oberleas, D., and Prasad, A. S. (1973). *Nutr. Rep. Int.* **7,** 309.

113. Calhoun, N. R., Smith, J. C., and Becker, K. L. (1974). *Clin. Orthop. Relat. Res.* **103,** 212.

114. Calhoun, N. R., McDaniel, E. G., Howard, M. P., and Smith, J. C. (1978). *Nutr. Rep. Int.* **17,** 299.

115. Callender, G. R., and Gentzkow, C. J. (1937). *Mil. Surg.* **80,** 67.

116. Campbell, J. K., and Mills, C. F. (1974). *Proc. Nutr. Soc.* **33,** 15A.

117. Campbell, J. K., and Mills, C. F. (1979). *Environ. Res.* **20,** 1.

118. Canfield, W. K., Menge, R., Walravens, P. A., and Hambidge, K. M. (1980). *J. Pediatr.* **97,** 87.

119. Canfield, W. K., Hambidge, K. M., and Johnson, L. (1982). *Am. J. Clin. Nutr.* **35,** 842.

120. Canfield, W. K., Hambidge, K. M., and Johnson, L. A. (1984). *J. Pediatr. Gastroenterol. Nutr.* **3,** 577.

121. Carpentieri, U., Smith, L., Daeschner, C. W., and Hagard, M. E. (1983). *Pediatrics* **72,** 88.

122. Casey, C. E. (1976). *Ph.D. Thesis, University of Otago, Dunedin, New Zealand.*

123. Casey, C. E., and Robinson, M. F. (1978). *Br. J. Nutr.* **39,** 639.

124. Casey, C. E., Walravens, P. A., and Hambidge, K. M. (1981). *Pediatrics* **68,** 394.

125. Casey, C. E., Walravens, P. A., and Hambidge, K. M. (1981). *Am. J. Clin. Nutr.* **34,** 1443.

126. Casey, C. E., Guthrie, B. E., and McKenzie, J. M. (1981). *In* "Proceedings of the New Zealand Workshop on Trace Elements in New Zealand" (J. V. Dunckley, ed.), p. 210. Univ. of Otago Press, Dunedin, New Zealand.

127. Casey, C. E., Guthrie, B. E., and McKenzie, J. M. (1982). *N.Z. Med. J.* **95,** 768.

128. Casey, C. E., Guthrie, B. E., and Robinson, M. F. (1982). *Biol. Trace Elem. Res.* **4,** 105.

129. Casey, C. E., Hambidge, K. M., and Neville, M. C. (1985). *Am. J. Clin. Nutr.* **41,** 1193.

130. Casey, C. E., and Hambidge, K. M. (1985). *In* "Vitamin and Mineral Requirements in Preterm Infants" (R. Tsang, ed.), p. 153. Dekker, New York.

131. Cassens, R. G., Briskey, E. J., and Hoekstra, W. G. (1963). *J. Sci. Food Agric.* **6,** 427.

132. Cassens, R. G., Hoekstra, W. G., Faltin, E. C., and Briskey, E. J. (1967). *Am. J. Physiol.* **212**, 688.

133. Castro-Magana, M., Collipp, P. J., Chen, S. Y., Cheruvanky, T., and Maddaiah, V. T. (1981). *Am. J. Dis. Child.* **135**, 322.

134. Catalanotto, F. A., and Frank, M. (1979). *Abstr. Soc. Neurosci.* **5**, 126.

135. Cavdar, A. O., Arcasoy, A., Cin, S., Babacan, E., and Gozdasoglu, S. (1983). *In* "Zinc Deficiency in Human Subjects" (A. S. Prasad, A. O. Cavdar. G. J. Brewer, and P. J. Aggett, eds.), p. 71. Alan R. Liss, Inc., New York.

136. Cavdar, A. O., Babacan, E., Asik, S., Arcasoy, A., Ertem, U., Himmetoglu, O., Baycu, T., and Akar, N. (1983). *In* "Zinc Deficiency in Human Subjects" (A. S. Prasad, A. O. Cavdar, G. J. Brewer, and P. J. Aggett, eds.), p. 99. Alan R. Liss, Inc., New York.

137. Cavell, P. A., and Widdowson, E. M. (1964). *Arch. Dis. Child.* **39**, 496.

139. Chandra, R. K., and Au, B. (1980). *Am. J. Clin. Nutr.* **33**, 736.

140. Chandra, R. K., and Kutty, K. M. (1980). *Acta Paediatr. Scand.* **69**, 25.

141. Chandra, R. K. (1980). *Pediatrics* **66**, 789.

142. Cheek, D. B., Smith, R. M., and Spargo, R. M. (1982). *In* "Clinical Applications of Recent Advances in Zinc Metabolism" (A. S. Prasad, I. E. Dreosti, and B. S. Hetzel, eds.), p. 151. Alan R. Liss, Inc., New York.

143. Cheek, D. B., Wishart, J., Phillipon, G., Field, J., and Spargo, R. (1985). *In* "Trace Elements in Nutrition of Children" (R. K. Chandra, ed.), Nestle Nutr. Workshop, Vol. 8, p. 209. Raven Press, New York.

144. Chen, R. W., Eakin, D. J., and Whanger, P. D. (1974). *Nutr. Rep. Int.* **10**, 195.

145. Cherry, F. F., Bennett, E. A., Bazzano, E. S., Johnson, L. K., Fosmire, G. J., and Batson, H. K. (1981). *Am. J. Clin. Nutr.* **34**, 2367.

146. Chesters, J. K., and Quarterman, J. (1970). *Br. J. Nutr.* **24**, 1061.

147. Chesters, J. K. (1972). *Biochem. J.* **130**, 133.

148. Chesters, J. K., and Will, M. (1973). *J. Nutr.* **30**, 555.

149. Chesters, J. K., and Will, M. (1978). *Br. J. Nutr.* **39**, 375.

150. Chesters, J. K., and Will, M. (1981). *Br. J. Nutr.* **46**, 111.

151. Chesters, J. K., and Will, M. (1981). *Br. J. Nutr.* **46**, 119.

152. Chesters, J. K. (1982). *In* "Clinical, Biochemical, and Nutritional Aspects of Trace Elements" (A. S. Prasad, ed.), p. 221. Alan R. Liss, Inc., New York.

153. Chesters, J. K. (1983). *J. Inherited Metab. Dis.* **6**, Suppl. 1, 34.

154. Chvapil, M., Weldly, P. L., Stankova, L., Clark, D. S., and Zukoski, C. F. (1975). *Life Sci.* **16**, 561.

155. Chvapil, M. (1976). *Med. Clin. North Am.* **60**, (4), 799.

156. Chvapil, M., Stankova, L., Zukoski, C., IV, and Zukoski, C., III (1977). *J. Lab. Clin. Med.* **89**, 135.

157. Chvapil, M., Montgomery, D., Ludwig, J. C., and Zukoski, C. F. (1979). *Proc. Soc. Exp. Biol. Med.* **162**, 480.

158. Clejan, S., Castro-Magana, M., Collipp, P. J., Jonas, E., and Maddaiah, V. T. (1982). *Lipids* **17**, 129.

159. Coble, Y. D., Schulert, A. R., and Farid, Z. (1966). *Am. J. Clin. Nutr.* **18**, 421.

160. Coble, Y. D., Bardin, C. W., Ross, G. T., and Darby, W. T. (1971). *J. Clin. Endocrinol. Metab.* **32**, 361.

161. Cohen, J. J., and Duke, R. C. (1984). *J. Immunol.* **132**, 1.

162. Colin, M. A., Taper, J., and Ritchey, S. J. (1983). *J. Nutr.* **113**, 1480.

163. Consumer Nutrition Center (1976). "Composition of Foods, Raw, Processed, Prepared." Handb. No. 8, Sects. 8-1 to 8-11. U.S. Dept. of Agriculture, Washington, D.C.

164. Cotzias, G. C., Borg, D. C., and Selleck, B. (1962). *Am. J. Physiol.* **202**, 359.
165. Cotzias, G. C., and Papavasiliou, P. S. (1964). *Am. J. Physiol.* **206**, 787.
166. Cousins, R. J. (1979). *Nutr. Rev.* **37**, 97.
167. Cousins, R. J. (1982). *In* "Clinical, Biochemical, and Nutritional Aspects of Trace Elements" (A. S. Prasad, ed.), p. 117. Alan R. Liss, Inc., New York.
168. Cox, D. H., and Harris, D. L. (1960). *J. Nutr.* **70**, 514.
169. Craig, W. J., Balbach, L., Harris, S., and Vyhmeister, N. (1984). *J. Am. Coll. Nutr.* **3**, 183.
170. Crawford, I. L., and Connor, J. D. (1972). *J. Neurochem.* **19**, 1451.
171. Creger, C. R. (1978). *Poult. Int.* **17**, 76.
172. Crofton, R. W., Clapham, M., Humphries, W. R., Aggett, P. J., and Mills, C. F. (1983). *Proc. Nutr. Soc.* **42**, 128A.
173. Crossley, L. G., Falchuk, K. H., and Vallee, B. L. (1982). *Biochemistry* **21**, 5359.
174. Cummings, F. J., Fardy, J. J., and Briggs, M. H. (1983). *Obstet. Gynecol.* **62**, 506.
175. Cunnane, S. C., and Wahle, K. W. J. (1981). *Lipids* **16**, 771.
176. Cunningham-Rundles, C., Cunningham-Rundles, S., Iwata, T., Incefy, G., Garofalo, J. A., Mendez-Botet, C., Lewis, V., Twomey, J. J., and Good, R. A. (1981). *Clin. Immunol. Immunopathol.* **21**, 387.
177. Cunningham-Rundles, S., Cunningham-Rundles, C., Dupont, B., and Good, R. A. (1980). *Clin. Immunol. Immunopathol.* **16**, 115.
178. Danford, D. E., Smith, J. C., and Huber, A. M. (1982). *Am. J. Clin. Nutr.* **35**, 958.
179. Dang, H. S., Jaiswal, D. D., Somasundaram, S., Deshpande, A., and DaCosta, H. (1984). *Sci. Total Environ.* **35**, 85.
180. Dardenne, M., Savino, W., Wade, S., Kaiserlain, D., Lemonnier, D., and Bach, J. F. (1984). *Eur. J. Immunol.* **14**, 454.
181. Dauncey, M. J., Shaw, J. C. L., and Urman, J. (1977). *Pediatr. Res.* **11**, 991.
182. Davies, N. T., and Williams, R. B. (1977). *Br. J. Nutr.* **38**, 417.
183. Davies, N. T., Soliman, H. S., Corrigall, W., and Flett, A. (1977). *Br. J. Nutr.* **38**, 153.
184. Davies, N. T., and Olpin, S. E. (1979). *Br. J. Nutr.* **41**, 590.
185. Davies, N. T. (1980). *Br. J. Nutr.* **43**, 189.
186. Davies, N. T., Carswell, A. J. P., and Mills, C. F. (1985). *In* "Trace Element Metabolism in Man and Animals-5" (C. F. Mills, P. J. Aggett, I. Bremner, and J. K. Chesters, eds.). CAB Publications, Farnham Royal, U.K. (in press).
187. Dawson, J. B., and Walker, B. E. (1969). *Clin. Chim. Acta* **26**, 465.
188. Deeming, S. B., and Weber, C. W. (1978). *Am. J. Clin. Nutr.* **31**, 1175.
189. Delezenne, C. (1919). *Ann. Inst. Pasteur, Paris* **33**, 68.
190. Demertzis, P. N. (1972). *Bull. Hell. Vet. Med. Soc.* **23**, 256.
191. Demertzis, P. N., and Mills, C. F. (1983). *Vet. Rec.* **92**, 219.
192. Dennes, E., Tupper, R., and Wormall, A. (1961). *Biochem. J.* **78**, 578.
193. DePasquale-Jardieu, P., and Fraker, P. J. (1979). *J. Immunol.* **109**, 1847.
194. DePasquale-Jardieu, P., and Fraker, P. J. (1980). *J. Immunol.* **124**, 2650.
195. Dewar, W. A., Teague, P. W., and Downie, J. N. (1974). *Br. Poult. Sci.* **15**, 119.
196. Dewar, W. A., Wight, P. A. L., Pearson, R. A., and Gentle, M. J. (1983). *Br. Poult. Sci.* **24**, 397.
197. Diamond, I., Swenerton, H., and Hurley, L. S. (1971). *J. Nutr.* **101**, 77.
198. Dinsdale, D., and Williams, R. B. (1977). *Br. J. Nutr.* **37**, 135.
199. DiSilvestro, R. A., and Cousins, R. J. (1983). *Annu. Rev. Nutr.* **3**, 261.
200. DiSilvestro, R. A., and Cousins, R. J. (1984). *Am. J. Physiol.* **247**, E436.
201. Dorea, J. G., and Pereira, S. E. (1983). *J. Nutr.* **113**, 2375.
202. Douglas, B. S., Lines, D. R., and Tse, C. A. (1976). *N. Z. Med. J.* **83**, 192.

203. Dreosti, I. E., Tsao, S., and Hurley, L. S. (1968). *Proc. Soc. Exp. Biol. Med.* **128**, 169.
204. Dreosti, I. E., Grey, P. C., and Wilkins, P. J. (1972). *S. Afr. Med. J.* **46**, 1585.
205. Dreosti, I. E., and Hurley, L. S. (1975). *Proc. Soc. Exp. Biol. Med.* **150**, 161.
206. Dreosti, I. E., Manuel, S. J., Buckley, R. A., Fraser, F. J., and Record, I. R. (1981). *Life Sci.* **28**, 2133.
207. Dreosti, I. E., McMichael, A. J., Gibson, G. T., Buckley, R. A., Hartshorne, J. M., and Colley, D. P. (1982). *Nutr. Res.* **2**, 591.
208. Dreosti, I. E. (1983). *In* "Neurobiology of the Trace Elements" (I. E. Dreosti and R. M. Smith, eds.), Vol. 1, p. 135. Humana Press, Clifton, New Jersey.
209. Drews, L. M., Keis, C., and Fox, H. M. (1979). *Am. J. Clin. Nutr.* **32**, 1893.
210. Duerre, J. A., Ford, K. M., and Sandstead, H. H. (1977). *J. Nutr.* **197**, 1082.
211. Duncan, J. R., and Hurley, L. S. (1978). *Proc. Soc. Exp. Biol. Med.* **159**, 39.
212. Duncan, J. R. (1984). *Nutr. Res.* **4**, 93.
213. Dvergsten, C., and Sandstead, H. H. (1980). *Fed. Proc., Fed. Am. Soc. Exp. Biol.* **39**, 431.
214. Dvergsten, C. L., Fosmire, G. J., Ollerich, D. A., and Sandstead, H. H. (1983). *Brain Res.* **271**, 217.
215. Dynna, P., and Havre, G. N. (1963). *Acta Vet. Scand.* **4**, 197.
216. Eckhert, C. D. (1983). *Exp. Eye Res.* **37**, 639.
217. Eggert, J. V., Siegler, R. L., and Edomkesmalee, G. (1982). *Int. J. Pediatr. Nephrol.* **3**, 21.
218. Elias, S. and Chvapil, M. (1973). *J. Surg. Res.* **15**, 59.
219. Elmes, M. E., and Jones, J. G. (1980). *J. Pathol.* **130**, 37.
220. Engel, R. W., Miller, R. F., and Price, N. O. (1966). *In* "Zinc Metabolism" (A. S. Prasad, ed.), p. 326. Thomas, Springfield, Illinois.
221. Erten, J., Arcasoy, A., Cavdar, A. O., and Cin, S. (1978). *Am. J. Clin. Nutr.* **31**, 1172.
222. Essatara, M. B., McClain, C. J., Levine, A. S., and Morley, J. E. (1984). *Physiol. Behav.* **32**, 479.
223. Essatara, M. B., Morley, J. E., Levine, A. S., Elson, M. K., Shafer, R. B., and McClain, C. J. (1984). *Physiol. Behav.* **32**, 475.
224. Essatara, M. B., Levine, A. S., Morley, J. E., and McClain, C. J. (1984). *Physiol. Behav.* **32**, 469.
225. Etzel, K. R., Shapiro, S. G., and Cousins, R. J. (1979). *Biochem. Biophys. Res. Commun.* **89**, 1120.
226. Etzel, K. R., and Cousins, R. J. (1981). *Proc. Soc. Exp. Biol. Med.* **167**, 233.
227. Etzel, K. R., and Cousins, R. J. (1983). *J. Nutr.* **113**, 1657.
228. Failla, M. L., and Cousins, R. J. (1978). *Biochim. Biophys. Acta* **543**, 293.
229. Failla, M. L., and Cousins, R. J. (1978). *Biochim. Biophys. Acta* **538**, 435.
230. Fair, W. R., and Heston, W. D. W. (1977). *In* "Zinc Metabolism: Current Aspects in Health and Disease" (G. J. Brewer and A. S. Prasad, eds.), p. 129. Alan R. Liss, Inc., New York.
231. Falchuk, K. H., Krishan, A., and Vallee, B. L. (1975). *Biochemistry* **14**, 3439.
232. Falchuk, K. H. (1977). *N. Engl. J. Med.* **296**, 1129.
233. Falchuk, K. H., Hardy, H., Ulpino, L., and Vallee, B. L. (1978). *Proc. Natl. Acad. Sci. U.S.A.* **75**, 4175.
234. Favier, A., Faure, H., and Arnaud, J. (1985). *In* "Trace Element Metabolism in Man and Animals-5" (C. F. Mills, P. J. Aggett, I. Bremner, and J. K. Chesters, eds.). CAB Publications, Farnham Royal, U.K. (in press).
235. Feaster, J. P., Hansard, S., McCall, J. T., Skipper, F. H., and Davis, G. K. (1954). *J. Anim. Sci.* **13**, 781.
236. Feeley, R. M., Eitenmiller, R. R., Jones, J. B., and Barnhart, H. (1983). *Am. J. Clin. Nutr.* **37**, 443.

237. Fell, B. F., Leigh, L. C., and Williams, R. B. (1973). *Res. Vet. Sci.* **14**, 317.
238. Fell, G. S., Cuthbertson, D. P., Morrison, C., Fleck, A., Queen, K., Bessent, R. G., and Husain, S. L. (1973). *Lancet* **1**, 280.
239. Fernandes, G., Nair, M., Onoe, K., Tanaka, T., Floyd, R., and Good, R. A. (1979). *Proc. Natl. Acad. Sci. U.S.A.* **76**, 457.
240. Fernandez-Madrid, F., Prasad, A. S., and Oberleas, D. (1973). *J. Lab. Clin. Med.* **82**, 951.
241. Ferry, F., and Donner, M. (1984). *Scand. J. Immunol.* **19**, 435.
242. Filteau, S. M., and Woodward, B. (1982). *J. Nutr.* **112**, 1974.
243. Flagstad, T. (1976). *Nord. Veterinaermed.* **28**, 160.
244. Flanagan, P. R. (1984). *J. Nutr.* **114**, 493.
245. Flynn, A., Pories, W. J., Strain, W. H., Hill, D. A., and Fratianne, R. B. (1971). *Lancet* **2**, 1169.
246. Flynn, A., Strain, W. H., and Pories, W. J. (1972). *Biochem. Biophys. Res. Commun.* **46**, 1113.
247. Flynn, A., Pories, W. J., Strain, W. H., and Hill, O. A. (1973). *Lancet* **1**, 780.
248. Follis, R. H., Day, H. G., and McCollum, E. V. (1941). *J. Nutr.* **22**, 23.
249. "Food and Nutrient Intakes of Individuals in 1 Day in the United States, Spring 1977" (1980). Nationwide Food Consumption Survey 1977–1978, Prelim. Rep. No. 2. USDA, Washington, D.C.
250. Foote, J. W., and Delves, H. T. (1983). *Analyst* **108**, 492.
251. Forbes, R. M., and Yohe, M. (1960). *J. Nutr.* **70**, 53.
252. Forbes, R. M. (1960). *Fed. Proc., Fed. Am. Soc. Exp. Biol.* **19**, 643.
253. Fox, M. R. S., and Harrison, B. N. (1964). *Proc. Soc. Exp. Biol. Med.* **116**, 256.
254. Fox, M. R. S., and Harrison, B. N. (1965). *J. Nutr.* **86**, 89.
255. Fraker, P. J., Haas, S. M., and Luecke, R. W. (1977). *J. Nutr.* **107**, 1889.
256. Fraker, P. J., DePasquale-Jardieu, P., Zwickl, C. M., and Luecke, R. W. (1978). *Proc. Natl. Acad. Sci. U.S.A.* **75**, 5660.
257. Fraker, P. J., Zwickl, C. M., and Luecke, R. W. (1982). *J. Nutr.* **112**, 309.
258. Fransson, G.-B., and Lönnerdal, B. (1983). *Pediatr. Res.* **17**, 912.
259. Fransson, G.-B., and Lönnerdal, B. (1984). *Am. J. Clin. Nutr.* **39**, 185.
260. Freeland, J. H., and Cousins, R. J. (1976). *J. Am. Diet. Assoc.* **68**, 526.
262. Freeland-Graves, J. H., Ebangit, M. L., and Hendrikson, P. J. (1980). *Am. J. Clin. Nutr.* **33**, 1757.
263. Freeland-Graves, J. H., Bodzy, P. W., and Eppright, M. A. (1980). *J. Am. Diet. Assoc.* **77**, 655.
264. Freeland-Graves, J. H., Ebangit, M. L., and Boszy, P. W. (1980). *J. Am. Diet. Assoc.* **77**, 648.
265. Freeland-Graves, J. H., Han, W.-H., Friedman, B. J., and Shorey, R. L. (1980). *Nutr. Rep. Int.* **22**, 285.
266. Freeland-Graves, J. H., Hendrickson, P. J., Ebangit, M. L., and Snowden, J. Y. (1981). *Am. J. Clin. Nutr.* **34**, 312.
267. Frommer, D. J. (1975). *Med. J. Aust.* **2**, 793.
268. Fujioka, M., and Lieberman, I. (1964). *J. Biol. Chem.* **239**, 1164.
269. Gallaher, D., and Hurley, L. S. (1980). *J. Nutr.* **110**, 591.
270. Geissler, C., Calloway, D. H., and Margen, S. (1978). *Am. J. Clin. Nutr.* **31**, 341.
271. Gentile, P. S. (1969). *Nuova Vet.* **45**, 113.
272. Gentile, P. S., Trentalange, M. J., and Coleman, M. (1981). *Pediatr. Res.* **15**, 123.
273. Ghavami-Maibodi, S. Z., Collipp, P. J., Castro-Magana, M. Stewart, C., and Chen, S. Y. (1983). *Ann. Nutr. Metab.* **27**, 214.
274. Ghishan, F. K., and Sobo, G. (1983). *Pediatr. Res.* **17**, 148.

275. Ghishan, F. K., and Greene, H. L. (1983). *Pediatr. Res.* **17,** 529.
276. Gibson, R. S., and Scythes, C. A. (1982). *Br. J. Nutr.* **48,** 241.
277. Gibson, R. S., Anderson, B. M., and Scythes, C. A. (1982). *Am. J. Clin. Nutr.* **37,** 37.
278. Gilbert, I. G. F., and Taylor, D. M. (1956). *Biochim. Bipohys. Acta* **21,** 546.
279. Giroux, E. L. (1975). *Biochem. Med.* **12,** 258.
280. Giroux, E., Schechter, P. J., and Schoun, J. (1976). *Clin. Sci. Mol. Med.* **51,** 545.
281. Giugliano, R., and Millward, D. J. (1984). *Br. J. Nutr.* **52,** 545.
282. Golden, B. E., and Golden, M. H. N. (1979). *Am. J. Clin. Nutr.* **32,** 2490.
283. Golden, B. E., and Golden, M. H. N. (1981). *In* "Trace Element Metabolism in Man and Animals-4" (J. M-C. Howell, J. M. Gawthorne, and C. L. White, eds.), p. 73, Aust. Acad. Sci., Canberra.
284. Golden, B. E., and Golden, M. H. N. (1981). *Am. J. Clin. Nutr.* **34,** 892.
285. Golden, B. E., and Golden, M. H. N. (1985). *In* "Trace Element Metabolism in Man and Animals-5" (C. F. Mills, P. J. Aggett, I. Bremner, and J. K. Chesters, eds.). CAB Publications, Farnham Royal, U.K. (in press).
286. Golden, M. H. N., Jackson, A. A., and Golden, B. E. (1977). *Lancet* **2,** 1057.
287. Golden, M. H. N., Golden, B. E., Harland, P. S. E. G., and Jackson, A. A. (1978). *Lancet* **1,** 1226.
288. Golden, M. H. N., and Golden, B. E. (1981). *Am. J. Clin. Nutr.* **34,** 900.
289. Goldiner, W. H., Hamilton, B. P., Hyman, P. D., and Russell, R. M. (1983). *J. Am. Coll. Nutr.* **2,** 157.
290. Golub, M. S., Gershwin, M. E., Hurley, L. S., Baly, D. L., and Hendrickx, A. G. (1984). *Am. J. Clin. Nutr.* **39,** 879.
291. Golub, M. S., Gerswhin, M. E., Hurley, L. S., Baly, D. L., and Hendrickx, A. G. (1984). *Am. J. Clin. Nutr.* **39,** 265.
292. Gombe, S., Apgar, J., and Hansel, W. (1973). *Biol. Reprod.* **9,** 415.
293. Gonzalez, J. R., Botet, M. V., and Sanchez, J. L. (1982). *Am. J. Dermatol.* **4,** 303.
294. Good, R. A., West, A., Day, N. K., Dong, Z. W., and Fernandes, G. (1982). *Cancer Res.* **42,** 737.
295. Gordon, E. F., Bond, J. T., and Denny, M. R. (1982). *Physiol. Behav.* **28,** 893.
296. Gordon, P. R., Browning, J. D., and O'Dell, B. L. (1983). *J. Nutr.* **113,** 766.
297. Gordon, P. R., and O'Dell, B. L. (1983). *J. Nutr.* **113,** 239.
298. Gormican, A. (1970). *J. Am. Diet. Assoc.* **56,** 397.
299. Grace, N. D. (1972). *N. Z. J. Agric. Res.* **15,** 284.
300. Grace, N. D., and Gooden, J. M. (1980). *N. Z. J. Agric. Res.* **23,** 293.
301. Grace, N. D. (1981). *In* "Proceedings of the New Zealand Workshop on Trace Elements in New Zealand" (J. V. Dunckley, ed.), p. 15. Univ. of Otago Press, Dunedin, New Zealand.
302. Grace, N. D. (1983). *N. Z. J. Agric. Res.* **26,** 59.
303. Grace, N. D. (1983).*Proc. N. Z. Soc. Anim. Prod.* **43,** 127.
304. Grant-Frost, D. R., and Underwood, E. J. (1958). *Aust. J. Exp. Biol. Med. Sci.* **36,** 339.
305. Greger, J. L. (1977). *J. Gerontol.* **32,** 549.
306. Greger, J. L., and Sciscoe, B. S. (1977). *J. Am. Diet. Assoc.* **70,** 37.
307. Greger, J. L., Zaikis, S. C., Abernathy, R. P., Bennett, O. A., and Huffman, J. (1978). *J. Nutr.* **108,** 1449.
308. Greger, J. L., Abernathy, R. P., and Bennett, O. A. (1978). *Am. J. Clin. Nutr.* **31,** 112.
309. Greger, J. L., Higgins, M. M., Abernathy, R. P., Kirksey, A., DeCorso, M. B., and Baligar, P. (1978). *Am. J. Clin. Nutr.* **31,** 269.
310. Greger, J. L., Marhefka, S., Huffman, J., Baligar, P., Peterson, T., Zaikis, S., and Sickles, V. (1978). *Nutr. Rep. Int.* **18,** 345.
311. Greger, J. L., and Geissler, A. H. (1978). *Am. J. Clin. Nutr.* **31,** 633.

312. Greger, J. L. and Sickles, V. S. (1979). *Am. J. Clin. Nutr.* **32,** 1859.
313. Greger, J. L., and Snedeker, S. M. (1980). *J. Nutr.* **110,** 2243.
314. Grimmett, R. E. R., McIntosh, I. G., and Wall, E. M. (1937). *N. Z. J. Agric.* **54,** 216.
315. Groppel., B., and Hennig, A. (1971). *Arch. Exp. Veterinaermed.* **25,** 817.
316. Gross, R. L., Osdin, N., Fong, L., and Newberne, P. M. (1979). *Am. J. Clin. Nutr.* **32,** 1260.
317. Guillard, O., Piriou, A., Gombert, J., and Reiss, D. (1979). *Biomedicine* **31,** 193.
318. Gunn, S. A., and Gould, T. C. (1958). *Am. J. Physiol.* **193,** 505.
319. Guthrie, B. E., and van Rij, A. M. (1975). *Proc. Univ. Otago Med. Sch.* **53,** 7.
320. Guthrie, B. E. (1975). *N. Z. Med. J.* **82,** 418.
321. Guthrie, B. E., and Robinson, M. F. (1977). *Br. J. Nutr.* **38,** 55.
322. Guthrie, B. E., and Robinson, M. F. (1978). *Fed. Proc., Fed. Am. Soc. Exp. Biol.* **37,** 254.
323. Guzelian, P. S., O'Connor, L., Fernandes, S., Chan, W., Giampietro, P., and Desnick, R. (1982). *Life Sci.* **31,** 1111.
324. Györkey, F., Min, K.-W., Huff, J. A., and Györkey, P. (1967). *Cancer Res.* **27,** 1348.
325. Haaranen, S. (1962). *Nord. Veterinaermed.* **14,** 265.
326. Haaranen, S. (1963). *Nord. Veterinaermed.* **15,** 536.
327. Habib, F. K. (1978). *J. Steroid Biochem.* **9,** 403.
328. Haeflein, K. A., and Rasmussen, A. I. (1977). *J. Am. Diet. Assoc.* **70,** 610.
329. Hagenfeldt, K., Plantin, L. O., and Diczfalusy, E. (1973). *Acta Endocrinol. (Copenhagen)* **72,** 115.
330. Halas, E. S., Reynolds, G. M., and Sandstead, H. H. (1977). *Physiol. Behav.* **19,** 653.
331. Halas, E. S., Heinrich, M. D., and Sandstead, H. H. (1979). *Physiol. Behav.* **22,** 991.
332. Halas, E. S., Eberhardt, M. J., Diers, M. A., and Sandstead, H. H. (1983). *Physiol. Behav.* **30,** 371.
333. Halas, E. S. (1983). *In* "Neurobiology of the Trace Elements" (I. E. Dreosti and R. M. Smith, eds.), Vol. 1, p. 213. Humana Press, Clifton, New Jersey.
334. Halas, E. S. (1985). *In* "Trace Element Metabolism in Man and Animals-5" (C. F. Mills, P. J. Aggett, I. Bremner, and J. K. Chesters. eds.). CAB Publications, Farnham Royal, U.K. (in press).
335. Hallböök, T., and Lanner, E. (1972). *Lancet* **2,** 780.
336. Halsted, J. A. (1968). *Am. J. Clin. Nutr.* **21,** 1384.
337. Halsted, J. A., Hackley, B. M., and Smith, J. C. (1968). *Lancet* **2,** 278.
339. Halsted, J. A. (1970). *Trans. Am. Clin. Climatol. Assoc.* **82,** 170.
340. Halsted, J. A., Ronaghy, H. A., Abadi, P., Haghshenass, M., Amirhakimi, G. H., Barakat, R. M., and Reinhold, J. G. (1972). *Am. J. Med.* **53,** 277.
341. Halsted, J. A., Smith, J. C., and Irwin, M. I. (1974). *J. Nutr.* **104,** 345.
342. Hambidge, K. M., Hambidge, C., Jacobs, M. A., and Baum, J. D. (1972). *Pediatr. Res.* **6,** 868.
343. Hambidge, K. M., and Silverman, A. (1973). *Arch. Dis. Child.* **48,** 567.
344. Hambidge, K. M., Walravens, P., Kumar, V., and Tuchinda, C. (1975). *In* "Trace Substances in Environmental Health-8" (D. D. Hemphill, ed.), p. 39. Univ. of Missouri Press, Columbia.
345. Hambidge, K. M., and Droegemueller, W. (1974). *Obstet. Gynecol.* **44,** 666.
346. Hambidge, K. M., Neldner, K. H., and Walravens, P. A. (1975). *Lancet* **1,** 577.
347. Hambidge, K. M., Walravens, P. A., Brown, R. M., Webster, J., White, S., Anthony, M., and Roth, M. L. (1976). *Am. J. Clin. Nutr.* **29,** 734.
348. Hambidge, K. M., Walravens, P. A., and Neldner, K. H. (1978). *In* "Zinc and Copper Clinical Medicine" (K. M. Hambidge and B. L. Nichols, eds.), p. 81. Spectrum Publ., New York.

349. Hambidge, K. M., Walravens, P. A., and Neldner, K. H. (1977). *In* "Zinc Metabolism: Current Aspects in Health and Disease" (G. J. Brewer and A. S. Prasad, eds.), p. 329. Alan R. Liss, Inc., New York.

350. Hambidge, K. M., Walravens, P. A., Neldner, K. H., and Daugherty, N. A. (1978). *In* "Trace Element Metabolism in Man and Animals-3" (M. Kirchgessner, ed.), p. 413. Tech. Univ. Munich, F.R.G., West Germany.

351. Hambidge, K. M., and Walravens, P. A. (1978). *In* "Trace Element Metabolism in Man and Animals-3" (M. Kirchgessner, ed.), p. 296. Tech. Univ. Munich, F.R.G., West Germany.

352. Hambidge, K. M., Nelder, K. H., Walravens, P. A., Weston, W. L., Silverman, A., Sabol, J. L., and Brown, R. M. (1978). *In* "Zinc and Copper in Clinical Medicine" (K. M. Hambidge and B. L. Nichols, eds.), p. 81. Spectrum Publ., New York.

353. Hambidge, K. M., and Mauer, A. M. (1978). *In* "Laboratory Indices of Nutritional Status in Pregnancy," p. 157. Committee on Nutrition of the Mother and the Preschool Child, Food and Nutrition Board, National Research Council, National Academy of Sciences, Washington, D.C.

354. Hambidge, K. M., Walravens, P. A., Casey, C. E., Brown, R. M., and Bender, C. (1979). *J. Pediatr.* **94,** 607.

355. Hambidge, K. M., Chavez, M. N., Brown, R. M., and Walravens, P. A. (1979). *Am. J. Clin. Nutr.* **32,** 2532.

356. Hambidge, K. M. (1980). *Pediatr. Clin. North Am.* **27,** 855.

357. Hambidge, K. M. (1981). *Philos. Trans. R. Soc. London, Ser. B* **294,** 129.

358. Hambidge, K. M. (1982). *Am. J. Clin. Nutr.* **36,** 943.

359. Hambidge, K. M., and Walravens, P. A. (1982). *Clin. Gastroenterol.* **11,** 87.

360. Hambidge, K. M., Krebs, N. F., Jacobs, M. A., Favier, A., Guyette, L., and Ikle, D. (1983). *Am. J. Clin. Nutr.* **37,** 429.

361. Hambidge, K. M., Jacobs, M. A., Kuether, K. O., and Barth, R. L. (1985). *In* "Trace Element Metabolism in Man and Animals-5" (C. F. Mills, P. J. Aggett, I. Bremner, and J. K. Chesters, eds.). CAB Publications, Farnham Royal, U.K. (in press).

362. Hambidge, K. M. (1985). *In* "Trace Elements in Nutrition of Children" (R. K. Chandra, ed.), Nestle Nutr. Workshop, Vol. 8, p. 1. Raven Press, New York.

363. Hambidge, K. M., Norris, D. A., Githens, J. H., Ambrusco, D., and Catalanotto, F. A. (1985). *J. Pediatr.* **106,** 450.

364. Hambidge, K. M., Krebs, N. F., and Walravens, P. A. (1985). *Nutr. Res.* Suppl. **1,** S-306.

365. Hamilton, D. L., Bellamy, J. E. C., Valberg, J. D., and Valberg, L. S. (1978). *Can. J. Physiol. Pharmacol.* **56,** 384.

366. Hamilton, E. J., Minski, M. J., and Cleary, J. J. (1972–1973). *Sci. Total Environ.* **1,** 341.

367. Hampton, D. L., Miller, W. J., Blackmon, D. M., Gentry, R. P., Neathery, M. W., and Stake, P. E. (1975). *Fed. Proc., Fed. Am. Soc. Exp. Biol.* **34,** 907.

368. Harland, B. F., Spivey-Fox, M. R., and Fry, B. E. (1975). *J. Nutr.* **105,** 1509.

369. Harland, B. F., and Peterson, M. (1978). *J. Am. Diet. Assoc.* **72,** 259.

370. Hartzer, G., and Kauer, H. (1982). *Am. J. Clin. Nutr.* **35,** 981.

371. Haumont, S. (1961). *J. Histochem. Cytochem.* **9,** 141.

372. Haumont, S., and Vincent, J. (1961). *Experientia* **17,** 296.

373. Haumont, S., and McLean, F. C. (1966). *In* "Zinc Metabolism" (A. S. Prasad, ed.), p. 169. Thomas, Springfield, Illinois.

374. Helgeland, K., Haider, T., and Jonsen, J. (1982). *Scand. J. Clin. Lab. Invest.* **42,** 35.

375. Heller, V. G., and Burke, A. D. (1927). *J. Biol. Chem.* **74,** 85.

376. Hendricks, D. G., and Mahoney, A. W. (1972). *J. Nutr.* **102,** 1079–1084.

377. Henkin, R. I., Graziadei, P. P. G., and Bradley, D. F. (1969). *Ann. Intern. Med.* **71,** 791.

378. Henkin, R. I., Meret, S., and Jacobs, J. B. (1969). *J. Clin. Invest.* **48,** 38a.
379. Henkin, R. I., and Bradley, D. F. (1970). *Life Sci.* **9** (Part 2), 701.
380. Henkin, R. I., and Smith, F. R. (1972). *Am. J. Med. Sci.* **264,** 401.
381. Henkin, R. I., Keiser, H. R., and Bronzert, D. (1972). *J. Clin. Invest.* **51,** 44a.
382. Henkin, R. I., Schulman, J. D., Schulman, C. B., and Bronzert, D. A. (1973). *J. Pediatr.* **82,** 831.
383. Henkin, R. I. (1974). *In* "Trace Element Metabolism in Animals-2" (W. E. Hoekstra, J. W. Suttie, H. E. Ganther, and W. Mertz, eds.), p. 652. University Park Press, Baltimore, Maryland.
384. Henkin, R. I. (1974). *In* "Trace Element Metabolism in Animals-2" (W. G. Hoekstra, J. W. Suttie, H. E. Ganther, and W. Mertz, eds.), p. 647. University Park Press, Baltimore, Maryland.
385. Henkin, R. I., Patten, B. M., Re, P. K., and Bronzert, D. A. (1975). *Arch. Neurol. (Chicago)* **32,** 745.
386. Henkin, R. I., Lippoldt, R. E., Bilstad, J., and Edelhoch, H. (1975). *Proc. Natl. Acad. Sci. U.S.A.* **72,** 448.
387. Henkin, R. I., Schechter, P. J., Friedewald, W. T., Demets, D. L., and Raff, M. (1976). *Am. J. Med. Sci.* **272,** 285.
388. Henkin, R. I. (1978). *In* "Zinc and Copper in Clinical Medicine" (K. M. Hambidge and B. L. Nichols, eds.), p. 35. Spectrum Publ., New York.
389. Henkin, R. I. (1984). *Biol. Trace Elem. Res.* **6,** 263.
390. Hennig, A. (1965). *Arch. Tierernaehr.* **15,** 31, 345, 353, 363, 377.
391. Henzel, J. H., DeWeese, M. S., and Lichti, E. L. (1970). *Arch. Surg. (Chicago)* **100,** 349.
392. Hesketh, J. E. (1981). *Int. J. Biochem.* **13,** 921.
393. Hess, F. M., King, J. C., and Margen, S. (1977). *J. Nutr.* **107,** 1610.
394. Hesse, G. W., Hesse, K. A., and Caralanotto, F. A. (1979). *Proc. Soc. Exp. Biol. Med.* **123,** 692.
395. Hetland, O., and Brubakk, E. (1973). *Scand. J. Clin. Lab. Invest.* **32,** 225.
396. Hidiroglou, M., and Spurr, D. T. (1975). *Can. J. Anim. Sci.* **55,** 31.
397. Hidiroglou, M., Williams, C. J., and Tryphonas, L. (1979). *Am. J. Vet. Res.* **40,** 103.
398. Hill, G. M., and Miller, E. R. (1983). *J. Anim. Sci.* **57,** 106.
399. Hill, G. M., Miller, E. R., and Ku, P. K. (1983). *J. Anim. Sci.* **57,** 123.
400. Hill, G. M., Miller, E. R., and Stowe, H. D. (1983). *J. Anim. Sci.* **57,** 114.
401. Hill, G. M., Miller, E. R., Whetter, P. A., and Ullney, D. E. (1983). *J. Anim. Sci.* **57,** 130.
402. Hinks, L. J., Colmsee, M., and Delves, H. T. (1982). *Analyst* **107,** 815.
403. Hinks, L. J., Clayton, B. E., and Lloyd, R. S. (1983). *J. Clin. Pathol.* **36,** 1016.
404. Hoefer, J. A., Miller, E. R., Ullrey, D. E., Ritchie, H. D., and Luecke, R. W. (1960). *J. Anim. Sci.* **19,** 249.
405. Hoekstra, W. G., Lewis, P. K., Phillips, P. H., and Grummer, R. H. (1956). *J. Anim. Sci.* **15,** 752.
406. Hoekstra, W. G., Faltin, E. C., Lin, C. W., Roberts, H. F., and Grummer, R. H. (1967). *J. Anim. Sci.* **26,** 1348.
407. Holden, J. M., Wolf, W. R., and Mertz, W. (1979). *J. Am. Diet. Assoc.* **75,** 2328.
408. Honda, K., and Tatsukawa, R. (1983). *Arch. Environ. Contam. Toxicol.* **12,** 543.
409. Hooper, P. L., Visconti, L., Garry, P. J., and Johnson, G. E. (1980). *JAMA, J. Am. Med. Assoc.* **244,** 1960.
410. Hove, E., Elvenhjem, C. A., and Hart, E. B. (1937). *J. Physiol. (London)* **119,** 768.
411. Hove, E., Elvehjem, C. A., and Hart, E. B. (1938). *Am. J. Physiol.* **124,** 750.
412. Hove, E., Elvehjem, C. A., and Hart, E. B. (1940). *J. Biol. Chem.* **136,** 425.
413. Hsu, F. S., Krook, L., Pond, W. G., and Duncan, J. R. (1975). *J. Nutr.* **105,** 112.
414. Hsu, J. M., Anthony, W. L., and Buchanan, P. J. (1969). *J. Nutr.* **99,** 425.

415. Hsu, J. M., and Anthony, W. L. (1971). *J. Nutr.* **101,** 445.
416. Hsu, J. M., and Anthony, W. L. (1975). *J. Nutr.* **105,** 26.
417. Huang, Y. S., Cunnane, S. C., Horrobin, D. F., and Davignon, J. (1982). *Atherosclerosis* **41,** 1932.
418. Hubbell, R. B., and Mendel, L. B. (1927). *J. Biol. Chem.* **75,** 567.
419. Huber, A. M., and Gershoff, S. N. (1973). *J. Nutr.* **103,** 1175.
420. Huber, A. M. and Gershoff, S. N. (1975). *J. Nutr.* **105,** 1486.
421. Hunt, I. F., Murphy, N. J., Gomez, J., and Smith, J. C., Jr. (1979). *Am. J. Clin. Nutr.* **32,** 1411.
422. Hunt, I. F., Murphy, N. J., Cleaver, A. E., Faraji, B., Swendseid, M. E., Coulson, A. H., Clark, V. A., Laine, N., Davis, C. A., and Smith, J. C. (1983). *Am. J. Clin. Nutr.* **37,** 572.
423. Hunt, I. F., Murphy, N. J., Cleaver, A. E., Faraji, B., Swendseid, M. E., Coulson, A. H., Clark, V. A., Browdy, B. L., Cabalum, M. D., and Smith, J. C., Jr. (1984). *Am. J. Clin. Nutr.* **40,** 508.
424. Hurley, L. S., and Swenerton, H. (1966). *Proc. Soc. Exp. Biol. Med.* **123,** 692.
425. Hurley, L. S., and Swenerton, H. (1971). *J. Nutr.* **101,** 597.
426. Hurley, L. S., and Shrader, R. E. (1972). *Int. Rev. Neurobiol., Suppl.* **1,** 7.
427. Hurley, L. S., and Tao, S. H. (1972). *Am. J. Physiol.* **222,** 322.
428. Hurley, L. S., and Mutch, P. B. (1973). *J. Nutr.* **103,** 649.
429. Hurley, L. S., and Cosens, G. (1974). *In* "Trace Element Metabolism in Animals-2" (W. G. Hoekstra, J. W. Suttie, H. E. Ganther, and W. Mertz, eds.), p. 516. University Park Press, Baltimore, Maryland.
430. Hurley, L. S., and Schrader, R. E. (1975). *Nature (London)* **254,** 427.
431. Hurley, L. S. (1981). *Physiol. Rev.* **61,** 249.
432. Hurley, L. S., Golub, M. S., Gershwin, M. E., and Hendricks, A. G. (1985). *In* "Trace Element Metabolism in Man and Animals-5" (C. F. Mills, P. J. Aggett, I. Bremner, and J. K. Chesters, eds.). CAB Publications, Farnham Royal, U.K. (in press).
433. Husain, S. L. (1969). *Lancet* **1,** 1069.
434. Hutton, C. W., and Hayes-Davis, R. B. (1983). *J. Am. Diet. Assoc.* **82,** 148.
435. Hwang, D. H., Chanmugam, P., and Wheeler, C. (1984). *J. Nutr.* **114,** 398.
436. Iniguez, C., Casa, J., and Carreres, J. (1978). *Acta Anat.* **101,** 120.
437. Ismail-Beigi, F., Reinhold, J. G., Faraji, B., and Abadi, P. (1977). *J. Nutr.* **107,** 510.
438. Iwata, T., Incefy, G. S., Tanaka, T., Fernandes, G., Mendez-Bote, C. J., Phi, K., and Good, R. A. (1979). *Cell. Immunol.* **47,** 100.
439. Iyengar, G. V., Kollmer, W. E., and Bowen, H. J. M. (1978). "The Elemental Composition of Human Tissues and Body Fluids." Verlag Chemie, Weinheim.
440. Jackson, M. J. (1977). *J. Clin. Pathol.* **30,** 284.
441. Jackson, M. J., Jones, D. A., and Edwards, R. H. T. (1982). *Clin. Physiol.* **2,** 333.
442. Jackson, M. J., Jones, D. A., Edwards, R. H. T., Swainbank, I. G., and Coleman, M. L. (1984). *Br. J. Nutr.* **51,** 199.
443. Jakinovich, W., and Osborn, D. W. (1981). *Am. J. Physiol.* **241,** R233.
444. Jameson, S. (1976). *Acta Med. Scand., Suppl.* **593.**
445. Jameson, S. (1980). *In* "Zinc in the Environment" (J. O. Nriagu, ed.), Part 2, p. 183. Wiley, New York.
446. Jameson, S. (1982). *In* "Clinical Applications of Recent Advances in Zinc Metabolism" (A. S. Prasad, I. E. Dreosti, and B. S. Hetzel, eds.), p. 39. Alan R. Liss, Inc., New York.
447. Janick, J., Zeitz, L., and Whitmore, W. F. (1971). *Fertil. Steril.* **22,** 573.
448. Jardine, N. J., and Leaver, J. L. (1978). *Biochem. J.* **169,** 103.
449. Jeffrey, D., Edwards, Y. H., Jackson, M. J., Jeffrey, S., and Carter, N. D. (1982). *Comp. Biochem. Physiol. B* **73B,** 971.
451. Johnson, P. E., and Evans, G. W. (1978). *Am. J. Clin. Nutr.* **31,** 416.

452. Jones, J. G., Elmes, M. E., Aggett, P. J., and Harries, J. T. (1983). *Pediatr. Res.* **17,** 354.
453. Judd, A. M., Macleod, R. M., and Login, I. S. (1984). *Brain Res.* **294,** 190.
454. Julius, R., Schulkind, M., Sprinkle, T., and Rennert, O. (1973). *J. Pediatr.* **83,** 1007.
455. Kalinowski, M., Wolf, G., and Markefski, M. (1983). *Acta Histochem.* **73,** 33.
456. Kane, E., Morris, J. G., Rogers, Q. R., Ihrke, P. J., and Cupps, P. T. (1981). *J. Nutr.* **111,** 488.
457. Kang, Y. J., Olsen, M. O. J., and Busch, H. (1974). *J. Biol. Chem.* **249,** 5580.
458. Karcioglu, Z. A. (1982). *Surv. Ophthalmol.* **27,** 114.
459. Kasperek, K., Feinendegen, L. E., Lombeck, I., and Bremer, H. J. (1977). *Eur. J. Pediatr.* **126,** 199.
460. Kay, R. G., and Tasman-Jones, C. (1975). *Aust. N. Z. J. Surg.* **45,** 325.
461. Kay, R. G., Tasman-Jones, C., Pybus, J., Whiting R., and Black, H. (1976). *Ann. Surg.* **183,** 331.
462. Kazimierczak, W., Adamas, B., and Maslinski, C. (1978). *Biochem. Pharmacol.* **27,** 243.
463. Keen, C. L., and Hurley, L. S. (1980). *Mech. Ageing Dev.* **13,** 161.
464. Keen, C. L., Lönnerdal, B., Clegg, M., and Hurley, L. S. (1981). *J. Nutr.* **111,** 226.
465. Keen, C. L., Lönnerdal, B., and Fisher, G. L. (1981). *Am. J. Vet. Res.* **42,** 347.
466. Keilin, D., and Mann, T. (1939). *Nature (London)* **144,** 442.
467. Keilin, D., and Mann, T. (1940). *Biochem. J.* **34,** 1163.
468. Kellokumpu, S., and Rajaniemi, H. (1981). *Biol. Reprod.* **24,** 298.
469. Kellokumpu, S., and Rajaniemi, H. (1982). *Biochim. Biophys. Acta* **718,** 26.
470. Kelly, R., Davidson, G. P., Townley, R. R. W., and Campbell, P. E. (1976). *Arch. Dis. Child.* **51,** 219.
471. Kelsay, J. L., and Prather, E. S. (1983). *Am. J. Clin. Nutr.* **38,** 12.
472. Kennedy, A. C., Bessent, R. G., Davis, P., and Reynolds, P. M. G. (1978). *Br. J. Nutr.* **40,** 115.
473. Kerr, W. K., Kerestechi, A. G., and Mayoh, H. (1960). *Cancer* **13,** 550.
474. Ketola, G. H. (1979). *J. Nutr.* **109,** 965.
475. Kiem, J., Borberg, H., Iyengar, G. V., Kasperek, K., Siegers, M., Feinendegen, L. E., and Gross, R. (1979). *Clin. Chem. (Winston-Salem, N.C.)* **25,** 705.
476. Kienholz, E. W., Turk, D. E., Sunde, M. L., and Hoekstra, W. G. (1961). *J. Nutr.* **75,** 211.
477. Kies, C., Young, E., and McEndree, L. (1983). *In* "Nutritional Bioavailability of Zinc" (G. E. Inglett, ed.), p. 115. Am. Chem. Soc., Washington, D.C.
478. Killerich, S., Christensen, M. S., Naestogft, J., and Christiansen, C. (1980). *Clin. Chim. Acta* **105,** 231.
479. Kincaid, R. L. (1981). *Nutr. Rep. Int.* **23,** 493.
480. King, J. C., Stein, T., and Doyle, M. (1981). *Am. J. Clin. Nutr.* **34,** 1049.
481. Kirchgessner, M., Merz, G., and Oelschlager, W. (1966). *Arch. Tierernäehr.* **10,** 414.
482. Kirchgessner, M., and Pallauf, J. (1972). *Z. Tierphysiol., Tierenäehr. Futtermittelkd.* **29,** 65 and 77.
483. Kirchgessner, M., Stadler, A. E., and Roth, H. P. (1975). *Bioinorg. Chem.* **5,** 33.
484. Kirchgessner, M., and Schwarz, W. A. (1976). *Arch. Tierernaehr.* **26,** 3.
485. Kirchgessner, M., Roth, H. P., and Schwarz, W. A. (1976). *Z. Tierphysiol., Tierernaehr. Futtermittelkd.* **36,** 175.
486. Kirchgessner, M., and Roth, H. P. (1980). *In* " Zinc in the Environment" (J. O. Nriagu, ed.), Part 2, p. 71. Wiley, New York.
487. Kirchgessner, M., and Weigand, E. (1983). *Met. Ions Biol. Syst.* **15,** 319.
488. Kirchgessner, M., and Roth, H. P. (1985). *In* "Trace Element Metabolism in Man and Animals-5" (C. F. Mills, P. J. Aggett, I. Bremner, and J. K. Chesters, eds.). CAB Publications, Farnham Royal, U.K. (in press).

489. Kirksey, A., Ernst, J. A., Roepke, J. L., and Tsai, T.-L. (1979). *Am. J. Clin. Nutr.* **32,** 30.
490. Klevay, L. M. (1973). *Am. J. Clin. Nutr.* **26,** 1060.
491. Koo, S. I., and Turk, D. E. (1977). *J. Nutr.* **107,** 896.
492. Kramer, L., Spencer, H., and Osis, D. (1981). *Am. J. Clin. Nutr.* **34,** 1372.
493. Kratzer, F. H., Vohra, P., Allred, J. B., and Davis, P. N. (1958). *Proc. Soc. Exp. Biol. Med.* **98,** 205.
494. Krebs, N. F., Hambidge, K. M., and Walravens, P. A. (1984). *Am. J. Dis. Child.* **138,** 270.
495. Krebs, N. F., Hambidge, K. M., Jacobs, M. A., and Oliva-Rasbach, J. (1985). *Am. J. Clin. Nutr.* **41,** 560.
496. Krebs, N. F., Hambidge, K. M., Jacobs, M. A., and Mylet, S. (1985). *J. Pediatr. Gastroenterol. Nutr.* **4,** 227.
497. Ku, P. K., Ullrey, D. E., and Miller, E. R. (1970). *In* "Trace Element Metabolism in Animals-1" (C. F. Mills, ed.), p. 158. Churchhill-Livingstone, Edinburgh and London.
498. Ku, P. K. (1971). *Diss. Abstr. Int. B* **32,** 6717.
500. Kumar, S. P., and Anday, E. K. (1984). *Pediatrics* **73,** 327.
501. Kynast, G., Wagner, N., Saling, E., and Herold, W. (1978). *J. Perinat. Med.* **6,** 231.
502. Kynast, G., Saling, E., and Wagner, N. (1979). *J. Perinat. Med.* **7,** 69.
503. Lang, V., Kirchgessner, M., and Voightlander, G. (1972). *Z. Acker-Pfanzenbau* **135,** 216.
504. Lantzsch, H.-J., and Berschauer, F. (1982). *Ann. Nutr. Metab.* **26,** 178.
505. Lappalainen, R., Knuttila, M., and Salminen, R. (1981). *Arch. Oral Biol.* **26,** 1.
506. Larson, D. L. (1974). *In* "Clinical Applications of Zinc Metabolism" (W. J. Pories *et al.,* eds.), p. 234. Thomas, Springfield, Illinois.
507. Lau, A. L., and Failla, M. L. (1984). *J. Nutr.* **114,** 224.
508. Law, J. S., Nelson, N., and Henkin, R. I. (1983). *Biol. Trace Elem. Res.* **5,** 219.
509. Lechartier, G., Bellamy, F., Raoult, F., and Breton, H. (1887). *C. R. Hebd. Seances Acad. Sci.* **84,** 867.
510. Legg, S. P., and Sears, L. (1960). *Nature (London)* **186,** 1061.
511. Lei, K. Y., Abbasi, A., and Prasad, A. S. (1976). *Am. J. Physiol.* **230,** (6), 1730.
512. Lennard, E., Bjornson, A. B., Petering, H., and Alexander, J. (1974). *J. Surg. Res.* **16,** 286.
513. LeRiche, H. H. (1968). *J. Agric. Sci. (Camb.)* **71,** 205.
514. Leure-Dupree, A. E., Rothman, R. J., and Fosmire, G. J. (1982). *Am. J. Anat.* **165,** 295.
515. Lewis, P. K., Hoekstra, W. G., Grummer, R. H., and Phillips, P. H. (1956). *J. Anim. Sci.* **15,** 741.
516. Lewis, P. K., Hoekstra, W. G., and Grummer, R. H. (1957). *J. Anim. Sci.* **16,** 578.
517. Lieberman, I., Abrams, R., Hunt, N., and Ove, P. (1963). *J. Biol. Chem.* **238,** 3955.
518. Lifschitz, M. D., and Henkin, R. I. (1971). *J. Appl. Physiol.* **31,** 88.
519. Likuski, H. J. A., and Forbes, R. M. (1965). *J. Nutr.* **85,** 230.
520. Lindeman, R. D., Clark, M. L., and Colmore, J. P. (1971). *J. Gerontol.* **26,** 358.
521. Lo, G. S., Settle, S. L., Steinke, F. H., and Hopkins, D. T. (1981). *J. Nutr.* **111,** 2223.
522. Lockitch, G., Godolphin, W., Pendray, M. R., Riddell, D., and Quigley, G. (1983). *J. Pediatr.* **102,** 304.
523. Login, I. S., Thorner, M. O., and MacLeod, R. M. (1983). *Neuroendocrinology* **37,** 317.
524. Lombeck, I., von Bassewitz, D. B., Becker, K., Tinschmann, P., and Kastner, H. (1974). *Pediatr. Res.* **8,** 82.
525. Lombeck, I., Schnippering, H. G., Ritzl, F., Feinendegen, L. E., and Bremar, H. J. (1975). *Lancet* **1,** 855.
526. Lombeck, I., Kasperek, K., Feinendegen, L. E., and Bremer, H. J. (1978). *Monogr. Hum. Genet.* **9,** 114.
527. Lönnerdal, B., Stanislowski, A. G., and Hurley, L. S. (1980). *J. Inorg. Biochem.* **12,** 71.

528. Lönnerdal, B., Keen, C. L., and Hurley, L. S. (1981). *In* "Trace Element Metabolism in Man and Animals-4" (J. Mc C. Howell, J. M. Gawthorne, and C. L. White, eds.), p. 249. Aust. Acad. Sci., Canberra.
529. Lönnerdal, B., Keen, C. L., Hurley, L. S., and Fisher, G. L. (1981). *Am. J. Vet. Res.* **42**, 662.
530. Lönnerdal, B., Keen, C. L., Ohtake, M., and Tamura, T. (1983). *Am. J. Dis. Child.* **137**, 433.
531. Lönnerdal, B., Cederblad, Å., Davidsson, L., and Sandström, B. (1984). *Am. J. Clin. Nutr.* **40**, 1064.
532. Lönnerdal, B., Keen, C. L., Glazier, C. E., and Anderson, J. (1984). *Pediatr. Res.* **18**, 911.
533. Lopez-Minares, M., and Munoz, J. (1962). *Rev. Clin. Esp.* **87**, 157.
534. Luecke, R. W., Rukson, B. E., and Baltzer, B. V. (1970). *In* "Trace Element Metabolism in Animals-1" (C. F. Mills, ed.), p. 471. Churchill-Livingstone, Edinburgh and London.
535. Luecke, R. W., Simonel, C. E., and Fraker, P. J. (1978). *J. Nutr.* **108**, 881.
536. Lukaski, H. C., Bolonchuk, W. W., Klevay, L. M., Milne, D. B., and Sandstead, H. H. (1984). *Am. J. Physiol.* **247**, E88.
537. Lunde, G. (1968). *J. Sci. Food Agric.* **19**, 432.
538. Lyon, T. D., Smith, H., and Smith. L. B. (1979). *Br. J. Nutr.* **42**, 413.
539. Macpinlac, M. P., Pearson, W. N., and Darby, W. J. (1966). *In* "Zinc Metabolism" (A. S. Prasad, ed.), p. 142. Thomas, Springfield, Illinois.
540. Macapinlac, M. P., Barney, G. H., Pearson, W. N., and Darby, W. J. (1967). *J. Nutr.* **93**, 499.
541. Macapinlac, M. P., Pearson, W. N., Barney, G. H., and Darby, W. J. (1967). *J. Nutr.* **93**, 511.
542. Macapinlac, M. P., Pearson, W. N., Barney, G. H., and Darby, W. J. (1968). *J. Nutr.* **95**, 569.
543. MacDonald, L. D., Gibson, R. S., and Miles, J. E. (1982). *Acta Paediatr. Scand.* **71**, 785.
544. Macy, I. G. (1942). "Nutrition and Chemical Growth in Childhood," Vol. 1, p. 198. Thomas, Springfield, Illinois.
545. Mahajan, S. K., Prasad, A. S., Rabbani, P., Briggs, W. A., and McDonald, F. D. (1979). *J. Lab. Clin. Med.* **94**, 693.
546. Mahajan, S. K., Prasad, A. S., Lambujon, J., Abbasi, A. A., Briggs, W. A., and McDonald, F. D. (1980). *Am. J. Clin. Nutr.* **33**, 1517.
547. Mahajan, S. K., Abbasi, A. A., Prasad, A. S., Rabbani, P., Briggs, W. A., and McDonald, F. D. (1982). *Ann. Intern. Med.* **97**, 357.
548. Mahalko, J. R., Sandstead, H. H., Johnson, L. K., and Milne, D. B. (1983). *Am. J. Clin. Nutr.* **37**, 8.
549. Mahmoud, O. M., Samani, F. E., Bakheit, A. O., and Hassan, M. A. (1983). *J. Comp. Pathol.* **93**, 591.
550. Malecki, J. C., and McCausland, I. P. (1982). *Res. Vet. Sci.* **33**, 192.
551. Maltin, C. A., Duncan, L., Wilson, A. B., and Hesketh, J. E. (1983). *Br. J. Nutr.* **50**, 597.
552. Margolith, E. J., Schenker, J. G., and Chevion, M. (1983). *Cancer* **52**, 868.
553. Marmar, J. L., Katz, S., Praiss, D. E., and DeBenedictus, T. J. (1980). *Urology* **14**, 478.
554. Martin, M. T., Licklider, K. F., Brushmiller, J. G., and Jacobs, F. A. (1981). *J. Inorg. Biochem.* **15**, 55.
555. Martinez, A., and Church, D. C. (1970). *J. Anim. Sci.* **31**, 982.
556. Mason, K. E., Burns, W. A., and Smith, J. C. (1982). *J. Nutr.* **112**, 1019.
557. Mason, R., Bakka, A.. Samarawickrama, G. P., and Webb, M. (1981). *Br. J. Nutr.* **45**, 375.
558. Masters, D. G., and Fels, H. (1980). *Biol. Trace Elem. Res.* **2**, 281.
559. Masters, D. G., and Somers, M. (1980). *Aust. J. Exp. Agric. Anim. Husb.* **20**, 20.

560. Masters, D. G., Keen, C. L., Lönnerdal, B., and Hurley, L. S. (1983). *J. Nutr.* **113,** 905.
561. Masters, D. G., and Moir, R. J. (1983). *Br. J. Nutr.* **49,** 365.
562. Matseshe, J. W., Phillips, S. F., Malagelada, J.-R., and McCall, J. T. (1980). *Am. J. Clin. Nutr.* **33,** 1946.
563. May, P. M., Smith, G. L., and Williams, D. R. (1982). *J. Nutr.* **112,** 1990.
564. Mazus, B., Falchuk, K. H., and Vallee, B. L. (1984). *Biochemistry* **3,** 42.
565. McBean, L., and Halsted, J. A. (1969). *J. Clin. Pathol.* **22,** 623.
566. McCaffrey, R., Smoler, D. F., and Baltimore, D. (1973). *Proc. Natl. Acad. Sci. U.S.A.* **70,** 521.
567. McClain, C., Soutor, C., and Zieve, L. (1980). *Gastroenterology* **78,** 272.
568. McClain, C. J., van Thiel, D. H., Parker, S., Badzin, L. K., and Gilbert, H. (1979). *Alcohol.: Clin. Exp. Res.* **3,** 135.
569. McClain, P. E., Wiley, E. R., Beecher, G. R., Anthony, W. L., and Hsu, J. M. (1973). *Biochim. Biophys. Acta* **304,** 457.
570. McCormick, C. C., Menard, M. P., and Cousins, R. J. (1981). *Am. J. Physiol.* **240,** E414.
571. McCoy, H., Kirby, A., Ercanli, F. G., Korslund, M., Liebman, M., Moak, S., Wakefield, T., and Ritchey, S. J. (1984). *J. Am. Diet. Assoc.* **84,** 1453.
572. McHargue, J. S. (1926). *Am. J. Physiol.* **77,** 245.
573. McKenney, J. R., McClennan, R. O., and Bustad, L. K. (1962). *Health Phys.* **8,** 411.
574. McKenzie, J. M., Fosmire, G. J., and Sandstead, H. H. (1975). *J. Nutr.* **105,** 1466.
575. McKenzie, J. M., Guthrie, B. E., and Prior, I. A. M. (1978). *Am. J. Clin. Nutr.* **31,** 422.
576. McKenzie, J. M. (1979). *Am. J. Clin. Nutr.* **32,** 570.
577. McMaster, D., Lappin, T. R. J., Halliday, H. L., and Patterson, C. C. (1983). *Biol. Neonate* **44,** 108.
578. McMichael, A. J., Dreosti, I. E., and Gibson, G. T. (1982). *In* "Clinical Applications of Recent Advances in Zinc Metabolism" (A. S. Prasad, I. E. Dreosti, and B. S. Hetzel, eds.), p. 53. Alan R. Liss, Inc., New York.
579. McMichael, A. J., Dreosti, I. E., Gibson, G. T., Hartshorne, J. M., Buckley, R. A., and Colley, D. P. (1982). *Early Hum. Dev.* **7,** 59.
580. McMillan, E. M., and Rowe, D. J. F. (1982). *Clin. Exp. Dermatol.* **7,** 629.
581. McPherson, E. A., Beattis, I. S., and Young, G. B. (1964). *Nord. Veterinaermed.* **16,** Suppl. 1, 533.
582. Meadows, N. J., Ruse, W., Smith, M. F., Day, J., Keeling, P. W. N., Scopes, J. W., Thompson, R. P. H., and Bloxam, D. L. (1981). *Lancet* **2,** 1135.
583. Meadows, N. J., Smith, M. F., Keeling, P. W. N., Ruse, W., Day, J., Scopes, J. W., Dervish, C., and Gillieson, M. (1982). *Lancet* **1,** 169.
584. Meadows, N. J., Grainger, S. L., Ruse, W., Keeling, P. W. N., and Thompson, R. P. H. (1983). *Br. Med. J.* **287,** 1013.
585. Meinel, B., Bode, J. C., Koenig, W., and Richter, F.-W. (1979). *Biol. Neonate* **36,** 225.
586. Meiners, C. R., Taper, L. J., Korslund, M. K., and Ritchey, S. J. (1977). *Am. J. Clin. Nutr.* **30,** 879.
587. Mellow, M. H., Layne, E. A., Lipman, T. O., Kaushik, M., Hostetler, C., and Smith, J. C. (1983). *Cancer* **51,** 1615.
588. Menard, M. P., McCormick, C. C., and Cousins, R. J. (1981). *J. Nutr.* **111,** 1353.
589. Mendel, L. B., and Bradley, H. C. (1905). *Am. J. Physiol.* **14,** 313.
590. Mendel, L. B., and Bradley, H. C. (1906). *Am. J. Physiol.* **17,** 167.
591. Mendelson, R. A., Anderson, G. H., and Bryan, M. H. (1982). *Early Hum. Dev.* **6,** 145.
592. Mesrobian, A. Z., and Shklar, G. (1968). *Periodontics* **6,** 224.
593. Metcoff, J., Cottilo, J. P., Crosby, W., Bentle, L., Sethachalam, D., Sandstead, H. H., Bodwell, C. E., Weaver, F., and McClain, P. (1981). *Am. J. Clin. Nutr.* **34,** 708.
594. Methfessel, A. H., and Spencer, H. (1974). *In* "Trace Element Metabolism in Animals-2"

(W. G. Hoekstra, J. W. Suttie, H. E. Ganther, and W. Mertz, eds.) p. 541. University Park Press, Baltimore, Maryland.

595. Meydani, S. N., and Dupont, J. (1982). *J. Nutr.* **112,** 1098–1104.
596. Meydani, S. N., Meydani, M., and Dupont, J. (1983). *J. Nutr.* **113,** 494.
597. Meyer, J., and Alvares, O. F. (1974). *Arch. Oral Biol.* **19,** 471.
598. Millar, M. J., Fisher, M. I., Elcoate, P. V., and Mawson, C. A. (1958). *Can J. Biochem. Physiol.* **36,** 557.
599. Millar, M. J., Elcoate, P. V., Fischer, M. I., and Mawson, C. A. (1960). *Can J. Biochem. Physiol.* **38,** 1457.
600. Millar, M. J., Vincent, N. R., and Mawson, C. A. (1961). *J. Histochem. Cytochem.* **9,** 111.
601. Miller, E. B., Sorscher, A., and Spencer, H. (1964). *Radiat. Res.* **22,** 216.
602. Miller, E. R., Luecke, R. W., Ullrey, D. E., Baltzer, B. V., Bradley, B. L., and Hoefer, J. A. (1968). *J. Nutr.* **95,** 278.
603. Miller, E. R., Liptrap, D. O., and Ullrey, D. E. (1970). *In* "Trace Element Metabolism in Animals-1" (C. F. Mills, ed.), p. 377. Churchill-Livingstone, Edinburgh and London.
604. Miller, J. K., and Miller, W. J. (1960). *J. Dairy Sci.* **43,** 1854.
605. Miller, J. K., and Miller, W. J. (1962). *J. Nutr.* **76,** 467.
606. Miller, J. K., and Cragle, R. G. (1965). *J. Dairy Sci.* **48,** 370.
607. Miller, J. K., and Jensen, L. S. (1966). *Poult. Sci.* **45,** 1051.
608. Miller, W. J., Clifton, C. M., and Camerson, N. W. (1963). *J. Dairy Sci.* **46,** 715.
609. Miller, W. J., Pitts, W. J., Clifton, C. M., and Schmittle, S. C. (1964). *J. Dairy Sci.* **47,** 556.
610. Miller, W. J., Clifton, C. M., Fowler, P. R., and Perkins, H. F. (1965). *J. Dairy Sci.* **48,** 450.
611. Miller, W. J., Powell, G. W., Pitts, W. J., and Perkins, H. (1965). *J. Dairy Sci.* **48,** 1091.
612. Miller, W. J., Pitts, W. J., Clifton, C. M., and Morton, J. D. (1965). *J. Dairy Sci.* **48,** 1329.
613. Miller, W. J., Morton, J. D., Pitts, W. J., and Clifton, C. M. (1965). *Proc. Soc. Exp. Biol. Med.* **118,** 427.
614. Miller, W. J., Blackmon, D. M., Gentry, R. P., Powell, G. W., and Perkins, H. E. (1966). *J. Dairy Sci.* **49,** 1446.
615. Miller, W. J., Powell, G. W., and Hiers, J. M. (1966). *J. Dairy Sci.* **49,** 1012.
617. Miller, W. J., Martin, Y. G., Gentry, R. P., and Blackmon, D. M. (1968). *J. Nutr.* **94,** 391.
618. Miller, W. J. (1969). *Am. J. Clin. Nutr.* **22,** 1323.
619. Miller, W. J., Neathery, M. W., Gentry, R. P., Blackmon, D. M., and Stake, P. E. (1974). *In* "Trace Element Metabolism in Animals-2" (W. G. Hoekstra, J. W. Suttie, H. E. Ganther, and W. Mertz, eds.), p. 550. University Park Press, Baltimore, Maryland.
620. Miller, W. J., Kincaid, R. L., Neathery, M. W., Gentry, R. P., Ansari, M. S., and Lassiter, J. W. (1978). *In* "Trace Element Metabolism in Man and Animals-3" (M. Kirchgessner, ed.), p. 175. Tech. Univ. Munich, F.R.G., West Germany.
621. Miller, W. J., and Neathery, M. W. (1980). *In* "Zinc in the Environment" (J. O. Nriagu, ed.), Part 2, p. 61. Wiley, New York.
622. Mills, C. F., Dalgarno, A. C., Williams, R. B., and Quarterman, J. (1967). *Br. J. Nutr.* **21,** 751.
623. Mills, C. F., Quarterman, J., Chesters, J. K., Williams, R. B., and Dalgarno, A. C. (1969). *Am. J. Clin. Nutr.* **22,** 1240.
624. Mills, C. F., and Dalgarno, A. C. (1972). *Nature (London)* **239,** 171.
625. Milne, D. B., Canfield, W. K., Mahalko, J. R., and Sandstead, H. H. (1983). *Am. J. Clin. Nutr.* **38,** 181.

626. Milne, D. B., Canfield, W. K., Mahalko, J. R., and Sandstead, H. H. (1984). *Am. J. Clin. Nutr.* **39,** 535.
627. Milne, D. B., Wallwork, J. C., Ralston, N. V. C., and Korynta, E. (1985). *Clin. Chem. (Winston-Salem, N.C.)* **31,** 65.
628. Milne, D. B., Wallwork, J. C., and Ralston, N. V. C. (1985). *In* "Trace Element Metabolism in Man and Animals-5" (C. F. Mills, P. J. Aggett, I. Bremner, and J. K. Chesters, eds.). CAB Publications, Farnham Royal, U.K. (in press).
629. Moeller, M. W., and Scott, H. M. (1958). *Poult. Sci.* **37,** 1227.
630. Molokhia, M. M., and Portnoy, B. (1969). *Br. J. Dermatol.* **81,** 759.
631. Molokhia, M., Sturniolo, G., Shields, R., and Turnberg, L. A. (1980). *Am. J. Clin. Nutr.* **33,** 881.
632. Montgomery, H. (1967). *In* "Dermatopathology," p. 81. Harper & Row, New York.
633. Montgomery, M. L., Sheline, G. E., and Chaikoff, I. L. (1943). *J. Exp. Med.* **78,** 151.
634. Moore, M. E. C., Moran, J. R., and Greene, H. L. (1984). *J. Pediatr.* **105,** 600.
635. Moore, R. (1978). *Br. Med. J.* 1, 754.
636. Moran, J. R., Vaughan, R., Stroop, S., Coy, S., Johnston, H., and Greene, H. L. (1983). *J. Pediatr. Gastroenterol. Nutr.* **2,** 629.
637. Morley, J. E., Gordon, J., and Hershman, J. M. (1980). *Am. J. Clin. Nutr.* **33,** 1767.
638. Morley, J. E., Russell, R. M., Reed, A., Carney, E. A., and Hershman, J. M. (1981). *Am. J. Clin. Nutr.* **34,** 1489.
639. Morris, E. R., and Ellis, R. (1980). *J. Nutr.* **110,** 1037.
640. Morris, E. R., and Ellis, R. (1983). *In* "Nutritional Bioavailability of Zinc" (G. E. Inglet, ed.), p. 159. Am. Chem. Soc., Washington, D.C.
641. Morris, E. R., and Ellis, R. (1985). *In* "Trace Element Metabolism in Man and Animals-5" (C. F. Mills, P. J. Aggett, I. Bremner, and J. K. Chesters, eds.). CAB Publications, Farnham Royal, U.K. (in press).
642. Morrison, S. A., Russell, R. M., Carney, E. A., and Oaks, E. V. (1978). *Am. J. Clin. Nutr.* **31,** 276.
643. Moser, P. B., and Gunderson, C. J. (1983). *Nutr. Res.* **3,** 279.
644. Moser, P. B., and Reynolds, R. D. (1983). *Am. J. Clin. Nutr.* **38,** 101.
645. Moser, P. B., and Allen, D. (1984). *J. Am. Diet. Assoc.* **84,** 42.
646. Moynahan, E. J., and Barnes, P. M. (1973). *Lancet* **2,** 676.
647. Moynahan, E. J. (1974). *Lancet* **2,** 399.
648. Moynahan, E. J. (1976). *Lancet* **1,** 91.
649. Mukherjee, M. D., Sandstead, H. H., Ratnaparkhi, M. V., Johnson, L. K., Milne, D. B., and Stelling, H. P. (1984). *Am. J. Clin. Nutr.* **40,** 496.
650. Murphy, E. W., Willis, B. W., and Watt, B. K. (1975). *J. Am. Diet. Assoc.* **66,** 345.
651. Murphy, J. V. (1970). *JAMA, J. Am. Med. Assoc.* **212,** 2119.
652. Murray, E. J., and Messer, H. H. (1981). *J. Nutr.* **111,** 1641.
653. Mutch, P. B., and Hurley, L. S. (1974). *J. Nutr.* **104,** 828.
654. Nassi, L., Poggini, G., and Vecchi, C. (1970). *Riv. Clin. Pediatr.* **8,** 69.
655. Nassi, L., Poggini, G., Vecchi, C., and Galvan, P. (1971). *Boll. Soc. Ital. Biol. Sper.* **48,** 86.
656. National Research Council (1978). "Zinc." Committee on Medical and Biological Effects of Environmental Pollutants, Nat. Acad. Sci., Washington, D.C.
657. National Research Council (1980). "Recommended Dietary Allowances," 9th ed. National Acad. Sci., Washington, D.C.
658. Naveh, Y., Lightman, A., and Zinder, O. (1983). *J. Pediatr.* **102,** 734.
659. Neathery, M. W., Miller, W. J., Blackmon, D. M., Gentry, R. P., and Jones, J. B. (1973). *J. Anim. Sci.* **37,** 848.

660. Neathery, M. W., Miller, W. J., Blackmon, D. M., Pate, F. M., and Gentry, R. P. (1973). *J. Dairy Sci.* **56,** 98.

661. Neathery, M. W., Miller, W. J., Blackmon, D. M., and Gentry, R. P. (1974). *J. Anim. Sci.* **38,** 854.

662. Neldner, K. H., Hagler, L., Wise, W. R., Stifel, F. B., Lufkin, E. G., and Herman, R. H. (1974). *Arch. Dermatol.* **110,** 711.

663. Neldner, K. H., and Hambidge, K. M. (1975). *N. Engl. J. Med.* **292,** 879.

664. Neville, M. C., Keller, R. P., Seacat, J., Casey, C. E., Allen, J. C., and Archer, P. (1984). *Am. J. Clin. Nutr.* **40,** 635.

665. Nevin, N. C. (1983). *In* "Prevention of Spina Bifida and Other Neural Tube Defects" (J. Dobbing, ed.), p. 127. Academic Press, London.

666. Nielsen, F. H., Sunde, M. L., and Hoekstra, W. G. (1964). *Proc. Soc. Exp. Biol. Med.* **116,** 256.

667. Nielsen, F. H., Sunde, M. L., and Hoekstra, W. G. (1965). *J. Nutr.* **86,** 89.

668. Nielsen, F. H., Sunde, M. L., and Hoekstra, W. G. (1966). *J. Nutr.* **89,** 24 and 35.

669. Nishi, Y., Hatano, S., Horino, N., Sakano, T., and Usui, T. (1981). *Hiroshima J. Med. Sci.* **30,** 65.

670. Nishi, Y., Hatano, S., Aihara, K., Okahata, H., Kawamura, H., Tanaka, K., Miyachi, Y., and Usui, T. (1984). *Pediatr. Res.* **18,** 232.

671. Nixon, G. S., Livingston, H. D., and Smith, H. (1967). *Arch. Oral Biol.* **12,** 411.

672. Norridin, R. W., Krook, L., Pond, W. G., and Walker, E. F. (1973). *Cornell Vet.* **63,** 264.

673. Norton, D. S., and Heaton, F. W. (1980). *J. Inorg. Biochem.* **13,** 1.

674. Oberleas, D., Muhrer, M. E., and O'Dell, B. L. (1966). *J. Nutr.* **90,** 56.

675. Oberleas, D., and Harland, B. E. (1981). *J. Am. Diet. Assoc.* **79,** 433.

676. Oberleas, D. (1983). *In* "Nutritional Bioavailability of Zinc" (G. E. Inglet, ed.), p. 145. Am. Chem. Soc. Washington, D.C.

677. O'Dell, B. L., and Savage, J. E. (1957). *Poult. Sci.* **36,** 489.

678. O'Dell, B. L., Newberne, F. M., and Savage, J. E. (1958). *J. Nutr.* **65,** 503.

679. O'Dell, B. L. (1969). *Am. J. Clin. Nutr.* **22,** 1315.

680. O'Dell, B. L., Reynolds, G., and Reeves, P. G. (1977). *J. Nutr.* **107,** 1222.

681. O'Dell, B. L., Browning, J. D., and Reeves, P. G. (1983). *J. Nutr.* **113,** 760.

682. O'Dell, B. L. (1984). *Nutr. Rev.* **42,** 301–308.

683. Oelshlegel, F. J., and Brewer, G. J. (1977). *In* "Zinc Metabolism: Current Aspects in Health and Disease" (G. J. Brewer and A. S. Prasad, eds.), p. 299. Alan R. Liss, Inc., New York.

684. Oestreicher, P., and Cousins, R. J. (1982). *J. Nutr.* **112,** 1978.

685. Oh, S.-H., Deagen, J. T., Whanger, P. D., and Weswig, P. H. (1978). *Am. J. Physiol.* **234,** E282.

686. Oh, S.-H., Nakaue, H., Deagen, J. T., Whanger, P. D., and Arscott, G. H. (1979). *J. Nutr.* **109,** 1720.

687. Ohtake, M. (1977). *Tohoku J. Exp. Med.* **123,** 265.

688. O'Leary, M. J., McClain, C. J., and Hegarty, P. V. J. (1979). *Br. J. Nutr.* **42,** 487.

689. Oleske, J. M., Westphal, M. L., Shore, S., Gorden, D., Bodgen, J. D., and Nahmias, A. (1979). *Am. J. Dis. Child.* **133,** 915.

690. O'Mary, C. C., Butts, W. T., Reynolds, R. A., and Bell, M. C. (1969). *J. Anim. Sci.* **28,** 268.

691. Oner, G., Bhaumick, B., and Bala, R. M. (1984). *Endocrinology (Baltimore)* **114,** 1860.

692. O'Rourke, J., Durrani, J., Benson, C., Bronzino, J., and Miller, C. (1972). *Arch. Ophthalmol. (Chicago)* **88,** 185.

693. Orro, H. F., and Weitz, H. (1972). *Beitr. Pathol.* **145,** 336.
694. Osis, D., Royston, K., Samachson, J., and Spencer, H. (1969). *Dev. Appl. Spectrosc.* **7A,** 227.
695. Osis, D., Kramer, L., Wiatrowski, E., and Spencer, H. (1972). *Am. J. Clin. Nutr.* **25,** 582.
696. Ott, E. A., Smith, W. H., Stob, M., and Beeson, W. M. (1964). *J. Nutr.* **82,** 41.
697. Ott, E. A., Smith, W. H., Stob, M., Parker, H. E., Harrington, R. B., and Beeson, W. M. (1965). *J. Nutr.* **87,** 459.
698. Ott, E. A., Smith, W. H., Harrington, R. B., and Beeson, W. M. (1966). *J. Anim. Sci.* **25,** 414.
699. Ott, E. A., Smith, W. H., Harrington, R. B., Stob, M., Parker, H. E., and Beeson, W. M. (1966). *J. Anim. Sci.* **25,** 432.
700. Pallauf, J., and Kirchgessner, M. (1971). *Int. J. Vitam. Nutr. Res.* **41,** 543.
701. Pallauf, J., and Kirchgessner, M. (1972). *Zentrabl. Veterinaermed., Reihe A* **19,** 594.
702. Pallauf, J., and Kirchgessner, M. (1973). *Zentrabl. Veterinaermed., Reihe A* **20,** 100.
703. Palm, R., and Hallmans, G. (1982). *J. Neurol., Neurosurg. Psychiatry* **45,** 685.
704. Palma, P. A., Seifert, W. E., Caprioli, R. M., and Howell, R. R. (1983). *J. Lab. Clin. Med.* **102,** 88.
705. Parizek, J., Boursnell, J. C., Hay, M. F., Babicky, A., and Taylor, D. M. (1966). *J. Reprod. Fertil.* **12,** 501.
706. Patrick, J., Golden, B. E., and Golden, M. H. N. (1980). *Am. J. Clin. Nutr.* **33,** 617.
707. Paul, A. A., and Southgate, D. A. T., eds. (1978). "McCance and Widdowson's The Composition of Foods." Elsevier/North-Holland Biomedical Press, Amsterdam.
708. Pedrosa, F. O., Pontremoli, S., and Horrecker, B. L. (1977). *Proc. Natl. Acad. Sci. U.S.A.* **74,** 2742.
709. Peirce, P., Jackson, M., Tomkins, A., and Millward, D. J. (1985). *Proc. Nutr. Soc.* **44,** 78A.
710. Pekarek, R. S., and Beisel, W. R. (1969). *Appl. Microbiol.* **18,** 482.
711. Pekarek, R. S., and Evans, G. W. (1975). *Proc. Soc. Exp. Biol. Med.* **150,** 755.
712. Pekarek, R. S., Sandstead, H. H., Jacob, R. A., and Barcome, D. F. (1979). *Am. J. Clin. Nutr.* **32,** 1466.
713. Pekas, J. C. (1966). *Am. J. Physiol.* **211,** 407.
714. Pensack, J. M., Henson, J. N., and Pogdonorf, P. D. (1958). *Poult. Sci.* **37,** 1232.
715. Perry, T. W., Beeson, W. M., Smith, W. H., and Mohler, M. T. (1968). *J. Anim. Sci.* **27,** 1674.
716. Persigehl, M., Schicha, H., Kasperek, K., and Klein, H. J. (1977). *Beitr. Pathol.* **161,** 209.
717. Petering, D. H., and Saryan, L. A. (1979). *Biol. Trace Elem. Res.* **1,** 27.
718. Picciano, M. F., and Guthrie, H. A. (1976). *Am. J. Clin. Nutr.* **29,** 242.
719. Piletz, J. E., and Ganschow, R. E. (1978). *Science* **199,** 181.
720. Pitts, W. J., Miller, W. J., Fosgate, O. T., Morton, J. D., and Clifton, C. M. (1966). *J. Dairy Sci.* **49,** 455.
721. Polanco, I., Nistal, M., Guerrero, J., and Vasquez, C. (1976). *Lancet* **1,** 430.
722. Pond, W. G., and Jones, J. R. (1964). *J. Anim. Sci.* **23,** 1057.
723. Pories, W. J., Schaer, E. W., Jordan, D. R., Chase, J., Parkinson, G., Whittaker, R., Strain, W. H., and Rob, C. G. (1966). *Surgery (St. Louis)* **59,** 821.
724. Pories, W. J., and Strain, W. H. (1966). *In* "Zinc Metabolism" (A. S. Prasad, ed.), p. 378. Thomas, Springfield, Illinois.
725. Pories, W. J., Mansour, E. G., Plecha, F. R., Flynn, A., and Strain, W. H. (1976). *In* "Trace Elements in Human Health and Disease" (A. S. Prasad and D. Oberleas, eds.), Vol. I, p. 115. Academic Press, New York.

726. Prasad, A. S., Halsted, J. A., and Nadimi, M. (1961). *Am. J. Med.* **31**, 532.
727. Prasad, A. S., Miale, A., Farid, Z., Schulert, A., and Sandstead, H. H. (1963). *J. Lab. Clin. Med.* **61**, 531.
728. Prasad, A. S., Miale, A., Farid, Z., Sandstead, H. H., Schulert, A., and Darby, W. J. (1963). *Arch. Intern. Med.* **111**, 407.
729. Prasad, A. S., Miale, A., Farid, Z., Sandstead, H. H., and Schulert, A. R. (1963). *J. Lab. Clin. Med.* **61**, 537.
730. Prasad, A. S., Schulert, A. R., Sandstead, H. H., Miale, A., and Farid, Z. (1963). *J. Lab. Clin. Med.* **62**, 84.
731. Prasad, A. S., Oberleas, D., and Halsted, J. A. (1965). *J. Lab. Clin. Med.* **66**, 508.
732. Prasad, A. S. (1966). *In* "Zinc Metabolism" (A. S. Prasad, ed.), p. 250. Thomas, Springfield, Illinois.
733. Prasad, A. S., Oberleas, D., Wolf, P., and Horwitz, J. P. (1967). *J. Clin. Invest.* **46**, 549.
734. Prasad, A. S., Oberleas, D., Wolf, P., and Horwitz, J. P. (1969). *J. Lab. Clin. Med.* **73**, 486.
735. Prasad, A. S., Oberleas, D., Wolf, P., Horwitz, J. P., Miller, E. R., and Luecke, R. W. (1969). *Am. J. Clin. Nutr.* **22**, 628.
736. Prasad, A. S., and Oberleas, D. (1974). *J. Lab. Clin. Med.* **83**, 634.
737. Prasad, A. S., Abbasi, A., and Ortega, J. (1977). *In* "Zinc Metabolism: Current Aspects in Health and Disease" (G. J. Brewer and A. S. Prasad, eds.), p. 211. Alan R. Liss, Inc., New York.
738. Prasad, A. S., Rabbani, P., Abassi, A., Bowersox, E., and Fox, M. (1978). *Ann. Intern. Med.* **89**, 483.
739. Prasad, A. S., Brewer, G. J., Schoomaker, E. B., and Rabbani, P. (1978). *JAMA, J. Am. Med. Assoc.* **240**, 2166.
740. Prasad, A. S., Abbasi, A. A., Rabbani, P., and DuMouchelle, E. (1981). *Am. J. Hematol.* **10**, 119.
741. Prask, J. A., and Plocke, D. J. (1971). *Plant Physiol.* **48**, 150.
742. Prema, K. (1980). *Indian J. Med. Res.* **71**, 534.
743. Price, N. O., Bunce, G. E., and Engel, R. W. (1970). *Am. J. Clin. Nutr.* **23**, 258.
744. Prohaska, J. R., Luecke, R. W., and Jasinski, R. (1974). *J. Nutr.* **104**, 1525.
745. Purcell, S. K., Jacobs, M. A., and Hambidge, K. M. (1984). *Fed. Proc., Fed. Am. Soc. Exp. Biol.* **43**, 687.
746. Purves, D., and MacKenzie, E. J. (1973). *Plant Soil* **39**, 361.
747. Quarterman, J., Mills, C. F., and Humphries, W. R. (1966). *Biochem. Biophys. Res. Commun.* **25**, 354.
748. Quarterman, J., and Florence, E. (1972). *Br. J. Nutr.* **28**, 75.
749. Quarterman, J. (1974). *In* "Trace Element Metabolism in Animals-2" (W. G. Hoekstra, J. W. Suttie, J. W. Ganther, and W. Mertz, eds.), p. 742. University Park Press, Baltimore, Maryland.
750. Quarterman, J., and Humphries, R. W. (1979). *Life Sci.* **24**, 177.
751. Quarterman, J., and Humphries, W. R. (1983). *J. Comp. Pathol.* **93**, 261.
752. Rahman, M. M., Davies, R. E., Deyoc, C. W., Reid, B. L., and Couch, J. R. (1960). *Poult. Sci.* **40**, 195.
753. Raper, J. T., and Curtin, L. V. (1953). *In* Proceedings of the Third Conference on Processing as Related to Nutritive Value of Cottonseed Meal.
754. Raulin, J. (1869). *Ann. Sci. Nat., Bot. Biol. Veg.* **11**, 93.
755. Raum, N. S., Stables, G. L., Pope, L. S., Harper, O. F., Waller, G. R., Renbarger, R., and Tillman, A. D. (1968). *J. Anim. Sci.* **27**, 1695.

756. Record, I. R., Dreosti, I. E., Manuel, S. J., Buckley, R. A., and Tulsi, R. S. (1985). *In* "Trace Element Metabolism in Man and Animals-5" (C. F. Mills, P. J. Aggett, I. Bremner, and J. K. Chesters, eds.). CAB Publications, Farnham Royal, U.K. (in press).

757. Reeves, P. G., Frissell, S. G., and O'Dell, B. L. (1977). *Proc. Soc. Exp. Biol. Med.* **156,** 500.

758. Reeves, P. G., and O'Dell, B. L. (1981). *J. Nutr.* **111,** 375.

759. Reeves, P. G., and O'Dell, B. L. (1983). *Br. J. Nutr.* **49,** 441.

760. Reeves, P. G., and O'Dell, B. L. (1985). *In* "Trace Element Metabolism in Man and Animals-5" (C. F. Mills, P. J. Aggett, I. Bremner, and J. K. Chesters, eds.). CAB Publications, Farnham Royal, U.K. (in press).

761. Reinhold, J. G., Kfoury, G. A., Chalambor, M. A., and Bennett, J. C. (1966). *Am. J. Clin. Nutr.* **18,** 294.

762. Reinhold, J. G., Kfoury, G. A., and Arslanian, M. (1968). *J. Nutr.* **96,** 519.

763. Reinhold, J. G., Nasr, K., Lahimgarzadeh, A., and Hedayati, H. (1973). *Lancet* **1,** 283.

764. Richards, B., Flint, D. M., and Wahlqvist, M. L. (1981). *Nutr. Rep. Int.* **23,** 939.

765. Richards, G. E., and Marshall, R. N. (1983). *J. Am. Coll. Nutr.* **2,** 133.

766. Richmond, C. R., Furchner, J. E., Trafton, G. A., and Langham, W. H. (1962). *Health Phys.* **8,** 481.

767. Rickard, B. F. (1975). *N. Z. Vet. J.* **23,** 41.

768. Riordan, J. R., and Richards, V. (1980). *J. Biol. Chem.* **255,** 5380.

769. Ritchey, S. J., Korslund, M. K., Gilbert, L. M., Fay, D. C., and Robinson, M. F. (1979). *Am. J. Clin. Nutr.* **32,** 799.

770. Robinson, L. K., and Hurley, L. S. (1981). *J. Nutr.* **111,** 858.

771. Rodin, A. E., and Goldman, S. (1969). *Am. J. Clin. Pathol.* **51,** 315.

772. Ronaghy, H. A., Fox, M. R. S., Garn, S. M., Israel, H., Harp, A., Moe, P. G., and Halsted, J. A. (1969). *Am. J. Clin. Nutr.* **22,** 1279.

773. Ronaghy, H. A., Reinhold, J. G., Mahloudji, M., Ghakami, P., Fox, M. R. S., and Halsted, J. A. (1974). *Am. J. Clin. Nutr.* **27,** 112.

774. Ronaghy, H. A., and Halsted, J. A. (1975). *Am. J. Clin. Nutr.* **28,** 831.

775. Root, A. W., Duckett, G., Sweetland, M., and Reiter, E. O. (1979). *J. Nutr.* **109,** 958.

776. Rosoff, B., and Martin, C. (1966). *Fed. Proc., Fed. Am. Soc. Exp. Biol.* **25,** 316.

777. Roth, H. P., and Kirchgessner, M. (1974). *Z. Tierphysiol., Tierernaehr. Futtermittelkd.* **32,** 289.

778. Roth, H. P., and Kirchgessner, M. (1974). *Z. Tierphysiol., Tierernaehr. Futtermittelkd.* **32,** 296.

779. Roth, H. P., and Kirchgessner, M. (1974). *Z. Tierphysiol., Tierernaehr. Futtermittelkd.* **33,** 57, 62, and 67.

780. Roth, H. P., and Kirchgessner, M. (1976). *Zentralbl. Veterinaermed., Reihe A* **23,** 578.

781. Roth, H. P., Schneider, U., and Kirchgessner, M. (1975). *Arch. Tierernaehr.* **25,** 545.

782. Rothbaum, R. J., Maur, P. R., and Farrell, M. K. (1982). *Am. J. Clin. Nutr.* **35,** 595.

783. Rubin, H. (1972). *Proc. Natl. Acad. Sci. U.S.A.* **69,** 712.

784. Saito, S., Zeitz, L., Bush, I. M., Lee, R., and Whitmore, W. F. (1967). *Am. J. Physiol.* **213,** 749.

785. Saito, S., Zeitz, L., Bush, I. M., Lee, R., and Whitmore, W. F. (1969). *Am. J. Physiol.* **217,** 1039.

786. Salem, S. I., Coward, W. A., Lunn, P. G., and Hudson, G. J. (1984). *Ann. Nutr. Metab.* **28,** 44.

787. Sandrock, B. C., Kern, S. R., and Bryan, S. E. (1983). *Biol. Trace Elem. Res.* **5,** 503.

788. Sandstead, H. H., Prasad, A. S., Schulert, A. R., Farid, Z., Miale, A., Jr., Bassilly, S., and Darby, W. J. (1967). *Am. J. Clin. Nutr.* **20**, 422.
789. Sandstead, H. H., and Shephard, G. H. (1968). *Proc. Soc. Exp. Biol. Med.* **128**, 687.
790. Sandstead, H. H., Gillespie, D. D., and Brady, R. N. (1972). *Pediatr. Res.* **6**, 119.
791. Sandstead, H. H. (1973). *Am. J. Clin. Nutr.* **26**, 1251.
792. Sandstead, H. H., Fosmire, G. J., McKenzie, J. M., and Halas, E. S. (1975). *Fed. Proc., Fed. Am. Soc. Exp. Biol.* **34**, 86.
793. Sandstead, H. H., Vo-Khactu, K. P., and Solomons, N. (1976). *In* "Trace Elements in Human Health and Disease" (A. S. Prasad and D. Oberleas, eds.), Vol. 1, p. 33. Academic Press, New York.
794. Sandstead, H. H., Strobel, D. A., Logan, G. M., Marks, E. O., and Jacob, R. A. (1978). *Am. J. Clin. Nutr.* **31**, 844.
795. Sandstead, H. H., Klevay, L. M., Jacob, R. A., Munoz, J. M., Logan, G. M., Reck, S. J., Dintzis, F. R., Inglett, G. E., and Shuey, W. C. (1979). *In* "Dietary Fibers: Chemistry and Nutrition" (I. Inglett-Falkehag, ed.), p. 147. Academic Press, New York.
796. Sandstead, H. H., (1981). *In* "Disorders of Mineral Metabolism" (F. Bronner and J. W. Coburn, eds.), Vol. 1, p. 94. Academic Press, New York.
797. Sandstead, H. H. (1985). *In* "Trace Element Metabolism in Man and Animals-5" (C. F. Mills, P. J. Aggett, I. Bremner, and J. K. Chesters, eds.). CAB Publications, Farnham Royal, U.K. (in press).
798. Sandström, B., Arvidsson, B., Cederblad, Å and Björn-Rasmussen, E. (1980). *Am. J. Clin. Nutr.* **33**, 739.
799. Sandström, B., and Cederblad, Å. (1980). *Am. J. Clin. Nutr.* **33**, 1778.
800. Sandström, B., Cederblad, Å., and Lönnerdal, B. (1983). *Am. J. Dis. Child.* **137**, 726.
801. Sandström, B., Davidsson, L., Cederblad, Å., and Lönnerdal, B. (1985). *In* "Trace Element Metabolism in Man and Animals-5" (C. F. Mills, P. J. Aggett, I. Bremner, and J. K. Chesters, eds.). CAB Publications, Farnham Royal, U.K. (in press).
802. Sanecki, R. K., Corbin, J. E., and Forbes, R. M. (1982). *Am. J. Vet. Res.* **43**, 1642.
803. Sann, L., Rigal, D., Galy, G., BienVenu, F., and Bourgeois, J. (1980). *Pediatr. Res.* **14**, 1040.
804. Sato, M., Mehra, R. K., and Bremner, I. (1984). *J. Nutr.* **114**, 1683.
805. Sato, N., and Henkin, R. I. (1973). *Am. J. Physiol.* **225**, 508.
806. Savlov, E. D., Strain, W. H., and Huegin, F. (1962). *J. Surg. Res.* **2**, 209.
807. Saylor, W. M., and Leach, R. M. (1980). *J. Nutr.* **110**, 448.
808. Schechter, P. J., Friedewald, W. T., Bronzert, D. A., Raff, M. S., and Henkin, R. I. (1972). *Int. Rev. Neurobiol.* **15**, 125.
809. Schechter, P. J., Giroux, E. L., Schlienger, J. L., Hoenig, V., and Sjoerdsma, A. (1976). *Eur. J. Clin. Invest.* **6**, 147.
810. Schechter, P. J., and Prakash, N. J. (1979). *Am. J. Clin. Nutr.* **32**, 1011.
811. Schelenz, R. F. W., and Harmuth-Hoene, A. E. (1985). *In* "Trace Element Metabolism in Man and Animals-5" (C. F. Mills, P. J. Aggett, I. Bremner, and J. K. Chesters, eds.). CAB Publications, Farnham Royal, U.K. (in press).
812. Schlage, C., and Wortberg, B. (1972). *Acta Paediatr. Scand.* **61**, 421.
813. Schlicker, S. A., and Cox, D. H. (1968). *J. Nutr.* **95**, 287.
814. Schneeman, B. O., Lönnerdal, B., Keen, C. L., and Hurley, L. S. (1983). *J. Nutr.* **113**, 1165.
815. Schoonees, R., de Klerk, J., and Murphy, G. (1968). *S. Afr. Med. J.* **42**, Suppl., 87.
816. Schraer, K. K., and Calloway, D. H. (1974). *Nutr. Metab.* **17**, 205.
817. Schroeder, H. A., Nason, A. P., Nason, B. A., Tipton, I. H., and Balassa, J. J. (1967). *J. Chronic Dis.* **20**, 179.

818. Schroeder, H. A. (1971). *Am. J. Clin. Nutr.* **24,** 562.
819. Schwarz, W. A., and Kirchgessner, M. (1975). *Arch. Tierernaehr.* **25,** 597.
820. Schwarz, W. A., and Kirchgessner, M. (1975). *Z. Tierphysiol., Tierernaehr. Futtermittelkd.* **35,** 9.
821. Scott, D. A. (1934). *Biochem. J.* **28,** 1592.
822. Scott, M. L., Holm, E. R., and Reynolds, R. E. (1959). *Poult. Sci.* **38,** 1344.
823. Scrutton, M. C., Wu, C. W., and Goldthwait, D. A. (1971). *Proc. Natl. Acad. Sci. U.S.A.* **68,** 2497.
824. Seal, C. J., and Heaton, F. W. (1983). *Br. J. Nutr.* **50,** 317.
825. Seelig, W., Feist, H., Grünert, A., Heinrich, H., and Luckner, L. (1982). *Magnesium-Bull.* **4,** 18.
826. Sempos, C. T., Johnson, N. E., Elmer, P. J., Allington, J. K., and Matthews, M. E. (1982). *J. Am. Diet. Assoc.* **81,** 35.
827. Sever, L. E., and Emanuel, I. (1973). *Teratology* **7,** 117.
828. Severy, H. (1923). *J. Biol. Chem.* **55,** 79.
829. Shanklin, S. H., Miller, E. R., Ullrey, D. E., Hoefer, J. A., and Luecke, R. W. (1968). *J. Nutr.* **96,** 101.
830. Shapcott, D. (1982). *In* "Clinical Applications of Recent Advances in Zinc Metabolism" (A. S. Prasad, I. E. Dreosti, and B. S. Hetzel, eds.), p. 121. Alan R. Liss, Inc., New York.
831. Shapiro, S. G., and Cousins, R. J. (1980). *Biochem. J.* **190,** 755.
832. Shatzman, A. R., and Henkin, R. I. (1981). *Proc. Natl. Acad. Sci. U.S.A.* **78,** 3867.
833. Shaw, J. C. L. (1979). *Am. J. Dis. Child.* **133,** 1260.
834. Shearer, T. R., Lis, E. W., Johnson, K. S., Johnson, J. R., and Prescott, G. H. (1979). *Nutr. Rep. Int.* **19,** 209.
835. Shin, Y. A., and Eichhorn, G. L. (1968). *Biochemistry* **7,** 1026.
836. Shingwekar, A. G., Mohanran, M., and Reddy, V. (1979). *Clin. Chim. Acta* **93,** 97.
837. Sluyser, M. (1977). *Trends Biochem.-Sci.* **2,** 202.
838. Smirnov, A. A. (1948). *Chem. Abstr.* **42,** 8302.
839. Smith, B. L., Embling, P. P., and Pearce, M. G. (1983). *Proc. N. Z. Soc. Anim. Prod.* **43,** 217.
840. Smith, H. (1967). *J. Forensic Sci. Soc.* **7,** 97.
841. Smith, J. C., Jr., and Halsted, J. A. (1970). *J. Nutr.* **100,** 973.
842. Smith, J. C., Jr., Brown, E. D., McDaniel, E. G., and Chan, W. (1976). *J. Nutr.* **106,** 569.
843. Smith, J. C., Jr., Zeller, J. A., Brown, E. D., and Ong, S. C. (1976). *Science* **193,** 496.
844. Smith, J. C., Jr. (1982). *In* "Clinical, Biochemical, and Nutritional Aspects of Trace Elements" (A. S. Prasad, ed.), p. 239. Alan R. Liss, Inc., New York.
845. Smith, J. E., Brown, E. D., and Smith, J. C., Jr. (1974). *J. Lab. Clin. Med.* **84,** 692.
846. Smith, K. T., Cousins, K. J., Silbon, B. L., and Failla, M. L. (1978). *J. Nutr.* **108,** 1849.
847. Smith, K. T., Failla, M. L., and Cousins, R. J. (1979). *Biochem. J.* **184,** 627.
848. Smith, K. T., and Cousins, R. J. (1980). *J. Nutr.* **110,** 316.
849. Smith, W. H., Plumlee, M. P., and Beeson, W. M. (1960). *Science* **128,** 1280.
850. Smythe, P., Schonland, M., Brereton-Stiles, G., Loovadia, H., Grace, H., Loening, W., Mafoyane, A., Parent, M., and Vos, G. (1971). *Lancet* **2,** 939.
851. Solomon, S. J., and King, J. C. (1983). *Fed. Proc., Fed. Am. Soc. Exp. Biol.* **42**(3), 391.
852. Solomons, N. W., and Jacob, R. A. (1981). *Am. J. Clin. Nutr.* **34,** 475.
853. Solomons, N. W., Janghorbani, M., Ting, B. T. G., Steinke, F. H., Christensen, M., Bijlani, R., Istfan, N., and Young, V. R. (1982). *J. Nutr.* **112,** 1809.
854. Solomons, N. W. (1982). *Am. J. Clin. Nutr.* **35,** 1048.
855. Solomons, N. W., Pineda, O., Viteri, F., and Sandstead, H. H. (1983). *J. Nutr.* **113,** 337.
856. Solomons, N. W., and Cousins, R. J. (1984). *In* "Absorption and Malabsorption of Mineral

Nutrients'' (N. W. Solomons and I. H. Rosenberg, eds.), p. 125. Alan R. Liss, Inc., New York.

857. Soltan, M. H., and Jenkins, D. M. (1982). *Br. J. Obstet. Gynaecol.* **89,** 56.
858. Soman, S. D., Panday, V. K., Joseph, K. T., and Raut, S. J. (1969). *Health Phys.* **17,** 35.
859. Somers, M., and Underwood, E. J. (1969). *Aust. J. Agric. Res.* **20,** 899.
860. Somers, M., and Underwood, E. J. (1969). *Aust. J. Biol. Sci.* **22,** 1229.
861. Somner, A. L., and Lipman, C. B. (1926). *Plant Physiol.* **1,** 231.
862. Somner, A. L. (1928). *Plant Physiol.* **3,** 231.
863. Souci, S. W., Fachmann, W., and Kraut, H. (1981). *In* "Food Composition and Nutrition Tables 1981/82." Wiss. Verlagsges. Stuttgart.
864. Speich, M., Métayer, C., Arnaud, P., Van Goc, N., and Boitear, H.-L. (1983). *Ann. Nutr. Metab.* **27,** 531.
865. Spencer, H., Rosoff, B., Feldstein, A., Cohn, S. H., and Gusmano, E. (1965). *Radiat. Res.* **24,** 432.
866. Spencer, H., Vankinscott, V., Lewin, I., and Samachsan, J. (1965). *J. Nutr.* **86,** 169.
867. Spencer, H., and Rostoff, B. (1966). *Health Phys.* **12,** 475.
868. Spencer, H., Osis, D., Kramer, L., and Norris, C. (1976). *In* "Trace Elements in Human Health and Disease" (A. S. Prasad and D. Oberleas, eds.), Vol. 1, p. 345. Academic Press, New York.
869. Spencer, H., Kramer, L., and Osis, D. (1982). *In* "Clinical, Biochemical and Nutritional Aspects of Trace Elements" (A. S. Prasad, ed.), p. 103. Alan R. Liss, Inc., New York.
870. Spray, C. M., and Widdowson, E. M. (1950). *Br. J. Nutr.* **4,** 332.
871. Spray, C. M. (1950). *Br. J. Nutr.* **4,** 354.
872. Stacey, N. H., and Klaassen, C. D. (1981). *Biochim. Biophys. Acta* **640,** 693.
873. Stake, P. E., Miller, W. J., Gentry, R. P., and Neathery, M. W. (1975). *J. Anim. Sci.* **40,** 132.
874. Stake, P. E., Miller, W. J., Neathery, M. W., and Gentry, R. P. (1975). *J. Dairy Sci.* **58,** 78.
875. Stankiewicz, A. J., Falchuk, K. H., and Vallee, B. L. (1983). *Biochemistry* **22,** 5150.
876. Stanwell-Smith, R., Thompson, S. G., Haines, A. P., Ward, R. J., Cashmore, G., Stedronska, J., and Hendry, W. F. (1983). *Fertil Steril.* **40,** 670.
877. Starcher, B. C., Hill, C. H., and Madaras, J. G. (1980). *J. Nutr.* **110,** 209.
878. Starcher, B. C., Glauber, J. G., and Madaras, J. G. (1980). *J. Nutr.* **110,** 1391.
879. Statter, M., and Krieger, J. (1983). *Pediatr. Res.* **17,** 239.
880. Steagand-Pedersen, K., Fredeus, K., and Larson, L. I. (1981). *Brain Res.* **212,** 230.
881. Steel, L., and Cousins, R. J. (1985). *Am. J. Physiol.* **248,** G46.
882. Stevenson, J. W., and Earle, I. P. (1964). *J. Anim. Sci.* **23,** 300.
883. Stober, M. (1971). *Dtsch. Tieraerztl. Wochenschr.* **78,** 257.
884. Strain, W. H., Dutton, A. M., Heyer, H. B., and Ramsey, G. H. (1953). "Experimental Studies on the Acceleration of Burn and Wound Healing," p. 18. University of Rochester, Rochester, New York.
885. Strobel, D. A., Sandstead, H. H., Zimmerman, L., and Reuter, A. (1979). *In* "Nursery Care of Nonhuman Primates" (G. C. Ruppenthal and D. J. Reese, eds.), p. 43. Plenum, New York.
886. Sturtevant, F. M. (1980). *Pediatrics* **65,** 610.
887. Sugawara, N. (1982). *Dev. Toxicol. Environ. Sci.* **9,** 155.
888. Sullivan, J. F., O'Grady, J., and Lankford, H. G. (1965). *Gastroenterology* **48,** 438.
889. Sullivan, J. F., Blotcky, A. J., Jetton, M. M., Hahn, H. K. J., and Burch, R. E. (1979). *J. Nutr.* **109,** 1432.
890. Sullivan, J. F., Williams, R. V., Wisecarver, J., Etzel, K., Jetton, M. M., and Magee, D. F. (1981). *Proc. Soc. Exp. Biol. Med.* **166,** 39.

891. Suttle, N. F. (1979). *Br. J. Nutr.* **42,** 89.
892. Suttle, N. F., Davies, H. L., and Field, A. C. (1982). *Br. J. Nutr.* **47,** 105.
893. Sutton, W. R., and Nelson, V. E. (1937). *Proc. Soc. Exp. Biol. Med.* **36,** 211.
894. Suwarnasarn, A., Wallwork, J. C., Lykken, G. I., Low, F. N., and Sandstead, H. H. (1982). *J. Nutr.* **112,** 1320.
895. Swanson, C. A., and King, J. C. (1982). *J. Nutr.* **112,** 697.
896. Swanson, C. A., Turnlund, J. R., and King, J. C. (1982). *J. Nutr.* **113,** 2557.
897. Swenerton, H., and Hurley, L. S. (1968). *J. Nutr.* **95,** 8.
898. Swenerton, H., and Hurley, L. S. (1980). *J. Nutr.* **110,** 575.
899. Swift, C. E., and Berman, M. D. (1969). *Food Technol.* **13,** 365.
900. Tamura, T., Shane, B., Baer, M. T., King, J. C., Margen, S., and Stokstad, E. L. R. (1978). *Am. J. Clin. Nutr.* **31,** 1984.
901. Tanabe, S. (1980). *Br. J. Nutr.* **44,** 355.
902. Tengrup, I., Ahonen, J., Rank, F., and Zederfeldt, B. (1980). *Acta Chir. Scand.* **146,** 243.
903. Tengrup, I., Ahonen, J., and Zederfeldt, B. (1980). *Acta Chir. Scand.* **143,** 195.
904. Terhune, M. W., and Sandstead, H. H. (1972). *Science* **177,** 68.
905. Thiers, R. E., and Vallee, B. L. (1957). *J. Biol. Chem.* **226,** 911.
906. Thomas, B., Thompson, A., Oyenuga, V. A., and Armstrong, R. H. (1952). *Emp. J. Exp. Agric.* **20,** 10.
907. Thomson, R. W., Gilbreath, R. L., and Bielk, E. (1975). *J. Nutr.* **105,** 154.
909. Todd, W. R., Elvehjem, C. A., and Hart, E. G. (1934). *Am. J. Physiol.* **107,** 146.
910. Tribble, H. M., and Scoular, F. I. (1954). *J. Nutr.* **52,** 209.
911. Tucker, H. F., and Salmon, W. D. (1955). *Proc. Soc. Exp. Biol. Med.* **88,** 613.
912. Tupper, R., Watts, R. W. E., and Wormall, A. (1954). *Biochem. J.* **57,** 245.
913. Turk, D. E., Sunde, M. L., and Hoekstra, W. G. (1959). *Poult. Sci.* **38,** 1256.
914. Turnlund, J., Costa, F., and Margen, S. (1981). *Am. J. Clin. Nutr.* **34,** 2641.
915. Turnlund, J. R., King, J. C., Keyes, W. R., Gong, B., and Michel, M. C. (1984). *Am. J. Clin. Nutr.* **40,** 1071.
916. Tyrala, E., Manser, J. I., Brodsky, N. L., and Tran, N. (1983). *Acta Paediatr. Scand.* **72,** 695.
917. Ullrey, D. E., Ely, W. T., and Covert, R. L. (1974). *J. Anim. Sci.* **38,** 1276.
918. Umoren, J., and Kies, C. (1982). *Nutr. Rep. Int.* **26,** 717.
919. Underwood, E. J. (1962). *World's Poult. Congr., Proc., 12th, 1962,* p. 216.
920. Underwood, E. J., and Somers, M. (1969). *Aust. J. Agric. Res.* **20,** 889.
921. Underwood, E. J. (1981). "The Mineral Nutrition of Livestock," 2nd ed. Commonw. Agric. Bur., England.
922. Urban, E., and Campbell, M. E. (1984). *Am. J. Physiol.* **247,** G88.
923. Vallee, B. L., and Neurath, H. (1954). *J. Am. Chem. Soc.* **76,** 5006.
924. Vallee, B. L., Wacker, W. E. C., Bartholomay, A. F., and Robin, E. D. (1956). *N. Engl. J. Med.* **255,** 403.
925. Vallee, B. L., and Williams, R. J. P. (1968). *Proc. Natl. Acad. Sci. U.S.A.* **59,** 498.
926. Vallee, B. L., and Falchuk, K. H. (1981). *Philos. Trans. R. Soc. London, Ser. B* **294,** 185.
927. Vallee, B. L. (1983). *In* "Zinc Enzymes" (T. G. Spiro, ed.), p. 1. Wiley, New York.
928. Vallee, B. L. (1983). *J. Inherited Metab. Dis.* **6,** Suppl. 1, 31.
929. Van Adrichem, P. M. W., van Leeuwen, J. M., and van Kluyve, J. J. (1970). *Tijdschr. Diergeneeskd.* **95,** 1170.
930. Van Campen, D. R., and Mitchell, E. A. (1965). *J. Nutr.* **86,** 120.
931. Van Rij, A. M., Godfrey, P. J., and McKenzie, J. M. (1979). *J. Surg. Res.* **26,** 293.
932. Van Rij, A. M., and Pories, W. J. (1980). *In* "Zinc in the Environment" (J. O. Nriagu, ed.), Part 2, p. 215. Wiley, New York.

933. Van Rij, A. M. (1982). *In* "Clinical, Biochemical and Nutritional Aspects of Trace Elements" (A. S. Prasad, ed.), p. 259. Alan R. Liss, Inc., New York.
934. Vaughan, L. A., Weber, C. W., and Kemberling, S. R. (1979). *Am. J. Clin. Nutr.* **32,** 2301.
935. Verheijden, J. H. M., Schotman, A. J. H., van Miert, A. S. J. P. A. M., and van Duin, C. T. M. (1983). *Am. J. Vet. Res.* **44,** 1637.
936. Versieck, J., Hoste, J., and Barbier, F. (1976). *Acta Gastro-enterol. Belg.* **39,** 340.
937. Versieck, J., Hoste, J., Barbier, F., Michels, H., and Rudder, J. (1977). *Clin. Chem. (Winston-Salem, N.C.)* **23,** 1301.
938. Versieck, J., and Cornelis, R. (1980). *Anal. Chim. Acta* **116,** 217.
939. Victery, W., Smith, J. M., and Vander, A. J. (1981). *Am. J. Physiol.* **241,** F532.
940. Vir, S. C., and Love, A. H. G. (1979). *Am. J. Clin. Nutr.* **32,** 1472.
941. Vuori, E., and Kuitunen, P. (1979). *Acta Paediatr. Scand.* **68,** 33.
942. Vuori, E., Mäkinen, S. M., Kara, R., and Kuitunen, P. (1980). *Am. J. Clin. Nutr.* **33,** 227.
943. Wacker, W. E. C., and Vallee, B. L. (1959). *J. Biol. Chem.* **234,** 3257.
944. Wada, L. L., Rosyner-Cohen, H., and King, J. C. (1983). *Fed. Proc., Fed. Am. Soc. Exp. Biol.* **42,** 390.
945. Wallwork, J. C., and Sandstead, H. H. (1981). *Fed. Proc., Fed Am. Soc. Exp. Biol.* **40,** 939.
946. Wallwork, J. C., Fosmire, G. J., and Sandstead, H. H. (1981). *Br. J. Nutr.* **45,** 127.
947. Wallwork, J. C., and Sandstead, H. H. (1983). *J. Nutr.* **113,** 47.
948. Walravens, P. A., and Hambidge, K. M. (1976). *Am. J. Clin. Nutr.* **29,** 1114.
949. Walravens, P. A., van Doorninck, W. J., and Hambidge, K. M. (1978). *J. Pediatr.* **93,** 535.
950. Walravens, P. A., and Hambidge, K. M. (1978). *In* "Zinc and Copper in Clinical Medicine (K. M. Hambidge and B. L. Nichols, eds.), p. 49. Spectrum Publ., New York.
951. Walravens, P. A., Krebs, N. F., and Hambidge, K. M. (1983). *Am. J. Clin. Nutr.* **38,** 195.
952. Wapnir, R. A., Khani, D. E., Bayne, M. A., and Lifshitz, F. (1983). *J. Nutr.* **113,** 1346.
953. Warkany, J., and Petering, H. G. (1972). *Teratology* **5,** 319.
954. Warkany, J., and Petering, H. G. (1973). *Am. J. Ment. Defic.* **77,** 645.
955. Warren, D. C., Lane, H. W., and Mares, M. (1981). *Biol. Trace Elem. Res.* **3,** 99.
956. Warren, H. V., and Delavault, R. E. (1971). *Mem.—Geol. Soc. Am.* **123.**
957. Watanabe, T., Sato, F., and Endo, A. (1983). *Yamagata Med. J.* **1,** 13.
958. Watkins, D. W., Antoniou, L. D., and Shalhoub, R. J. (1981). *Can. J. Physiol. Pharmacol.* **59,** 562.
959. Watson, A. R., Stuart, A., Wells, F. E., Houston, I. B., and Addison, G. M. (1983). *Hum. Nutr.: Clin. Nutr.* **37C,** 219.
960. Watts, T. (1969). *J. Trop. Pediatr.* **15,** 155.
961. Wegner, T. N., Ray, D. E., Lox, C. D., and Scott, G. H. (1973). *J. Dairy Sci.* **56,** 748.
962. Weigand, E., and Kirchgessner, M. (1976). *Nutr. Metab.* **20,** 314.
963. Weigand, E., and Kirchgessner, M. (1977). *Z. Tierphysiol., Tierernaehr. Futtermittelkd.* **39,** 84.
964. Weigand, E., and Kirchgessner, M. (1980). *J. Nutr.* **110,** 469.
965. Weigand, E., and Kirchgessner, M. (1982). *Z. Tierphysiol., Tierernaehr. Futtermittelkd.* **47,** 1.
966. Weismann, K., and Flagstad, T. (1976). *Acta Derm.-Venereol.* **56,** 151.
967. Weissman, K., Hoe, S., Knudson, L., and Sorenson, S. S. (1979). *Br. J. Dermatol.* **101,** 573.
968. Weitzel, G., and Fretzdorff, A.-M. (1953). *Hoppe-Seyler's Z. Physiol. Chem.* **292,** 221.
969. Weitzel, G., Strecker, F.-J., Roester, U., Buddecke, E., and Fretzdorff, A.-M. (1954). *Hoppe-Seyler's Z. Physiol. Chem.* **296,** 19.

970. Welch, R. R., House, W. A., and Allaway, W. H. (1974). *J. Nutr.* **104,** 733.
971. Wester, P. O. (1974). *Atherosclerosis* **20,** 207.
972. Westmoreland, N., First, N. L., and Hoekstra, W. G. (1967). *J. Reprod. Fertil.* **13,** 223.
973. Westmoreland, N. (1971). *Fed. Proc., Fed. Am. Soc. Exp. Biol.* **30,** 1001.
974. Weston, W. L., Huff, J. C., Humbert, J. R., Hambidge, K. M., Neldner, K. H., and Walravens, P. A. (1977). *Arch. Dermatol.* **113,** 422.
975. Whanger, P. D., Oh, S.-H., and Deagen, J. T. (1981). *J. Nutr.* **111,** 1196.
976. Whanger, P. D., and Ridlington, J. W. (1982). *Dev. Toxicol. Environ. Sci.* **9,** 263.
977. White, H. S. (1969). *J. Am. Diet. Assoc.* **55,** 38.
978. White, H. S., and Gynne, T. N. (1971). *J. Am. Diet. Assoc.* **59,** 27.
979. White, H. S. (1976). *J. Am. Diet. Assoc.* **68,** 243.
980. Whitehouse, R. C., Prasad, A. S., Rabbani, P. I., and Cossack, Z. T. (1982). *Clin. Chem. (Winston-Salem, N.C.)* **28,** 475.
981. Widdowson, E. M., and Spray, C. M. (1951). *Arch. Dis. Child.* **26,** 205.
982. Widdowson, E. M., McCance, R. A., and Spray, C. M. (1951). *Clin. Sci.* **10,** 113.
983. Widdowson, E. M., and Dickerson, J. W. T. (1964). *In* "Mineral Metabolism: An Advanced Treatise" (C. L. Comar and F. Bronner, eds.), Vol. 2, Part A, p. 1. Academic Press, New York.
984. Widdowson, E. M., Chan, H., Harrison, G. E., and Milner, R. D. G. (1972). *Biol. Neonate* **20,** 360.
985. Widdowson, E. M., Dauncey, J., and Shaw, J. C. L. (1974). *Proc. Nutr. Soc.* **33,** 275.
986. Williams, R. B., and Mills, C. F. (1970). *Br. J. Nutr.* **24,** 989.
987. Williams, R. B., and Chesters, J. K. (1970). *Br. J. Nutr.* **24,** 1053.
988. Williams, R. B., Demertzis, P., and Mills, C. F. (1973). *Proc. Nutr. Soc.* **32,** 3A.
989. Williams, R. B., McDonald, I., and Bremner, I. (1978). *Br. J. Nutr.* **40,** 377.
990. Wilson, M. C., Fischer, T. J., and Riordan, M. M. (1982). *Ann. Allergy* **48,** 288.
991. Wirshing, L., Jr. (1962). *Acta Ophthalmol.* **40,** 567.
992. Wolman, S. L., Anderson, H., Marliss, E. B., and Jeejeebhoy, K. N. (1979). *Gastroenterology* **76,** 458.
993. World Health Organization (1973). "Trace Elements in Human Nutrition," Report of a WHO Expert Committee, Tech. Rep. Ser. No. 532. WHO, Geneva.
994. Wright, A. L., King, J. C., Baer, M. T., and Citron, L. J. (1981). *Am. J. Clin. Nutr.* **34,** 848.
995. Yassur, Y., Snir, M., Melamed, S., and Ben-Sira, I. (1981). *Br. J. Ophthalmol.* **65,** 184.
996. Young, R. J., Edwards, H. M., and Gillis, M. B. (1958). *Poult. Sci.* **37,** 1100.
997. Yunice, A. A., King, R. W., Kraikitpanitch, S., Haygood, C. C., and Lindeman, R. D. (1978). *Am. J. Physiol.* **235,** F40.
998. Zanzonico, P., Fernandes, G., and Good, R. A. (1981). *Cell. Immunol.* **60,** 203.
999. Ziegler, E. E., Edwards, B. B., Jensen, R. L., Filer, L. J., and Fomon, S. J. (1978). *In* "Trace Element Metabolism in Man and Animals-3" (M. Kirchgessner, ed.), p. 292. Tech. Univ. Munich, F.R.G., West Germany.
1000. Ziegler, T. R., Scott, M. L., McEvoy, R., Greenlaw, R. H., Huegin, F., and Strain, W. H. (1962). *Proc. Soc. Exp. Biol. Med.* **109,** 239.
1001. Zimmerman, A. W., Hambidge, K. M., Lepow, M. L., Greenberg, R. D., Stover, M. L., and Casey, C. E. (1982). *Pediatrics* **69,** 176.
1002. Zimmerman, A. W., Dunham, B. S., Nochimson, D. J., Kaplan, B. M., Clive, J. M., and Kunkel, S. L. (1984). *Am. J. Obstet. Gynecol.* **149,** 523.
1003. Zook, E. G., Greene, F. E., and Morris, E. R. (1974). *Cereal Chem.* **51,** 788.
1004. Zwickl, C. M., and Fraker, P. J. (1980). *Immunol. Commun.* **9,** 611.

2

Iodine

BASIL S. HETZEL

Division of Human Nutrition, CSIRO
Adelaide, South Australia, Australia

GLEN F. MABERLY

Department of Medicine, Endocrine Unit, Westmead Hospital
Westmead, New South Wales, Australia

I. INTRODUCTION

Iodine (I) is an essential trace element that is widely distributed in nature and is of particular importance to human nutrition. Its biological importance arises from the fact that it is a constituent of the thyroid hormones thyroxine 3,5,3′,5′-tetraiodothyronine (T_4) and 3,5,3′-triiodothyronine (T_3). The major role of iodine in nutrition arises from the importance of thyroid hormones in the growth and development of humans and animals.

A. The Cycle of Iodine in Nature

There is an iodine cycle in nature (161). Iodine was present during the primordial development of the earth, but large amounts were leached from the surface soil by snow and rain and were carried by wind, rivers, and floods into the sea (108). Iodine was also removed from the surface soil by glaciation (195). The ocean has become the primary source of iodine. Iodide ions are oxidized by sunlight (wavelengths up to 560 nm) to elemental iodine, which is volatile (108). Every year some 400,000 tons of iodine escape from the surface of the oceans (161,203). The concentration of iodine in seawater is about 50 to 60 μg/liter, about the same as in human serum. The average iodine concentration in the earth is 300 μ/kg, and in the air it is ~0.7 μg/m^3 (108).

Atmospheric iodine can be increased by combustion of fossil fuels. Rain (1.8–8.5 μg/liter) contains more iodine than air, so that rain enriches the superficial layers of soil with iodine, thus completing the cycle.

Iodine-deficient areas arise from the effects of heavy rain on steep mountain slopes and from floods of snow water and rain, as in the Ganges plains in India. All crops grown on these soils are iodine deficient. As a result, the animal and human populations will also become iodine deficient if they are totally dependent on such crops. This accounts for the occurrence of severe iodine deficiency in vast populations in Asia (as in China and India) who are living within a system of subsistence agriculture.

The iodine content of water reflects the iodine content of the adjacent soil. In general, iodine-deficient areas have water iodine levels <2 μg/liter; for example, in goitrous areas in India, Nepal, and Ceylon, ranges of 0.1 to 1.2 μg/liter were found compared with levels of 9.0 μg/liter in the nongoitrous area of Delhi (151).

The iodine content of terrestrial plants varies in relation to the iodine content of the soil. The content averages 1 mg/kg dry weight, but in iodine-deficient soils the concentration may drop to 10 μg/kg. By contrast, plankton has a concentration of 30 to 1500 μg/kg (256).

B. The Biological Significance of Iodine

The biological significance of iodine arises from the fact that it is a major constituent of the thyroid hormones. This probably arises from the chemical affinity of tyrosine for iodine. In the iodine-rich marine environment the formation of monoiodotyrosine (MIT) and diiodotyrosine (DIT) would readily occur in many tissues and organisms. The coupling of iodinated tyrosines would be facilitated by proteins.

The thyroid gland occurs in all vertebrates as a development from the midventral pharyngeal floor. The thyroid has evolved a special facility for the concentration of inorganic iodine and also for the biosynthesis of thyroid hormones.

The role of the thyroid differs in different species. In fish, the thyroid plays a role in seasonal changes. However, in amphibians the thyroid hormone provides the primary signal that is necessary for metamorphosis. This probably depends on an interaction with receptors. Such an interaction provides a major opportunity for study of the mechanisms involved. However, the evidence available indicates that amphibian metamorphosis is a uniquely thyroid hormone-dependent phenomenon, and general extrapolation of the constituent mechanisms involved is not justified (112).

However, a euthyroid state dependent on an adequate iodine intake is necessary for normal growth, differentiation, and development. In severe iodine deficiency gross retardation of fetal and postnatal development occurs. A full discussion is provided in Section V, where the human effects, iodine deficiency disorders (IDD), are described, and in Section VI where a series of animal models are reviewed. The effect of iodine deficiency on fetal development is mediated by both maternal and fetal thyroid glands. Correction of the iodine deficiency will prevent the retardation.

Iodine deficiency is a continuing major human public health problem, particularly in Third World countries and especially in Asia, where some 400 million are affected (63). The eradication of the effects of iodine deficiency on human growth and development is feasible and now constitutes a major challenge in the field of international nutrition (133).

II. IODINE IN ANIMAL TISSUES AND FLUIDS

A. General Distribution

The healthy human adult body contains a total of 15 to 20 mg iodine, of which 70–80% is present in the thyroid gland. Since the normal thyroid weighs only 15–25 g, or ~0.03% of the whole body, this represents a unique degree of concentration of any trace element in a single organ (244). The iodine concentration in the skeletal muscles is <1/1000 of that of the thyroid, but because of their large mass they contain the next largest proportion of total-body iodine. Hamilton et al. (123) give the following mean levels in a range of human tissues in micrograms of iodine per gram wet weight: muscle 0.01 ± 0.001, brain 0.02 ± 0.002, testis 0.02 ± 0.003, lymph nodes 0.03 ± 0.01, kidneys 0.04 ± 0.01, lungs 0.07 ± 0.03, ovaries 0.07 ± 0.03, and liver 0.20 ± 0.06. Other workers have found the ovaries, pituitary gland, bile, and salivary glands to be appreciably higher in iodine than most other extrathyroidal tissues (37,187,280). Significant iodine concentrations also occur in parts of the eye, notably the orbital fat and the orbicular muscle. In one study of a small range of samples, this muscle averaged close to 0.25 μg iodine per gram wet weight (110).

Iodine in the tissues occurs in both inorganic and organically bound forms. The former is normally present in extremely low concentrations, of the order of 0.01 μg/g (249). In the saliva the iodine is almost entirely in the inorganic form, even in conditions when organic iodine compounds are secreted in the urine (3,219). The salivary iodine concentration is proportional to the plasma inorganic iodine concentration at physiological levels (124) and at plasma concentrations up to 100 μg/dl, or ~500 times normal (125). Such increases can be achieved by the administration of iodine in the prophylaxis of simple goiter, and particularly at the much higher levels used in the therapy of exophthalmic goiter (31).

Most of the small amounts of organic iodine in the extrathyroidal tissues consists of thyroxine bound to protein, together with widely distributed low concentrations of other compounds, including T_3. The solubility of muscle iodine differs from that of T_4 added to tissue extracts, and its distribution is not uniform between myosin and actin. However, muscle iodine levels decrease in hypothyroidism and increase in hyperthyroidism (250).

B.   Iodine in the Thyroid Gland

The total concentration of iodine in the thyroid varies with the iodine intake and age of the animal and with the activity of the gland. Variation among species is small, except that the thyroids of sea fish are richer and those of rats slightly poorer than those of most mammalian species. The normal healthy thyroid of mammals contains 0.2–0.5% iodine (dry basis), giving a total of 8 to 12 mg in the adult human gland. This amount can be reduced to ≤1 mg in endemic goiter, with an even greater reduction in concentration because of the hyperplastic changes that characterize the disease. Many years ago Marine and co-workers (183,185) showed that hyperplastic changes are regularly found when the iodine concentration falls below 0.1%. This has been confirmed by later studies with several species. Thus sheep thyroids with marked follicular hyperplasia contained 0.01% iodine, those with moderate hyperplasia 0.04%, and pig thyroids showing very slight hyperplasia 0.11% (dry basis) (7, 12). Further investigations of neonatal mortality in lambs associated with goiter also indicate that a thyroid iodine level of 0.1% or slightly higher is a critical level below which the gland cannot function properly (255,266).

Iodine exists in the gland as inorganic iodine, MIT, DIT, T_4, T_3, polypeptides containing T_4, thyroglobulin, and probably other iodinated compounds (246). The iodinated amino acids are bound with other amino acids in peptide linkage to form thyroglobulin, the unique iodinated protein of the thyroid. Thyroglobulin, the chief constituent of the colloid filling the follicular lumen, is a glycoprotein with a molecular weight of 650,000. It constitutes the storage form of the thyroid hormones and normally represents some 90% of the total iodine of the gland. The amounts and proportions of the various iodine-containing components of the thyroid vary with the supply of iodine to the gland (87), with the presence of goitrogens that can inhibit the iodine-trapping mechanism or the process of hormonogenesis, and with the existence of certain disease states and metabolic defects of genetic origin. These questions are discussed further in Sections IV and XI.

C.   Iodine in Blood

Iodine exists in blood in both inorganic and organic forms. The normal range of plasma inorganic iodide (PII) is stated by Wayne *et al.* (308) to be 0.08–0.60

μg/dl, with values <0.08 suggesting iodine deficiency and values >1.0 pointing to exogenous iodine administration. Karmarkar *et al.* (151) give mean PII values of 0.096 ± 0.02, 0.088 ± 0.017, and 0.089 ± 0.013 for individuals from goitrous areas in India, Nepal, and Ceylon, respectively, compared with 0.137 ± 0.018 μg/dl for normal controls.

The organic iodine of the blood, which does not occur in the erythrocytes, is present mainly as T_4 bound to the plasma proteins. Only a very small proportion, normally ~0.5%, is free in human serum (274). Up to 10% of the organic iodine of the plasma is made up of several iodinated substances, including T_3 and DIT (117,247). Thyroglobulin occurs only in pathological states involving damage to the thyroid gland, but the iodotyrosines do appear in the peripheral circulation, following thyroid-stimulating hormone (TSH) stimulation and in hyperthyroidism (313).

The levels of several different iodine-containing components of blood have been estimated in attempts to develop convenient and satisfactory indices of thyroid function. The protein-bound iodine of the serum (PBI), or the butanol-extractable iodine (BEI) of the serum, has had a considerable vogue and corresponds reasonably well with the level of thyroid activity in humans (121,308,317) and farm animals (147,232). In adult humans the limits of normality have been placed at 4 to 8 or 3 to 7.5 μg/dl, with a "mean" close to 5 to 6 μg/dl (121,308,311,312). Lower PBI norms (3–4 μg/dl) have been found for mice, rats, and dogs (290), adult sheep (310), and beef cattle (291). Still lower mean serum PBI levels have been recorded in studies with the domestic fowl (329) and horses (147).

Estimation of total T_4 in serum, based on competitive protein-binding analysis as introduced by Ekins (86), correlates well with thyroid function (86) but is open to the objection that changes in T_4-binding proteins can invalidate the results. For example, as a consequence of raised T_4-binding globulin (TBG) concentration, euthyroid women taking oral contraceptives or pregnant women may have falsely elevated serum T_4 levels (86,109). The concept developed by Robbins and Rall (245) that the small amount of unbound or free T_4 is the factor determining the true thyroid status of the individual has received strong support from later studies (144,206,311). Free-T_4 (FT_4) assays on plasma can sharply differentiate hypothyroid and thyrotoxic patients from euthyroid individuals (79,149,156,296,311).

In recent years, commercial companies have invested major resources in simplifying the methods for the measurement of thyroid hormones and TSH in serum (143). Most use radioimmunoassay techniques, where the labeled analogs are competing with the hormone for a specific antibody. These methods have been shown to perform satisfactorily in most clinical situations (48,100,278,312). Other methods use nonisotopic immunoassay techniques where the analog reacts with an enzyme or where the bound analog emits a different spectrum of polarized light. Techniques of separation of bound from free hormones have been greatly simplified, thus improving the ease of performance of assays. Many assay systems have been linked to computer systems that can automatically store and plot quality control data.

D. Iodine in Milk

Cow's colostrum has much higher iodine levels than true milk, and there is a fall in concentration in late lactation. Kirchgessner (156) reports a mean value of 264 ± 100 μg for colostrum, compared with 98 ± 82 for true milk. Lewis and Ralston (1,167) found the colostrum iodine level of five cows to range from 200 to 350 μg/liter compared with 72 to 136 μg/liter in the later milk of the same animals. Salter (249) quotes values of 50 to 240 μg/liter for the iodine content of human colostrum and 40–80 μg/liter for human milk once lactation is established.

In ruminants, iodine is present in the milk entirely in the form of iodine, since only iodide has been detected after iodine-131 (^{131}I) administration and no thyroactive compounds can be found by chromatographic procedures (24, 107,323). This is in agreement with biological tests (27) of bovine milk and with the finding that the normal bovine mammary gland is impervious to T_4 (243). In the milk of the rat, rabbit (36), and dog (197), an iodine-containing protein can be detected after ^{131}I administration.

The iodine concentration of milk is greatly influenced by dietary iodine intakes. Blom (29) increased the iodine level in cow's milk from a "normal" 20–70 μg/liter to 510 to 1070 μg/liter by feeding a daily supplement of 100 mg potassium iodide; Kirchgessner (156) demonstrated rising milk iodine concentrations from increments of dietary iodine, and these findings have been confirmed by others (130).

Teat dipping with iodophors in the treatment of mastitis also increased milk iodine levels. Funke *et al.* (102) reported iodine levels of 89 and 94 μg/liter for controls and 127 and 152 μg/liter for the milk of cows treated after each milking. This question is further discussed in Section VIII.

III. IODINE METABOLISM

A. Iodine Absorption

In food and water, iodine occurs largely as inorganic iodide and is absorbed from all levels of the gastrointestinal tract (57). Orally administered iodide is rapidly and almost completely absorbed, with little appearing in the feces (301). Iodinated amino acids are well absorbed, though more slowly and less completely than iodide. A proportion of their iodine may be lost in the feces as organic metabolites, with the remainder being broken down to iodide. Other forms of iodine are reduced to iodide prior to absorption (153).

In the ruminant, the rumen is the major site of absorption of iodide and the abomasum the major site of endogenous secretion (i.e., for the reentry of circulating iodine into the digestive tract). Net absorption also occurs from the small intestine and from the remainder of the gastrointestinal tract (19,201). It is apparent that considerable absorption and endogenous secretion takes place throughout the ruminant digestive tract.

The normal range of PII in humans was reported by Wayne *et al.* (308) to be 0.08–0.60 µg/dl, with values <0.08 suggesting iodine deficiency and values >1 µg pointing to administration of exogenous iodine.

B. Iodine Excretion

The only known and key physiological role of iodine is linked to thyroid hormone synthesis and action. In fact Wolff (318) described the major function of the thyroid as "an efficient collector of this rare element." As such, the thyroid must compete mainly with the kidneys, which have no mechanism to conserve iodine and hence are the major source of iodine excretion. The level of urinary iodine excretion correlates well with plasma iodide concentrations and ^{131}I thyroid uptakes (99,153).

Follis (99) has set a urinary iodine level of 50 µg/g creatinine as the "tentative lower limit of normal" for adolescents, with 32.5 µg/g creatinine as the corresponding figure for children 5–10 years of age and 75 µg/g creatinine for adult men. Koutras (160) considers that a urinary iodine excretion <40 µg/day is suggestive of iodine deficiency in humans, if renal clearance is normal. Stanbury *et al.* (271) have suggested that iodine excretion of <50 µg/hr or 50 µg/g creatinine in randomly obtained specimens in a fair sample of the population is indicative of the existence of endemic goiter in a community.

In addition, iodine appears in the milk, feces, and sweat. In tropical areas of low dietary iodine status, losses in the sweat could impose a significant drain on a limited iodine supply. Vought and co-workers (301) found fecal iodine excretion in normal adults to range from 6.7 to 42.1 µg/day. Koutras (160) gives 5–20 µg/day as the normal range of fecal iodine excretion in humans.

C. The Iodine Pool

Iodide ions resemble chloride ions in that they permeate all tissues. The total iodide pool therefore consists of the iodide present in the whole extracellular space, together with the red blood cells and certain areas of selective concentration, namely, the thyroid, salivary, and gastric glands. Equilibration within the total pool is reached rapidly (26). In the rat ~52% of either absorbed or intraperitoneally injected ^{131}I is excreted, mainly in the urine, with a half-life of 6 to

7 hr (129). This half-life represents the turnover of the inorganic iodide phase. Metabolic equilibrium between the retained ^{131}I and the whole-body iodide pool was achieved within 14 days, and the half-life of this iodine reached 9.5 days and remained constant at that level. This is considered to represent the turnover of the organic iodine pool.

Despite its high iodine content and the efficiency with which it traps iodine, the thyroid gland contributes little to the iodine pool, because the binding into the organic form is normally so rapid. Significant quantities of iodide are also trapped by the salivary glands (37), apparently by mechanisms similar to those of the thyroid (126,127). Since salivary gland iodide is not converted into its organic form, and is normally reabsorbed, this process represents little net loss to the iodine pool.

The iodide pool is replenished continuously, exogenously from the diet, and endogenously from the saliva, the gastric juice, and the breakdown of thyroid hormones. Gastric clearance of iodide exceeds that of chloride in dogs by 10 to 50 times (140), and in dairy calves the net gastric (abomasal) secretion of iodine exceeds that of chloride by 15 times (198). It has been suggested that the iodine-concentrating action of the abomasum may promote iodine conservation by creating an extravascular iodine pool, thus preventing its excessive loss in the urine (201).

Iodine is continuously lost from the iodide pool by the activities of the thyroid, kidneys, salivary glands, and gastric glands, which compete for the available iodine. Koutras (160) conceives of iodine metabolism as a metabolic cycle consisting of three principal pools. These are PII, the intrathyroidal iodine, and the pool comprising the hormonal iodine or PBI of the plasma and tissues, The rate of removal of iodide from the first of these pools by the thyroid and kidneys are expressed as thyroid and renal clearances, calculated as organ accumulation of iodide per unit time divided by plasma iodide concentration. In normal humans total clearance from the iodide pool occurs at the rate of ~50 ml/min, and renal iodide clearance is constant at ~35 ml/min over all ranges of plasma iodide examined (47). Thyroid clearance, by contrast, is sensitive to changes in plasma iodide concentration and varies with the activity of the gland. In normal individuals the thyroid clears an average of 10 to 20 ml/min, whereas in exophthalmic goiter or Graves' disease, a clearance of 100 ml/min is usual (249) and >1000 ml/min is possible (26).

IV. THE THYROID HORMONES

A. Iodine Metabolism in the Thyroid Gland

The human thyroid must trap ~60 μg of iodine daily, to ensure an adequate supply of hormone. The gland, which normally weighs 15–25 g (or 0.28% of total body weight) can achieve this because of its vascularity and extremely active iodide-trapping mechanism. Under normal physiological conditions, the iodide trap or pump mechanism maintains a gradient from the extracellular fluid to the thyroid cell cytoplasm of 100 : 1. Under certain conditions, such as iodine deficiency or Graves' disease, this gradient may exceed 400 : 1 (318,320). It is this trapping mechanism that regulates a more or less constant iodine supply to the thyroid over a wide range of plasma iodide levels (260).

The steps of thyroid hormone biosynthesis and release are represented diagrammatically in Fig. 1. The iodide-trapping mechanism is an active transport mechanism linked to Na^+,K^+-ATPase activity, which is ouabain sensitive (154,155). This mechanism is also regulated by TSH, which is released from the pituitary to regulate thyroidal activity (132). Large inorganic anions can act as competitive inhibitors of iodine transport, and these include halides such as astatide (At) and bromide (Br), or pseudohalides such as thiocyanate (SCN) and selenocyanide (SeCN), or nonhalogens such as perrhenate (ReO_4) and pertechnetate (TcO_4) (289). The most clinically important is SCN, a goitrogen derived from hydrogen cyanide (HCN)-containing foods (33) (see later section). Pertechnetate (8,254) in the form of $^{99m}TcO_4$ is also important for its diagnostic use in thyroid scanning. Its short half-life (6 hr), absence of β-emission, and 140-KeV γ-ray radiation, makes it ideal for this purpose.

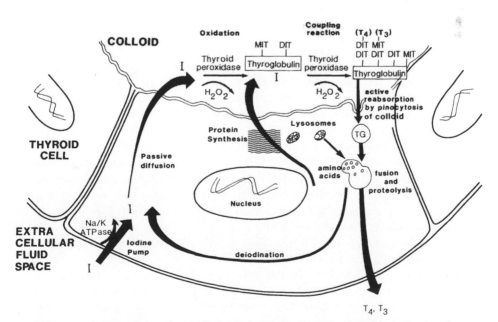

Fig. 1. Diagram showing pathways of synthesis of thyroid hormones from iodine within the thyroid gland.

After the iodide is released by thyroid cells into the colloid space, it is oxidized using hydrogen peroxide by the thyroid peroxidase enzyme (1,286,287). Iodine is then able to combine with the tyrosine residues associated with the thyroglobulin protein to form MIT and DIT. The oxidation process proceeds further under the influence of the thyroid peroxidase enzyme to couple MIT and DIT to form various iodothyronines (289). The most common of these is T_4. This organification phase of iodine metabolism by the thyroid is much less efficient than the trapping mechanism and can be blocked by a large number of reducing agents, including drugs such as propylthiouracil and carbimazole, and goitrogen substances of the goitrin group (288,289).

Finally, iodinated thyroglobulin and thyroid hormones are reabsorbed into the thyroid cells and exposed to proteolytic enzymes (42,82,322). Much of the protein and iodinated tyrosines are lysed and returned as substrates for the whole process to start again. At the same time, some thyroid hormones are released into the circulation.

B. Regulation of Thyroid Hormones

The regulation of the action of thyroid hormones is a complex interaction among neurotransmitters, hormones, and enzymes in the central nervous system, the pituitary, the thyroid gland, the circulation, and peripheral tissues. This is schematically summarized in Fig. 2. The central nervous system has input to adjust thyroid hormone activity via neuronal pathways from the thalamus and its associated nuclei to the hypothalamus. The interactions among these pathways are still controversial. Current evidence would indicate that the adrenergic and nonadrenergic (catecholamine), serotonin (monoamine), and opioid transmitters are stimulatory (150,267) for thyrotrophin-releasing hormone (TRH) release from the hypothalamus (118), whereas dopamine (64) exerts a suppressive effect on TRH release. Apart from the day-to-day normal control (149), a mechanism exists for the thyroid axis to respond to abnormal stresses, such as illness, shock, starvation, and cold (141,194,309). When TRH is released into the portal hypothalamic–pituitary circulation, its action is directed to influencing the synthesis and release of TSH, which is produced by the thyrotrophic cells in the anterior pituitary. It is currently not clear whether feedback of either TSH or thyroid hormones affects the release of TRH.

It has become clear that TRH (a tripeptide) is not confined in its action to the control of TSH. It is found in high concentrations in other parts of the central nervous system, especially in the pineal region, and also in the pancreas (52,166). This widespread distribution of TRH leads to the concept that TRH may also act as a neurohormonal regulator in systems other than the thyroid system. Without continued stimulation from TRH, however, the thyrotrophic

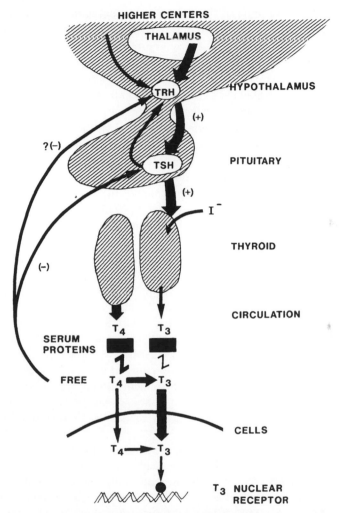

Fig. 2. Schematic representation of the thyroid hormonal interactions from the higher centers in the brain, hypothalamus, pituitary, thyroid, circulation, and peripheral cells.

cells would cease to produce an adequate level of TSH and result in thyroid underactivity (hypothyroidism).

Without the normal inhibitory-feedback effect of thyroid hormones on the pituitary, the stimulatory effect of TRH would lead to excessive TSH and hence thyroid overactivity. The thyrotrophic cells, by altering TSH synthesis and release in response to circulating levels of thyroid hormones, determine the appropriate level of thyroid activity. The central nervous system, in response to stress and other factors, can modulate but does not directly control the level of thyroid activity.

Thyroid-stimulating hormone is a glycoprotein with a molecular weight of ~28,000 (59). It consists of two polypeptide-subunit (α and β) chains, and the combination of these subunits is necessary for full expression of hormonal activity. This basic structure is similar to other pituitary hormones, luteinizing hormone (LH) and follicle-stimulating hormone (FSH) and the placental human chorionic gonadotrophin (HCG). As with all these hormones within a species, the α-subunit (i.e., α-TSH or α-LH) has essentially identical amino acid sequences. These sequences, however, do vary from species to species and account for the major differences in size and structure of TSH among the different species. The amino acid sequence for the β-subunit differs for each hormone and confers unique biological activity for the hormone. The β-TSH subunit is essentially the same across all species (59,228).

Under normal circumstances, without TSH the thyroid gland will fail to produce adequate amounts of thyroid hormone for normal tissue requirements. TSH action is initiated by the β-subunit binding to specific receptor sites in the membrane of thyroid cells. The precise subsequent events are not fully understood, but it is clear that the major effects are exerted via activation of the cAMP mechanism (42). The initial binding event may result in a Ca^{2+} and Mg^{2+} influx and activation of adenylate cyclase (cAMP), which in turn initiates activation and phosphorylation of numerous proteins and enzymes within the cell. In addition to the direct effects on metabolism, activation of these enzymes causes phosphorylation of the nuclear nonhistone chromosomal proteins permitting exposure of certain gene structures for RNA replication and subsequent further protein synthesis (39–41,65,152).

TSH influences all phases of iodine metabolism in the thyroid from iodide transport to secretion of T_4 and T_3. Iodide transport is directly linked to oxidative phosphorylation (82,83,283,297). In addition to causing an overall increase in organification of iodide, TSH increases the coupling rate of MIT and DIT to form T_4 and T_3 (248). TSH influences the rate at which colloid is pinocytosed and fused with liposomes to be released as thyroid hormones in the circulation (42). It has been suggested that this whole process may be largely regulated by TSH having its effect via modulation of the calcium calmodulin mechanism, which controls the microtubules and microfilaments within the thyroid cell (318,321).

C. Circulating Thyroid Hormones

Under normal conditions most of the thyroid hormone released from the thyroid gland is in the form of T_4 (~80 μg/day; ref. 51) and only a small amount as

T_3 ($\sim$3 µg/day; ref. 52). These hormones rapidly combine with predominantly three serum proteins, which in the order of their affinity for thyroid hormones are TBG, T_4-binding prealbumin (TBPA), and albumin.

The serum-binding proteins in the subhuman primates and horses are similar to humans. In cattle, sheep, pigs, goats, and dogs there is a protein of similar structure to TBG, but little or no TBPA. Pigeons, chickens, and kangaroos have a TBPA-like protein and albumin, but lack TBG. Guinea pigs and rats lack both TBG and TBPA with only albumin to bind thyroid hormones (93,242). In humans, 99.97% of T_4 and 99.7% of T_3 is bound to these binding proteins. Differences in binding characteristics of these proteins help to explain the differences in serum half-lives of T_4 (8 days) and T_3 (8 hr). The small amounts of FT_4 or FT_3 are available for interaction with the tissues.

As thyroid hormone passes certain tissues, especially the kidneys, liver, lungs, and muscle, deiodination of T_4 occurs making more T_3 available in the circulation (35,229). In fact, 80% of circulating T_3 is derived from T_4 via this peripheral conversion pathway, so that adult human serum total T_4 levels are normally between 70 and 160 nmol/liter and total serum T_3 levels between 1.2 and 2.8 nmol/liter (51,229,273).

D. Mechanism of Action of Thyroid Hormones in Cells

The mechanism of thyroid hormone action at the molecular level has been the subject of considerable interest and debate since 1950 (21,25,214,215). In clinical evaluation of the thyroidal status of a patient, experience has shown that serum concentrations of thyroid hormones are not the absolute reference point. Among the many hypotheses advanced to explain the action of thyroid hormones on cells, there appear to be two schools of thought: (1) that thyroid hormone initiates its action at the level of the nucleus (284), or (2) that thyroid hormones act at the plasma membrane or with extranuclear organelles (275). Since the mid-1960s a considerable body of circumstantial evidence has accumulated supporting the concept that the basic unit of thyroid hormone action was the T_3 nuclear receptor complex. The evidence supporting this concept is that nuclear T_3 receptors are of high affinity and low capacity, so that they may respond to physiological concentrations of thyroid hormones and amplify the message within the responsive cells. The nuclear binding protein is a nonhistone protein, a class of protein believed to be important in gene expression. The T_3 nuclear binding sites are distributed appropriately in hormone-responsive tissue, and there is a good correlation between T_3 nuclear occupancy in many tissues and biological responses, as measured by thyroid hormone-dependent enzymes such as α-glycerophosphate dehydrogenase and malic enzyme (215).

The evidence for T_3 nuclear binding sites being the principal locus for thyroid hormone action does not exclude extranuclear, subcellular sites of action or preclude other iodothyronines from mediating thyromimetic effects. Furthermore, many physiological and pathophysiological states, characterized by subnormal plasma and tissue T_3 levels and euthyroidism, are not readily explained by this unifying nuclear T_3 hypothesis.

Other proposed sites for thyroid hormone action include the mitochondrion and cell membrane. Sterling and co-workers (276) have described high-affinity, T_3-binding sites on the inner mitochondrial membrane, but this finding awaits more extensive study and the demonstration of these sites by *in vivo* techniques. The cell membrane is a potential site of hormone action in view of the large number of effects induced by thyroid hormones acting apparently at this level, and which are not blocked by inhibitors of protein synthesis (253). In addition, there have been several reports of high-affinity binding sites for thyroid hormones on rat hepatic membrane preparations (4,10,106,128,182,231). The function of these binding sites remains unknown; however, the physicochemical characteristics of binding and specificity suggest they may be involved in mediating hormone transport or action at the membrane level.

Deiodination of T_4 by the tissues to various thyroid hormone analogs appears to be another important mechanism for locally controlling the effects of thyroid hormones. Two monodeiodinase enzymes are present in tissue, and a schematic view of a possible cascade of metabolism is shown in Fig. 3. Outer-ring deiodination of T_4 produces metabolically active T_3 ($3,5,3'$-T_3), which by binding to the nuclear receptor is able to enhance thyromimetic activity. Inner-ring deiodination, however, leads to reverse-T_3 ($3,3',5'$-T_3, or rT_3), which is rapidly further degraded to inactive metabolites.

In the rat it has been demonstrated that certain tissues metabolize and utilize thyroid hormones differently (35,68,263). This is represented in Fig. 4. In the cerebral cortex, cerebellum, and pituitary, most of the T_3 that eventually occupies the nuclear receptor is generated locally by an active intracellular outer-ring T_4 monodeiodinase enzyme. In these tissues little T_4 is wasted in the production of inactive metabolites such as rT_3, so the nuclear occupancy is high. Conversely, in more peripheral tissues such as liver and kidneys, the inner-ring intracellular T_4 monodeiodinase enzyme activity predominates. This means that much of the intracellular T_4 (derived from serum T_4) is metabolized to rT_3 rather than T_3. Most of the nuclear T_3 must thus find its way to the nucleus directly from the plasma pool. In peripheral tissues, therefore, T_3 nuclear occupancy is usually lower than in central (brain and pituitary) tissue (35; see Table I).

E. Thyroidal Activity at Various Stages of Life

Thyroidal activity varies at different stages of life. The mammalian hypothalamic–pituitary–thyroid system in general, and the human system in particular, develop in several phases (139). The initial phase, during the first 12 weeks of

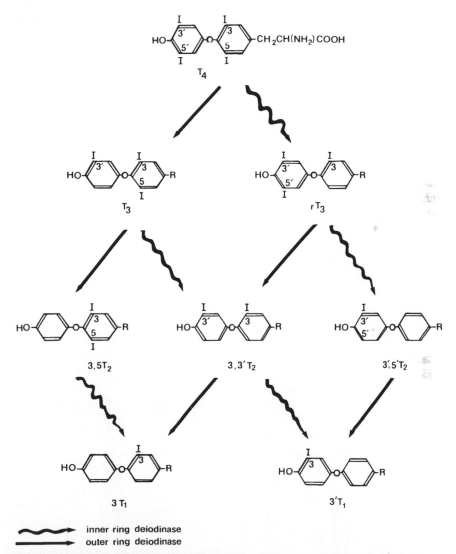

Fig. 3. Diagrammatic representation of the possible cascade of metabolism of T_4, depending on which (inner or outer ring) deiodinase enzyme is active.

gestation, is the formation of the thyroid gland from the ectoderm of the primitive buccal cavity to a solid mass of thyroid tissue able to concentrate iodide and release iodothyronines (96,259).

The anterior pituitary gland also derived from the buccal cavity (Rathke's pouch) develops simultaneously and by 12 weeks is secreting TSH (97,101).

Maturation of the hypothalamus develops from the tenth to thirtieth week of gestation (239). Maturation of neuroendocrine control occurs between the twentieth week of gestation and 3–4 weeks postpartum. Between 18 and 22 weeks of gestation TSH concentrations rise abruptly to a mean level of between 7 and 9 mU/ml, and this is followed by a rise in both total T_4 and FT_4 concentrations (96,97,239). During the initial phase of thyroid development (from 12 weeks), the inner-ring monodeiodinase enzyme is active so rT_3 is formed from T_4, and this is reflected in the serum thyroid hormone levels (49,50). Fetal serum T_3 concentrations are low. Just prior to birth the outer-ring monodeiodinase enzyme activity matures and the inner-ring deiodinase activity declines (139,324). This results in a rapid rise in serum T_3 and a fall in serum rT_3, so that T_3 is higher in concentration compared with rT_3 by the second week of extrauterine life. During the latter half of pregnancy the fetal thyroid system is independent of the maternal thyroid system, as maternal TSH, T_4, and T_3 are either blocked or degraded by the placenta (68,97). Events, such as iodine deficiency or excess, immunological thyroid blocking or stimulation, or interference in thyroid metabolism by goitrogens or drugs will however be common to both the mother and the fetus. The placenta is no barrier to these interfering factors (139).

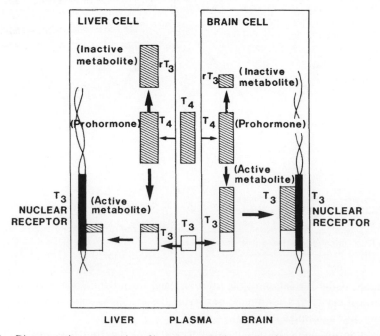

Fig. 4. Diagrammatic representation of how the metabolism of thyroid hormones by liver and brain cells differs and how this contributes to differences in T_3 nuclear occupancy.

Table I

Relative Contribution of Plasma T_3 and Locally Produced T_3 from T_4 to
Nuclear T_3 in Various Tissues of the Rat[a]

Tissue	$T_3(T_3)$[b]	$T_3(T_4)$[c]	Nuclear occupancy (%)
Liver	75	25	50
Kidney	85	15	50
Pituitary	50	50	80
Cerebellum	50	50	60
Cerebral cortex	20	80	90

[a]Modified from Silva (263).
[b]Tissue T_3 from plasma T_3.
[c]Tissue T_3 from plasma T_4.

In pregnancy or any other condition that causes a rise in estrogen levels (such as many oral contraceptive pills), the level of TBG in serum rises. This causes the total circulating thyroid hormones, especially T_4, to increase to maintain free-hormone levels within the normal range (85,86,109).

In the neonate, the abrupt increase in serum T_3 and T_4 immediately after birth is followed by a sustained elevation of both T_3 and T_4 levels throughout childhood, with a slow decline to adult levels during sexual maturity (66,82,222). In elderly people there is a gradual decline in total serum T_3 levels with age without any obvious changes in total T_4 or TBG concentrations (82). The mechanism for this is not well understood but may be related to a decline in tissue outer-ring deiodinase enzyme activity with a reduction in conversion of T_4 to T_3.

F. Biological Effects of Thyroid Hormones

The biological effects of thyroid hormones vary with different species (e.g., metamorphosis in tadpoles; 230), as well as at different stages of development within a species (fetal versus adult life; 139) and among different tissues (liver versus brain; 263). In the fetus, neonate, and child, thyroid hormones exert a major influence on cellular differentiation, growth, and development (257). This is probably mediated by affecting gene expression and manifested via increased synthesis of new proteins and enzymes or activation of existing enzymes (215,285). Thyroid hormones also induce stimulation of oxygen consumption and the basal metabolic rate in tissues. These effects may be related to new synthesis as well as activation of ouabain-sensitive Na^+,K^+-ATPase in the cell membrane. This effect seems to be more important in certain tissues (i.e., liver, kidney, and muscle) and less in others (i.e., spleen, testes, and adult brain). Thyroid hormones cause a reduction in amino acid efflux from cells. Certain proteins appear to be more specifically regulated by thyroid hormones, including those associated with skin epidermis and hair production and also cartilage metabolism. The conversion of carotene to vitamin A appears to be regulated by thyroid hormones, and they also have an effect on stimulation of erythropoietin-mediated erythropoiesis (52).

There appears to be a close link in the interaction between thyroid hormones and other hormones, especially the catecholamines. Their actions often appear synergistic within cells. Research indicates that cell surface receptors for these hormones may be closely linked, such that when one receptor is occupied the other membrane receptor will have an enhanced sensitivity (9,10).

V. EFFECTS OF IODINE DEFICIENCY IN HUMANS

The best known effect of iodine deficiency is endemic goiter. Goiter is a swelling of the thyroid gland that was well known to the ancient world and has continued to excite interest over the centuries. Iodine deficiency is the major primary etiological factor in endemic goiter.

Extensive reviews of its global geographical occurrence have been published, notably by Kelly and Snedden in the 1960 WHO monograph (54), and also in the 1980 monograph (272).

Iodine deficiency is demonstrated by determination of urine iodine excretion using either 24-hr samples or more conveniently casual samples with determination of iodine content per gram of creatinine. This question is further discussed in Section XI. Most of the data have become available since 1960; in general, in endemic goiter areas the intake is well below 100 μg/day and goiter is usually seen when the level is <50 μg/day (236). The rate increases as the iodine excretion falls, so that goiter may be almost universal at levels <10 μg/day. As already noted, the iodine content of drinking water is also low in association with endemic goiter (151).

Goiter arises from causes other than iodine deficiency; these causes are reviewed in Section X. In general they are usually of secondary importance to iodine deficiency as etiological factors in endemic goiter.

Iodine depletion causes depletion of thyroid iodine stores with reduced daily production of thyroid hormone (T_4). A fall in the blood level of T_4 triggers the secretion of increased amounts of pituitary TSH, which increases thyroid activity with hyperplasia of the thyroid. An increased efficiency of the thyroid iodide pump occurs with faster turnover of thyroid iodine. This can be demonstrated by an increased thyroidal uptake of radioactive iostopes ^{131}I and ^{125}I. These features were first demonstrated in the field in the classical observations of Stanbury *et al.* (270) in the Andes in Argentina. (See also Sections III and IV.)

Chronic severe iodine deficiency is associated with gross thyroid hyperplasia and goiter. Chronicity may also be associated with nodule formation, which is more characteristic of goiter in older age groups. Nodules probably reflect alternating stimulation and involution of the gland reacting heterogeneously to a growth stimulus (22).

Sequestration of iodine occurs in chronic endemic goiter, with transfer of iodine into pools not available for hormone secretion. There are quite large quantities of long half-life iodocompounds of low molecular weight that are resistant to enzyme digestion. They may be fragments of normal thyroglobulin (22,279).

These changes have been confirmed in experimental studies of iodine deficiency in animals (see Section VI).

Apart from goiter itself, subsequent work on the effects of iodine deficiency in humans has revealed a great variety of effects on human growth and development. In the light of this knowledge it has been suggested that the term iodine deficiency disorders (IDD) be used instead of "goiter" to denote the effects of iodine deficiency (133).

Iodine deficiency disorders are best described in relation to four different phases of life (Table II).

Table II
Iodine Deficiency Disorders (IDD)

Age group	Disorder	Clinical features
Fetus	Abortions	
	Stillbirths	
	Congenital anomalies	
	Increased perinatal mortality	
	Increased infant mortality	
	Neurological cretinism	Mental deficiency
		Deaf-mutism
		Spastic diplegia
		Squint
	Myxedematous cretinism	Dwarfism
		Mental deficiency
	Psychomotor defects	
	Fetal hypothyroidism	
Neonate	Neonatal hypothyroidism	
	Neonatal goiter	
Child and adolescent	Juvenile hypothyroidism	
	Goiter	
	Impaired mental function	
	Retarded development	
	Cretinism (myxedematous; neurological)	Mental deficiency
		Dwarfism
Adult	Goiter with its complications	
	Hypothyroidism	
	Impaired mental function	

Table III
Comparative Clinical Features in Neurological and Hypothyroid Cretinism[a]

Clinical feature	Neurological cretin	Hypothyroid cretin
Mental retardation	Present, often severe	Present, less severe
Deaf-mutism	Usually present	Absent
Cerebral diplegia	Often present	Absent
Stature	Usually normal	Severe growth retardation usual
General features	No physical signs of hypothyroidism	Course dry skin, husky voice
Reflexes	Excessively brisk	Delayed relaxation
ECG	Normal	Small-voltage QRS complexes and other abnormalities of hypothyroidism
Appearance of limbs on X-ray film	Normal	Epiphyseal dysgenesis
Effect of thyroid hormones	No effect	Improvement

[a]Modified from Hetzel and Potter (137).

A. Iodine Deficiency in the Fetus

Iodine deficiency of the fetus is the result of iodine deficiency in the mother. The condition is associated with a greater incidence of stillbirths, abortions, and congenital abnormalities, which can be reduced by iodization (193,225).

Another major effect of fetal iodine deficiency is the condition of endemic cretinism (226,238). This condition, which occurs with an iodine intake of <20 µg/day in contrast to a normal intake of 80 to 150 µg/day, is still widely prevalent, affecting for example up to 10% of the populations living in severely iodine-deficient areas in India (218), Indonesia (80), and China (173). In its most common form, it is characterized by mental deficiency, deaf-mutism, and spastic diplegia, which is referred to as the neurological type, in contrast to the less common myxedematous type characterized by hypothyroidism with dwarfism.

These two conditions were first described in modern medical literature by McCarrison in 1908 (188), and the differences are summarized in Table III. The condition still exists in the same areas of the Karakoram Mountains and the Himalayas (218). Neurological, myxedematous, and mixed types occur in the Hetian District of Sinkiang, China (Fig. 5) some 300 km east of Gilgit, where McCarrison made his original observations (173). In both China and India, the

Fig. 5. Photograph of a dwarfed cretin woman together with a barefoot doctor of the same age from the Hetian District in Sinkiang, China. Courtesy of Dr. Ma Tai of Tianjin, China. Reprinted with permission from the Editor of the *Lancet*(133).

condition occurs most frequently below the mountain slopes in the fertile silt plains that have been leached of iodine by snow waters and glaciation.

Apart from its prevalence in Asia and Oceania (Papua New Guinea), it also occurs in Africa (Zaire) and in South America in the Andean region (Ecuador, Peru, Bolivia, and Argentina) (226). In all these situations, with the exception of Zaire, neurological features are predominant (226). In Zaire the myxedematous form is more common, possibly because of the high intake of cassava (90).

The common form of endemic cretinism is not usually associated with severe clinical hypothyroidism as in the case of the so-called sporadic cretinism, although mixed forms with both the neurological and myxedematous features do occur. However, the neurological features are not reversed by the administration of thyroid hormones as they are with hypothyroidism (94).

The apparent spontaneous disappearance of endemic cretinism in southern Europe raised considerable doubts as to the relation of iodine deficiency to the condition. Such a spontaneous disappearance without iodization was noted by Costa *et al.* (67) in northern Italy and by Konig and Veraguth (159) in Switzerland. In Yugoslavia, Ramzin *et al.* (240) noted a fall in cretinism from 13% before 1930 to 7% following economic development, and then a disappearance following iodine prophylaxis after 1954.

In Switzerland, Wespi (315) reported a decline in deaf-mutism from 1.2 to 1.7 per 1000 births between 1915 and 1922, to a level of 0.4 per 1000 births in 1925—associated with the introduction of iodized salt (by 1925, 23% of all the salt in Switzerland had been iodized). There was, however, a discrepancy between the decline of deaf-mutism in individual cantons and the degree of salt iodization. For example, in the canton of Bern, deaf-mutism fell from 1.7 per 1000 in 1916 through 1920 to 0.5 per 1000 in 1926 through 1930, although only 4–6% of the salt was iodized. As Trotter (298) has pointed out, no cause–effect relationship between correction of iodine deficiency and the decline of deaf-mutism can be deduced from these observations. There was active social development in Switzerland at the time of iodization, associated with greater social mobility, greater dietary diversification, and improved hygiene, which raised many other possible causal factors (137).

It was in these circumstances that it was decided in 1966 to set up a controlled trial in the western highlands of Papua New Guinea to see whether endemic cretinism could be prevented by iodization. This study, carried out in collaboration with the public health department, was based on the use of iodized oil in a single intramuscular injection of 4 ml of Lipiodol, which provided 2.15 g of iodine. This dose had previously been shown (44) to provide satisfactory correction of severe iodine deficiency for a period of 4.5 years. Iodized oil or saline injections were given to alternate families in the Jimi River District at the time of the first census (1966). Each child born subsequently was examined for evidence of motor retardation, as assessed by the usual milestones of sitting, standing, or walking, and for evidence of deafness. Examination was carried out without knowledge as to whether the mother had received iodized oil injection or saline. Infants showing a full syndrome of hearing and speech abnormalities together with abnormalities of motor development with or without squint were classified as suffering from endemic cretinism. Later follow-up confirmed the diagnoses of cretinism in these cases.

Full details are published (224), and the results of the follow-up are shown in Table IV.

It was concluded that an injection of iodized oil given prior to pregnancy could prevent the occurrence of the neurological syndrome of endemic cretinism in the infant. The occurrence of the syndrome in infants born to women who were already pregnant at the time of oil injection indicated that the damage probably occurred during the first half of pregnancy, possibly in the first trimester.

In the light of later experimental findings (212), it is most likely that this effect is due to absence of transfer of maternal thyroid hormones across the placental barrier and not iodine deficiency itself as originally suggested (224). This possibility is supported by other evidence from Papua New Guinea indicating a relationship between maternal T_4 levels and psychomotor development in the child (227).

Still later studies in Papua New Guinea and Indonesia have demonstrated the existence of a coordination defect in otherwise normal children (28,60) exposed to severe iodine deficiency during fetal development. Lesser degrees of neurological damage are also observed (isolated deaf-mutism and mental deficiency), which probably reflect a less severe fetal iodine deficiency. In China, these less severe forms are called "cretinoids" (173).

B. Iodine Deficiency in the Neonate

The availability of methods for neonatal screening in developed countries (43) has led to their application in developing countries such as India and Zaire. In

Table IV

Children Born in Jimi River Subdistrict, Classified According
to Treatment Received by Mother[a]

Treatment received by mother	Total number of new births	Number of children examined	Number of deaths recorded	Number of endemic cretins
Iodized oil	498	412	66	7[b]
Untreated	534	406	97	26[c]

[a]From Pharoah et al. (224).
[b]Six already pregnant when injected with oil.
[c]Five already pregnant when injected with saline solution.

Table V
Effect of Injection of Iodized Oil Given during Pregnancy[a]

Parameter	Not treated	Treated
Birth weight (g)	2634 ± 552 (98)[b]	2837 ± 542[c] (112)
Perinatal mortality per 1000	188 (123)	98[c] (129)
Infant mortality per 1000	250 (263)	167[c] (252)
Developmental quotient	104 ± 24 (66)	115 ± 16[c] (72)

[a]Modified from Thilly (292).
[b]Number of subjects in parentheses.
[c]Difference significant.

India observations on cord blood in iodine-deficient areas indicate as many as 4% of neonates with serum T_4 levels <2 μg/dl (158). In Zaire up to 10% of neonates have been observed with low levels (89). These frequencies should be compared with 0.02% in developed countries with normal iodine nutrition (43).

In a further study from Zaire the effect of the injection of iodized oil on birth weight, perinatal and infant mortality, and development quotient was assessed by comparison with an untreated group (292). The findings are shown in Table V. They indicate substantial improvements in birth weight, with reductions in perinatal and infant mortality as well as improvement in development quotient in the infants. These findings indicate the necessity of iodine and normal thyroid function for general fetal development and neonatal health.

C. Iodine Deficiency in Childhood and Adolescence

Iodine deficiency in this period is characteristically associated with endemic goiter. The prevalence increases with age, reaching a maximum after the first decade. The condition can be effectively prevented by iodization using various methods following the original demonstration by Marine and Kimball in schoolchildren in Akron, Ohio, in 1921 (185).

Studies from China (306) in children indicate a higher general prevalence of lowered intellectual performance (as measured by IQ and other tests modified for use in China) in iodine-deficient areas compared to areas without iodine deficiency.

A study from a mountain village in Bolivia suggests that improved intelligence in school-age children followed the administration of oral iodized oil in a double-blind study (20). The improvement was related to reduction of goiter and was particularly evident in girls.

Iodization programs have been shown to increase the level of circulating thyroid hormones in children in India (268) and in China (327).

These changes occur whether or not the child is goitrous and indicate a mild degree of hypothyroidism without any apparent symptoms.

As already pointed out (Section IV), the major determinant for brain (and pituitary) T_3 is serum T_4 and not serum T_3 (as is true of the liver, kidneys, and muscle) (70). Low levels of brain T_3 have been demonstrated in the iodine-deficient rat in association with reduced levels of serum T_4, and these have been restored to normal with correction of iodine deficiency (75).

These findings provide a rationale for suboptimal brain function in subjects with endemic goiter and lowered serum T_4 levels, and its improvement following correction of iodine deficiency (95).

D. Iodine Deficiency in Adults

The common effect of iodine deficiency in adults is endemic goiter. Characteristically there is an absence of classical clinical hypothyroidism in adults with endemic goiter. However, laboratory evidence of hypothyroidism with reduced T_4 levels is common. This is often accompanied by normal T_3 levels and raised TSH levels (111,175,221,327).

Iodine administration in the form of iodized salt (327), iodized bread (54), or iodized oil (44) have all been demonstrated to be effective in the prevention of goiter in adults. Iodine administration may also reduce existing goiter in adults. This is particularly true of iodized oil injections. This obvious effect leads to ready acceptance of the measure by people living in iodine-deficient communities.

A rise in circulating T_4 can be readily demonstrated in adult subjects following iodization. As already pointed out, this could mean a rise in brain T_3 levels with improvement in brain function. In northern India a high degree of apathy has been noted in populations living in iodine-deficient areas. This may even affect domestic animals such as dogs. It is apparent that reduced mental function is widely prevalent in iodine-deficient communities without effects on their capacity for initiative and decision making (157).

In general, social effects of iodization are clearly demonstrated by the Chinese village of Jixian (near Jamusi in Heilongjiang Province, northeastern China) (170). In 1978 there were 1313 people with a goiter incidence of 65%, with 11.4% cretins. The cretins included many severe cases, which caused the villagers to be known locally as "the village idiots." The economic development of the village was retarded; for example, no truck driver or teacher was available. Girls from other villages did not want to marry and live in the village. The intelligence of the student population was known to be low: children 10 years of age had a mental development level equivalent to those aged 7.

Iodized salt was introduced in 1978, after which the goiter rate in 1982 had dropped to 4.25%. No cretins had been born since 1978. The attitude of the people had changed greatly: they were much more positive in their approach to life in contrast to their attitude before iodization. The average income had increased from 43 yuan per head in 1981 to 223 yuan in 1982 and 414 yuan in 1984, which was higher than the average caput income in the district. In 1983 cereals were exported for the first time. Before iodization no family had a radio; now 55 families have a TV set. Now 44 girls have come from other villages to marry boys in Jixian. Seven men had joined the People's Liberation Army, whereas they had been rejected before with goiter. All these effects are mainly due to the correction of hypothyroidism by iodized salt.

This section indicates a great variety of effects of iodine deficiency, much more extensive than goiter, even though this familiar feature is the obvious one. The designation of the effects of iodine deficiency by the term goiter is clearly no longer an adequate reflection of current knowledge. The term iodine deficiency disorders (IDD) has been well accepted (see Section XIV). It denotes the range of disorders described in this section, all of which can be prevented by correction of iodine deficiency.

VI. EFFECTS OF IODINE DEFICIENCY IN ANIMALS

The significant role of iodine deficiency in the etiology of endemic goiter has been confirmed by extensive studies in animals. The major morphological and functional abnormalities can be readily reproduced, as originally shown in 1909 by Marine and co-workers (183,184).

Experimental iodine-deficient goiter is usually diffuse, but with long-standing deficiency nodules are seen in the rat that show an increased number of cells and follicles. As Marine originally showed, when the iodine deficiency subsides the goiter becomes of colloid type and does not return to normal.

As summarized by Beckers and Delange (22), subsequent studies have shown that animals subjected to chronic iodine deficiency with intermittent periods of normal iodine intake show

1. Low cellular iodine content with large total iodine stores
2. Heterogeneous activity of the thyroid gland as indicated by iodine kinetic studies
3. Qualitative change in thyroid secretion with an enhanced synthesis of T_3
4. Impaired iodination of thyroglobulin associated with impaired proteolysis and a decreased deiodinating action of iodotyrosines
5. In the goitrous gland, a significant escape of nonhormonal iodine

In the last decade, systematic experimental studies of the effects of iodine deficiency on development, particularly fetal development, have been carried out. The most extensive studies have been done in sheep, but the effects in the marmoset (New World primate) (*Callithrix jacchis jacchis*) have also been studied. These studies have been particularly concerned with fetal brain development because of its relevance to the human problem of endemic cretinism and other forms of brain damage due to fetal iodine deficiency (135) (see Section V). Animal models in the rat have been established in China with diets closely resembling those consumed in endemic areas. These various models are now reviewed. Finally, observations in farm animals are considered.

A. Iodine Deficiency in Sheep

Severe iodine deficiency has been produced in sheep (234) with a low-iodine diet of crushed maize and pelleted pea pollard (8–15 μg iodine per kilogram), which provided 5–8 μg iodine per day. After a period of 5 months, although body weights were maintained, iodine deficiency was evident with the appearance of goiter, low plasma T_4 and T_3 values, elevated TSH levels, and low daily urinary excretion of iodine. Control animals received the same diet but were supplemented with 2 mg sodium iodide administered by subcutaneous injections each week. The ewes were mated with normal fertile rams, dates of conception established, and fetuses delivered at 56, 70, 98, and 140 days' gestation by hysterotomy (234).

Table VI

Effect of Severe Dietary Iodine Deficiency on Maternal and Fetal Thyroid Function in Sheep[a]

Functional parameter		Gestational age			
		56 days	70 days	98 days	140 days
Maternal plasma T_4	ID[b]	37* (5)[c]	17* (6)	15* (5)	19* (7)
(nmol/liter)	Control	126 (3)	134 (4)	141 (5)	137 (3)
Maternal plasma TSH	ID	54	120*	125**	109*
(ng/ml)	Control	6	7	5	11
Fetal thyroid weight (g)	ID	0.01	0.37**	4.30*	12.69*
	Control	0.05	0.08	0.28	0.99
Fetal thyroid iodine (μg)	ID	0.06	1.15***	8.7***	29**
	Control	0.24	8.79	146.9	2636
Fetal plasma T_4 (nmol/liter)	ID	3**	4*	4*	6*
	Control	10	25	125	216
Fetal plasma TSH (ng/ml)	ID	56	165*	170*	211*
	Control	12	11	13	12

[a]Modified from Potter *et al.* (233).
[b]ID, Severe dietary iodine deficiency.
[c]Number of observations shown in parentheses.
[d]*$p < .00$; **$p < .01$; ***$p < .05$ (two-tailed *t* test).

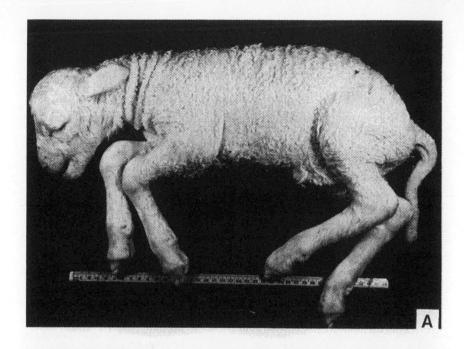

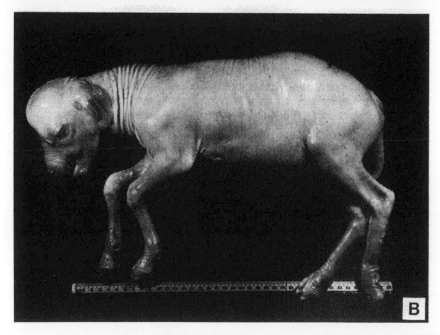

Fig. 6. A 140-day-old control fetal lamb (A) compared to an iodine-deficient fetal lamb (B) of the same gestational age. Note absence of wool coat, subluxation of leg joints, and domelike appearance of cranium.

Goiter was evident from 70 days in the iodine-deficient fetuses, and thyroid histological findings revealed hyperplasia from 56 days' gestation. The increase in thyroid weight was associated with a reduction in fetal thyroid iodine content, reduced plasma T_4 values, and increased plasma TSH (Table VI).

The iodine-deficient fetuses at 140 days were grossly different in physical appearance from the control fetuses (Fig. 6). There was reduced weight, absence of wool growth, goiter, varying degrees of subluxation of the foot joints, and deformation of the skull. There was also delayed bone maturation, as indicated by delayed appearance of epiphyses in the limbs.

There were subnormal brain weight and brain DNA as early as 70 days, indicating a reduction in cell number probably due to slowed neuroblast multiplication, which normally occurs from 40 to 80 days in the sheep (137). Although brain protein was reduced in the deficient fetuses, the ratio of protein to DNA remained unchanged, indicating no significant changes in the size of the brain cells. After 80 days, neuroglial development also appeared to be slowed in iodine deficiency, as both the DNA and protein content were reduced to less than normal in the 98-day and 140-day fetal brains (Table VII).

Retardation of fetal brain development in severe dietary iodine deficiency was revealed also by histological studies at 140 days' gestation (234). Delayed maturation of the cerebellum was shown by reduced migration of cells from the external granular layer to the internal granular layer and increased density of Purkinje cells. The greater density of Purkinje cells indicates a reduction in Purkinje cell arborization within the molecular layer. In the cerebral hemispheres the cells were more densely packed in the motor and visual areas while the

Table VII
Effect of Severe Dietary Iodine Deficiency on Fetal Brain Development in Sheep[a]

Developmental parameter		Gestational age			
		56 days	70 days	98 days	140 days
Body weight (g)	ID[b]	31.7 (5)[c]	101*** (7)	662 (5)	2930** (7)
	Control	32.2 (3)	129 (4)	753 (5)	3820 (6)
Brain weight (g)	ID	1.79	4.20***	19.0	46.4**
	Control	1.68	5.01	22.1	53.8
Cell number (mg DNA)	ID	8.86	14.2***	27.8**	62.6*
	Control	8.37	16.2	32.5	74.5
Cell size	ID	9.31	12.4	25.2	40.3
(protein : DNA)	Control	8.95	12.8	26.6	44.1

[a] Modified from Potter et al. (233).
[b] ID, Severe dietary iodine deficiency.
[c] Number of observations shown in parentheses.
[d] $*p < .001$; $**p < .01$; $***p < .05$ (two-tailed t test).

pyramidal neurons in the hippocampus were denser in the CA1 and CA4 regions, indicating severe retardation in neuropil growth in both subfields (234).

Evidence of retarded myelination in the cerebral hemispheres and brain stem was provided by lowered cholesterol–DNA ratios, and an increased water content in the brain at 140 days was further confirmation of brain retardation in iodine deficiency (234).

The effect of iodine on this retarded fetal brain development due to iodine deficiency has been investigated with a single intramuscular injection of iodized oil containing 500 mg iodine given at 100 days' gestation (137). In injected animals, the difference between iodine-deficient and control fetal brain weights was reduced from 10.8% to 6% by the iodized oil injection. The difference in body weight was also reduced, and maternal and fetal plasma T_4 values were restored to normal (Table VIII).

The effects of iodine deficiency and the iodized oil administration on cerebellum and cerebral hemisphere are summarized in Fig. 7.

The effects of severe iodine deficiency on fetal brain development in the sheep are more severe but similar to those of fetal thyroidectomy carried out at 50 to 60 days or at 98 days. Maternal thyroidectomy carried out some 6 weeks before pregnancy had a significant effect on fetal brain development in midgestation. The combination of maternal thyroidectomy and fetal thyroidectomy at 98 days produces more severe effects than that of iodine deficiency (Fig. 7) (192).

The findings following maternal, fetal, and combined thyroidectomy suggest that the effect of iodine deficiency on fetal brain development is mediated by the combination of reduced maternal and fetal thyroid secretion and not by a direct effect of iodine (138). The effect of reduced maternal secretion occurs in the first half of pregnancy and the effect of reduced fetal secretion in the latter half of pregnancy. The conclusion is consistent with evidence in the rat of the passage of maternal T_4 across the placental barrier early in pregnancy (212).

B. Iodine Deficiency in Marmosets

Severe iodine deficiency has been produced in the marmoset (*Callithrix jacchis jacchis*) with a mixed diet of maize (60%), peas (15%), torula yeast (10%), and dried iodine-deficient mutton (10%) derived from the iodine-deficient sheep produced in the study already described (235). There was a gross reduction in maternal T_4 levels (Table IX), with grossly reduced thyroid iodine. After a year on the diet the animals were allowed to become pregnant, and the newborn animals were studied following the first pregnancy and then again following the second pregnancy. The effects on brain development are summarized in Table IX. Significant effects were apparent in the first pregnancy, with more striking effects apparent in the second pregnancy.

There are other evidence of hypothyroidism in the form of impaired hair

Table VIII

Effect of Iodized Oil Given to the Iodine-Deficient Ewe at 100 Days' Gestation on Fetal Development as Assessed at 140 Days[a,b]

Experimental group		Fetal thyroid iodine (μg)	Plasma T$_4$ (nmol/liter)		Body weight (kg)	Total brain		
			Maternal	Fetal		Weight (g)	Cell number (mg DNA)	Cell size (protein : DNA)
Control ($n = 10$)	Mean	1,758[ab]	126[a]	203[a]	3.66[abc]	53.88[ab]	73.50[ab]	45.05[a]
	SEM	±458	±8	±13	±0.15	±0.93	±1.65	±1.15
Iodine deficient ($n = 12$)	Mean	38[acd]	17[acd]	18[acd]	2.73[ad]	48.05[a]	62.04[ac]	41.93[acd]
	SEM	±7	±4	±6	±0.12	±0.79	±1.40	±0.66
Iodized oil given at 100 days ($n = 8$)	Mean	10,434[bc]	155[c]	208[c]	3.23[bd]	50.67[b]	72.32[c]	44.35[c]
	SEM	±2,782	±14	±30	±0.11	±1.22	±2.82	±0.95

[a]Modified from Hetzel and Potter (137).

[b]Values within a column with the same superscript differ significantly ($p < .05$, two-tailed t test).

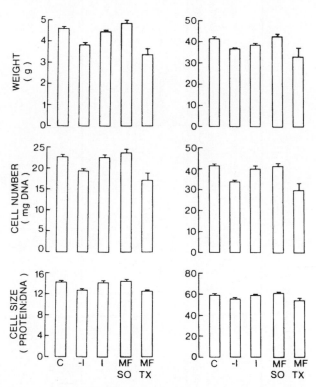

Fig. 7. Comparison of brains of control (C), iodine-deficient (−I), iodine at 100 days (I), maternal + fetal sham-operated (MFSO), and thyroidectomized (MFTX) sheep fetuses at 140 days' gestation. Reprinted with permission from the Editor of the *Lancet* (133).

growth and some skull deformity, but there were no striking effects on epiphyseal development. In general the findings in this primate resemble those in the sheep.

C. Iodine Deficiency in Rats

Studies in rats using diets consumed in two endemic areas in China have been carried out. In both instances fetal hypothyroidism has been produced.

The most extensive studies have been carried out using the diet consumed by the people of Jixian village (near Jamusi) in Heilongjiang Province by Li *et al.* (168). This village, as already noted (Section V), was severely iodine deficient with an endemic cretin rate of 11%. The diet included available main crops

Table IX

Effect of Iodine Deficiency on Fetal Brain Development in the Marmoset[a,b]

		Plasma T_4 (nmol/liter)		Brain		
Experimental group		Maternal	Fetal	Weight (g)	Cell number (mg DNA)	Cell size (protein : DNA)
Control (n = 6)	Mean	182[aaa,bb]	>400[aaa,bbb]	3.4[bb]	7.90[ab]	31.3[bb]
	SEM	±15		±0.08	±0.10	±0.7
Iodine deficiency (first pregnancy)	Mean	16[aaa]	75[aaa]	3.50[ccc]	7.40[a]	30.9[c]
(n = 8)	SEM	±3	±17	±0.06	±0.15	±0.9
Iodine deficiency (second pregnancy)	Mean	11[bb]	40[bbb]	2.97[bb,ccc]	7.17[b]	27.8[bb,c]
(n = 4)	SEM	±3	±8	±0.11	±0.28	±0.7

[a]Modified from Potter et al. (235).

[b]Values within a column with the same superscript differ significantly. Single letters, $p < .05$; double letters, $p < .01$; triple letters, $p < .001$.

(maize, wheat), vegetables, and water (1.0 μg iodine per liter) from the area with an overall iodine content of 4.5 μg/kg. The rats were actually housed and fed in a specially constructed animal laboratory in the village. Control animals received the same diet with the addition of iodine, providing 54.7 μg/kg. Special attention was paid to the fetal thyroid and brain during gestation (16–20 days) and birth (1–60 days). After the mother had received the diet for 4 months, There was obvious neonatal goiter (T_4 3.6 μg/dl, compared to controls 10.4 μg/dl, higher ^{125}I uptake, with reduced brain weight. The density of brain cells was increased in the cerebral hemispheres. The cerebellum showed delayed disappearance of the external granular layer with reduced incorporation of [³H]leucine in comparison to the control group.

Similar effects have been reported with the iodine-deficient diet (mainly rice) being consumed in South China in Guijou Province (326a).

Extensive studies have also been carried out in the mouse using the iodine-deficient diet being consumed in Inner Mongolia (Chemeng County). There were definite effects on learning capacity (174) observable in >10 generations.

So far in these Chinese models, the effects observed, as in the sheep and the marmoset, are those of fetal hypothyroidism. Despite the predominant occurrences of neurological cretinism in the two endemic areas studied in northeastern and southern China, this condition has not yet been reproduced in the rat model. This may yet occur in succeeding generations. However, the question raised is whether there are some other factors in the environment (presumably dietary) that together with iodine deficiency produce the condition.

D. Iodine Deficiency in Farm Animals

Reproductive failure is often the oustanding manifestation of iodine deficiency and consequent impairment of thyroid activity in farm animals. The birth of weak, dead, or hairless young in breeding stock has long been recognized in goitrous arcas (91). Fetal development may be arrested at any stage, leading either to early death and resorption, abortion and stillbirth, or the live birth of weak young, often associated with prolonged gestation and parturition and retention of fetal membranes (4,207). Allcroft et al. (4) demonstrated subnormal serum PBI levels in herds showing a high incidence of aborted, stillborn, and weakly calves. Falconer (92) has shown that thyroidectomy of ewes some months before conception severely reduced both the prenatal and postnatal viability of the lambs, despite the presence of an apparently adequate thyroid in the lamb itself. Neonatal mortality in lambs from ewes fed goitrogenic kale, responsive to iodine administration during pregancy, has frequently been observed (6,258).

In addition to the reproductive disturbances just described, irregular or suppressed estrus in dairy cattle causing infertility has been associated with goiter and shown to respond to iodine therapy. Moberg (204), working with 190 herds totaling 1572 cows in goitrous areas in Finland, obtained a significant improvement from iodine therapy in first-service conception rate and in the number of cows with irregular estrus incidence. McDonald *et al.* (191), working in an iodine-deficient area in Canada, also obtained a marked improvement in first-service conception rate by feeding an organic iodine preparation, beginning 8–10 days before the cows came into estrus.

The thyroid gland plays an equally important part in the maintenance of male fertility (181). A decline in libido and a deterioration in semen quality have been associated with iodine deficiency in bulls and stallions, and a seasonal decline in semen quality in rams has been related to a mild hypothyroid state. This condition is responsive in part to doses of thyroactive proteins, but not to iodine except in known iodine-deficient areas.

VII. IODINE REQUIREMENTS

Calculations based on average daily losses of iodine in the urine give an adult human daily requirement of 100 to 200 μg (72), while the results of balance studies indicate that equilibrium or positive balance can be achieved at daily intakes ranging from 44 to 162 μg (58,251). From an assessment of many studies Elmer (88) placed the daily requirements of humans at 100 to 200 μg, and Wayne *et al.* (308) state that 160 μg/day is "the minimum certainly safe amount of iodine which must be available in the individual's diet if iodine deficiency goiter is to be avoided." They suggested that it might be advisable to raise this figure to 200 μg for children and pregnant women.

In 1980 the Food and Nutrition Board, National Academy of Sciences, National Research Council, United States, recommended a daily iodine intake of 40 μg for children aged 0–6 months, 50 μg from 6 to 12 months, 70–120 μg from 1 to 10 years, and 120–150 μg from 11 years onward. The recommended rates during pregnancy and lactation were, respectively, 175 and 200 μg. The recommendations applied equally to both sexes. (241).

The daily nutritional iodine requirement of the rat has been given as 0.7 to 0.9 μg on the basis of whole-body iodine turnover rates (129). This would give a requirement of 20 to 40 μg iodine per 1000 kcal of food consumed. If this reasoning is applied to human adults consuming 2800 kcal daily, the daily iodine requirement would be 56–112 μg.

Mitchell and McClure (202) have calculated the minimum iodine requirements of different classes of farm animals on the basis of their heat production rather than their energy production (Table X).

These estimates, except for the cow where they are much lower, compare well with the minimum intakes given by Orr and Leitch (216) for those species in

Table X
Iodine Requirements of Different Farm Animals[a]

Animal	Body weight (lb)	Heat production (kcal)	Iodine requirements (µg/day)
Poultry	5	225	5–9
Sheep	110	2,500	50–100
Pigs	150	4,000	80–160
Cow, in milk (40 lb/day)	1000	20,000	400–800

[a]From Underwood (298a).

nongoitrous areas. Hartmans (128) cites 0.6 mg/kg of ration as the iodine requirement of high-yielding cows, with a tolerance of 0.4 mg iodine per kilogram for short periods. A requirement of 0.8 mg iodine per kilogram for pregnant and lactating livestock and 0.12 mg/kg for other animals in the feed dry matter is given by the Agricultural Research Council of Great Britain (210). Normal growth in chickens has been reported on diets as low as 0.07 mg iodine per kilogram, although 0.3 mg/kg was required for completely normal thyroid structure (71). The dietary iodine requirements of pheasants and quail are no greater than 0.3 mg/kg, either for growth or the development of normal thyroid glands (252). The results of experiments with growing pigs suggest that their iodine requirement is not greater than 0.14 mg/kg of the diet, and on corn–soybean meal rations approximately 0.086–0.132 mg/kg (262).

VIII. SOURCES OF IODINE

A. Iodine in Water

The level of iodine in the drinking water reflects the iodine content of the rocks and soils of a region and hence of the locally grown foods or feeds. Four studies in widely separated areas illustrate this relationship. Kupzis (162) found the water iodine content in goitrous areas in Latvia to range from 0.1 to 2.0 µg/liter, compared with 2 to 15 µg/liter for nongoitrous areas. Young et al. (326) reported that the drinking water iodine level in English villages with a goiter incidence assessed at 56% averaged 2.9 µg/liter, compared with 8.2 µg/liter in other villages where the goiter incidence was only 3%. Karmarkar et al. (151) found the iodine content of water from goitrous areas in India, Nepal, and Ceylon to range from 0.1 to 1.2 µg/liter, compared with a nongoitrous area in Delhi of 9.0 µg/liter. In a study of goiter incidence in Egyptian oases, the iodine content of waters from goitrous villages ranged from 7 to 18 µg/liter, compared with the very high levels of 44 and 100 μg/liter in two samples from villages with no goiter (56). These last figures indicate that in some circumstances the water supply can contribute significant amounts of iodine to the daily diet, although in most areas the proportion of the total intake from this source is very low.

B. Iodine in Human Foods and Dietaries

The iodine concentrations in human foods of all types are exceedingly variable, mainly because of differences in the content and availability of iodine in the soil, and in the amount and nature of the fertilizers applied. Chilean nitrate of soda, the only mineral fertilizer naturally rich in iodine (203), can double or triple the iodine content of food crops (and pastures) when applied in the amounts required to meet their nitrogen needs. Gurevich (120) showed that the iodine level in a range of vegetables and cereals were increased 10–100 times or more by applications of seaweed and by-products of the fish, crab, and whale processing industries. Barakat *et al.* (18), in a study of iodine in the edible portion of 15 vegetables, found large variations between samples of the same vegetables, presumably reflecting differences in the availability of iodine in the soils. Iodine in cabbage, for example, was reported to range from 0 to 0.95 μg/g.

Similarly, animal products such as milk or eggs are much richer in iodine when they come from animals that have consumed iodine-enriched rations than from animals not so treated. The magnitude of this effect on cow's milk was considered in Section II. On average rations, hen's eggs contain 4–10 μg iodine, most of which is located in the yolk (182). Feeding the hens large amounts of iodine either as iodized salt, sodium iodide, or seaweed can raise this amount 100-fold (118,119,148) or 1000-fold (182). The significance to humans of adventitious iodine sources, such as iodophor antiseptics and iodates in bread, is considered below.

Foods of marine origin are so much richer in iodine than any other class of normal foodstuff that differences among other foods are usually of relatively minor importance. The edible flesh of sea fish and shellfish may contain 300–3000 μg/kg of iodine on the fresh basis, compared with 20 to 40 μg/kg for freshwater fish (146). The iodine content of composites of food categories in the United States taken from the work of Vought and London (302) is given in Table XI. Highly refined products such as sugar contain negligible amounts of iodine. Hamiltion and Minski (122) found Barbados brown sugar to contain 30 μg iodine per kilogram, compared with <1 μg/kg in refined white sugar. These workers also reported a mean daily iodine intake of 220 ± 51 μg from English adult diets. In 1969 the estimated average daily intakes for the United States ranged from 238 μg in the Northeast to 738 μg in the Southwest (213). The daily dietary iodine intakes by Japanese people are suggested as ~300 μg, with a basic diet contain-

Table XI
Iodine Content of Composites of Food Categories[a]

| Food category | Number of samples | Iodine (μg/kg wet wt.) | |
		Mean ± SE	Median
Seafoods	7	660 ± 180	540
Vegetables	13	320 ± 100	280
Meat products	12	260 ± 70	175
Eggs	11	260 ± 80	145
Dairy products	18	130 ± 10	139
Bread and cereal	18	100 ± 20	105
Fruits	18	40 ± 20	18

[a]From Vought and London (302).

ing a minimum of seaweed, with incidental consumption of seaweed to a maximum of 10 mg/day (325).

Overall iodine intakes are determined more by the source of foods composing dietaries than by the choice or proportion of different foods, except for those of marine origin. Residents of an endemic goiter area can only obtain sufficient iodine for their needs by the consumption of substantial amounts of marine foods, or foods imported from iodine-rich areas elsewhere, or by the use of iodized salt, bread, or some other form of iodine-fortified material. The consumption of seaweeds, which may contain as much as 0.4 to 0.6% iodine on the dry basis (145), is believed to be one reason for the low incidence of goiter in Japan. The supply of sea salt thrown up in the spray into the country is suggested as a further factor contributing to a relatively high iodine content in Japanese-grown foods (325).

C. Iodine in Animal Feeds

The iodine levels in pasture species were reported in earlier studies to range between 300 and 1500 μg/kg on the dry basis (119,216). Hartmans (128), in a study of the factors affecting herbage iodine content in the Netherlands, found that dicotyledonous species had up to 14 times higher iodine contents than grasses, while the iodine content of grass species varied over a two-fold range. Samples of white clover (*Trifolium repens*) ranged from 160 to 180 μg/kg and pasture grasses from 60 to 140 μg/kg (dry basis). The mean levels in a range of Welsh pasture grasses ranged from 200 to 310 μg/kg (2). Roughages are usually substantially higher in iodine than cereal grains, or the oilseed meals commonly employed as protein supplements in farm rations (146). Protein concentrates of animal origin, other than fish meal, cannot be relied on as significant sources of dietary iodine, unless the animals from which they are obtained are ingesting exceptionally large amounts of the element.

D. Adventitious Sources of Iodine

Iodophors are also used as antiseptic cleansing agents for teat dips, udder washes, milking machines, and tanks for handling, storing, and transportation of milk and milk products. Iodophor solutions for sanitizing equipment release iodine in concentrations of 0.001 to 0.0025%. The use of iodophors was considered to be a major factor in the occurrence of iodine-induced thyrotoxicosis in Tasmania (see Section XII).

Another widely used agent, ethylenediaminedihydroiodide, is given in daily doses of 50 and 200 mg per cow for the prevention and therapy of foot rot and soft tissue lumpy jaw, respectively. Its iodine content is 5 mg/dl. ^{131}I-labeled ethylenediaminedihydroiodide is absorbed equally but retained better in cartilage and soft tissue than sodium iodide. The iodide content of milk during administration of 200 mg of this compound per day was 1559 ± 771 μg/liter (98).

Another important artificial source of iodine is the iodine-containing coloring additive 2,4,5,7-tetraiodofluorescein (erythrosine) to drugs, beverages, foods, and cosmetics. It contains 58% iodine. In the gastrointestinal tract of the rat erythrosine is partially digested, and about 10% iodine becomes available for utilization by the thyroid gland (300,304). The metabolism of erythrosine is probably similar in humans and rats. Many preparations in multivitamins, minerals, and antacids are coated and colored red with erythrosine. The multivitamin capsules may contain 4–625 μg of iodine, and 233 μg of iodine may be present in one antacid tablet (300,304). Numerous beverages, fruits, and maraschino cherries are colored with erythrosine. Erythrosine also serves for coloring some cake mixes and cereals. Some brands of dry cereal contain in 20 g as much as 850 μg of iodine in the form of erythrosine (300).

It is clear that the iodine content of bread, meat, milk, and eggs and their products from different regions and countries will depend on the natural iodine content as well as increasingly on iodine supplements in the form of iodized salt, iodine-enriched animal feeds, iodine-containing veterinary medicinals, sanitizing agents, and coloring substances (300).

The results of detailed studies of the iodine content of the American (220) and British (314) diets have been published. The data indicate contents far in excess of RDAs mostly because of adventitious sources of iodine. The studies in Britain revealed a daily intake of 92 μg from liquid milk, meat and meat products 36 μg, and cereal products 31 μg. Fish and fish products, though rich in iodine, contributed only 5% to the total intake. In the United States, figures from the "market basket" survey in 1978 were for milk 534 μg/day, grain and cereal products 152 μg/day, and meat, fish, and poultry 103 μg/day (220).

IX. IODINE TOXICITY

Iodine toxicity has been critically studied in humans, laboratory species, poultry, pigs, and cattle. Wolff (319) has defined four degrees of iodide excess in humans:

1. Relatively low levels that lead to temporary increases in the absolute iodine uptake by the thyroid and the formation of organic iodine, until such time as the thyroid is required to reduce iodide clearances
2. A larger amount that can inhibit iodine release from the thyrotoxic human thyroid or from thyroids in which iodine release has been accelerated by TSH
3. A slightly greater intake that leads to inhibition of organic iodine formation and that probably causes iodide goiter (the so-called Wolff–Chaikoff effect)
4. Very high levels of iodide that saturate the active-transport mechanism for this ion; the acute pharmacological effects of iodide can usually be demonstrated before saturation becomes significant

This worker has also suggested that human iodine intakes of 2000 μg/day should be regarded as an excessive or potentially harmful level of intake. Normal diets composed of natural foods are unlikely to supply as much iodine as 2000 μg/day and most would supply <1000 μg/day, except where the diets are exceptionally high in marine fish or seaweed, or where foods are contaminated with iodine from adventitious sources. Inhabitants of the coastal regions of Hokkaido, the northern island of Japan, whose diets contain large amounts of seaweed, have astonishingly high iodine intakes amounting to 50,000 to 80,000 μg/day (282). Urinary iodine excretion in five patients exhibiting clinical signs of iodide goiter exceeded 20 mg/day or ~100 times normal. Similar findings have been reported from two Chinese villages on the Yellow Sea coast in association with the consumption of large amounts of kelp (328).

Connolly (61) showed that the use of iodophor antiseptics in milking machines, storage vats, and bulk milk tankers leads to a marked increase in the iodine content of milk, ice cream, and confections made from dairy products. The iodine concentration of milk from four areas where iodophors were not used ranged from 13 to 23 μg/liter, compared with 113 to 346 μg/liter in milk from five areas where iodophor bactericidal agents were used. The use of potassium iodate instead of bromate as a dough conditioner or "improver" in bread making represents a further abnormal source of iodine to humans. Where these two sources of extra iodine were concurrently present, as in parts of Tasmania, a marked increase in the incidence of thyrotoxicosis occurred, mainly in women in the 40- to 80-year age group with preexisting goiter (Section XII). The normal thyroid is tolerant of iodine intakes well above those provided by most diets, but it is clear that toxic or potentially toxic intakes can occur as just described, where abnormal amounts are included as a consequence of various technological developments involving industrial use of the element.

Large doses of stable iodine are known to reduce radioiodine uptake by the thyroid and decrease retention of this radionuclide, an effect of particular interest with respect to protection against radioactive fallout involving ^{131}I. Driever et al. (81) found that administration to calves of stable potassium iodide decreased whole-body retention of ^{131}I given subsequently by about one-half in 23 days.

Significant species differences exist in tolerance to high iodine intakes. In all species studied the tolerance is high (i.e., relative to normal dietary iodine intakes), pointing to a wide margin of safety for this element. Thus adult female rats fed 500, 1000, 1500, and 2000 mg/kg of iodine as potassium iodide from 0 to 35 days prepartum revealed increasing neonatal mortality of the young with increasing levels of iodine, but the effects of the lowest level of supplemental iodine fed (500 mg/kg) were slight when compared with those receiving no supplemental iodine (5). Examination of mammary gland tissue from females fed iodine indicated that milk secretion was absent or markedly reduced. The fertility of male rats fed 2500 mg iodine per kilogram from birth to 200 days of age appeared to be unimpaired (5). In subsequent studies with rats, rabbits, hamsters, and pigs (11), rabbits fed ≥250 mg iodine per kilogram for 2 to 5 days in late gestation showed significantly higher mortality of the young than controls receiving no supplemental iodine. Hamsters were unaffected except for a slightly reduced feed intake and a decreased weaning weight of the young.

In a series of experiments with poultry, Arrington and co-workers (11,12,223) demonstrated profound effects on egg production and hatchability. When laying hens were fed 312–5000 mg iodine per kilogram as potassium iodide in a practical laying ration, egg production ceased within the first week at the highest level and was reduced at the lower levels (223). The fertility of the eggs produced was not affected, but early embryonic death, reduced hatchability, and delayed hatching resulted. Within 7 days after cessation of iodine feeding the hens resumed egg production, indicating that the adverse effects of the excess iodine are only temporary. Similar results were obtained subsequently with pullets and hens fed supplementary iodine in amounts ranging from 625 to 5000 mg/kg for 6 weeks, but the effects were much smaller for sexually mature pullets than for the mature hens (12).

The mechanism by which excess iodine affects egg production and embryonic mortality, or reproduction in female rats and rabbits, is not understood. Preliminary experiments conducted by Marcilese et al. (182) indicate that T_4 production is not impaired in hens fed high iodine levels. However, the growing ova were shown to have a marked ability to concentrate iodine from the high doses administered, a finding in keeping with the earlier demonstration of the specific incorporation of orally administered radioiodine into hens' eggs and follicles

(182,295). Thus iodine in the eggs of hens given 100 mg iodine daily as sodium iodide increased linearly for 10 days and reached a plateau of 3 mg per egg at that time. When hens were given 500 mg iodine daily, the level in the eggs increased rapidly to an average of 7 mg iodine per egg by 8 days, at which time most hens ceased production. Ova continued to develop in hens not laying, and many ova were found to be regressing (182). It was suggested that when a threshold amount of iodine reaches the ova, development ceases and regression takes place.

The minimum toxic iodine intakes, as calcium iodate, for calves (80–112 kg body weight) were shown by Newton *et al.* (208,209) to be close to 50 mg/kg, although some experimental animals were adversely affected at lower levels. At 50, 100, and 200 mg iodine per kilogram, weight gain and feed intake were depressed and signs of toxicity, including coughing and profuse nasal discharge, became evident. At the higher iodine levels, heavier thyroid and adrenal glands were apparent at the end of the 104- to 112-day treatment periods, and blood hemoglobin levels were slightly depressed. With milking cows, no signs of toxicity were observed at iodine intakes of 50 mg/kg of the diet (as sodium or potassium iodide, or as ethylenediaminedihydroiodide), although extremely high iodine levels were present in the blood serum and in milk, urine, and feces (200,209).

Pigs are much more tolerant of excess iodine than cattle. Newton and Clawson (208), in experiments with growing–finishing swine, fed graded increments of calcium iodate to give iodine additions from 10 to 1600 mg/kg of the diet, found the minimum toxic level to lie between 400 and 800 mg/kg. Growth rate, feed intake, hemoglobin levels, and liver iron concentrations were depressed at 800 to 1600 mg iodine per kilogram. Liver iron levels were also significantly depressed at 400 mg/kg, suggesting that the minimum toxic iodine level over extended periods could be <400 mg/kg. The interaction between iron and iodine demonstrated in these studies with pigs is of particular interest. The effects of the elevated iodine intakes on growth, feed intake, and hemoglobin levels were found to be reduced by supplementary iron, whether given orally or intramuscularly.

X. GOITROGENIC SUBSTANCES

Goitrogens are substances capable of producing thyroid enlargement by interfering with thyroid hormone synthesis. The pituitary responds by increasing its output of TSH, which induces hypertrophy in the gland in an effort to increase thyroid hormone production. The extent to which the increased thyroid tissue mass compensates for the inhibition or blocking of thyroid hormonogenesis depends on the dose of goitrogen and, in some circumstances, on the level of iodine intake by the animal.

The first clear evidence of a goitrogen in food was obtained by Chesney and co-workers in 1928 (46). Rabbits fed a diet of fresh cabbage developed goiters that could be prevented by a supplement of 7.5 mg iodine per rabbit per week. "Cabbage goiter" was subsequently demonstrated in other animal species (30), and goitrogenic activity was reported for a wide range of vegetable foods, including virtually all cruciferous plants (16,30,269). About this time workers in New Zealand showed that rapeseed goiter arises from interference with the process of thyroid hormonogenesis and, unlike cabbage goiter, is only partially controlled by supplemental iodine (115,116,237). Astwood and co-workers (17,113,114) subsequently isolated and identified a new compound named goitrin (L-5-vinyl-2-thiooxalidone) from rutabagas and showed that the goitrogenic activity of *Brassica* seeds could largely be accounted for by the presence of this compound in combined form (progoitrin). Plants containing goitrins include cabbage, kale, brussels sprouts, cauliflower, broccoli, kohlrabi, turnips, metabagas, white and black mustard, rape and rapeseed, horseradish, and garden cress. Small aliphatic disulfides, the major volatile components of onions and garlic, have also been identified as having marked antithyroid effects. Most of these foods, however, are not consumed in sufficient quantities by any population to justify a major role to them in the etiology of endemic goiter in humans (16).

Bourdoux *et al.* (32–34) and Delange *et al.* (77,78) have identified other food families that are even more goitrogenic. These include staple foods from the Third World such as cassava, maize, bamboo shoots, sweet potatoes, lima beans, and millets. These foods contain cyanogenic glucosides, which are capable of liberating large quantities of cyanide by hydrolysis. Not only is the cyanide toxic, but the metabolite is predominantly SCN. With the exception of cassava, these glycosides are located in the inedible portions of the plants, or if in the edible portion, in small quantities that they do not cause a major problem. Cassava, on the other hand, is cultivated extensively in developing countries and represents an essential source of calories for more than 200 million people living in the tropics (69). The role of cassava with iodine deficiency in the etiology of endemic goiter has now been established by Delange (76) from their studies in Zaire and in Sicily. These observations have also been confirmed by Maberly *et al.* (178) in Sarawak, Malaysia.

Abnormalities in thyroid metabolism induced by SCN from cassava include a decrease in thyroidal iodine reserves, alteration in intrathyroidal hormone production (increased MIT : DIT ratio) and decreased T_4 synthesis, a drop in serum T_4, an elevation in TSH, and an increase in thyroid size. At higher concentrations of SCN (5 mg/day), an increased renal iodine wastage could be demonstrated (33). It could be concluded that with chronic cassava consumption, development of goiter is critically related to the balance between supplies of iodine and SCN. Endemic goiter was found to develop when the urinary iodine–SCN ratio (in micrograms per milligram) was below a critical threshold of 3 and became hyperendemic when this ratio was <2. Factors that affected this ratio were the level of iodine intake in the diet, the HCN content of fresh cassava roots and leaves, the efficiency of the detoxification processes used during the preparation of cassava foods, and the frequency and quantity of consumption of these foods (78).

An important finding was that newborn children were found to be more sensitive to the antithyroid action of cassava than the rest of the population. The human placenta is permeable to SCN. In the population where the iodine–SCN ratio was only slightly decreased and the thyroid function in adults was unaffected, TSH and T_4 concentrations in cord blood showed clear shifts toward high and low values, respectively. When the iodine–SCN ratio was <2 in the general population, the changes in the newborn were dramatic, with 10% of them exhibiting a characteristic biochemical picture of severe hypothyroidism. Experimental data gained from rat studies concluded that the severe brain retardation in the fetus supplemented with HCN or SCN was probably mediated by the induction of thyroid failure rather than direct toxic effects and was dependent on a critical supply of iodine (78).

Since ancient times, poor-quality water has been thought to be a cause of goiter. McCarrison (188) attributed goitrogenic situations in the Himalayan areas of India to bacterial contamination of the water. In bacteriological studies of some villages in Greece, the drinking water in the goiter-prevalent villages was found to be polluted with *Escherichia coli* and other coliforms significantly more than water from nongoitrous regions (180). Vought *et al.* (303,305), in the U.S. state of Virginia, had previously reported a similar relationship where goiter existed despite adequate iodine supplementation. Vought *et al.* (303) later demonstrated antithyroid activity in cell-free ultrafiltration fraction of both cultures of four species of *E. coli* isolated from an area of high endemia. In 1980 Gaitan (103) reported no correlation between *E. coli* and goiter prevalence in western Colombia. Gaitan found, however, an inverse relationship between *Klebsiella pneumoniae* water contamination and endemic goiter. The researchers in western Colombia hypothesized that the lower goiter prevalence may be associated with this organism biodegrading other goitrogenic organic contaminants. In their multivariate analysis, the variable that had the most significant association with high goiter prevalence as a geological one indicating sedimentary rocks in the waterbed. High mineral content of water, particularly magnesium or calcium salts or fluoride- or sulfur-bearing organic compounds of humic, high molecular weight polymetric, or complex nonprotein organic substances have all been implicated as goitrogens (74,104,179,189,316). Iodide excess is also a well-documented goitrogen in coastal Japanese communities consuming large amounts of seafood (282). Endemic goiter has also been reported in villages in China with high seaweed intake or consumption of iodine-rich water taken from deep wells around oil fields (328).

XI. ASSESSMENT OF IODINE NUTRITIONAL STATUS

The assessment of iodine nutritional status is important in relation to public health programs in which iodine supplementation is carried out. The problem is therefore one of assessment of a population or group living in an area or region that is suspected to be iodine deficient. Only a brief outline can be given here; for additional information the reader is referred to comprehensive references elsewhere (54,272).

The data required include

1. Total population including the number of children <15 years of age where the effects of iodine deficiency are so important
2. Goiter incidence including the incidence of palpable or visible goiter classified according to accepted criteria (294)
3. Incidences of cretinism and cretinoids in the population
4. Urine iodine excretion
5. Level of iodine in the drinking water
6. Determination of the level of serum T_4 in various age groups; particular attention is now focused on the levels in the neonate because of the importance of the level of T_4 for early brain development

1. Basic population data are usually available and make a reference point of obvious importance in developing an iodization program if it is to be comprehensive. There are difficulties in reaching the entire iodine-deficient population, predominantly because of the remoteness of many of these communities. Observations of schoolchildren have obvious advantages of access and convenience and have been extensively reported.

2. A classification of goiter severity has been adopted by the World Health Organization (294). There are still minor differences in technique between different observers. In general, visible goiter is more readily verified than palpable goiter. However, the extensive assessment of goiter incidence, while desirable, is not essential. It is time-consuming and costly, and limited samples of the population sufficient to verify its presence may be adequate.

3. The incidences of cretinism and cretinoids may be difficult to determine. Observations of schoolchildren will not detect those most severely affected, who are likely not to be attending school. Studies of IQ, as reported above from China, provide additional important evidence justifying programs.

4. The determination of urine iodine excretion can be carried out appropriately on 24-hr samples. However, the difficulties of such collections may be insurmountable. For this reason, as originally suggested by Follis (99), determinations can be carried out on casual samples from a group of ~30 subjects. The iodine levels are expressed as micrograms per gram of creatinine excretion and the range plotted out as a histogram. This provides a reference point for the level

of iodine excretion that is also a good index of the level of iodine nutrition. The availability of modern automated equipment (autoanalyzer) is making the analysis of large numbers of samples quite feasible. Methods have been improved so that reliable results can be obtained (23,105).

It has been suggested that there are three grades of severity of iodine deficiency in a population based on the urinary iodine excretion (238):

Grade I.   Goiter endemics with an average iodine excretion of >50 μg/g creatinine. At this level, thyroid hormone supply adequate for normal mental and physical development can be anticipated.

Grade II. Goiter endemics with an average urinary iodine excretion of between 25 and 50 μg/g creatinine. In these circumstances, adequate thyroid hormone formation may be impaired. This group is at risk for hypothyroidism but not for overt cretinism.

Grade III. Goiter endemics with an average urinary iodine excretion <25 μg/g creatinine. Endemic cretinism is a serious risk in such a population.

5. The level of iodine in drinking water indicates the level of iodine in the soil, which determines the level of iodine in the crops and animals in the area. Iodine levels in iodine-deficient areas are usually <2 μg/liter.

6. The determination of the level of serum T_4 provides an indirect measure of iodine nutritional status. The availability of radioimmunoassay methods with automated equipment has greatly facilitated this approach. Particular attention should be given to levels of T_4 in the neonate; levels <4 μg/dl have to be regarded as prejudicial to brain development (43).

In the developed countries of the world, where iodine deficiency in humans normally does not exist, all babies born are screened to ensure they have adequate thyroid hormone levels. These screening programs use blood from heel pricks of neonates, spotted onto filter paper, which is dried and sent to a regional laboratory. Blood levels of either T_4 or TSH or both are measured by immunoassay techniques. The detection rate of neonatal hypothyroidism requiring treatment is ~1 per 3500 babies screened. This rate varies little among developed countries (43).

Neonatal hypothyroid screening has been initiated in several less developed iodine-deficient regions. As already noted in Section V, Kochupillai et al. in India (158) and Ermans et al. in Zaire (89) have reported severe biochemical hypothyroidism (T_4 concentrations <2 μg/dl) occurring in 4 and 10% of neonates, respectively. It is evident from this and from other reports (78) that within an iodine-deficient population, serum T_4 levels are lowest at birth and lower in children than in the adult population. In addition, goitrogens such as cassava seem to be much more potent in reducing serum T_4 levels in neonates and children than in adults. This may be a critical factor, since T_4 levels are lowest at the most crucial time of development, especially brain development. There is a strong argument, then, to extend neonatal hypothyroid screening beyond the developed countries to regions where it is considered that iodine deficiency may be a problem.

To summarize, the most critical evidence is that available from measurement of urine iodine and from the measurement of serum T_4 in the neonate. The results of these two determinations indicate the severity of the problem. They can also be used to assess the effectiveness of remedial measures.

XII. CORRECTION OF IODINE DEFICIENCY
IN HUMANS

A. Iodized Salt

Many methods have been used to increase iodine intake. These include the addition of iodine as iodide or iodate to various foods such as bread, water, and milk. The addition of iodide to sweets has been used in Mexico. However, the addition of iodine to salt is by far the most widely used and simplest method available. Unlike various foods and water, the intake of salt as a condiment tends to be more constant from day to day. However, the problem of acceptability remains with non-salt-eating populations or populations (as in Thailand and China) who prefer their own non-iodine-containing salt.

However, there is an even bigger problem in ensuring the production of iodized salt of adequate quality in sufficient amounts for large iodine-deficient populations as in Indonesia, India, and China. Cost is also a problem whether or not it is provided by governments.

Recommended levels of iodization vary in different countries. In India $\frac{1}{40000}$ has been recommended, in Finland $\frac{1}{25000}$ was found to be required for effective control of goiter. On the assumption of an intake of 5 g of salt per day, a level of $\frac{1}{100000}$ would provide 50 µg. If the salt intake is lower than this, as it often is in populations in developing countries, a higher level of supplementation is required. In general a level of intake of 100 µg/day can be regarded as adequate for the prevention of goiter and cretinism. In view of the significance of salt intake for the development of hypertension, lower levels of salt intake can be anticipated with a higher concentration of iodine.

The most popular and simple method of iodization consists in applying a solution of an iodine compound on salt by a drip or a spray. If the salt is then well dried, if it is free of impurities (in particular ferric ion) and has a slightly alkaline pH, potassium iodide will be quite stable. On the other hand, if the salt retains moisture (or acquires moisture in a humid climate), contains impurities, or has a pH <7.5, iodide may be oxidized to I_2 and evaporate (net loss), or it may move in the water film downward or into the fabric of the container (segregation). In such instances potassium iodate is the preferred iodinating compound. Iodate has a remarkable stability even in impure salts and under adverse climatic conditions. Surveillance of an iodization program should include monitoring of salt sales, checks on the iodine content of salt in the retail stores, and analysis of the urinary iodine excretion (38).

Iodine is mainly produced from brine wells in the United States and Japan. China is at present totally dependent on Japan for its iodine supplies. The price is also rising. Chile is no longer so important a source, as there is competition with the fertilizer industry (53).

Potassium iodide is used extensively for supplementation of refined (table) salt. It is highly soluble but easily oxidized with higher temperature and humidity. This can be reduced by the addition of stabilizers such as 0.1% sodium thiosulfate and 0.1% calcium hydroxide combined. Iodate is now used almost exclusively for crude salt treatment in developing countries (142). No stabilizer is required. Solubility is sufficient for spray treatment. Dry mixing is, however, required in very cold climates such as that of Heilongjiang Province in northeastern China. Here calcium iodate is more suitable.

The difficulties in the production and distribution of salt to the millions who are iodine deficient, especially in Asia, are vividly demonstrated in India, where there has been a breakdown in supply. The difficulties have been discussed in a detailed report prepared by the Nutrition Foundation of India (211). The gap between the actual production (106,000 metric tons 1978–1979) and the annual requirement (700,000 metric tons; Table XII) is so great that the Indian government is considering alternatives such as the use of iodized oil on a mass scale.

In Asia, the cost of iodized salt production and distribution at present is considered to be 2–3 cents per person per year (UNICEF 1984). This must be considered cheap in relation to the social benefits that have been described in the previous section.

Table XII
Production of Iodized Salt in India[a]

	Iodized salt amount (metric tons)
Production category	
Total annual requirement of iodized salt	700,000
Total installed capacity of 12 iodination plants already set up	376,000
Total annual quota fixed for production for use in India	220,000
Production quota in the Sambhar Lake and Khargoda area for supply to Nepal	60,000
Total actual production for both India and Nepal together	
1974–1975	122,000
1978–1979	106,000

[a]Nutrition Foundation of India (211).

However, there is still the problem of the salt actually reaching the iodine-deficient subject. There may be a problem with preservation of the iodine content; it may be left uncovered or exposed to heat. It should be added after cooking to reduce the loss of iodine.

Finally there is the difficulty of actual consumption of the salt. While the addition of iodine makes no difference to the taste of the salt, the introduction of a new variety of salt to an area where salt is already available and familiar and much appreciated as a condiment is likely to be resisted. In the Chinese provinces of Sinjiang and Inner Mongolia, the strong preference of the people for desert salt of very low iodine content led to a mass iodized oil injection program in order to prevent cretinism (173).

B. Iodized Oil by Injection

The value of iodized oil injection in the prevention of endemic goiter and endemic cretinism was first established in New Guinea with controlled trials involving the use of saline injection as a control. These trials established the value of the oil in the prevention of goiter (131,190) and in the prevention of cretinism (225). Experience in South America (136,236) has confirmed the value of the measure. The quantitative correction of severe iodine deficiency by a single intramuscular injection (2–4 ml) has been demonstrated (44) (Table XIII).

The injection of iodized oil can be carried out through local health services where they exist, or by special teams. In New Guinea (136) the injection of a population in excess of 100,000 was carried out by public health teams along with the injection of triple antigen (immunization procedure).

Iodized oil is singularly appropriate for the ioslated village community so characteristic of mountainous endemic goiter areas. It has been used in pilot studies in the Himalayan region of Pakistan (45), Burma, and Nepal (136), and in Zaire (293).

In a suitable area the oil should be administered to all females ≤40 years of age and all males ≤20 years of age (dosage schedule recommended is shown in Table XIV). A repeat injection would be required in 3 to 5 years depending on the dose given and the age. In children the need is greater than in adults, and the recommended dose should be repeated in 3 years if severe iodine deficiency persists.

It is now clear that iodized oil is suitable for use in a mass program. In Indonesia (80) some 1,036,828 injections were given between 1974 and 1978 together with the massive distribution of iodized salt. A further 6 million injections were given by specially trained paramedical personnel in the period 1978–1983. In China in Sinjiang Province, 707,000 injections were given by barefoot doctors between 1978 and 1981, and a further 300,000–400,000 were being given in 1982. There are considered to be advantages to the use of injections

Table XIII

Effect of Iodized Oil on Thyroid Function in New Guinea Subjects[a,b]

Group	Urinary iodine (μg/24 hr)	^{131}I uptake (% at 24 hr)	Serum PBI (μg/100 ml)	T_3 resin uptake (% of normal)
Untreated	11.5 ± 12.4 (91)	70 ± 19 (181)	4.1 ± 2.1 (204)	91 ± 12.1 (195)
Treated 18 months and 3 years before	67 ± 83 (47)	33 ± 20 (94)	8.0 ± 2.0 (79)	97 ± 14.8 (77)
Treated 3 months before	258 ± 109 (8)	6.0 ± 3 (20)	44.7 ± 18.4 (20)	110 ± 15.5 (20)
Treated 18 months before	119 ± 114 (18)	31 ± 20 (51)	8.2 ± 2.6 (27)	97 ± 15.8 (27)
Treated 3 years before	35 ± 25 (29)	37 ± 19 (43)	7.8 ± 1.6 (52)	97 ± 14.7 (50)
Treated 4.5 years before	23 ± 21 (11)	44 ± 18 (67)	6.4 ± 2.4 (43)	99 ± 16.0 (43)
Australian normal range	70–140	16–40	3.6–7.2	70–110

[a]Butfield and Hetzel (44).

[b]Statistical analysis showed highly significant differences between the treated and untreated groups in urinary iodine, ^{131}I uptake, and serum PBI ($p < .001$). There was no significant difference in the ^{131}I uptake or serum PBI between subjects treated 3 years and 18 months before. Figures given are for the mean ± SD with the number of subjects tested shown in parentheses.

Table XIV
Recommended Dosages of Ethiodized Oil
Containing 37% Iodine[a,b]

Age	Iodine (mg)	Dose (ml)
0–6 months	95.0–180.0	0.2–0.4
6–12 months	142.5–285.0	0.3–0.6
6 months–6 years	232.5–465.0	0.5–1.0
6–45 years	475.0–950.0	1.0–2.0

[a]From Stanbury et al. (271).
[b]The dosage should be reduced to 0.2 ml for all persons with nodular goiters or with single thyroid nodules without goiter.

because of the association of injections with the successful smallpox eradication campaign.

The disadvantages to the use of injections are the immediate discomfort produced and the infrequent development of abscesses at the site of injection. Sensitivity phenomena have not been reported. These side effects are not serious.

However, the major problem of injections is their cost (271), although this has been reduced with mass packaging. Clearly costs can be reduced if the syringes can be used for other purposes and staff is available from the primary health care center. An injection program could be phased in with the immunization program involving triple-antigen injections as occurred in Papua New Guinea.

C. Iodized Oil by Mouth

There is evidence available of the effectiveness of a single oral administration of iodized oil for 1 to 2 years in South America (307) and in Burma (163).

Iodized walnut oil and iodized soya bean oil are new preparations that have been developed in China since 1981 to avoid the need for foreign exchange. They are cheaper than the Western products, which means they could have wider application. Preliminary reports on the use of these preparations were given to the Second Asia and Oceania Thyroid Association Meeting in Tokyo (171,217).

Subsequent unpublished studies in India and China reveal that oral iodized oil lasts only half as long as a similar dose given by injection.

Experimental studies in China in guinea pigs indicate that oral oil is stored in adipose tissue, while the injection is stored at the intramuscular site of the injection. There is a slower absorption from the muscle than from the adipose tissue. There is also a greater loss of iodine from oral administration due to deiodination in the stomach. This iodine is absorbed but rapidly excreted through the kidney as inorganic iodide would be. In general with oral oil, there is an 85%

loss within 72 hr, with an intramuscular injection 30% in 2 weeks, and with oral iodized salt a 96% loss within 72 hr (169).

The production of oral iodized oil is clearly a different proposition from that of iodized oil for injection. The addition of an antioxidant would be feasible in the case of an oral preparation.

However, there is an urgent need to increase the production of iodized oil.

An investigation of possible chemical methods for iodizing oil has been carried out by Dr. Trevor Morton of the CSIRO Division of Applied Organic Chemistry, Melbourne, Australia (205). This study revealed that a variety of vegetable oils with linoleic acid as a major constituent can readily be iodinated. Studies with safflower oil (76.9–80.5% linoleic acid) indicated that satisfactory and cheap iodination could be achieved with iodide and phosphoric acid. There is no reason large quantites of iodized oil cannot be produced, particularly for oral use.

D. Iodized Bread

Iodized bread has been used in the Netherlands and in Australia. Detailed observations are available from the island of Tasmania in Australia.

Since 1949 the Tasmanian population has received a number of intended and unintended iodine dietary supplements. These began in 1949 with the use of weekly tablets of potassium iodide (10 mg) given to infants, children, and pregnant women through baby health clinics, schools, and antenatal clinics whenever possible.

The prevalence of endemic goiter fell progressively over these 16 years but was not eliminated. This was traced to lack of cooperation by a number of schools in the distribution of the iodide tablets. The distribution through the child health centers to infants and preschool children was also ineffective because of their irregular attendance.

For this reason a decision was made to change the method of prophylaxis from iodide tablets to iodization of bread. The use of potassium iodate up to 20 mg/kg as a bread improver was authorized by the National Health and Medical Research Council of Australia in May 1963, and the necessary legislation was passed by the Tasmanian Parliament in October 1964.

In the early 1960s the average consumption of bread was estimated by the Commonwealth Bureau of Census and Statistics to be ~3.3 lb (1500 g) per head per week (55). The improver commonly used at that time was potassium bromate. Since the estimated requirement of iodine for an adult is of the order of 150 μg/day, the use of iodate as the sole improver would have yielded much more iodine than was required to prevent goiter. It was decided to permit the addition of 2 to 4 mg of potassium iodate per kilogram, and the firm that supplied bread improver to Tasmania agreed to substitute the required amount of potassium iodate for potassium bromate in the improver shipped to Tasmanian bakers. It was estimated that the range of bread consumed by the various age groups would supply a significant percentage of the estimated daily requirements of iodine.

By the end of April 1966, practically all areas were being supplied with the new improver, and the distribution of tablets of potassium iodide to infants and children was gradually phased out, so that by the end of 1967 this method of prophylaxis had been abandoned.

The effect of bread iodization was followed by a series of surveys of palpable goiter incidence in schoolchildren. A definite effect on visible goiter incidence was apparent by 1969 (55). Studies or urinary iodide excretion and PII in May 1967 revealed no excessive intake of iodide. Correction of iodine deficiency was confirmed by evidence of a fall of 24-hr radioiodine uptake levels in hospital subjects as well as normal plasma inorganic iodine concentration and urine iodine excretion (277).

It may be concluded that bread iodization was effective in correcting iodine deficiency in the 1960s. However, there is now concern for possible excessive iodine intake from diverse sources such as milk, ice cream, and poultry. A transient increase in thyrotoxicosis was observed (see below).

E. Water Iodination

Reduction in goiter incidence from 61 to 30%, with 79% of goiters showing visible reduction, has been demonstrated following water iodization in Sarawak, Malaysia (176). Significant rises in serum T_4 and falls in TSH were also shown. Urinary iodine excretions were variable due to intermittent obstruction of the iodinator, but eventually the levels indicated iodine repletion.

Similar results have been obtained with preliminary studies in Thailand by Dr. Romsai Suwanik and his group at the Siriraj Hospital, Bangkok.

It is suggested that iodinated water may be more convenient than iodized salt and there may be less likelihood of iodine-induced thyrotoxicosis. This method is appropriate at the village level if a specific source of drinking water can be identified; otherwise there is a heavy cost for <1% of a general water supply used for drinking purposes. The antiseptic benefit is also significant with this method.

F. Other Methods

In Bangkok, Dr. Romsai Suwanik has also developed iodized fish sauce and iodized soya sauce as additional iodized condiments. These sauces are also being used for iron supplementation (281).

In the past iodide was used in the original controlled trial by Marine and Kimball (185) in Ohio schoolchildren (1917–1922) and in Tasmanian schoolchildren (1949–1965), as already noted (Section XII,D).

Marine and Kimball (185) used 200 mg sodium iodide in water daily for 10 days in the spring and repeated this in the autumn. Satisfactory regression of goiter was observed. In Tasmania, Clements (55) used 10 mg of potassium iodide weekly through schools and child health centers—with some (but an irregular) regression of goiter. Failure of reduction of goiter was attributed to inadequate cooperation in the schools and health centers, and led to the introduction of bread iodization in 1966.

The use of oral iodide certainly should be considered as an option in schoolchildren for moderate iodine deficiency, where compliance can be more readily enforced than in adults. However, it cannot be expected to meet the demand of severe iodine deficiency.

G. Iodine-Induced Thyrotoxicosis

A mild increase in incidence of thyrotoxicosis has now been described following iodized salt programs in Europe and South America and following iodized bread in the Netherlands and Tasmania (62,227). A few cases have been noted following iodized oil administration in South America. No cases have yet been described in New Guinea, India, or Zaire. This is probably because of the scattered nature of the population in small villages and limited opportunities for observation (165). The condition is largely confined to those >40 years of age— a smaller proportion of the population in developing countries than in developed countries. Detailed observations are available from the island of Tasmania (62,277).

It was clear that the rise in incidence of thyrotoxicosis in Tasmania had occurred following the rise in iodine intake from below normal to normal levels (277), following the introduction of iodized bread in April 1966.

However, careful scrutiny of records in northern Tasmania revealed a rise in the incidence of thyrotoxicosis as early as 1964, associated with a rise in food imports and the introduction during 1963 of iodophors to the dairy industry (61,135a). There was a much steeper increase in incidence from 1966 following the iodization of bread.

A cohort effect was demonstrated, because the peak passed—the peak being mainly composed of those over the age of 40 with life-long iodine deficiency and autonomous thyroid glands that continued the rapid turnover of iodine after the increase in iodine intake.

It was clear that the increase was mainly caused by patients with toxic autonomous goiter and not Graves' disease (135a,299). The findings were consistent with the original concepts of Plummer and support the old view of two types of thyrotoxicosis: Graves' disease and Plummer's disease (299).

The condition was readily controlled with antithyroid drugs or radioiodine. In general iodization should be avoided in those over the age of 40 because of the risk of thyrotoxicosis (177).

Apart from the question of thyrotoxicosis, the risk of iodism or iodide goiter seems to be very small. An increase in lymphocytic thyroiditis (Hashimoto's disease) has been claimed following iodization, but this is still disputed (271).

H. Summary

In reviewing the various methods available for iodization, it is apparent they are of two types: (1) measures for the whole population (iodized salt and iodized bread) and (2) measures suitable for an at-risk segment of the population (children and women of reproductive age) in the form of iodized oil and iodide tablets.

The great advantage of the prescriptive approach is that it can be carried out through the health care system. It does not, as in the case of population measures, require cooperation and enforcement measures involving other government departments and private industry.

The importance of *quantitative* correction of severe iodine deficiency, particularly in women and children, in order to prevent mental deficiency requires a much more critical approach than in the past, particularly with severe iodine deficiency.

The complication of iodine-induced thyrotoxicosis is unavoidable for population measures. Its incidence mainly depends on the proportion of the population over the age of 40 years, which is a lower proportion in a developing country than in developed countries.

In the case of the prescriptive measures of iodized oil and iodide tablets, the susceptible older age group for thyrotoxicosis can be avoided. Occasional cases will still occur in those under the age of 40, but the risk is much less.

There is therefore a good case for increasing the use of prescriptive measures in order to increase the efficiency of the cover provided and to reduce the incidence of thyrotoxicosis.

In China an iodized oil injection has been introduced for iodine-deficient young women just before marriage. This will cover them for 3 years, which would confer protection for the single pregnancy that is now in force as family planning policy in China.

Oral administration of iodized oil to children could be carried out through the baby health centers and schools; periods of 18 months could be covered at present (2-ml dose). This may well be increased as a result of further research investigation. The injection will last longer (2 ml for 3 years) and may be appropriate in some circumstances when it can be carried out by existing primary health care center staff. Cheaper production of iodized oil is readily achievable and should be provided in India and other countries with large populations at risk. The advantages of a single administration of iodized oil against multiple administration of iodide tablets are obvious.

The advantages and disadvantages of prescriptive versus population measures have to be considered in every different region and within any national program. In the massive iodine-deficient populations in Asia, the great advantages of salt iodization as a population measure in the prevention of large-scale mental deficiency outweigh the disadvantages of iodine-induced thyrotoxicosis.

XIII. CORRECTION OF IODINE DEFICIENCY IN ANIMALS

With farm animals, the best iodine supplementation method to adopt depends on the conditions of husbandry. Stall-fed animals are usually provided with iodized salt licks or pellets, or the iodine is incorporated into the mineral mixtures or concentrates provided to supply other nutritional needs. Salt licks containing potassium iodide lose iodine readily from volatilization or leaching if exposed for any length of time to hot or humid conditions (73). However, potassium iodate, which is more stable than iodide and is nontoxic at the levels required, can be used (265). Inclusion of this compound into salt licks or mineral mixtures for stock at a level of 0.01% iodine (172) is recommended to ensure adequate intakes in goitrous areas. In a comparison of potassium iodide, calcium iodate, and 3,5-diiodosalicylic acid as iodine sources for livestock, Shuman and Townsend (261) found the first two to be rapidly lost from the surface layer of salt blocks when exposed to outdoor weather conditions, whereas the diiodosalicylic acid remained constantly on the surface and was obtained by the animals in a normal manner. Aschbacher and co-workers (14,15) found this compound to be an unsatisfactory source of supplemental iodine for calves and dairy cattle. Subsequently Aschbacher (13) showed that allowing ewes free access to salt containing 0.007% iodine as diiodosalicylic acid did not prevent iodine deficiency in their newborn lambs, whereas access to salt containing the same iodine concentration as potassium iodide successfully achieved this aim. These workers (199) then compared the nutritional availability of iodine from calcium iodate, pentacalcium orthoperiodate (PCOP), and sodium iodide to pregnant cows. All three forms were found to supply the fetal thyroid with equal efficiency. Since PCOP has greater physical stability than sodium iodide and calcium iodate under field conditions (196), this compound is clearly a valuable form of supplemental iodine for use in livestock salt blocks or mineral mixes.

With sheep and cattle under permanent grazing in goitrous areas, different forms of treatment are ncessary. Iodized fertilizers cannot be relied on to maintain satisfactory iodine levels in the herbage for long periods after application (128), and the provision of iodized salt licks is subject to the hazard, as with all trace elements, of spasmodic and uncertain consumption by the grazing animal. Unless such stable and insoluble compounds as PCOP are used, there is also the problem of physical loss by volatilization and leaching. Regular dosing or "drenching" with inorganic iodine solutions is effective but can be costly and time-consuming. This form of treatment, consisting of two oral doses of 280 mg potassium iodide or 360 mg potassium iodate, given at the beginning of the fourth and fifth months of pregnancy, is satisfactory for the prevention of the neonatal mortality and associated goiter in lambs that arise when ewes are grazed on goitrongenic kale (*Brassica oleracea*) (264). This condition can also be controlled by intramuscular injections of an iodized poppyseed oil preparation, containing some 40% iodine by weight. A single 1-ml injection of this preparation into ewes 2 months before lambing raised the iodine concentration in the thyroid glands of lambs from kale-fed ewes to normal levels and prevented marked goiter and high neonatal mortality. Similar injections given 1 month later only partially prevented the thyroid enlargement, although the iodine concentrations in these glands were normal and the death rate was reduced (6,264).

A plastic capsule containing solid iodine has been developed by Laby (87a,164), which provides a sustained release of iodine into the rumen of sheep and cattle over several years. Other data indicate that this could be a cheap and satisfactory means of providing supplemental iodine to grazing ruminants in goitrous areas, especially as the iodine release rate and the lifetime can be varied over a wide range by choice from a variety of capsule dimensions and types of plastic.

Pregnant ewes in a goiter-prone area of northern Tasmania so treated showed reduced thyroid enlargement as determined by palpation and histological scores, as well as greatly increased iodine concentration in milk (186).

XIV. ERADICATION OF IODINE DEFICIENCY DISORDERS

Two monographs (54,272), published in 1960 and 1980, have summarized the information available regarding the global occurrence and control of endemic goiter and endemic cretinism. The hope expressed in 1960 for total prevention and eradication of these disorders has not been fulfilled. As stated in the preface to the 1980 monograph,

The reasons for the persistence of endemic goiter and its attendant results—infant mortality, intellectual and physical retardation, and deaf-mutism—are not deficits in our scientific information, but are socio-economic, cultural and political. The means for eradicating the disease are in hand. Iodization of salt is easy, efficient, and so cheap that it should not raise the price of the salt to the consumer. Iodized oil is efficient, long-lasting, effective, and although somewhat more expensive on a per capita per year basis than iodized salt, is a remarkably small item in a national budget when measured against the economic benefits, and can reach populations where social, economic, or geographic factors are not compatible with salt as the vehicle of distribution. Governments must contribute to their own national life by eliminating endemic goiter once and for all, for there is no disease which impedes national development and the quality of life so much and which can be eliminated for so little.

The relevance of economic, cultural, social, and political factors to this lag in application of available knowledge has been analyzed (293a). The geographical isolation of the iodine-deficient communities and hence their lack of political weight is an important factor.

However, one special reason for neglect of this problem at the social and political level is its designation simply as goiter. Goiter is often a cause of disfigurement; it may produce obstruction of the trachea or esophagus and may be associated with an increased incidence of thyroid carcinoma (272). However, goiter alone hardly constitutes a serious health problem in developing countries when compared with bacterial or parasitic infection.

However, as pointed out in Section V, the human effects of iodine deficiency extend far beyond those of goiter. Research since the mid-1960s has established the significant impact of iodine deficiency on brain function in the fetus, the newborn, and the child. It has become apparent that there is a spectrum of disorders that constitute a significant threat to the genetic potential of millions living in iodine-deficient areas, which are usually, though not only in developing countries. In fact the impairment of mental function brought about by iodine deficiency may well be a significant barrier to the general social development of iodine-deficient communities.

In the light of these advances, it has been suggested that the iodine deficiency problem be redesignated in the future as iodine deficiency disorders, with the acronym IDD (133). This suggestion was supported in an editorial in the *Lancet* entitled "From Endemic Goiter to Iodine Deficiency Disorders" (84).

A series of resolutions at several international meetings has pointed out the feasibility of the prevention and eradication of iodine deficiency disorders. These include the Regional WHO/UNICEF Committee for South-East Asia (1981, 1982), the International Nutrition Congress (Rio de Janiero 1978), the Asia and Oceania Thyroid Congress (Tokyo 1982), and the Fourth Asian Congress of Nutrition (Bangkok 1983). The international symposium at the Fourth Asian Congress of Nutrition in Bangkok made the following recommendations to the various international agencies, national aid bodies, and national governments with responsibility for the problem (63):

1. The iodine deficiency problem should in the future be designated as iodine deficiency disorders (IDD) and the term goiter be discontinued, as it no longer reflects the state of existing knowledge.

2. Reliable estimates indicate that 400 million people in Asia are currently suffering from iodine deficiency.

3. The major effect of iodine deficiency is on the brain. This includes the fetal and neonatal brain, as well as mental function in childhood and adult life.

4. Correction of iodine deficiency by the use of iodized salt or iodized oil has been shown to prevent the mental deficiency that might otherwise occur.

5. The eradication of iodine deficiency with prevention of mental deficiency is feasible and effective at modest cost in comparison to the cost of the disability.

6. The serious effects of iodine deficiency on human potential have now been adequately demonstrated. The constraints placed on iodine-deficient communities are such as to limit seriously their social life and development. The eradication of iodine deficiency in Asia has become mandatory.

7. Iodized oil (by injection or by mouth) offers an effective emergency method for the correction of severe iodine deficiency until iodized salt can be distributed. Such a measure can be carried out through the primary health care system.

8. Iodization programs with iodized salt and/or iodized oil should be monitored by determinations of urine iodine in regional laboratories. Those could be in India (which might also serve Sri Lanka, Nepal, Bhutan, and Bangladesh), Burma, Thailand, Indonesia, China, and Pakistan. These could be supported by WHO and UNICEF.

9. We urge international agencies and national governments in the region to reappraise existing programs with the aim of eradication of iodine deficiency in Asia in the next 10 years. We offer our expertise and knowledge to national governments and the international agencies with responsibility for this problem.

10. We strongly recommend regular meetings in the Asian region to evaluate programs and monitor progress in the eradication of iodine deficiency.

The evidence reviewed in a number of sections indicates clearly that a great opportunity exists for a major success in the field of international nutrition. A proposed strategy is being drawn up (134). It is to be hoped that this opportunity will be seized with benefits to many millions of people at present suffering from the deadening effects of iodine deficiency on their lives.

REFERENCES

1. Ahn, C. S., and Rosenberg, I. N. (1968). *Proc. Natl. Acad. Sci. U.S.A.* **60**, 830.
2. Alderman, G., and Jones, D. I. H. (1967). *J. Sci. Food Agric.* **18**, 197.

3. Alexander, W. D., Papdopoulos, S., Harden, R. McG., Macfarlane, S., Mason, D. K., and Wayne, E. (1966). *J. Lab. Clin. Med.* **67,** 808.
4. Allcroft, R., Scarnell, J., and Hignett, S. L. (1954). *Vet. Rec.* **66,** 367.
5. Ammerman, C. B., Arrington, L. R., Warnick, A. C., Edwards, J. L., Shirley, R. L., and Davis, G. J. (1964). *J. Nutr.* **84,** 107.
6. Andrews, E. D., and Sinclai', D. P. (1962). *Proc. N.Z. Soc. Anim. Prod.* **22,** 123.
7. Andrews, F. N., Shrewsbury, C. L., Harper, C., Vestal, C. M., and Doyle, L. P. (1948). *J. Anim. Sci.* **7,** 298.
8. Andros, G., Harper, P. V., Lathrop, K. A., and McArdle, R. J. (1965). *J. Clin. Endocrinol. Metab.* **25,** 1067.
9. Arnott, R. D., Eastman, C. J., and Waite, K. V. (1982). *Proc. Asia Oceania Congr. Endocrinol. Abstracts,* p. 57.
10. Arnott, R. D., and Eastman, C. J. (1983). *J. Recept. Res.* **3,** 393.
11. Arrington, L. R., Taylor, R. N., Ammerman, C. B., and Shirley, R. L. (1965). *J. Nutr.* **87,** 394.
12. Arrington, L. R., Santa Cruz, R. A., Harms, R. H., and Wilson, H. R. (1967). *J. Nutr.* **92,** 325.
13. Aschbacher, P. W. (1968). *J. Anim. Sci.* **27,** 127.
14. Aschbacher, P. W., Miller, J. K., and Craggle, R. G. (1963). *J. Dairy Sci.* **46,** 114.
15. Aschbacher, P. W., Cragle, R. G., Swanson, E. W., and Miller, J. K. (1966). *J. Dairy Sci.* **49,** 1042.
16. Astwood, E. B. (1940). *Ann. Intern. Med.* **30,** 1087.
17. Astwood, E. B., Greer, M. A., and Ettlinger, M. G. (1949). *J. Biol. Chem.* **181,** 121.
18. Barakat, M. Z., Bassiouni, M., and El-Wakil, M. (1972). *Bull. Acad. Pol. Sci.* **20,** 531.
19. Barua, J., Cragle, R. G., and Miller, J. K. (1964). *J. Dairy Sci.* **47,** 539.
20. Bautista, S., Barker, P. A., Dunn, J. T., Sanchez, M., and Kaiser, D. L. (1982). *Am. J. Clin. Nutr.* **35,** 127.
21. Baxter, J. D., Eberhart, N. L., Apriletti, J. W., Johnson, L. K., Ivarie, R. D., Scacher, B. S., Morris, J. A., Seeburg, P. H., Goodman, H. M., Latham, K. R., Polansky, J. R., and Martial, J. A. (1979). *Recent Prog. Horm. Res.* **35,** 97.
22. Beckers, C., and Delange, F. (1980). *In* "Endemic Goiter and Endemic Cretinism" (J. B. Stanbury and B. S. Hetzel, eds.), p. 199. Wiley, New York.
23. Belling, G. B. (1983). *Analyst* **108,** 763.
24. Bernal, J., and Refetoff, S. (1977). *Clin. Endocrinol. (Oxford)* **6,** 227.
25. Berson, S. A. (1956). *Am. J. Med.* **20,** 653.
26. Berson, S. A., and Yalow, R. A. (1954). *J. Clin. Invest.* **33,** 1533.
27. Blaxter, K. L. (1952). *Vitam. Horm. (N.Y.)* **10,** 217.
28. Bleichrodt, N., Drenth, P. J. D., and Querido, A. (1980). *Am. J. Phys. Anthropol.* **53,** 55.
29. Blom, I. J. B. (1934). *Onderstepoort J. Vet Sci. Anim. Ind.* **2,** 139.
30. Blum, F. (1971). *Schweiz. Med. Wochenschr.* [N.S.] **70,** 1301.
31. Bogard, R., and Mayer, D. T. (1946). *Am. J. Physiol.* **147,** 320.
32. Bourdoux, P., Delange, F., Gerard, M., Mafuta, A., Hanson, A., and Ermans, A. M. (1978). *J. Clin. Endocrinol. Metab.* **46,** 613.
33. Bourdoux, F., Delange, M. G., Mafuta, M., Hanson, A., and Ermans, A. M. (1980). *In* "Role of Cassava in the Etiology of Endemic Goiter and Cretinism" (A. M. Ermans, N. M. Moulameko, F. Delange, and R. Alhuwalia, eds.), p. 61. International Development Research Centre, Ottawa, Canada.
34. Bourdoux, P., Mafuta, M., Hanson, A., and Ermans, A. M. (1980). *In* "Role of Cassava in the Etiology of Endemic Goiter and Cretinism" (A. M. Ermans, N. M. Moulameko, F.

Delange, and R. Alhuwalia, eds.), p. 15. International Development Research Centre, Ottawa, Canada.
35. Braverman, L. E., Ingbar, S. H., and Sterling, K. (1970). *J. Clin. Invest.* **49,** 855.
36. Brown-Grant, K., and Galton, V. A. (1958). *Biochim. Biophys. Acta* **27,** 422.
37. Brown-Grant, K. (1961). *Physiol. Rev.* **41,** 189.
38. Burgi, H., and Rutishauser, R. (1983). *In* Fifth Meeting of PAHO/WHO Technical Group on Control of Endemic Goiter and Cretinism, Lima, Peru. Pan Am. Health Organ., Washington, D.C. (in press).
39. Burke, G. (1970). *Am. J. Physiol.* **218,** 1445.
40. Burke, G. (1970). *J. Clin. Endocrinol. Metab.* **32,** 76.
41. Burke, G. (1970). *Biochim. Biophys. Acta* **220,** 30.
42. Burke, G. (1971). *Acta Endocrinol. (Copenhagen)* **66,** 558.
43. Burrow, G. N., eds. (1980). "Neonatal Thyroid Screening." Raven Press, New York.
44. Buttfield, I. H., and Hetzel, B. S. (1967). *Bull. W.H.O.* **36,** 243.
45. Chapman, J. A., Grant, I. S., Taylor, G., Mahmud, K., Mulk, S. U., and Shahid, M. A. (1972). *Philos. Trans. R. Soc. London, Ser. B* **263,** 459.
46. Chesney, A. M., Clawson, T. A., and Webster, B. (1928). *Bull. Johns Hopkins Hosp.* **43,** 261.
47. Childs, D. S., Jr., Keating, F. R., Rall, J. E., Williams, M. M., and Power, M. H. (1950). *J. Clin. Invest.* **29,** 726.
48. Chopra, I. J., Solomon, D. H., and Ho, R. S. (1971). *J. Clin. Endocrinol. Metab.* **33,** 865.
49. Chopra, I. J., Sack, J., and Fisher, D. A. (1975). *Endocrinology (Baltimore)* **97,** 1080.
50. Chopra, I. J., Sack, J., and Fisher, D. A. (1975). *J. Clin. Invest.* **55,** 1137.
51. Chopra, I. J. (1981). *In* "Triiodothyronines in Health and Disease: Monographs on Endocrinology," p. 102. Springer-Verlag, Berlin and New York.
52. Chopra, I. J. (1981). *In* "Triiodothyronines in Health and Disease: Monographs on Endocrinology," Vol. 18, p. 118. Springer-Verlag, Berlin and New York.
53. Claridge, G. G. C., and Campbell, I. B. (1968). *Nature (London)* **217,** 428.
54. Clements, F. W. (1960). "Endemic Goitre." World Health Organ., Geneva.
55. Clements, F. W., Gibson, H. B., and Coy, J. F. (1970). *Lancet* **1,** 489.
56. Coble, Y., Davis, J., Schulert, A., Hetz, F., and Award, A. Y. (1968). *Am. J. Clin. Nutr.* **21,** 277.
57. Cohn, B. N. E. (1932). *Arch. Intern. Med.* **49,** 950.
58. Cole, V. V., and Curtis, G. M. (1935). *J. Nutr.* **19,** 493.
59. Condliffe, P. G., and Weintraub, B. D. (1979). *In* "Hormones in Blood" (C. H. Gray and V. H. T. James, eds.), Vol. 1, p. 499. Acaddmic Press, London.
60. Connolly, K. J., Pharoah, P. O. D., and Hetzel, B. S. (1979). *Lancet* **2,** 1149.
61. Connolly, R. J. (1971). *Med. J. Aust.* **2,** 1191.
62. Connolly, R. J., Vidor, G. I., and Stewart, J. C. (1970). *Lancet* **1,** 500.
63. "Control of Iodine Deficiency in Asia" (1983). *Lancet* **2,** 1244.
64. Cooper, D. S., Klibanski, A., and Ridgway, E. C. (1983). *Clin. Endocrinol. (Oxford)* **18,** 265.
65. Cooper, E., and Spaulding, S. W. (1983). *Endocrinology (Baltimore)* **112,** 1816.
66. Corcoran, J. M., Eastman, C. J., Carter, J. N., and Lazarus, L. (1977). *Arch. Dis. Child.* **52,** 716.
67. Costa, A., Cottino, F., Mortara, M., and Vogliazzo, U. (1964). *Panminerva Med.* **6,** 250.
68. Courrier, R., and Aron, M. (1929). *C.R. Seances Soc. Biol. Ses Fil.* **100,** 839.
69. Coursey, O. G., and Haynes, P. H. (1970). *World Crops* **22,** 261.
70. Crantz, F. R., and Larsen, P. R. (1980). *J. Clin. Invest.* **65,** 935.

71. Creek, R. D., Parker, H. E., Hauge, S. M., Andrews, E. N., and Carrick, C. W. (1954). *Poult. Sci.* **33**, 1052.
72. Curtis, G. M., Puppel, I. D., Cole, V. V., and Matthews, N. L. (1937). *J. Lab. Clin. Med.* **22**, 1014.
73. Davidson, W. M., and Watson, C. J. (1948). *Sci. Agric.* **28**, 1.
74. Day, T. K., and Powell-Jackson, P. R. (1972). *Lancet* **1**, 1135.
75. de Escobar, G. M., Obregon, M. J., and Escobar, F. (1982). *Proc. Asia Oceania Congr. Endocrinol.*
76. Delange, F. (1974). *In* "Monographs in Pediatrics" (F. Faulkner et al., eds.), Vol. 2, p. 6. S. Karger.
77. Delange, F., Camus, M., and Ermans, A. M. (1972). *J. Clin. Endocrinol. Metab.* **34**, 891.
78. Delange, F., Iteke, F. B., and Ermans, A. M. (1982). "Nutritional Factors Involved in the Goitrogenic Action of Cassava," p. 1. International Development Research Centre, Canada.
79. D'Haene, E. G. M., Crombag, F. J. L., and Tertoolen, J. F. W. (1974). *Br. Med. J.* **3**, 708.
80. Djokomoeljanto, R., Tarwotjo, I., and Maspaitella, F. (1983). *In* "Current Problems in Thyroid Research" (N. Ui, K. Torizuka, S. Nagataki, and K. Miyai, eds.), p. 403. Excerpta Medica, Amsterdam.
81. Driever, C. W., Christian, J. E., Bousquet, W. F., Plumlee, M. P., and Andrews, F. N. (1965). *J. Dairy Sci.* **48**, 1088.
82. Dumont, J. (1971). *Vitam. Horm. (N.Y.)* **29**, 287.
83. Eastman, C. J. (1977). *Aust. Fam. Physician* **6**, 119.
84. Editorial (1983). *Lancet* **2**, 1121.
85. Eggo, M. C., and Burrow, G. N. (1982). *Endocrinology (Baltimore)* **111**, 1663.
86. Ekins, R. P. (1960). *Clin. Chim. Acta* **5**, 453.
87. Ekpechi, O. L. V., and van Middlesworth, L. (1973). *Endocrinology (Baltimore)* **92**, 1376.
87a. Ellis, K. J., George, J. M., and Laby, R. H. (1983). *Aust. J. Exp. Agric. Anim. Husb.* **23**, 369.
88. Elmer, A. W. (1938). "Iodine Metabolism and Thyroid Function." Oxford Univ. Press, London and New York.
89. Ermans, A. M., Bourdoux, P., Lagasse, R., Delange, F., and Thilly, C. (1980). *In* "Neonatal Thyroid Screening (G. N. Burrow, ed.), p. 61. Raven Press, New York.
90. Ermans, A. M., Moulameko, N. M., Delange, F., and Alhuwalia, R., eds. (1980). "Role of Cassava in the Etiology of Endemic Goiter and Cretinism." International Development Research Centre, Ottawa, Canada.
91. Evvard, J. M. (1928). *Endocrinology (Baltimore)* **12**, 539.
92. Falconer, I. R. (1965). *Nature (London)* **205**, 703.
93. Farer, L. S., Robins, J., Blumberg, B. S., and Rall, J. E. (1962). *Endocrinology (Baltimore)* **70**, 686.
94. Fierro-Benitez, R., Stanbury, J. B., Querido, A., De Groot, L., Alban, R., and Endova, J. (1970). *J. Clin. Endocrinol. Metab.* **30**, 228.
95. Fierro-Benitez, R., Ramirez, I., Estrella, E., and Stanbury, J. B. (1974). *In* "Endemic Goiter and Cretinism: Continuing Threats to World Health" (J. T. Dunn and G. A. Medeiros-Neto, eds.), Publ. No. 292, p. 135. Pan Am. Health Organ., Washington, D.C.
96. Fisher, D. A., and Dussault, J. H. (M. A. Greer and D. H. Solomon, eds.), (1974). *In* "Handbook of Physiology" Sect. 7, Vol. III, Chapter 21. Am. Physiol. Soc., Washington, D.C.
97. Fisher, D. A., Dussault, J. H., Sack, J., and Chopra, I. J. (1977). *Recent Prog. Horm. Res.* **33**, 59.
98. Fisher, K. D., and Carr, C. J. (1974). "Iodine in Foods: Chemical Methodology and Source

of Iodine in the Human Diet,'' FDA 71-294. Life Sciences Research Office, Bethesda, Maryland.

99. Follis, R. H., Jr. (1963). *Am. J. Clin. Nutr.* **14,** 253.
100. Franklin, J. A., Shepherd, M. C., Ramsden, D. B., Wilkinson, R., and Hoffenberg, R. (1984). *Clin. Endocrinol. (Oxford)* **20,** 107.
101. Fukuchi, M., Inoure, T., Abe, H., and Kumahara, Y. (1970). *J. Clin. Endocrinol. Metab.* **31,** 565.
102. Funke, H., Iwarsson, K., Olsson, S. O., Salomonsson, P., and Strandberg, P. (1975). *Nord. Veterinaermed.* **27,** 270.
103. Gaitan, E. (1980). *In* "Endemic Goiter and Endemic Cretinism" (J. B. Stanbury and B. S. Hetzel, eds.), p. 219. Wiley, New York.
104. Gaitan, E., Medina, P., DeRowen, T., and Zia, M. S. (1980). *J. Clin. Endocrinol. Metab.* **51,** 957.
105. Garry, P. J., Lashley, D. W., and Owen, G. M. (1973). *Clin. Chem. (Winston-Salem, N.C.)* **19,** 950.
106. Gharbi, J., and Torresani, J. (1979). *Biochem. Biophys. Res. Commun.* **88,** 170.
107. Glascock, R. F. (1954). *J. Dairy Res.* **21,** 318.
108. Goldschmidt, V. M. (1954). *In* "Geochemistry" (A. Muir, ed.). Oxford Univ. Press (Clarendon), London and New York.
109. Goolden, A. W. G., Gartside, J. M., and Sanderson, G. (1967). *Lancet* **1,** 12.
110. Gorge, F. B. de, and Jose, N. K. (1967). *Nature (London)* **214,** 491.
111. Goslings, B. M., Djokomoeljanto, R., Docter, R., van Hardeveld, C., Hennemann, G., Smeenk, D., and Querido, A. (1977). *J. Clin. Endocrinol. Metab.* **44,** 481.
112. Greenberg, A. A., Najjar, S., and Blizzard, R. M. (1974). *In* "Handbook of Physiology" (M. A. Greer and D. H. Solomon, eds.), Sect. 7, Vol. III, Chapter 22, p. 377. Am. Physiol. Soc., Washington, D.C.
113. Greer, M. A., Ettlinger, M. G., and Astwood, E. B. (1949). *J. Clin. Endocrinol.* **9,** 1069.
114. Greer, M. A., Studer, H., and Kendall, J. W. (1967). *Endocrinology (Baltimore)* **81,** 623.
115. Griesbach, W. E., Kennedy, T. H., and Purves, H. D. (1941). *Br. J. Exp. Pathol.* **22,** 245.
116. Griesbach, W. E., and Purves, H. D. (1943). *Br. J. Exp. Pathol.* **24,** 174.
117. Gross, J., and Pitt-Rivers, R. (1952). *Lancet* **1,** 439.
118. Guillemin, R., Yamazaki, E., and Gard, D. A. (1963). *Endocrinology (Baltimore)* **73,** 564.
119. Gurevich, G. P. (1960). *Nutr. Abstr. Rev.* **30,** 697.
120. Gurevich, G. P. (1964). *Fed. Proc., Fed. Am. Soc. Exp. Biol.* **23,** Trans. Suppl., T511.
121. Hallman, B. L., Bondy, P. K., and Hagewood, M. A. (1951). *Arch. Intern. Med.* **87,** 817.
122. Hamilton, E. I., and Minski, M. J. (1972/73). *Sci. Total Environ.* **1,** 375.
123. Hamilton, E. I., Minski, M. J., and Cleary, J. J. (1972/73). *Sci. Total Environ.* **1,** 341.
124. Harden, R. McG., Mason, D. K., and Alexander, W. D. (1966). *Q. J. Exp. Physiol. Cogn. Med. Sci.* **51,** 130.
125. Harden, R. McG., Hilditch, T., Kennedy, I., Mason, D. K., Papadopoulos, S., and Alexander, W. D. (1967). *Clin. Sci.* **32,** 49.
126. Harden, R. McG., Alexander, W. D., Shimmins, J., and Robertson, J. (1968). *Q. J. Exp. Physiol. Cogn. Med. Sci.* **53,** 227.
127. Harden, R. McG., Alexander, W. D., Shimmins, J., Kostalas, N., and Mason, D. K. (1968). *J. Lab. Clin. Med.* **71,** 92.
128. Hartmans, J. (1974). *Neth. J. Agric. Sci.* **22,** 195.
129. Heinrich, H. C., Gabbe, E. E., and Whang, D. H. (1964). *Atomkernenergie* **9,** 279.
130. Hemkin, R. W., Vandersall, J. H., Oskarsson, M. A., and Fryman, L. R. (1972). *J. Dairy Sci.* **55,** 931.

131. Hennessy, W. B. (1964). *Med. J. Aust.* **1,** 505.
132. Herveg, J. P., Beckers, C., and De Vissher, M. (1966). *Biochem. J.* **100,** 540.
133. Hetzel, B. S. (1983). *Lancet* **2,** 1126.
134. Hetzel, B. S. (1985). "Toward a Global Strategy for the Eradication of Iodine Deficiency Disorders (IDD)," Report submitted to the Administrative Committee on Coordination-Sub-committee on Nutrition, United Nations, Nairobi.
135. Hetzel, B. S., and Hay, I. D. (1979). *Clin. Endocrinol. (Oxford)* **11,** 445.
135a. Hetzel, B. S., and Hales, I. B. (1980). *In* "Endemic Goiter and Endemic Cretinism" (J. B. Stanbury and B. S. Hetzel, eds.), p. 123. Wiley, New York.
136. Hetzel, B. S., Thilly, C. H., Fierro-Benitez, R., Pretell, E. A., Buttfield, I. H., and Stanbury, J. B. (1980). *In* "Endemic Goiter and Endemic Cretinism" (J. B. Stanbury and B. S. Hetzel, eds.), p. 513. Wiley, New York.
137. Hetzel, B. S., and Potter, B. J. (1983). *In* "Neurobiology of the Trace Elements" (I. Dreosti and R. M. Smith, eds.), Vol. I, p. 45. Humana Press, Clifton, New Jersey.
138. Hetzel, B. S., Potter, B. J., Mano, M., Belling, B., McIntosh, G., and Cragg, B. G. (1984). *In* "Endocrinology" (F. Labrie and L. Proulx, eds.), pp. 731–734. Proc. Int. Congr. Endocrinol., 7th, Quebec. Excerpta Medica, Amsterdam.
139. Hollingsworth, D., Fisher, D. A., and Pretell, E. A. (1980). *In* "Endemic Goiter and Endemic Cretinism" (J. B. Stanbury and B. S. Hetzel, eds.), p. 423. Wiley, New York.
140. Howell, G. L., and van Middlesworth, L. (1956). *Proc. Soc. Exp. Biol. Med.* **93,** 602.
141. Hughes, J. N., Reinberg, A., Jordan, D., Seabauoun, J., and Modigliani, E. (1982). *Acta Endocrinol. (Copenhagen)* **101,** 403.
142. Hunnikin, C., and Wood, F. O. (1980). *In* "Endemic Goiter and Endemic Cretinism" (J. B. Stanbury and B. S. Hetzel, eds.), p. 497. Wiley, New York.
143. Hunter, W. M., and Corrie, J. E. T. (1983). "Immunoassays in Clinical Chemistry." Churchill-Livingstone, Edinburgh and London.
144. Ingbar, S. H., Braverman, L. E., Dawber, N. A., and Lee, G. Y. (1965). *J. Clin. Invest.* **44,** 1679.
145. "Iodine and Plant Life" (1950). Chilean Iodine Educational Bureau, London.
146. "Iodine Content of Foods" (1952). Chilean Iodine Education Bureau, London.
147. Irvine, C. H. G. (1967). *Am. J. Vet. Res.* **28,** 1687.
148. Johnson, J. M., and Butler, G. W. (1957). *Physiol. Plant.* **10,** 100.
149. Jordan, D., Rousset, B., Perrin, F., Fournier, M., and Orgiazzi, J. (1980). *Endocrinology (Baltimore)* **107,** 1245.
150. Judd, A. M., and Hedge, G. A. (1983). *Endocrinology (Baltimore)* **113,** 706.
151. Karmarkar, M. G., Deo, M. G., Kochupillai, N., and Ramalingaswami, V. (1974). *Am. J. Clin. Nutr.* **27,** 96.
152. Kawano, H., Daikoku, S., and Saito, S. (1983). *Endocrinology (Baltimore)* **112,** 951.
153. Keating, F. R., Jr., and Albert, A. (1949). *Recent Prog. Horm. Res.* **4,** 429.
154. Kendall-Taylor, P. (1972). *J. Endocrinol.* **52,** 533.
155. Kendall-Taylor, P. (1972). *J. Endocrinol.* **54,** 137.
156. Kirchgessner, M. (1959). *Z. Tierphysiol., Tierernaehr. Futtermittelkd.* **14,** 270 and 278.
157. Kochupillai, N. (1983). Personal communication, All India Institute of Medical Sciences, Delhi.
158. Kochupillai, N., Pandav, C. S., and Karmarkar, M. G. (1984). *Indian J. Med. Res.* **80,** 293.
159. Konig, M. P., and Veraguth, P. (1961). *In* "Advances in Thyroid Research" (R. Pitt-Rivers, ed.), p. 294. Pergamon, Oxford.
160. Koutras, D. A. (1968). *In* "Activation Analysis in the Study of Mineral Metabolism in Man," IAEA, Vienna.

161. Koutras, D. A., Matovinovic, J., and Vought, R. (1980). *In* "Endemic Goiter and Endemic Cretinism" (J. B. Stanbury and B. S. Hetzel, eds.), p. 185. Wiley, New York.
162. Kupzis, J. (1932). *Z. Hyg. Infektionskr.* **113**, 551.
163. Kywe-Thein, Tin-Tin-oo, Khin-Maung-Niang, Wrench, J., and Buttfield, I. H. (1979). *In* "Current Thyroid Problems in Southeast Asia and Oceania" (B. S. Hetzel, M. L. Wellby, and R. Hoschl, eds.), Proc. Asia Oceania Thyroid Assoc. Workshops Endemic Goiter Thyroid Test. p. 78.
164. Laby, R. H. (1975). Private communiation.
165. Larsen, P. R., Silva, J. E., Hetzel, B. S., and McMichael, A. J. (1980). *In* "Endemic Goiter and Endemic Cretinism" (J. B. Stanbury and B. S. Hetzel, eds.), p. 551. Wiley, New York.
166. Lechan, R. M., Molitch, M. E., and Jackson, I. M. D. (1983). *Endocrinology (Baltimore)* **112**, 877.
167. Lewis, R. C., and Ralston, N. P. (1953). *J. Dairy Sci.* **36**, 33.
168. Li, J., Wang, X., Yan, Y., Wang, K., Qin, D., Xin, Z., and Wei, J. (1985). *Neuropathol. Appl. Neurobiol.* (in press).
169. Li, J., and Jun, W. (1985). *Nutr. Rep. Int.* **31**, 1085.
170. Li, J., He, Q., and Wang, X. (1981). *J. Endemic Correspondence* **3**, 1.
171. Liu, Z. (1983). *In* "Current Problems in Thyroid Research" (N. Ui, K. Torizuka, S. Nagataki, and K. Miyai, eds.), p. 410. Excerpta Medica, Amsterdam.
172. Loosli, J. K., Becker, R. B., Huffman, C. F., Phillips, P. H., and Shaw, J. C. (1956). "Nutrient Requirements of Dairy Cattle." Natl. Res. Counc., Washington, D.C.
173. Ma, T., Lu, T., Tan, U., Chen. B., and Chu, H. I. (1982). *Food Nutr. Bull.* **4**, 13.
174. Ma, T., Chen, X. X., Bai, G., Chen, Z. P., Zu, M. Q., Suen, S. X., Zuen, X. R., Yang, L. C., Yang, H. X., Li, C. G., Suen, J. Y., and Ou, Y. C. (1983). *In* "Current Problems in Thyroid Research" (N. Ui, K. Torizuka, S. Nagataki, and K. Miyai, eds.), p. 366. Excerpta Medica, Amsterdam.
175. Maberly, G. F., Eastman, C. J., and Corcoran, J. M. (1978). *In* "Current Thyroid Problems in South-east Asia and Oceania" (B. S. Hetzel, M. L. Wellby, and R. Hoschl, eds.), Proc. Asia Oceania Thyroid Assoc. Workshops Endemic Goiter Thyroid Test., p. 21.
176. Maberly, G. F., Eastman, C. J., and Corcoran, J. (1981). *Lancet* **2**, 1270.
177. Maberly, G. F., Corcoran, J. M., and Eastman, C. J. (1982). *Clin. Endocrinol. (Oxford)* **17**, 253.
178. Maberly, G. F., Waite, K. V., Eastman, C. J., and Corcoran, J. (1983). *In* "Current Problems in Thyroid Research" (N. Ui, K. Torizuka, A. Nagataki, and K. Miyai, eds.), p. 341. Excerpta Medica, Amsterdam.
179. Malamos, B., and Koutras, D. A. (1962). *Proc. Int. Congr. Intern. Med., 7th, 1960* Vol. 2, p. 678.
180. Malamos, P., Koutras, D. A., Rigopoulos, G. A., and Papapetrou, P. D. (1971). *J. Clin. Endocrinol. Metab.* **32**, 130.
181. Maqsood, M. (1952). *Biol. Rev. Cambridge Philos. Soc.* **27**, 281.
182. Marcilese, N. A., Harms, R. H., Valsechhi, R. M., and Arrington, L. R. (1968). *J. Nutr.* **94**, 117.
183. Marine, D., and Williams, W. W. (1908). *Arch. Intern. Med.* **1**, 349.
184. Marine, D., and Lenhart, C. H. (1909). *Arch. Intern. Med.* **3**, 66.
185. Marine, D., and Kimball, P. O. (1921). *JAMA, J. Am. Med. Assoc.* **77**, 1068.
186. Mason, R. S., and Laby, R. (1978). *Aust. J. Exp. Agric. Anim. Husb.* **18**, 653.
187. Maurer, E., and Ducrue, H. (1928). *Biochem. Z.* **193**, 356.
188. McCarrison, R. (1908). *Lancet* **2**, 1275.
189. McClelland, J. (1935). *Trans. Med. Phys. Soc. (Calcutta)* **7**, 145.

190. McCullagh, S. F. (1963). *Med. J. Aust.* **1,** 769.
191. McDonald, R. J., McKay, G. W., and Thomson, J. D. (1962). *Proc.—Int. Congr. Anim. Reprod. Artif. Insemin., 4th, 1961* Vol. 3, p. 679.
192. McIntosh, G. H., Potter, B. J., Mano, M. T., Hua, C. H., Cragg, B. G., and Hetzel, B. S. (1983). *Neuropathol. Appl. Neurobiol.* **9,** 215.
193. McMichael, A. J., Potter, J. D., and Hetzel, B. S. (1980). *In* "Endemic Goiter and Endemic Cretinism" (J. B. Stanbury and B. S. Hetzel, eds.), p. 445. Wiley, New York.
194. Mendewics, J. (1982). *J. Psychiatr. Res.* **7,** 388.
195. Merke, F. (1965). *Schweiz. Med. Wochenschr.* **95,** 1183.
196. Meyer, R. J., Internal Report. Morton Salt Co., Chicago, Illinois (unpublished).
197. Middlesworth, L. van (1956). *J. Clin. Endocrinol. Metab.* **16,** 989.
198. Miller, J. K. (1966). *Proc. Soc. Exp. Biol. Med.* **121,** 291.
199. Miller, J. K., Moss, B. R., Swanson, E. W., Aschbacher, P. W., and Cragle, R. G. (1968). *J. Dairy Sci.* **51,** 1831.
200. Miller, J. K., and Swanson, E. W. (1973). *J. Dairy Sci.* **56,** 378.
201. Miller, J. K., Swanson, E. W., Lyke, W. A., Moss, B. R., and Byrne, W. F. (1974). *J. Dairy Sci.* **57,** 193.
202. Mitchell, H. H., and McClure, F. J. (1937). *Bull. Natl. Res. Counc. (U.S.)* **99.**
203. Miyake, Y., and Tsunogai, S. (1963). *J. Geophys. Res.* **68,** 3989.
204. Moberg, R. (1959). *Proc. World Congr. Fertil. Steril.,* p. 71.
205. Morton, T., and Guillaume, H. (1984). CSIRO Internal Report.
206. Murphy, B. E. P., Pattee, C. J., and Gold, A. (1966). *J. Clin. Endocrinol. Metab.* **26,** 247.
207. Mussett, M. V., and Pitt-Rivers, R. (1954). *Lancet* **2,** 1212.
208. Newton, G. L., and Clawson, T. A. (1974). *J. Anim. Sci.* **39,** 879.
209. Newton, G. L., Barrick, E. R., Harvey, R. W., and Wise, M. B. (1974). *J. Anim. Sci.* **38,** 449.
210. "Nutrient Requirements of Farm Livestock" (1966). No. 2, p. 104. Agricultural Research Council of Great Britain, H. M. Stationery Office, London.
211. Nutrition Foundation of India (1983). The National Goitre Control Program. *Sci. Rep.* **1.**
212. Obregon, M. J., Mallot, J., Paston, R., Morreale de Escobar, G., and Escobar del Rey, F. (1984). *Endocrinology (Baltimore)* **114,** 305.
213. Oddie, T. H., Fisher, D. A., McConahey, W. M., and Thompson, C. S. (1970). *J. Clin. Endocrinol. Metab.* **30,** 659.
214. Oppenheimer, J. H. (1979). *Science* **203,** 971.
215. Oppenheimer, J. H., Schwartz, H. L., Surks, M. I., Koerner, D., and Dillman, W. H. (1976). *Recent Prog. Horm. Res.* **32,** 529.
216. Orr, J. B., and Leitch, I. (1929). *Med. Res. Counc. (G.B.), Spec. Rep. Ser.* **SRS-123.**
217. Ouyang, A., Wang, P. O., Liu, Z. T., Lin, F. F., and Wang, H. M. (1983). *In* "Current Problems in Thyroid Research" (N. Ui, K. Torizuka, S. Nagataki, and K. Miyai, eds.), p. 418. Excerpta Medica, Amsterdam.
218. Pandav, C. S., and Kochupillai, N. (1982). *Indian J. Pediatr.* **50,** 259.
219. Papadopoulos, S., MacFarlane, S., Harden, R. McG., Mason, D. K., and Alexander, W. D. (1966). *J. Endocrinol.* **36,** 341.
220. Park, Y. K., Harland, B. F., Vanderveen, J. E., and Shank, F. R.(1981). *J. Am. Diet. Assoc.* **79,** 17.
221. Patel, Y., Pharoah, P. O. D., Hornabrook, R., and Hetzel, B. S. (1973). *J. Clin. Endocrinol. Metab.* **37,** 783.
223. Penny, R., Spencer, C. A., Frasier, D., and Nicoloff, J. T. (1983). *J. Clin. Endocrinol. Metab.* **56,** 177.

223. Perdomo, J. T., Harms, R. H., and Arrington, L. R. (1966). *Proc. Soc. Exp. Biol. Med.* **122,** 758.
224. Pharoah, P. O. D., Buttfield, I. H., and Hetzel, B. S. (1971). *Lancet* **1,** 308.
225. Pharoah, P. O. D., Lawton, N. F., Ellis, S. M., Williams, E. S., and Ekins, R. P. (1973). *Clin. Endocrinol. (Oxford)* **2,** 193.
226. Pharoah, P. O. D., Delange, F., Fierro-Benitez, R., and Stanbury, J. B. (1980). *In* "Endemic Goiter and Endemic Cretinism" (J. B. Stanbury and B. S. Hetzel, eds.), p. 395. Wiley, New York.
227. Pharoah, P. O. D., Connolly, K. J., Ekins, R. P., and Harding, A. G. (1984). *Clin. Endocrinol. (Oxford)* **21,** 265.
228. Pierce, J. G. (1971). *Endocrinology (Baltimore)* **89,** 1331.
229. Pittman, C. S., Chambers, J. B., Jr., and Reed, V. H. (1971). *J. Clin. Invest.* **50,** 1187.
230. Pitt-Rivers, R., and Tata, J. R. (1959). "The Thyroid Hormones." Pergamon, Oxford.
231. Pliam, N. G., and Goldfine, I. H. (1977). *Biochem. Biophys. Res. Commun.* **79,** 166.
232. Post, T. B., and Mixner, J. P. (1961). *J. Dairy Sci.* **44,** 2265.
233. Potter, B. J., McIntosh, G. H., and Hetzel, B. S. (1981). *In* "Fetal Brain Disorders" (B. S. Hetzel and R. M. Smith, eds.), p. 119. Elsevier, Amsterdam.
234. Potter, B. J., Mano, M. T., Belling, G. B., McIntosh, G. H., Hua, C., Cragg, B. G., Marshall, J., Wellby, M. L., and Hetzel, B. S. (1982). *Neuropathol. Appl. Neurobiol.* **8,** 303.
235. Potter, B. J., Mano, M., Belling, B., and Hetzel, B. S. (1984). *Proc. Endocr. Soc. Aust.* **27,** 26.
236. Pretell, E. A., Torres, T., Zenteno, V., and Cornejo, M. (1972). *In* "Human Development and the Thyroid Gland: Relation to Endemic Cretinism" (J. B. Stanbury and R. L. Kroc, eds.), p. 249. Plenum, New York.
237. Purves, H. D. (1943). *Br. J. Exp. Pathol.* **24,** 171.
238. Querido, A., Delange, F., Dunn, J. T., Fierro-Benitez, R., Ibbertson, H. K., Koutras, D. A., and Perinetti, H. (1974). *In* "Endemic Goitre and Cretinism: Continuing Threats to World Health (J. T. Dunn and G. A. Medeiros-Neto, eds.), Publ. No. 292, pp, 267–272. Pan Am. Health Organ., Washington, D.C.
239. Raiha, H., and Hjelt, L. (1957). *Acta Paediatr. Scand* **72,** 610.
240. Ramzin, S., Kicic, M., Dordevic, S., and Todorovic, H. (1968). *Acta Med. Iugosl.* **22,** 77.
241. "Recommended Dietary Allowances" (1980). 9th ed. Natl. Acad. Sci., Washington, D.C.
242. Refetoff, S., Robin, N. I., and Fang, V. S. (1970). *Endocrinology (Baltimore)* **86,** 793.
243. Reineke, R. P., and Turner, C. W. (1944). *J. Dairy Sci.* **27,** 793.
244. Riggs, D. S. (1952). *Pharmacol. Rev.* **4,** 282.
245. Robbins, J., and Rall, J. E. (1960). *Physiol. Rev.* **40,** 415.
246. Roche, J., Michel, R., and Wolff, W. (1955). *C.R. Hebd. Seances Acad. Sci.* **240,** 251 and 921.
247. Roche, J., Michel, R., Truchot, R., and Wolff, W. (1955). *C.R. Seances Soc. Biol. Ses Fil.* **149,** 1219.
248. Rosenberg, L. L., Dimick, M. K., and Laroche, G. (1963). *Endocrinology (Baltimore)* **72,** 749.
249. Salter, W. T. (1950). *Hormones* **2,** 181.
250. Salter, W. T., and Johnson, McA. W. (1948). *J. Clin. Endocrinol.* **8,** 911.
251. Sceffer, L. (1933). *Biochem. Z.* **259,** 11.
252. Scott, M. L., van Tienhoven, A., Holm, E. R., and Reynolds, R. E. (1960). *J. Nutr.* **71,** 282.
253. Segal, J., and Ingbar, S. H. (1980). *In* "Endocrinology" (I. A. Cursming *et al.*, eds.), p. 405. Aust. Acad. Sci., Canberra.

254. Selby, J. B., Caldwell, J. G., Magoun, S. E., and Beihn, R. M. (1975). *Radiology* **114**, 107.
255. Setchell, B. P., Dickinson, D. A., Lascelles, A. K., and Bonner, R. B. (1960). *Aust. Vet. J.* **36**, 159.
256. Shacklette, H. T., and Cuthbert, M. E. (1967). *Spec. Pap.—Geol. Soc. Am.* **90**, 30.
257. Shambaugh, Gh. E., III, (1978). *In* "The Thyroid: A Fundamental and Clinical Text" (S. C. Werner and S. H. Ingbar, eds.), p. 115. Harper & Row, New York.
258. Shand, A. (1952). *Br. Vet. Assoc. Publ.* **23.**
259. Shepard, T. H. (1967). *J. Clin. Endocrinol. Metab.* **27**, 945.
260. Sherwin, J. R., and Tong, W. (1974). *Endocrinology (Baltimore)* **94**, 1465.
261. Shuman, A. C., and Townsend, D. P. (1963). *J. Anim. Sci.* **22**, 72.
262. Sihombing, D. T. H., Cromwell, G. L., and Hays, V. W. (1974). *J. Anim. Sci.* **39**, 1106.
263. Silva, J. E. (1983). *Prog. Clin. Biol. Res.* **116**, 23.
264. Sinclair, D. P., and Andrews, E. D. (1958). *N.Z. Vet. J.* **6**, 87.
265. Sinclair, D. P., and Andrews, E. D. (1959). *N.Z. Vet. J.* **7**, 39.
266. Sinclair, D. P., and Andrews, E. D. (1961). *N.Z. Vet. J.* **9**, 96.
267. Smythe, G. A., Bradshaw, J. E., Cai, W. Y., and Symons, R. G. (1982). *Endocrinology (Baltimore)* **111**, 1181.
268. Sooch, S. S., Deo, M. G., Karmarkar, M. G., Kochupillai, N., Ramachandrau, K., and Ramalingaswami, V. (1973). *Bull. W.H.O.* **49**, 307.
269. Srinivasan, V., Moudgal, N. R., and Sarma, P. S. (1957). *J. Nutr.* **61**, 87.
270. Stanbury, J. B., Brownell, G. L., Riggs, D. S., Perinetti, H., Itoiz, J., and Del Castillo, E. B. (1954). *In* "The Adaptation of Man to Iodine Deficiency," p. 1. Harvard Univ. Press, Cambridge, Massachusetts.
271. Stanbury, J. B., Ermans, A. M., Hetzel, B. S., Pretell, E. A., and Querido, A. (1974). *WHO Chron.* **28**, 220.
272. Stanbury, J. B., and Hetzel, B. S., eds. (1980). "Endemic Goiter and Endemic Cretinism: Iodine Nutrition in Health and Disease." Wiley, New York.
273. Sterling, K. (1970). *Recent Prog. Horm. Res.* **26**, 249.
274. Sterling, K., and Brenner, M. A. (1966). *J. Clin. Invest.* **45**, 155.
275. Sterling, K., and Milch, P. O. (1975). *Proc. Natl. Acad. Sci. U.S.A.* **72**, 3225.
276. Sterling, K., Lazarus, J. H., Milch, P. O., Sakurada, T., and Brenner, M. A. (1978). *Science* **201**, 1126.
277. Stewart, J. C., Vidor, G. I., Buttfield, I. H., and Hetzel, B. S. (1971). *Aust. N.Z. J. Med.* **1**, 203.
278. Stockight, J. R., Stevens, V., White, V. L., and Barlow, J. W. (1983). *Clin. Chem. (Winston-Salem, N.C.)* **29**, 1408.
279. Studer, H., Kohler, H., and Burgi, H. (1974). *In* "Handbook of Physiology" (M. A. Greer and D. H. Solomon, eds.), Sect. 7, Vol. III, p. 303. Am. Physiol. Soc., Washington, D.C.
280. Sturm, A., and Buchholz, B. (1928). *Dtsch. Arch. Klin. Med.* **161**, 227.
281. Suwanik, R. (1982). "The R & D Group, Iodine, Iron and Water Project." Siriraj Hospital, Mahidol Univ., Bangkok.
282. Suzuki, H. (1980). *In* "Endemic Goitre and Endemic Cretinism" (J. B. Stanbury and B. S. Hetzel, eds.), p. 237. Wiley, New York.
283. Tao, M. (1970). *Proc. Natl. Aca. Sci. U.S.A.* **67**, 408.
284. Tata, J. R. (1962). *Recent Prog. Horm. Res.* **18**, 221.
285. Tata, J. R., and Widnell, C. C. (1966). *Biochem. J.* **98**, 604.
286. Taurog, A. (1970). *Recent Prog. Horm. Res.* **26**, 189.
287. Taurog, A. (1974). *In* "Handbook of Physiology" Endocrinology" (R. O. L. Greep and E. B. Astwood, eds.), Sect. 7, Vol. III, p. 101. Am. Physiol. Soc., Washington, D.C.

288. Taurog, A. (1976). *Endocrinology (Baltimore)* **98,** 1031.
289. Taurog, A. (1978). *In* "The Thyroid: A Fundamental and Clinical Text" (S. C. Werner and S. H. Ingbar, eds.), p 31. Harper & Row, New York.
290. Taurog, A., and Chaikoff, I. L. (1946). *J. Biol. Chem.* **163,** 313.
291. Taurog, A., Chaikoff, I. L., and Feller, D. D. (1947). *J. Biol. Chem.* **171,** 189.
292. Thilly, C. H. (1981). *Bull. Acad. Med. Belg.* **136,** 389.
293. Thilly, C. H., Delange, F., Ramioul, L., Lagasse, R., Luvivila, K., and Ermans, A. M. (1977). *Int. J. Epidemiol.* **6,** 43.
293a. Thilly, C. H., and Hetzel, B. S. (1980). *In* "Endemic Goiter and Endemic Cretinism" (J. B. Stanbury and B. S. Hetzel, eds.), p. 475. Wiley, New York.
294. Thilly, C. H., Delange, F., and Stanbury, J. B. (1980). *In* "Endemic Goiter and Endemic Cretinism" (J. B. Stanbury and B. S. Hetzel, eds.), p. 157. Wiley, New York.
295. Thorell, C. B. (1964). *Acta Vet. Scand.* **5,** 224.
296. Thorson, S. C., Mincey, E. K., McIntosh, H. W., and Morrison, R. T. (1972). *Br. Med. J.* **2,** 67.
297. Tong, W. (1974). *In* "Handbook of Physiology" (M. A. Greer, and D. H. Solomon, eds.), Sect. 7, Vol. III, p. 255. Am. Physiol. Soc., Washington, D.C.
298. Trotter, W. R. (1960). *Br. Med. Bull.* **16,** 92.
298a. Underwood, E. J. (1977). "Trace Elements in Human and Animal Nutrition," 4th ed. Academic Press, New York.
299. Vidor, G. I., Stewart, J. C., Wall, J. R., Wangel, S., and Hetzel, B. S. (1973). *J. Clin. Endocrinol. Metab.* **37,** 901.
300. Vought, R. L. (1972). *In* "Trace Substance in Environmental Health-5" (D. D. Hemphill, ed.), p. 307. Univ. of Missouri Press, Columbia.
301. Vought, R. L., London, W. T., Lutwak, L., and Dublin, T. P. (1963). *J. Clin. Endocrinol. Metab.* **23,** 1218.
302. Vought, R. L., and London, W. T. (1964). *Am. J. Clin. Nutr.* **14,** 186.
303. Vought, R. L., London, W. T., and Stebbing, G. E. T. (1967). *J. Clin. Endocrinol. Metab.* **27,** 1381.
304. Vought, R. L., Brown, F. A., and Wolff, J. (1972). *J. Clin. Endocrinol. Metab.* **34,** 747.
305. Vought, R. L., Brown, F. A., and Sibiniovic, K. H. (1974). *J. Clin. Endocrinol. Metab.* **38,** 861.
306. Wang Dong, Chen Zu-pai, Lu T. Zhang, and Ma, T. (1984). *Chin. J. Endemic Dis.* (in press).
307. Watanabe, T., Moran, D., El Tamer, E., Staneloni, L., Salvaneschi, J., Altschuler, N., DeGrossi, O., and Niepominiszcse, H. (1974). *In* "Endemic Goiter and Cretinism: Continuing Threats to World Health" (J. T. Dunn and G. A. Medeiros-Neto, eds.), Publ. No. 292, p. 231. Pan Am. Health Organ. Sci., Washington, D.C.
308. Wayne, E. J., Koutras, D. A., and Alexander, W. D. (1964). "Clinical Aspects of Iodine Metabolism." Blackwell, Oxford.
309. Weeke, J., and Grunderson, B. J. (1983). *Acta Physiol. Scand.* **117,** 33.
310. Weeks, M. H., Katz, J., and Farnham, N. C. (1952). *Endocrinology (Baltimore)* **50,** 511.
311. Wellby, M. L., and O'Halloran, M. W. (1966). *Br. Med. J.* **2,** 668.
312. Wellby, M. L., Guthrie, L., and Riley, C. P. (1981). *Clin. Chem. (Winston-Salem, N.C.)* **27,** 2022.
313. Wellby, M. L., and O'Halloran, M. W. (1969). *Biochem. J.* **11,** 543.
314. Wenlock, R. W., Buss, D. H., Moxan, R. E., and Bunton, N. G. (1982). *Br. J. Nutr.* **47,** 381.
315. Wespi, H. J. (1945). *Schweiz. Med. Wochenschr.* **75,** 625.

316. Wilson, D. C. (1941). *Lancet* **1**, 211.
317. Woeber, K. A. (1968). *In* "Textbook of Endocrinology" (R. H. Williams, ed.). Saunders, Philadelphia, Pennsylvania.
318. Wolff, J. (1964). *Physiol. Rev.* **44**, 45.
319. Wolff, J. (1969). *Am. J. Med.* **47**, 101.
320. Wolff, J. (1972). *In* "The Thyroid and Biogenic Amines" (J. E. Rall and I. J. Kopin, eds.), p. 115. North-Holland Publ., Amsterdam.
321. Wolff, J., and Williams, J. A. (1973). *Recent Prog. Horm. Res.* **29**, 229.
322. Wollman, S. H. (1969). *In* "Lysosomes in Biology and Pathology" (J. T. Dingle and H. B. Fell, eds.), p. 483. North-Holland Publ., Amsterdam.
323. Wright, W. E., Christian, J. E., and Andrews, F. N. (1955). *J. Dairy Sci.* **38**, 131.
324. Wu, S. Y., Klein, A. H., Chopra, I. J., and Fisher, D. A. (1977). *Trans. Am. Thyroid Assoc. Meet.*, p. 15.
325. Yamagata, N., and Yamagata, T. (1972). *J. Radiat. Res.* **13**, 81.
326. Young, M., Crabtree, M. G., and Mason, I. M. (1936). *Med. Res. Counc. (G.B.), Spec. Rep. Ser.* **SES-217.**
326a. Zhong, F.-G., Cao, X. M., and Liu, J. L. (1983). *Chin. J. Pathol.* **12**, 205.
327. Zhu, X. Y (Chu, H. I.) (1983). *In* "Current Problems in Thyroid Research" (N. Ui, K. Torizuka, S. Nagataki, and K. Miyai, eds.), p. 13. Excerpta Medica, Amsterdam.
328. Zhu, X. Y., Lu, T. Z., Song, X. K., Li, X. T., Gao, E. M., Yang, H. M., Ma, T., Li, Y. Z., and Zhang, F. Q. (1983). *In* "Current Problems in Thyroid Research" (N. Ui, K. Torizuka, S. Nagataki, and K. Miyai, eds.), p. 360. Excerpta Medica, Amsterdam.
329. Zyl, A. van, and Kerrich, J. E. (1955). *S. Afr. J. Med. Sci.* **20**, 9.

3

Selenium

ORVILLE A. LEVANDER

U.S. Department of Agriculture
Agricultural Research Service
Beltsville Human Nutrition Research Center
Beltsville, Maryland

I. SELENIUM IN TISSUES AND FLUIDS

A. General Distribution

Selenium (Se) occurs in all cells and tissues of the body in concentrations that vary with the tissue and the amount and chemical form of selenium in the diet. Although the highest concentrations occur in the liver and kidneys, the largest total amount of selenium is in the muscle mass (20). Cardiac muscle is consistently higher in selenium than skeletal muscle (95,153,241). Selenium concentrations in the tissues reflect the level of dietary selenium over a wide range. In a study of the selenium level in the longissimus muscle of pigs from 13 locations in the United States with known differing natural dietary selenium intakes from 0.027 to 0.493 µg/g, a highly significant ($p < .01$) linear correlation of 0.95 between dietary selenium and tissue selenium concentration was established (209). Selenium deposition in the blood, muscle, liver, kidneys, and skin of chicks and poults has similarly been shown to bear a direct relationship to the inorganic selenium of the diet up to dietary levels of 0.2 to 0.3 µg/g (376). Increasing the dietary inorganic selenium further up to 0.8 µg/g resulted in higher levels in the liver and kidneys, but there was no appreciable increase in blood or muscle selenium. Increasing the dietary selenium to 0.67 µg/g by addition of organic selenium in soybean meal, fish meal, and wheat induced higher levels in muscle and blood than equivalent levels of dietary selenium as selenite. A similar superiority of natural selenium over selenite–selenium in increasing the levels in muscle and liver has been demonstrated in laying hens (216).

In rats fed a torula yeast (low-selenium) diet for 4 weeks, the selenium levels in the kidneys fell from 1.0 to 0.3 µg/g and in the liver from 0.7 to 0.1 µg/g (fresh basis) (40). Similar levels occur in these tissues in normal and selenium-deficient sheep (74,158,167,336), cattle (25), and pigs (241). Rat studies, however, have shown that certain other tissues, such as the testes and adrenals, tend to maintain their selenium levels during depletion (19), and this suggests important but as yet unknown functions for selenium in these tissues.

At toxic intakes of selenium—that is, 10–100 times or more greater than those normally ingested—much higher tissue selenium concentrations are usual than those just given. Levels as high as 5–7 µg/g in liver and kidneys and 1–2 µg/g in the muscles may be reached. Beyond these tissue levels excretion begins to keep pace with absorption (74,280,336,401). Even higher levels may be reached in the hair and hooves of severely affected animals, as discussed in Section I,D. The tissue selenium levels cited for selenotic animals are well above those that occur in animals treated at recommended rates to prevent selenium deficiency. Food products from such treated animals do not contain undesirably or dangerously high selenium concentrations (9,74,88,153,214,318).

Comparatively few data have appeared on the selenium levels in normal human tissues, other than blood. Schroeder et al. (365) reported the following values for autopsy specimens from six U.S. subjects: kidneys 0.61–1.84 (mean 1.09), liver 0.28–0.81 (mean 0.54), and muscle 0.11–0.38 (mean 0.24) µg selenium/g of wet tissue. Thus, it can be calculated (227) that the skeletal muscle mass would contain almost half of the total-body selenium content of 14.6 mg thought typical of North Americans (365). Similar mean values for the selenium content of human liver (0.44) and muscle (0.37) were found in Canadian autopsy material (82), but tissue samples from England were somewhat lower, 0.30 and 0.11 µg/g, respectively (152). Skeletal muscle samples taken from New Zealanders contained a mean of only 0.06 µg selenium/g recalculated on a fresh-weight basis (55), and this is consistent with their low dietary selenium intake and the observation that their total-body selenium content is less than half that of North Americans (405). The mean selenium content of adult human livers collected in three different surveys in New Zealand (55) was 0.23 µg/g (recalculated on a fresh-weight basis), which is less than half that reported in North Americans (365) but still more than four times the level considered by Andrews and co-workers as indicative of marginal selenium deficiency in sheep (9).

B. Selenium in Blood

The concentration of selenium in blood is highly responsive to changes in the le⁻el in the diet over a wide range. For example, the blood of sheep fed highly toxic amounts of selenium can contain selenium at levels as high as 1.34 to 3.1

$\mu g/ml$ (336,346). In sheep suffering from various selenium-responsive diseases, . selenium levels as low as 0.01 to 0.02 $\mu g/ml$ occur (74,158,167). Similar extreme values have been reported in human samples (474), since blood selenium levels ranged from 1.3 to 7.5 (mean 3.2) $\mu g/ml$ in persons residing in an area of China with endemic selenosis to 0.021 ± 0.010 (mean ± SD) $\mu g/ml$ found in people from an area with Keshan disease, a selenium-responsive cardiomyopathy (see Section III,I). The latter value is less than the 0.05 $\mu g/ml$ thought to be "satisfactory" for sheep by Hartley (158).

Less extreme values are usually found in human blood from other countries. For example, the selenium levels in whole human blood from 210 donors in 19 sites in the United States were reported to range from 0.10 to 0.34 $\mu g/ml$ (7). Some evidence was obtained of a geographical pattern reflecting established regional differences in the selenium levels in crops. Later more extensive surveys have verified this pattern, since mean whole-blood selenium levels in the United States varied from 0.161 $\mu g/ml$ in Ohio, a low-selenium area (402), to 0.265 $\mu g/ml$ in South Dakota, a high-selenium area (177). Lower values have been reported in countries with low-selenium soils. For example, in Finland selenium levels in whole blood ranged from 0.056 to 0.087 $\mu g/ml$ (454), whereas samples from Dunedin on the South Island of New Zealand averaged 0.062 $\mu g/ml$ (421). Still lower mean blood selenium levels, 0.048–0.060 $\mu g/ml$, were found in children from New Zealand (421).

C. Forms of Selenium in Blood and Tissues

Selenium occurs in animal tissues partly bound to proteins in a manner incompletely understood, partly incorporated into proteins as selenium analogs of the sulfur-containing amino acids, and as the main functioning form of selenium, glutathione peroxidase (GSH-Px). Some of the less well-characterized possible forms of selenium in tissue proteins include selenotrisulfide linkages, S—Se—S (116,191) and an acid-labile form presumed to be selenide (87). The ability of the animal to convert inorganic selenium into selenoamino acids, as occurs in plants, has been questioned (77,191) since McConnell and Wabnitz first reported that selenomethionine and selenocystine were present in dog liver protein (260). It has been suggested that nonruminant animals cannot synthesize selenomethionine *de novo* from inorganic selenium compounds (77,311). Fuss and Godwin reported the conversion of small though significant amounts of intravenously injected selenite–selenium into the selenomethionine and selenocystine of tissue proteins of sheep within 48 hr (107). Others have found that rats and rabbits can convert selenite into selenocysteine (129,311), the form of selenium at the active site of GSH–Px (215,480). However, the mechanism by which the selenium is inserted into the enzyme is still unresolved (161,409).

GSH-Px was identified as a selenoprotein by Rotruck *et al.* (349) in 1973, and in the same year Flohe and co-workers (103) showed that the selenium in the enzyme was present in stoichiometric amounts with 4 gram atoms selenium per mole. GSH-Px catalyzes the reduction of hydrogen peroxide or lipid peroxides according to the following general reaction:

$$ROOH + 2GSH \xrightarrow{\text{GSH-Px}} ROH + GSSG + H_2O$$

The biochemical properties of this enzyme have been reviewed (68). GSH-Px activity has been demonstrated in a wide range of body tissues, fluids, cells, and subcellular fractions at levels that vary greatly with the species, tissue, and selenium status of the animal (119). The highest GSH-Px activity commonly occurs in the liver, moderately high activity in the erythrocytes, heart muscle, lung, and kidneys, and smaller activity in the intestinal tract and skeletal muscles. The dramatic dependence of the GSH-Px activity of the tissues on dietary selenium intakes is evident from numerous studies with several species (62,131,148,317,375). For example, Scott (375) found that the level of GSH-Px in the tissues of chicks made selenium deficient dropped to ~10% of the normal value in 5 days. Hafeman and co-workers (148) observed a fall in liver GSH-Px activity to "undetectable" levels within 24 days of feeding an unsupplemented torula yeast diet to weanling rats. The response of erythrocyte GSH-Px activity to the lack of dietary selenium was smaller in magnitude and more gradual, but in both tissues rapid elevation of GSH-Px activity followed selenium supplementation. Godwin and co-workers (131) examined the tissues of 2- to 3-week-old lambs from selenium-deprived and selenium-supplemented ewes. Even at this early stage significant reductions in GSH-Px activities in the erythrocytes and muscles and a small reduction in the plasma were evident in the lambs from the deficient ewes. In ovine skeletal and cardiac muscle (131), rat liver (136), and chicken liver (302), the major proportion of the GSH-Px is localized in the soluble or cytosolic fraction. Most of the noncytosolic, particulate portion of liver GSH-Px is present in the mitochondria (102) or in the nuclear and mitochondrial fractions (302).

This relationship between GSH-Px activity and dietary selenium intake has also been demonstrated in humans since Thomson et al. (425) showed a good correlation between human whole-blood GSH-Px activities and selenium concentrations in New Zealanders of low-selenium status (Fig. 1). However, the activity of the enzyme did not increase appreciably when blood selenium levels exceeded 0.10 µg/ml, and it was suggested that intakes that maintained this concentration were sufficient for function as judged by the GSH-Px activity. Others have also found no correlation between blood GSH-Px activity and blood selenium level in subjects more fully replete with selenium (360).

It appears that there are significant variations in the amount of tissue selenium

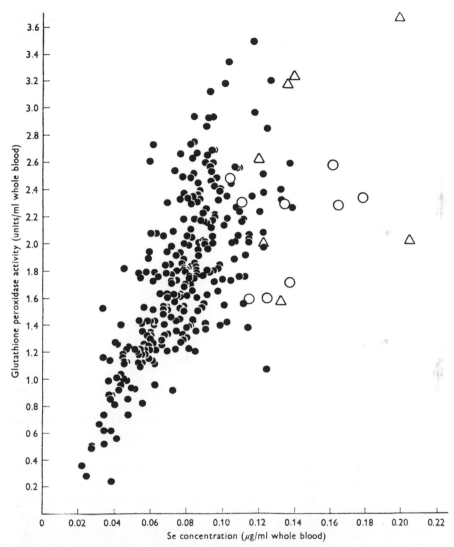

Fig. 1. Relationship of glutathione peroxidase activity and selenium content of human whole blood. From Thomson *et al.* (425). Values from 264 New Zealand residents (●); 9 New Zealand residents returned from overseas visits (○); and 7 new settlers to New Zealand (△).

that is associated with GSH-Px among various species and even among different tissues for the same species. For example, Beilstein and Whanger found that <10% of the erythrocyte selenium is associated with GSH-Px in humans, whereas >70% was associated with GSH-Px in rats and sheep (21). Behne and Wolters showed that the amount of selenium bound to GSH-Px in rat erythrocytes, liver, lungs, heart, kidneys, muscle, brain, and testes was 82, 63, 22, 20, 11, 5, 3, and <1% of the total tissue selenium, respectively (20). Such great species and tissue variations, of course, raise serious questions about the use of GSH-Px activity as a tool for assessing selenium status.

Many other less well-defined selenium-containing proteins have been described in animals. For example, the presence of a selenium-containing cytochrome in ovine muscle has been suggested (459), but this material has not been further characterized. Several reports have now appeared concerning a selenoprotein of 15 to 20 kilodaltons in bovine or rat sperm (50,261,319) that might be of significance in reproduction given the distinctive metabolism of selenium in the testis (19). A number of workers have found selenoproteins in rat plasma that are distinct from GSH-Px and that might have a role in selenium transport (see Section II,B).

D. Selenium in Keratinous Tissues

The selenium level of the hair of cattle is a useful indicator of both deficiency and toxicity. In one study cows with hair selenium levels between 0.06 and 0.23 $\mu g/g$ produced calves sick or dead from white muscle disease (WMD), whereas no WMD was observed in calves from dams with hair selenium levels >0.25 $\mu g/g$ (166). By contrast, the hair of yearling cattle on a seleniferous range averaged >10 $\mu g/g$, with values as high as 30 $\mu g/g$, compared with 1 to 4 $\mu g/g$ for the hair of cattle from unaffected areas (313). It was concluded that hair selenium values consistently <5 $\mu g/g$ indicate that the diet is unlikely to contain sufficient selenium to induce clinical signs of selenosis. On the other hand, Wahlstrom et al. (444) reported that the selenium content of hair from poisoned swine is related to its color and may not be a satisfactory index of dietary selenium intake.

Hair selenium content has been used to assess selenium status in humans in the People's Republic of China, where an excellent correlation was found between hair and blood levels (60). Hair levels <0.16 $\mu g/g$ were associated with Keshan disease, whereas mean levels of 32.2 $\mu g/g$ were seen in areas with endemic human selenosis (474). However, Gallagher et al. found no significant correlation between hair and whole-blood selenium values in eight North American subjects before and after selenium supplementation (109). Hair selenium levels in Western societies can also be greatly affected by the use of selenium-containing shampoos (81).

The high selenium content of human fingernails compared with other normal tissues has been pointed out by Hadjimarkos and Shearer (147). The range reported for 16 individuals was 0.70–1.69 and the mean 1.14 ± 0.06 μg selenium per gram. Morris and co-workers have advocated the use of toenails for assessing human selenium status and have presented some data showing dif-

ferences in selenium levels in toenails from areas rich and poor in the element
(276).

The feathers of chicks fed low-selenium diets for 64 weeks contained ~0.3 µg
selenium per gram compared with 4 to 5 times this level when the diets were
supplemented with 2 µg/g of selenium as selenite and 10 times this level (i.e.,
3.3–3.4 µ/g) when the diet was similarly supplemented at a level of 8 µg/g (14).

E. Selenium in Milk

The concentration of selenium in cow's milk varies greatly with the selenium
intake of the animal. In a study of pasteurized milk from different areas of New
Zealand, a three-fold variation from the highest to the lowest areas was found.
reflecting the selenium status of the soils and pastures of these areas (270). The
actual range reported was 2.9 ± 0.7 to 9.7 ± 0.7 ng/ml, with a mean of samples
from eight areas of 4.9 ng/ml. Allaway et al. (7) reported that cow's milk from a
low-selenium area in Oregon contained <20 ng/ml, compared with 50 ng/ml
from a high-selenium area in South Dakota. Higher levels, ranging between 160
and 1270 ng/ml, have been reported for cow's milk from high-selenium rural
areas in the United States (348). However, when milk from such areas is sent to
centralized dairies for processing, the high-selenium milks appear to be diluted
out in the distribution system so that the range of selenium content reported in
South Dakota milk was 32–138 ng/ml (312).

The selenium level in the milk is readily raised by supplementation of the
animal's diet. Grant and Wilson (132) obtained substantial milk selenium increases over a period of 3 to 4 weeks from cows receiving a single oral or
subcutaneous dose of 50 mg selenium as selenate, while the levels in the untreated cows remained steady at 3 to 4 ng/ml. The selenium concentration of the
milk of ewes fed a low-selenium diet was tripled by supplementing this diet with
2.25 mg selenium per day (122), and that of sows was doubled by supplementing
a low-selenium diet (0.03 and 0.05 µg/g) with sodium selenite at a selenium
level of 0.1 µg/g (250). The actual selenium levels in the milk were 13–15
ng/ml in the control sows and 25–29 ng/ml in the selenium-supplemented sows.
The levels of selenium in the colostrum of these sows were 43 and 47 ng/ml and
80 and 106 ng/ml, respectively. The selenium in milk is largely associated with
the protein fractions (255,269).

It was reported that in a low-selenium country like New Zealand, human milk
contains about twice as much selenium (11.5–14.5 ng/ml) as normal cow's milk
(5 ng/ml) despite its lower protein content (269). However, mean selenium
levels as low as 7.6 ng/ml were observed in human milk samples collected on the
South Island of New Zealand (468). A survey in the United States showed that
the mean selenium content for mature human milk from 241 subjects from 17
sites was 18 ng/ml (388). Geographical variations were apparent in this study,

but most of the individual values fell within the relatively narrow range of 7 to 33 ng selenium/ml. Some workers have observed a decline in the selenium content of human breast milk as lactation progresses. For example, the milk levels decreased from 20 to 15 ng/ml in the United States and from 10.7 to 5.8 ng/ml in Finland during the first and third months of lactation, respectively (211,234). However, others found no effect of the stage of lactation on the selenium content of breast milk from North American mothers (395). The most extreme values for the selenium content of human milk were 2.6 and 283 ng/ml for Keshan disease and endemic human selenosis areas in China, respectively (475). Studies carried out in Finland with mothers of low selenium status showed that it was possible to raise the selenium content of human milk and thereby increase the serum levels of the breast-fed infant by maternal supplementation (212).

F. Selenium in the Avian Egg

Under normal dietary conditions a hen's egg contains a total of 10 to 12 µg selenium, most of which is present in the yolk (416). The total amount and the proportions present in the yolk and white are markedly and rapidly influenced by the selenium status of the hen's diet and by the chemical form or forms in which the dietary selenium is supplied. Thus, Cantor and Scott (51) found the selenium concentration in the eggs from groups of hens receiving 20, 40, 60, and 80 ng/g of dietary selenium to be approximately equal to the level in the diet within 2 weeks of feeding the respective diets. Similarly, Arnold et al. (13), employing much higher dietary selenium levels, observed that the egg selenium concentration increased from ~50 ng/g for hens receiving a practical diet containing 500 ng/g of naturally occurring selenium to a maximum of 1.7 µg/g in 12 days when this diet was supplemented with 8 µg/g selenium as selenite. Within 8 days of withdrawing the supplemental selenium, the egg selenium levels returned to preexperimental values. The superiority of natural dietary selenium over selenite–selenium with respect to transfer to the egg is apparent from the work of Latshaw (216). This worker fed hens for 180 days a diet that contained either 0.10 µg/g natural selenium, 0.10 µg/g natural selenium plus 0.32 µg/g selenite–selenium, or 0.42 µg/g natural selenium. The selenium concentrations in the liver, breast muscle, and eggs were significantly greater in those fed the 0.42 µg/g natural selenium than in those given the equivalent selenite–selenium diet. The whole-egg selenium concentrations for the three diets as just given, were 0.32, 0.74, and 1.23 µg/g, respectively. It was further found that (1) a higher proportion of the natural than the selenite–selenium passed into the white than the yolk, (2) more of the selenite–selenium was present in the yolk than in the white, and (3) the selenium in the white resulting from selenite feeding could be removed by dialysis but not that in the yolk. A difference in the chemical forms of selenium present in the two parts is apparent from these findings. The nature of these forms is not yet known. Additional evidence for the differential deposition of selenium in eggs depending on the form of selenium fed was provided by Latshaw and Biggert (217), who fed laying hens diets supplemented with either selenite or selenomethionine and found that although both forms of selenium caused similar increases in the egg yolk content, selenomethionine was superior to selenite in increasing the selenium content of egg white. The increased selenium levels in egg white proteins after feeding selenomethionine appeared to parallel the methionine content of the proteins. Various dosing protocols have been developed to label chicken eggs with stable selenium isotopes for use in human tracer studies (393,413).

At highly toxic selenium intakes, extremely high levels can occur in hen's eggs, particularly in the white. Many years ago, Moxon and Poley (282) observed increases in yolk selenium from 3.6 to 8.4 $\mu g/g$ and in white selenium from 11.3 to 41.3 $\mu g/g$ (dry basis), when the selenium level in the hen's rations was raised from 2.5 to 10.0 $\mu g/g$. Arnold and co-workers (14) later reported five- to nine-fold increases in the selenium concentrations in eggs up to nearly 2 $\mu g/g$, when 8 $\mu g/g$ selenium as selenite was added to various low- or moderate-selenium diets of hens for 42 to 62 weeks. The further addition of 8 or 15 $\mu g/g$ arsenic as sodium arsenite to these diets reduced but did not eliminate the increases in egg selenium levels.

II. SELENIUM METABOLISM

Selenium metabolism (absorption, transport, distribution, excretion, retention, and metabolic transformation) is highly dependent on the chemical form and amount of the element ingested and on the presence or absence of numerous interacting dietary factors. There are also differences between species regarding various aspects of metabolism, especially between monogastric and ruminant animals.

A. Absorption

Soluble selenium compounds are very efficiently absorbed from the gastrointestinal tract, since rats absorbed 92, 91, and 81% of tracer doses of selenite, selenomethionine, and selenocystine, respectively (422,424). Studies with selenium-75 (^{75}Se) at physiological levels indicate that the duodenum is the main site of selenium absorption and that there is no absorption from the rumen or abomasum of sheep or the stomach of pigs (471). Likewise, the absorption of selenite or selenomethionine in rats occurred mainly from the duodenum, with slightly less from the jejunum or ileum and practically none from the stomach (460). There appears to be no homeostatic control of the absorption of selenite by rats, since at least 95% was absorbed over a range of dietary selenium intakes from deficient to mildly toxic levels (33). When rations containing 0.35–0.50 μg selenium per gram were ingested, total net absorption represented some 85% of the ingested isotope in pigs but only 35% in sheep (471). This relatively low intestinal selenium absorption in ruminants may be related to a reduction of selenite to insoluble forms in the rumen (44,74).

Numerous balance studies conducted with humans have shown that the apparent absorption of dietary selenium ranges between 55 and 70% (231,235,405). Studies carried out in New Zealand (141,423) with radiolabeled selenium compounds demonstrated that the absorption of selenomethionine by young women was much better than that of selenite (95–97% versus 44–70%). Radiotracers also allowed comparison (405) of the true and apparent absorption of dietary selenium in New Zealanders (79 versus 55%, respectively). Later work with stable isotope tracers has confirmed the relatively poor absorbability by humans of selenium as selenite compared to other forms of the element (63,257). The absorption of selenium in eggs from chickens fed stable-labeled selenium as selenite was ~80% by pregnant and nonpregnant women (412). As in the case of animals, humans appear to have little or no homeostatic control over the absorption of selenium compounds from the gastrointestinal tract. New Zealand women given 1 mg of selenium as sodium selenite or selenomethionine absorbed 93 and 97% of the dose, respectively (420,426). The selenium status of an individual also appears to have no effect on selenium absorption, since the apparent absorption of a 200-μg dose of stable-labeled sodium selenite by New Zealand women was 57 and 70% before and after a course of supplementation with high-selenium bread even though plasma levels rose almost three-fold during the intervention period (93). It was suggested that the absorption of selenium could be enhanced by feeding a high-protein diet, but the different forms of selenium given to the high- and low-protein groups in this experiment complicate its interpretation (138). A sudden decrease in selenium intake reduced apparent absorption, presumably because of a continued relatively high rate of endogenous fecal loss (235).

B. Transport and Distribution

Several reports have suggested the existence of possible selenium transport proteins in blood plasma of animals (39,165,279). A selenocysteine-containing plasma protein (called selenoprotein P) has been postulated to have a possible role in selenium transport in the rat (278). Selenoprotein P, synthesized in the liver, was thought to transfer selenium to extrahepatic tissues. The mechanism by which selenium, firmly bound in this protein as selenocysteine, could be removed and transferred into tissue cells is not clear. No selenium transport protein has been identified in humans.

The distribution of selenium in the internal organs was studied by Behne and Wolters (20), who found that the concentration of selenium in tissues from rats fed commercial rat pellets containing 0.3 μg selenium per gram was as follows: kidneys > liver > testes > adrenals > erythrocytes > spleen > pancreas > plasma > lungs > heart > thymus > muscle > brain. A similar tissue distribution of selenium was found (365) in autopsy samples from North Americans: kidneys > liver > spleen > pancreas > testes > heart > muscle > lungs > brain. On the other hand, Burk *et al.* (41) showed that a tracer dose of selenite concentrated in the testes, brain, thymus, and spleens of rats fed a diet deficient in selenium. The importance of the testes was also recognized by Behne and Hofer-Bosse (19), who observed that this tissue lost the least selenium when rats were fed a depletion diet. The total-body selenium content of North Americans was calculated to be 14.9 mg (range 13.0–20.3 mg), whereas that of New Zealanders was estimated to be only 3.0–6.1 mg (365,405). This geographical difference is thought to reflect differences in the dietary selenium intake between the two countries (see Section IV,C).

C. Excretion

Radioselenite experiments with rats have indicated the existence of a threshold level of dietary selenium as selenite (0.054–0.084 μg/g) above which urinary selenium excretion is directly related to its dietary level and below which it is not (42). In humans, the amount of selenium excreted in the urine is closely related to the dietary intake, and balance studies have shown that over a range of intakes from 9 to 226 μg/day the urine accounts for 50 to 60% of the total amount excreted (231,235,451). Rat studies also suggest that the shutting down of urinary selenium excretion below the dietary threshold is an important mechanism for selenium conservation during periods of low selenium intake (42). The importance of the kidneys in the homeostasis of selenium in humans was emphasized by work from New Zealand that demonstrated that persons of low selenium status had low renal plasma clearances of selenium (344).

In ruminants, fecal excretion of selenium assumes greater importance than in monogastric species. Sheep and swine given oral doses of radioselenite excreted 66 and 15% of the dose via this pathway, respectively (471). Rumen microorganisms were thought to reduce the selenite to an unavailable form, so the increased fecal excretion was due to poor absorption rather than elevated endogenous excretion. Human balance studies show that fecal excretion accounts for a relatively constant fraction of the total excretory output over a wide range of dietary intakes. For example, Chinese men residing in a Keshan disease area and consuming 8.8 μg/day in their diet excreted 3.4 μg/day or 48% of their total excretory output in their feces (451), whereas young North American males adjusted to a dietary selenium intake of 226 μg/day excreted 86 μg/day or 43%

of their total excretory output in their feces (235). If it is assumed that true gastrointestinal absorption was 80% in both cases, endogenous fecal loss would contribute about 40 μg/day in the latter case but only 2 μg/day in the former case. This suggests that adjustments in the fecal loss of selenium can, like urinary losses, play a role in the homeostasis of selenium.

The pulmonary excretion of volatile selenium compounds in rats assumes significance only when subacute doses of soluble selenium salts are injected (315). Negligible losses of selenium via the pulmonary or dermal routes were observed when humans took microgram oral doses of selenite or selenomethionine (141,423), and no pulmonary excretion was detected in a subject given 1 mg of selenium as sodium selenite (420). Human sweat from North Americans was found to contain an average of 1.3 μg selenium/liter (236). Thus, sweat losses would represent a minor pathway of selenium excretion in North Americans but might constitute a more important excretory route in persons with low dietary intakes, if indeed the selenium concentration in their sweat were similar.

D. Retention

The whole-body retention curve of a single dose of radioselenite injected into rats consists of two or more components (41,98). The apparent whole-body retention of radioselenium is the summation of several different processes, since each tissue within the body has its own discrete rate of selenium turnover. The biological half-lives of radioselenium in rat skeletal muscle, whole body, and kidneys were 74, 55, and 38 days, respectively (422). The whole-body turnover of selenium in humans can also be expressed by a three-term exponential curve (141). The biological half-life of an oral dose of radiolabeled selenomethionine during the third phase was 234 days in New Zealanders of low selenium status, whereas that of selenite was only 103 days (141). This is in agreement with the finding that selenomethionine is more effective than selenite in raising blood selenium levels in persons of low selenium status (see below). Stable selenium isotopes given as selenite have also been used to estimate whole-body selenium retention in North Americans, and the half-life calculated for the longest exponential component was 162 days (190). The longer half-life observed in the North Americans compared to New Zealanders may have resulted because the former subjects were undergoing selenium depletion.

In early studies of the metabolism of selenium at toxic levels, marked differences were seen in its retention in the tissues of rabbits depending on whether it was given in the form of seleniferous oats (presumably protein bound) or sodium selenite (401). For example, the skeletal muscle from rabbits receiving the oats contained 29 times as much selenium as muscle from rabbits getting similar doses of selenite. In selenium toxicity studies with mice, an interesting contrast was noted in the metabolism of selenomethionine and selenium–methylselenocysteine, since feeding the former resulted in markedly elevated tissue residues of selenium whereas the latter, like selenite, raised tissue selenium levels to a much lesser extent (253). Differences in the retention of selenium depending on the form fed are also seen at nutritionally relevant levels, since Latshaw showed that the tissues of hens fed a diet containing 0.42 μg/g selenium derived from natural sources contained more selenium than tissues of hens fed a diet containing 0.10 μg natural selenium plus 0.32 μg selenium added as selenite per gram (216). Animal studies have also shown that the retention of selenium is dependent on selenium status, since rats fed a low-selenium diet excreted less of a single dose of radioselenite in the urine and retained more in the carcass than rats fed diets high in selenium (175). Moreover, the long-term turnover of selenium was faster in rats fed diets higher in selenium content (98).

Striking differences in the retention of selenium as influenced by the form administered have also been observed in studies with human subjects. Selenium as selenomethionine was much more effective in raising blood selenium levels in New Zealanders of low selenium status than was the sodium selenite form (427). Supplementation with selenium-rich wheat or yeast raised plasma and red cell levels in Finnish men with low initial plasma level more efficiently than supplementation with sodium selenate (238). Moreover, once the supplements were discontinued, the subjects given the wheat or yeast had higher plasma selenium levels some weeks later than the group given selenate.

E. Metabolic Transformation

The metabolic transformations of selenium are complex and depend on the chemical form supplied to the body. Selenate is reduced to selenite, which in turn is reduced further to selenide (113). At this oxidation state (-2), selenium is presumably introduced into selenocysteine, the form at the active site of GSH-Px, but the mechanism for this conversion is not clear (409). Selenium as selenide can also be methylated to form dimethylselenide, which is excreted via the lungs, or trimethylselenonium ion, which is excreted via the kidney. Dimethylselenide is formed only when animals are challenged with toxic doses of selenium (315). Trimethylselenonium ion was originally thought to be a major urinary metabolite under conditions of both low and high selenium exposure (321), but later work suggests that it becomes more important under conditions of high exposure (292). Selenomethionine can be incorporated into tissue proteins as such (305), which probably accounts for the tendency of this form of selenium to be retained in the body (141,427). Selenomethionine can also be catabolized to selenide (409) and then follow the methylation pathway typical of this form of selenium. The metabolic transformations of selenium compounds have not been well studied in humans, but selenium metabolism in people appears to be similar to that of animals, since garlicky breath (thought due to dimethylselenide) is noted in industrial workers overexposed to selenium (125), and trimethylselenonium ion has been characterized in human urine samples (293).

The metabolism of selenium can be strongly influenced by the concurrent exposure to other elements, particularly arsenic and heavy metals. Subacute doses of arsenic salts administered to rats given toxic doses of selenium block the pulmonary excretion of volatile selenium compounds but still protect against selenium poisoning, presumably by promoting the biliary excretion of selenium instead (222). This increased biliary excretion of selenium by arsenic is also thought to be the mechanism by which arsenic protects against chronic selenium toxicity, since arsenic decreases the hepatic retention of selenium in chronically poisoned rats (228). However, arsenic does not exacerbate selenium deficiency in animals fed low-selenium diets (222). Cadmium, mercury, and thallium all decreased the pulmonary excretion of selenium, but none increased biliary selenium excretion (228). Rather, selenium retention in the tissues was increased by these heavy metals. Both silver and inorganic mercury decrease GSH-Px activity in rats, but only silver accelerated the development of liver necrosis (457). Silver also exacerbates selenium–vitamin E deficiency in turkeys (197) and swine (435), and it has been suggested that silver may make selenium unavailable for the snythesis of GSH-Px (114). Like arsenic, cyanide decreases the pulmonary excretion of selenium and protects against its acute toxicity, apparently by forming selenocyanate anion, $SeCN^-$ (320). Moreover, drenching with potassium cyanide has been reported to increase the incidence of nutritional myopathy in lambs (350).

F.   Placental and Mammary Transmission

Selenium, in either inorganic or organic forms, is transmitted through the placenta to the fetus (154,244,259,455,470). Such transmission is also evident from reports that selenium administration to the mother during pregnancy prevents WMD in lambs and calves. The placenta nevertheless presents something of a barrier to the transfer of selenium in inorganic forms. Jacobson and Oksanen (185) showed that the ^{75}Se in lambs was higher when their mothers were injected with [^{75}Se]selenomethionine or [^{75}Se]selenocystine than when [^{75}Se]selenite was injected. Similar results have been obtained by others in mice (154) and sheep (192). Furthermore, several workers have shown that following injection of the ewe with [^{75}Se]selenite, the ^{75}Se concentrations in the blood and most organs of the fetus are lower than those in the mother (43,470). It is apparent that these inorganic selenium compounds pass the placental barrier less readily than the selenoamino acids. Whether the mammary barrier is similarly selective is not entirely clear. Thus Fuss and Godwin (107) found a greater entry of ^{75}Se into the milk of ewes when the iostope was given as selenomethionine than when it was given as selenite, whereas Jenkins and Hidiroglou (192) reported data indicating that the ewe made better use of selenite than of selenomethionine for secreting higher selenium levels in milk. On the other hand, selenium in the form of selenium-rich yeast was much more effective than selenate–selenium in raising selenium levels in the breast milk of Finnish mothers of low selenium status (212). Injection of acute doses of inorganic mercury into pregnant or lactating rats decreased the passage of selenium across both the placenta and the mammary gland, and it was suggested that heavy-metal exposure of the mother might compromise the selenium nutrition of the young (326).

III. SELENIUM DEFICIENCY AND FUNCTIONS

Selenium is necessary for growth and fertility in animals and for the prevention of various disease conditions that show a variable response to vitamin E. These are liver necrosis in rats and other species, exudative diathesis (ED) and pancreatic fibrosis in poultry, muscular dystrophy (WMD) in lambs, calves, and other species, and hepatosis dietetica in pigs. The resorption sterility in rats (156) and encephalomalacia in chicks (80) that occur on vitamin E-deficient diets do not respond to selenium. Shortly after the original demonstration by Schwarz and Foltz (370) that selenium prevents liver necrosis in rats and by Patterson and co-workers (329) that this element prevents ED in chicks, growth and fertility responses to selenium in farm animals, greater than could be achieved by tocopherol, were observed in Oregon and New Zealand. Unequivocal evidence that selenium is a dietary essential independent of or additional to its function as a substitute for vitamin E was obtained by Thompson and Scott (417), who showed that chicks consuming a purified diet containing ≤ 0.005 μg/g selenium exhibited poor growth and high mortality even when 200 μg/g of d-α-tocopherol was added. Higher tocopherol levels prevented mortality, but even with 1000 μg/g growth was inferior to that obtained with selenium and no tocopherol. Comparable findings obtained with rats by McCoy and Weswig (262) and Wu and co-workers (472) clearly established that selenium does not function merely as a substitute for vitamin E. The association between selenium and Keshan disease has now furnished evidence that selenium is also essential for humans (see Section III,I).

A. Muscular Dystrophy

Nutritional muscular dystrophy is a degenerative disease of the striated muscles that occurs, without neural involvement, in a wide range of animal species. It was described in calves in Europe in the 1880s and has been observed as field occurrences in many countries in lambs, calves, foals, rabbits, and even marsupials (8). In some countries the incidence is low and sporadic. Less than 1% of the block or herd may be affected, and in some seasons or years the disease may not appear at all. In other countries, notably in parts of New Zealand, Turkey, and Estonia (200), the incidence of muscular dystrophy or white muscle disease (WMD) is higher and more consistent unless appropriate selenium treatment is undertaken.

White muscle disease is a complex disease with a multifactorial etiology that responds not only to vitamin E and selenium, but also to the synthetic antioxidant ethoxyquin (461). In Canadian trials, treatment with both vitamin E and selenite was the most successful in controlling WMD (172). The most important ingredient of the selenium–vitamin E mixture preventing WMD in northern Europe was thought to be the α-tocopherol (408). Some workers found that tissue activities of GSH-Px are not related to the incidence of WMD and concluded that while GSH-Px activity may be used to assess the selenium status of animals, it cannot be used to assess the probability of WMD (462).

WMD rarely occurs in mature animals. In lambs it can occur at birth (congenital muscular dystrophy), or at any age up to 12 months. It is most common between 3 and 6 weeks of age. Lambs affected at birth usually die within a few days. The deep muscles overlying the cervical vertebrae are particularly affected with the typical chalky-white striations. Lambs affected later in life (delayed muscular dystrophy) show a stiff and stilted gait and an arched back. They are disinclined to move about, lose condition, become prostrate, and die. Animals with severe heart involvement may die suddenly without showing any such signs. Clinical signs of WMD may not appear until the lambs are driven or moved about. Mildly affected animals may recover spontaneously.

These disabilities are associated with a noninflammatory degeneration or necrosis of varying severity of the skeletal or the cardiac musculature, or of both. A bilaterally symmetrical distribution of the skeletal muscle lesions is characteristic of WMD in lambs. The symmetry frequently extends beyond a simple bilateral involvement of paired muscles to the distribution of lesions within the muscles (478). The lesions are usually most readily discernible in the thigh and shoulder muscles. The lesions in the cardiac muscle are commonly confined to the right ventricle but may occur in other compartments. They are seen either as subendocardial grayish-white plaques or as more diffuse lesions of a similar color extending up to 1 mm into the myocardium (120). Alterations in the electrophoretic pattern of affected muscles are a consistent feature of the disease (478). Characteristic abnormalities in the electrocardiograms have also been demonstrated by Godwin (126) in lambs with WMD. The changes in the ECG pattern develop early and become very marked as death approaches. Similar changes in ECG pattern just before death have been observed in rats and lambs fed torula yeast-based diets deficient in both selenium and vitamin E (127,128).

WMD is characterized biochemically by abnormally high levels of serum aspartate aminotransferase (AspAT) (formerly called glutamic oxaloacetic transaminase, or GOT) and lactic dehydrogenase (330,458). In normal lambs and calves, AspAT activity rarely exceeds 200 units/ml and is usually only about half that level. AspAT concentrations 5–10 times higher occur in animals with WMD, with the increase above normal being roughly proportional to the amount of muscle damage (27). AspAT determinations are therefore of some value in the diagnosis of WMD, although individual variability is high in both healthy and affected animals. Oksanen (307) has further pointed out that "the SGOT [AspAT] value may be only moderately increased in animals with marked clinical symptoms caused by extensive subacute or chronic degenerative processes," and "a degeneration in the myocardium, even an acute one, may also produce only a moderate increase in the SGOT [AspAT] value, although this often ends in sudden death."

WMD has received most attention in lambs and calves because of its economic importance and natural occurrence, but similar degenerative changes occur in foals (159), and in association with hepatosis dietetica in pigs (307) and ED in chicks (297). Muscular dystrophy in chicks is characterized by degeneration of the skeletal muscle fibers, especially the pectoral muscles, and is most efficiently prevented by feeding both selenium and vitamin E (373). Selenium alone or at lower levels in combination with vitamin E is effective against myopathy of the gizzard in turkey poults (377). Muscular dystrophy in rabbits is prevented by vitamin E but not selenium (176).

B.   Exudative Diathesis

The disease ED first appears as an edema on the breast, wing, and neck and later has the appearance of massive subcutaneous hemorrhages, arising from abnormal permeability of the capillary walls and accumulation of fluid throughout the body. The greatest accumulation occurs under the ventral skin, giving it a greenish-blue discoloration. Plasma protein levels are low in affected chicks, and an anemia develops probably as a consequence of the hemorrhages that occur (298). Growth rate is subnormal and mortality can be high. In the outbreaks of the disease that occur in commercial flocks as a result of consuming low-selenium grain, chicks are most commonly affected between 3 and 6 weeks of age. They become dejected, lose condition, show leg weakness, and may become prostrate and die (9,159).

ED can be completely prevented by either selenium or vitamin E (418). Noguchi and co-workers (302) have produced evidence indicating that vitamin E and selenium prevent ED by two different mechanisms and that the effectiveness of vitamin E is not due to a simple antioxidant action. For example, the synthetic antioxidant ethoxyquin, which will substitute for vitamin E in the prevention of encephalomalacia, has no preventive effect on the incidence or severity of ED in vitamin E- and selenium-deficient chicks. These workers showed that the GSH-Px level of chick plasma is directly related to the selenium level in the diet (Fig.

2) and to the effectiveness of selenium in prevention of ED (Fig. 3). Dietary vitamin E and selenium were both found necessary for protection of hepatic mitochondrial and microsomal membranes from ascorbic acid-induced lipid peroxidation *in vitro*. On the basis of these results, the hypothesis was put forward that "the plasma GSH-Px present when the diet contains adequate selenium acts to prevent ED by destroying peroxides that may form in the plasma and/or cytosol of the capillary wall. Vitamin E appears to prevent ED by acting within the lipid membrane where it neutralizes free radicals, thereby preventing a chain-reactive autoxidation of the capillary membrane lipids." However, the concentration of GSH in plasma is so low that it would not seem sufficient to act as an electron donor for the GSH-Px-catalyzed reduction of peroxides.

WMD affecting particularly the breast muscles often develops concurrently with ED. A congenital myopathy, characterized by the hatching of dead chicks or chicks dying 3–4 days after hatching, is another selenium-responsive condition in this species (352). The chicks show extensive pale areas in the gizzard and sometimes in the hindlimb skeletal musculature.

C. Nutritional Pancreatic Atrophy

Nutritional pancreatic atrophy (NPA) can be produced in chicks (418) by feeding diets extremely deficient in selenium (<10 ng/g), even in the presence of

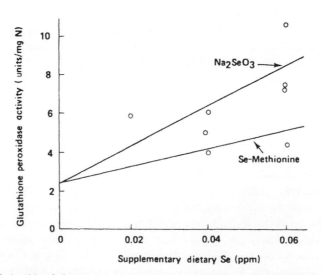

Fig. 2. Relationship of plasma glutathione peroxidase activity in chicks to dietary selenium levels supplied by sodium selenite ($y = 0.0222 + 1.03x$; $r = .889$) or DL-Se-methionine ($y = 0.0242 + 0.548x$; $r = .809$). From Noguchi *et al.* (302).

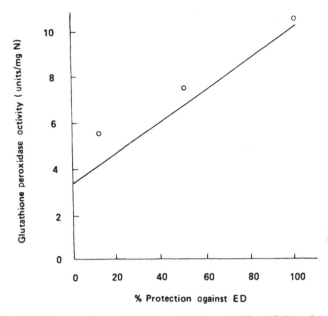

Fig. 3. Correlation between plasma glutathione peroxidase activities at 7 days of age and prevention of exudative diathesis (ED) measured at 13 days of age ($y = 0.0342 + 0.684x$; $r = .938$). From Noguchi *et al.* (302).

supplemental vitamin E in the normal nutritional range (100 IU/kg). Although long regarded as a specific severe deficiency of selenium, it is now known that NPA may be prevented in the selenium-deficient chick either by a high level of vitamin E (≥ 300 IU/kg) or by other antioxidants (500 μg/g of butylated hydroxytoluene, diphenyl-*p*-phenylenediamine, ascorbic acid, or ethoxyquin), despite the very low activities of pancreatic GSH-Px seen in such chicks (466). Since NPA can be prevented by dietary selenium levels of 20 to 50 ng/g, the highly purified basal diet used in these studies was based on crystalline amino acids rather than intact protein. NPA has been observed in chicks fed a practical-type diet containing only 7 ng selenium/g that was based on corn and soybean meal grown in a markedly selenium-deficient area of northeastern China (70). However, the NPA produced in the chicks fed this severely selenium-deficient practical diet was delayed in onset and reduced in severity compared to that which occurred in chicks fed the purified diet, thereby suggesting that uncharacterized factors partially protective against NPA existed in the practical ration.

The initial deficiency lesions of NPA are apparent at 6 days of age and consist of vacuolization and hyaline body formation in the acinar cells, with loss of zonation of the acini. Fibroblasts are later observed in increasing numbers in the interacinar spaces while the acinary cytoplasm is shrinking basally, thus enlarging the central lumen of the acinus. Finally, the acini appear as rings of cells composed mainly of small dense-staining nuclei, completely surrounded by the fibrotic tissue (139). Within 2 weeks of selenium supplementation the pancreas returns to normal in appearance (139). The degeneration of the selenium-deficient pancreas has been further characterized by the detailed studies of Noguchi *et al.* (303), demonstrating a decrease in zymogen activity and an increase in the activities of the lysosomal enzymes cathepsin and acid phosphatase. The invasion of macrophages accounted for the increase in lysosomal enzymes, and no evidence of lysosomal disruption during the course of the pancreatic degeneration was obtained. It was concluded that the role of selenium in preventing pancreatic fibrotic degeneration was not in the protection of lysosomal membranes of the acinar cells. Perhaps a decreased number of functional acinar mitochondria may be involved (465).

D. Hepatosis Dietetica in Pigs

Hepatosis dietetica has been produced on vitamin E-free diets based on torula yeast (94) or soybean meal (133), and occurs spontaneously in New Zealand (9,159) and Scandinavia (307) when pigs are fed grain rations naturally low in selenium. The disease is most common at 3 to 15 weeks of age and results in high mortality. Severe necrotic liver lesions are apparent at postmortem examination. There is also deposition of ceroid pigment in adipose tissue, giving a yellowish-brown color to the body fat and a generalized subcutaneous edema.

A complex nutritional disease in swine called the vitamin E and selenium deficiency syndrome was described by Hakkarainen *et al.* (149) and consisted of hepatosis dietetica, mulberry heart, and skeletal muscle degeneration. The syndrome was produced in weaned pigs fed a diet containing only 8 ng selenium and 1.4 μg α-tocopherol per gram. Supplementation of the diet with 45 (but not 15) μg α-tocopheryl acetate per gram conferred complete protection against the condition. Addition of as much selenium as 135 ng/g as selenite gave only partial protection. However, a combined supplement of 5 μg α-tocopheryl acetate plus 135 ng selenium per gram as selenite, supplementation levels that were inadequate when administered separately, prevented the syndrome.

The mortality and liver lesions characteristic of hepatosis dietetica are associated with a marked depletion of tissue selenium concentrations and an increase in the level of the liver-specific enzyme ornithine carbamyltransferase (OCT) in the blood (307). Since the degenerative changes that may occur in the muscles do not affect the OCT level, blood OCT levels can be useful aids in the diagnosis of hepatosis dietetica. However, the interpretation of plasma enzyme activities may be complicated by other factors, and cautions about their proper use in the diagnosis of selenium–vitamin E deficiency in swine have been presented (408).

E. Selenium-Responsive Unthriftiness in Sheep and Cattle

In parts of New Zealand a serious condition known as "ill thrift" occurs in lambs at pasture and can occur in beef and dairy cattle of all ages (9,158). The condition varies from a subclinical growth deficit (158,263) to clinical unthriftiness with rapid loss of weight and sometimes mortality (91,159). No characteristic microscopic lesions are apparent, there is no increase in AspAT levels, and the condition may or may not be associated with WMD and infertility. Another selenium-responsive syndrome in lactating ewes, characterized by rapid weight loss and scouring in association with heavy parasitic infestation, has been described (263).

Ill thrift can be prevented by selenium treatment, with striking increases in growth and wool yield in some instances (9,91,159,199). Significant increases in wool yield from selenium supplements have also been observed in Canadian (394) and Scottish (340) experiments, in the growth of lambs and calves in the western United States (308), and in lambs from farms in northeastern Scotland (26). In most parts of the world the responses to selenium are less spectacular than those reported in New Zealand. In one series of New Zealand trials, lambs 5 months old were divided into two groups, the controls and those receiving 5 mg selenium per lamb as selenite, given orally at commencement and again at 2 and 6 weeks. The mortality was reduced from 27 to 8% by the treatment, and highly significant weight gains were observed in the selenium-treated groups. Smaller oral doses—namely 1 mg selenium at docking, 1–5 mg at weaning, and 1–5 mg at 3-month intervals thereafter—were later found adequate. Neither vitamin E nor ethoxyquin had any effect on the unthriftiness (158).

F. Reproductive Disorders

Early work established that selenium was of no value in preventing resorption gestation in vitamin E-deficient rats (156) and also could not improve reproductive performance in chickens or turkeys deficient in vitamin E (75,196). Generally, rats fed selenium-deficient diets reproduce normally, but their offspring are hairless, grow poorly, and fail to reproduce if maintained on a low-selenium diet (262). In five of the eight untreated males, immotile sperm, with separation of heads from tails, were observed, with no spermatozoa in the remaining untreated males. A selenium supplement of 0.1 μg/g diet as Na_2SeO_3 restored hair coat, growth, and fertility, but only one or two pups were delivered per litter, all of which died in a few days. The motility of spermatozoa from male rats born to dams fed a selenium-deficient diet was markedly reduced, and many sperm exhibited breakage of the tail near the midpiece (472,473). High levels of dietary vitamin E or various other antioxidants did not prevent these manifestations of selenium deficiency. Similar experiments carried out with mice showed that selenium deficiency, over the course of three generations, resulted in progressive decreases in body, testis, and epididymal weights, as well as impaired sperm production (445). The percentage of morphologically abnormal sperm increased in successive generations, and as in the rats, most of the sperm defects occurred in the midpiece of the tail. Fertility declined progressively during the first two generations of selenium deficiency.

The mechanism by which selenium deficiency affects the morphology of sperm from rodents is unknown, but much of an injected dose of radioselenite injected into selenium-deficient rats tends to localize in the midpiece (32). Calvin (49) isolated a selenopolypeptide of 17 kDa from rat sperm that may play a role in the normal assembly of the sperm tail. Progressively increasing abnormalities in the morphology and arrangement of midpiece mitochondria were noted in sperm from mice fed a selenium-deficient diet for three successive generations (446), and it was suggested that selenium may be required for the normal development or stabilization of mitochondrial shape during spermiogenesis.

Induction of severe selenium deficiency in chickens by feeding a low-selenium (20 ng/g) but high-vitamin E diet to 1-day-old chicks until they were 12.5 months old resulted in depressed egg production and hatchability that could be improved by supplementing the diet with selenium at 0.1 μg/g as sodium selenite (51). Likewise, feeding a diet supplemented with 0.2 μg selenium per gram as sodium selenite to hen turkeys that had previously been fed a low-selenium diet for 8 months improved hatchability from 50 to 59% (54).

A high seasonal incidence of infertility in ewes occurs in parts of New Zealand in association with WMD and unthriftiness. In certain of these areas, 30% of ewes may be infertile and losses of lambs are high. The fertility of these ewes is dramatically increased by selenium administration before mating, and further selenium treatment reduces the lamb losses and incidence of WMD (157). The recommended doses are 5 mg selenium as sodium selenite by mouth before mating and a similar dose 1 month before lambing. Hartley (157) has shown that estrus, ovulation, fertilization, and early embryonic development proceed normally in the affected flocks. The infertility results from a high embryonic mortality occurring between 3 and 4 weeks after conception, that is, at about the time of implantation. This mortality can be prevented by selenium but not by either vitamin E or an antioxidant. In ewes fed a selenium-deficient purified diet, satisfactory reproductive performance was obtained only when selenium and vitamin E were administered in combination (35). A combination of selenium and vitamin E injected a month before mating was similarly shown to improve ewe prolificacy in selenium-deficient areas of Scotland (286). The total number of uterine contractions in ewes at the onset of estrus as well as the number of contractions moving toward the oviduct were increased by injecting the animals with a preparation containing both selenium and vitamin E (378).

A sodium selenite–vitamin E mixture injected into cows a month before calving completely prevented losses from the birth of premature, weak, or dead calves in parts of California (247) and greatly reduced the incidence of retained placentas in a herd of cows in Scotland (428). In this last study selenate alone was less effective than the vitamin E–selenate combination. In the United States, supplemental selenium eliminated retained placenta in Ohio, a low-selenium area, but had no effect in South Dakota, a high-selenium area (408). Optimum fertilization of ova from beef cattle occurred in cows receiving supplemental selenium and vitamin E and an adequate plane of nutrition (380). Injection of yearling Angus bulls with 50 mg of selenium as selenite followed by 30-mg injections at 3-week intervals for 150 days had no apparent effect on sperm cell viability, number, or selenium concentration, even though serum selenium levels increased from 10 to 80 ng/ml (379).

In a study of commercial Australian piggeries, a significant response in far-rowing percentage has been observed following a single subcutaneous injection of 2.5 mg selenium as selenite given just prior to mating (160). The farrowing percentages were 82.5 and 77.0 (mean 79.3) in the two groups of treated gilts, compared with 68 and 62% in two groups of untreated controls. The selenium response appeared to be due very largely to decreased neonatal losses. No abnormalities in the viability or morphology of sperm were seen in 210-day-old boars that had been fed a low-selenium diet for 4.5 months (381), but reproductive disorders in boars have been described in Keshan disease areas of China (242).

The remarkable protective action of selenium injections against the testicular necrosis caused by cadmium is discussed in Chapter 18.

G. Anemia

The role of vitamin E as an erythropoietic factor in certain species such as swine and some nonhuman primates is well established (92), but there is little work to suggest that any anemias are due to selenium deficiency. However, Morris *et al.* (275) described an anemia in cattle that was associated with the presence of Heinz bodies and suboptimal blood selenium levels. Supplementation with selenium corrected the anemia, prevented formation of Heinz bodies, increased body weight, and raised blood selenium levels. This was thought to be the first recorded example of anemia due to selenium deficiency.

H. Immunological Responses

Early work showed that the humoral immune response of several different species to bacterial and viral antigens could be improved by dietary vitamin E in excess of nutritionally required levels (301). Later work indicated that vitamin E deficiency caused impaired mitogen response in spleen cells from rats (22).

Likewise, early work with selenium showed that levels above those generally accepted as adequate (0.1 μg/g) enhanced the primary immune response in mice as measured by the plaque-forming cell test and by hemagglutination (403). Similarly, later work indicated that selenium deficiency caused impaired mitogen response in cultures of spleen cells from mice (287). Selenium deficiency in rats and cattle had no effect on the phagocytic ability of neutrophils but significantly decreased microbicidal activity (29,383). On the other hand, Bass *et al.* (18) found no difference in the bactericidal activity of polymorphonuclear leukocytes from rats fed normal versus selenium-deficient diets. Both vitamin E and selenium were found to enhance independently humoral antibody production in weanling swine after an antigenic challenge with sheep red blood cells (333). Both vitamin E and selenium were also reported necessary for optimum immune function in the chick, but the effect of selenium was influenced by antigen concentration, sex, and ontogenic state of the animal (251). Attempts to demonstrate effects of selenium on immune function in humans have been less successful than in animals. GSH-Px activity of granulocytes from a patient on total parenteral nutrition (TPN) for 18 months was <20% of control values, but no difference was seen in bactericidal activity of patient and control granulocytes against *Staphylococcus aureus* 502A (17). Supplementation of middle-aged Finnish men of low selenium status had no effect on phagocytosis, chemotactic factor generation, antibody or leukocyte migration inhibitory factor production by lymphocytes, or proliferative responses to phytohemagglutinin or concanavalin A, even though plasma selenium levels were more than doubled, from 74 to 169 ng/ml (15). However, intracellular killing of *S. aureus* by granulocytes was slightly lower in the control versus the supplemented group (77.2 versus 85.2%). The cardiotoxic effect of Coxsackie B_4 virus was found to be more severe in selenium-deficient mice compared to mice fed adequate-selenium diets or given oral doses of selenite (16), and it has been suggested that this decreased resistance to viral infection during selenium deficiency may play some role in the etiology of Keshan disease (see next section).

I. Keshan Disease

Keshan disease is an endemic cardiomyopathy that has been known in China for >100 years (123). The disease derives its name from a violent outbreak that occurred in Keshan County of northern China in 1935. It affects primarily children and women of child bearing age and is regionally distributed in a long belt extending from northeast to southwest China (123). The susceptible population consists mainly of peasants who live in rural areas. The susceptibility of the peasants is thought to be due to the monotonous and limited nature of their diet (406,479). There is a seasonal variation in the incidence of the disease with the peak occurring during the winter in the North and during the summer in the

South. There is also a great yearly variation in the incidence of the disease. During the peak years of 1959, 1964, and 1970 the annual incidence exceeded 40 per 100,000 of the affected population, with more than 8000 cases, and 1400 to >3000 deaths recorded each year. The incidence has since dropped markedly to <5 per 100,000, with only ~1000 new cases reported annually.

Clinically, Keshan disease can be divided into four types: acute, subacute, chronic, and latent. Although the disease cannot be identified by any specific signs or symptoms, criteria for the diagnosis and subtyping of Keshan disease have been presented (60). Sudden acute insufficiency of heart function characterizes acute cases, whereas chronic cases usually exhibit moderate to severe heart enlargement with varying degrees of insufficient heart function. Latent cases generally have normal heart function with mild heart enlargement, while subacute cases show clinical features that are a combination of those of acute and chronic types. Gross examination often reveals a highly expanded ball-shaped heart in chronic cases, but in acute cases the heart often appears generally normal.

The main pathological characteristics of Keshan disease are multifocal necrosis and fibrous replacement of the myocardium. Myocytolysis was present in most cases. Details of the disease process have been summarized elsewhere (123,143). The histopathological picture appears somewhat similar to the lesions seen in WMD in calves but is different from that of idiopathic congestive cardiomyopathy or viral myocarditis. Ultrastructurally, the cellular organelles with the most conspicuous morphological changes were the mitochondria, which showed marked dilatation with a rather clear matrix. A triple membrane structure was present in the cristae of the swollen mitochondria with the third membrane usually occurring at the widened interspace of the cristal membranes. Some hearts of the Keshan disease victims also contained three kinds of specific granules of unknown properties that have not been described in other reports of cardiomyopathy or myocarditis.

Investigators in China have compiled an impressive array of evidence that links an increased incidence of Keshan disease with poor selenium status. The disease has been associated with low dietary selenium intakes and low blood and hair selenium levels (203). Keshan disease generally did not occur in those areas in which the selenium content of grains was ≥ 0.04 µg/g. Hair selenium levels were shown to be directly related to blood levels, and the average selenium content of hair was <0.12 µg/g in areas affected with Keshan disease. Because of these relationships between selenium status and Keshan disease, a large-scale intervention trial was carried out to determine whether selenium had any possible protective effect against the disease in a high-risk population of 1- to 9-year-old children (202). In a trial that started in 1974, 4510 children received tablets containing sodium selenite whereas 3985 children got placebo tablets. The dose given to the treated children was 0.5 mg of sodium selenite per week if 1–5 years

Table I
Protective Effect of Selenium against Keshan Disease[a]

Treatment	Year	Subjects	Cases	Deaths
Placebo	1974	3,985	54	27
	1975	5,445	52	26
Sodium selenite	1974	4,510	10	0
	1975	6,767	7	1
	1976	12,579	4	2
	1977	12,747	0	0

[a] Adapted from Keshan Disease Research Group (202).

old or 1.0 mg/week if 6–9 years old. The effectiveness of the selenium supplement is shown by the data in Table I. No side effects were seen due to the selenium supplementation except for a few isolated instances of nausea that could be avoided by consuming the tablet postprandially. No liver damage was detected by physical examination or function tests after constant consumption of the supplement for 3 to 4 years.

Despite the successful selenium intervention and the many associations between poor selenium nutrition and the increased risk for Keshan disease, certain evidence indicates that factors other than selenium status may be involved in the etiology of the disease (162). For example, the seasonal variation in the incidence of the disease is difficult to rationalize on the basis of differences in selenium exposure alone. A viral involvement in the disease was suggested on the basis of experiments that showed the decreased resistance of selenium-deficient mice to the cardiotoxic effects of a Coxsackie B_4 virus that had been isolated from a patient with Keshan disease (see Section III,H). Whatever the involvement of other factors, it is nonetheless clear that low selenium status is the underlying condition that predisposes an individual to the disease.

J. Kashin–Beck Disease

Kashin–Beck disease is an endemic osteoarthropathy that occurs in eastern Siberia, northern China, and North Korea (299). The disease derives its name from the Russian investigators who first described (N. I. Kashin in 1861) and later gave detailed clinical accounts (E. B. and A. N. Beck, husband and wife, in 1899) of the condition (299). Kashin–Beck disease is characterized as a chronic, disabling, degenerative, generalized osteoarthrosis that involves both the peripheral joints and the spine. There is disturbed endochondral ossification and deformity of the affected joints but no systemic or visceral manifestations.

The most prominent pathological feature is multiple focal chondronecrosis widely involving the hyaline cartilage tissue, as in the cartilaginous epiphysis, epiphyseal growth plate, and articular cartilage (64,272). Epiphyseal impairment is often followed by disturbances in bone development such as shortened fingers and toes and in extreme cases, dwarfism. The pathological changes in the articular cartilage progress slowly and lead to secondary osteoarthrosis with enlarged, deformed joints.

Although the etiology and pathogenesis of the disease remain obscure, animal studies revealed acute massive liver necrosis in rats given grain and drinking water from an endemic area (272). These dietary components were found to be low in selenium by analysis. Chinese children <5 years of age, 6–10 years of age, and nursing mothers residing in an acutely endemic area were supplemented with 0.5, 1.0, and 2.0 mg of sodium selenite per week, respectively, over a period of 6 years (240). X-Ray records showed that the incidence of Kashin–Beck disease declined from 42% (20 of 47) to only 4% (2 of 49) in children 3–10 years old after the selenium intervention. Although these findings suggest a role for selenium in Kashin–Beck disease, it should be pointed out that other hypotheses (e.g., contaminated drinking water or the presence of mycotoxins in food) have been advanced concerning the etiology of the disease. A tibial dyschondroplasia has been produced in chickens that may or may not be related to Kashin–Beck disease by feeding the birds a mycotoxin produced by the mold *Fusarium graminarum* (220). Further investigation is required to clarify the relationship between these hypotheses and Kashin–Beck disease (272).

K. Mode of Action of Selenium

Two major steps forward have been made in our understanding of the mode of action of selenium and its relationship to vitamin E since liver necrosis in rats and ED in chicks were first shown to be prevented by small amounts of selenite, as described above. The first of these was the demonstration of an essential requirement for dietary selenium by chicks and rats, in the presence of adequate vitamin E. The second major advance was the discovery that GSH-Px is a selenium-containing enzyme with an activity in the blood and tissues of rats and chicks directly related to the selenium level in the diet. These important findings gave a great impetus to the antioxidant role for selenium compounds fostered by Tappel (415) but open to some doubt through the researches of Green and Diplock and their group and others (135).

Both selenium and vitamin E protect biological membranes from oxidative degradation in the prevention of ED (see Section III,B). Noguchi *et al.* (302) have put forward the hypothesis that vitamin E functions as a specific lipid-soluble antioxidant in the membrane and that selenium functions as a component of cytosolic GSH-Px that reduces peroxides. In other words, the GSH-Px is considered to be of primary importance, acting to destroy peroxides before they can attack the cellular membranes, while vitamin E acts within the membrane itself in preventing the chain-reactive autoxidation of the membrane lipids. Although GSH-Px can attack free fatty acid hydroperoxides in the cytoplasm directly, the enzyme apparently is unable to reduce hydroperoxy fatty acids esterified in phospholipids as they might occur in cell membranes (142).

"Respiratory decline" of liver mitochondria from selenium–vitamin E-deficient rats was a phenomenon described in some of the early work of Schwarz (368), and several workers have observed various effects of selenium deficiency on the oxidative processes of the tricarboxylic acid cycle or on other aspects of mitochondrial function (36,130,233,456). Sies and Moss (391) later suggested that GSH-Px may have a role in the regulation of mitochondrial oxidation. Perhaps such a regulatory role for GSH-Px could provide an explanation for many of the earlier reports linking selenium deficiency with different mitochondrial processes.

It is now apparent that two different GSH-Px activities are present in the tissues, one that depends on dietary selenium and another that is independent of it (38). The so-called non-selenium-dependent GSH-Px activity is due to a family of enzymes known as the GSH S-transferases. These enzymes have an important role in the detoxification of certain xenobiotics by conjugating them with GSH (201), but they also can catalyze the reduction of organic hydroperoxide substrates by GSH in a reaction similar to that catalyzed by GSH-Px. However, the K_m of this reaction is much higher for the transferases than for GSH-Px, and the transferases have no activity toward hydrogen peroxide. Nonetheless, 50–100% increases in GSH S-transferase activity have been observed in male rats deficient in selenium (218), so the non-selenium-dependent GSH-Px activity contributed by the transferases must be considered when using GSH-Px activity measurements to assess selenium status.

Further elucidation of the precise mode of action of selenium and confirmation or otherwise of the hypotheses already advanced will no doubt emerge as further research is undertaken. Several selenium-containing enzymes have been reported in microbial systems (404), and meanwhile evidence is accumulating that selenium may serve functions in mammalian and avian tissues unrelated to that of GSH-Px. For example, Whanger et al. (459) have isolated a selenoprotein from the heart and muscle of lambs given adequate selenium that resembles cytochrome c and have shown that this compound is not present in these tissues in the selenium-deficient animal. Diplock studied the distribution and various intracellular forms of selenium in rat liver and suggested that selenium, particularly as selenide in hypothetical nonheme iron selenide proteins, may have a role in the electron transfer functions associated with mitochondria and smooth endoplasmic reticulum (57). However, the chemical stability of the forms of selenium in these tissue fractions is poor, and little progress has been made in their characterization (85,86). Some workers have produced evidence that selenium may be functioning in the metabolism of sulfhydryl compounds (31). Protein sulfhydryl groups were decreased in lamb muscle and rat liver from selenium-deficient animals, whereas nonprotein sulfhydryl groups increased. Hill and Burk have reported that the interorgan metabolism of GSH is affected in selenium-deficient rats. The release of GSH from rat liver is increased in selenium deficiency, and this results in higher levels of plasma GSH (170). Under such conditions, the removal of plasma GSH by the kidneys is also increased (171). In certain strains of chicks, selenium deficiency has been shown to influence another aspect of sulfur metabolism, namely the conversion of methionine to cysteine (37,150). Although concentrations of free cysteine and cystathionine in the plasma were depressed, no elevation of free methionine occurred in the plasma, thereby suggesting that selenium deficiency may cause a greater portion of the methionine to be metabolized via an alternate (e.g., transaminative) pathway in these birds.

Burk and co-workers (38) have conducted an extensive series of experiments on the effects of selenium on xenobiotic metabolism and have found that the induction of cytochrome P_{450} in rat liver microsomes by phenobarbital is impaired by a moderate selenium deficiency, whereas a more severe selenium deficiency causes a decrease in cytochrome P_{450} even in the uninduced state. Pascoe *et al.* (328) have found that dietary selenium is also needed to maintain cytochrome P_{450} levels in rat small intestinal mucosa. Feeding a selenium-deficient diet to normal adult rats just for a single day dramatically reduced intestinal cytochrome P_{450} content. The biochemical mechanism by which selenium affects cytochrome P_{450} levels is not clear, but Correia and Burk (73) favor the concept of a defect in the attachment of heme to the apocytochrome P_{450}. Whatever the mechanism, it apparently does not involve GSH-Px, since it was possible to correct the abnormality in heme metabolism by injecting selenium into selenium-deficient rats before any detectable GSH-Px activity appeared (73). The relationship between selenium deficiency and drug metabolism has also been carefully investigated by Reiter and Wendel (342,343), who found many changes in the activities of various drug-metabolizing enzymes in mice that were deprived of selenium for long periods of time. These workers emphasized that many processes unrelated to GSH-Px occur in a later phase of selenium deficiency in mouse liver and that a study of these processes will be required to understand fully the effects of selenium status on drug metabolism.

Another possible metabolic role for selenium might be in the detoxification of heavy metals. Early short-term acute toxicity experiments by Parizek and Ostadalova demonstrated that selenium could protect against mercury toxicity (325), and Parizek advanced the hypothesis that nutritional levels of selenium might protect against traces of toxic metals in the environment (326). Work by Ganther and associates showed that selenium added to the diet in the nutritional range protected against chronic methylmercury poisoning in rats and quail (118), and they suggested that the selenium that occurs naturally in seafoods might have some beneficial effect against any mercury residues in such fish (117). Despite much research, the biochemical mechanism by which selenium can diminish mercury toxicity is unknown (114). However, vitamin E and certain synthetic antioxidants also protect against mercury poisoning, even in the absence of dietary selenium (452), so the beneficial effect of selenium against mercury might somehow be related to its function as a biological antioxidant (112). It should be pointed out that although inorganic selenium compounds decrease mercury toxicity, dimethylselenide, a volatile selenium metabolite produced under conditions of selenium overexposure (see Section II,E), potentiates the toxicity of mercuric chloride (326). Again, the biochemical basis of this interaction is not known.

Selenium protects against the testicular damage caused by cadmium when salts of both elements are injected into animals, presumably by diverting the cadmium to less sensitive protein-binding sites (59,326). Such diversion, however, does not occur when the salts are given orally, so the significance of this effect under environmental conditions is yet to be established (457). Selenium may have some beneficial effect against lead under some experimental conditions, but vitamin E status seems more important in this regard (223). Even though large doses of selenate prevent acute thallium poisoning in rats, tissue retention of thallium is increased by such treatment and the mechanism of the protective effect of selenium is unknown (228).

IV. SELENIUM SOURCES AND REQUIREMENTS

A. Selenium Requirements

The minimum selenium requirements of animals vary with the form of the element ingested and the nature of the rest of the diet. A dietary selenium intake of 0.1 µg/g is thought to provide a satisfactory margin of safety against any dietary variables or environmental stresses likely to be encountered by grazing sheep and cattle (3). A minimum requirement of 0.06 µg/g selenium for the prevention of WMD in lambs is indicated by the work of Oldfield et al. (308), while New Zealand experience indicates that lambs can grow normally and remain free from clinical signs of selenium deficiency on pastures containing 0.03–0.04 µg/g selenium (159). The U.S. National Research Council concluded that 0.02 µg/g is the critical level of dietary selenium below which deficiency signs are observed in ruminants (407). However, Gardiner and Gorman (121) showed that WMD can occur in lambs grazing Australian pastures estimated to contain 0.05 µg/g selenium, and Whanger et al. (462,463) have suggested that 0.1 µg/g dietary selenium may not be sufficient to meet the physiological requirements of sheep under some conditions.

Some of the above differences probably reflect analytical uncertainties, in-

cluding variable selenium losses in drying and storing of samples, while variations in the herbage levels of tocopherol and of other substances affecting selenium metabolism can be important. In fact, a direct rather than an inverse
relationship between the selenium content of alfalfa and the occurence of WMD
in calves has been reported, despite evidence of the effectiveness of selenium
supplementation (366). This can probably be explained by the unusually high
sulfate levels in the alfalfa. High-sulfate intakes are known to reduce selenium
availability to animals at high selenium intakes (115), so that selenium requirements are likely to be greater when sulfate intakes are high than when they are
low.

The minimum selenium requirements of poultry have been greatly illuminated
by the critical studies of Scott and collaborators. Nesheim and Scott (297) found
that a torula yeast diet containing 0.056 μg/g selenium and 100 IU of vitamin E
per pound did not sustain maximum growth rates in chicks unless it was supplemented with 0.04 μg/g selenium as selenite. Subsequently, Thompson and Scott
(417) demonstrated a selenium requirement for growth close to 0.05 μg/g when
chicks were fed a purified diet with no vitamin E. With 10 μg/g of added vitamin
E the selenium requirement was >0.02 μg/g, and with 100 μg/g of vitamin E it
was <0.01 μg/g. The dietary selenium requirement of the vitamin E-adequate
chick for the prevention of *in vitro* ascorbic acid-stimulated hepatic microsomal
peroxidation (69), and for complete protection of the chick pancreas from fibrotic degeneration (418), is ~0.06 μg/g when the selenium is provided as sodium
selenite. The U.S. National Research Council suggested that 0.03–0.05 μg/g is
the critical level of dietary selenium below which deficiency signs are observed
in poultry (407). However, the selenium requirements of turkey poults are higher
than those of chicks on the basis of experiments carried out by Scott *et al.* (377).
These workers reported a selenium requirement of 0.17 μg/g for the prevention
of gizzard and heart myopathies when the diet was well supplied with vitamin E,
and ~0.28 μg/g in diets marginal in this vitamin and in sulfur amino acids.

The above requirements do not necessarily apply when the selenium is provided in other forms or when the diets contain abnormal amounts of elements
with which selenium interacts metabolically. For example, Cantor *et al.* (52)
established a dose–response curve for the effectiveness of sodium selenite in the
prevention of ED and in the maintenance of plasma GSH-Px activity. The effectiveness for these purposes of the selenium in various feedtsuffs and selenium
compounds was then compared with the selenite at equivalent selenium dose
levels. The selenium in most of the feeds of plant origin was highly available,
ranging from 60 to 90%, whereas that in the animal products tested was <25%
available. On the basis of those results, it would seem that the selenium requirements of chicks are higher when such animal products are included as protein
sources than when plant products are similarly employed. On the other hand,
Gabrielson and Opstvedt (108) criticized the statistical interpretation of these

results and found that the selenium in capelin or mackerel fish meal was about twice as effective as that in soybean or corn gluten meal for restoring plasma GSH-Px activity in selenium-depleted chicks. Further research is required to resolve these discrepancies.

The selenium requirements of chicks are also increased when the diet contains abnormally high dietary levels of silver, copper, and zinc. Jensen (194) observed high mortality and a high incidence of ED and muscular dystrophy in chicks fed a diet containing 0.2 $\mu g/g$ selenium supplemented with 800 or 1600 $\mu g/g$ copper or with 2100 to 4100 $\mu g/g$ zinc. A selenium supplement of 0.5 $\mu g/g$ completely prevented the deficiency signs and markedly reduced mortality. In other words, the selenium requirement of the chicks was raised from ≤ 0.2 $\mu g/g$ to 0.7 $\mu g/g$ by the treatments imposed. Similarly, Peterson and Jensen (335) observed a marked growth depression and high mortality, mostly due to ED, in chicks fed a diet marginal in vitamin E and containing 0.2 $\mu g/g$ selenium, when 900 $\mu g/g$ silver as silver nitrate was added. Including either 1 $\mu g/g$ selenium as selenite or 100 IU vitamin E per kilogram prevented these signs of selenium deficiency.

Studies with weanling male rats fed torula yeast-based diets (0.01 $\mu g/g$ selenium supplemented with 0, 0.05, 0.1, 0.5, 1.0, and 5.0 $\mu g/g$ selenium as Na_2 SeO_3 revealed optimal growth to 134 days with the 0.05 $\mu g/g$ supplement (i.e., 0.06 $\mu g/g$ total dietary selenium). This level did not satisfy the GHS-Px requirements for selenium. This requirement was apparently met by 0.1 $\mu g/g$ selenium, since increasing dietary selenium from 0.1 to 0.5 or 1.0 $\mu g/g$ caused only comparatively small increases in GSH-Px activities in erythrocytes and liver (148).

Supplemental selenium at the rate of 0.1 $\mu g/g$ as selenite has similarly been found adequate to meet the selenium requirements of sows throughout gestation and lactation through two reproductive cycles when corn–soybean meal diets containing 0.02 or 0.03 $\mu g/g$ selenium were fed (250). On the other hand, mortality of baby pigs attributed to vitamin E and selenium deficiency continued in certain areas of the United States, even though selenium was added to swine diets at the rate of 0.1 $\mu g/g$ (249). This problem generally occurred within 2 weeks postweaning. Meyer et al. (267) fed weanling pigs different levels of dietary selenium and on the basis of breakpoint regression analysis of both plasma and liver GSH-Px activities concluded that the selenium requirement of young swine was 0.35 $\mu g/g$ (total concentration in the diet).

Certain animal experiments have suggested that dietary selenium may have some beneficial effects at levels above those generally accepted as adequate. For example, Binnerts and El Boushy (23) reported that chicks fed a vitamin E-adequate diet supplemented with 0.25 μg selenium per gram for 7 weeks were ~3% heavier than chicks fed the same diet supplemented with 0.10 μg selenium per gram. The favorable effect of elevated dietary levels of selenium on some immune responses of mice is discussed in Section III,H, and this phenomenon may have some relationship to the beneficial response of certain experimentally induced cancers in animals to high dietary intakes of selenium (see Section V,A).

Selenium deficiency characterized by loss of weight, listlessness, and alopecia has been produced in subhuman primates by feeding adult squirrel monkeys (*Saimiri sciureus*) a low-selenium torula yeast diet adequate in vitamin E for 9 months (291). Monkeys injected with small doses (0.04 mg) of selenium as sodium selenite at 2-week intervals recovered rapidly. Untreated monkeys died and showed cardiac and skeletal muscle degeneration and hepatic necrosis. Some of these signs are more typical of vitamin E or combined vitamin E–selenium deficiency, but no changes in plasma tocopherol levels occurred as selenium deficiency progressed (408). In contrast, feeding selenium-deficient diets to adult rhesus monkeys (*Macaca mulata*) for 16 months produced no physiological signs of selenium deficiency in this species despite lowered blood selenium levels and erythrocyte and plasma GSH-Px activities (45,46). It was suggested that one reason for this species difference between squirrel and rhesus monkeys might be the different distribution of selenium and GSH-Px activity among the blood fractions from these animals (21).

Several different approaches have now been taken in an attempt to estimate human selenium requirements, including extrapolation from animal experiments, determination of the amount needed to maintain metabolic balance, comparison of dietary intakes in those areas with and without deficiency disease (Keshan disease), and measurement of functional responses during depletion–repletion studies. Regarding the first approach, the preceding discussion reveals that the selenium requirement of animals can vary from 0.02 to 0.35 $\mu g/g$ in the dry diet depending on the species, the age of the animal, the criterion of deficiency selected, and various other factors. If it is assumed that an adult human consumes about 500 g of food daily on a dry basis and that a nutritionally generous level of selenium for most animal species is 0.1 $\mu g/g$ in the dry diet, this would result in a daily dietary selenium intake of 50 μg, which is right at the lower limit of the safe and adequate daily dietary selenium intake for adult humans as defined by the U.S. National Research Council in 1980 (295). If the critical dietary level needed to prevent selenium deficiency in ruminants (0.02 $\mu g/g$) is used as the basis for extrapolation, this would translate into an adult human intake of only 10 $\mu g/day$, or somewhat less than that needed to prevent Keshan disease (see below). On the other hand, if one estimated the daily human adult selenium requirement by extrapolating on the basis of how much is needed by young swine (0.35 $\mu g/g$), one arrives at 175 μg, an intake still within that judged safe and adequate (50–200 $\mu g/day$, see ref. 295) and somewhat below that sometimes reported in healthy adults in the U.S. state of South Dakota (see Section IV,C). Such extrapolations do not take into account species differences and at any rate are not particularly useful in pinpointing a human selenium requirement, because the range obtained thereby is so great. The marked influence of cross-species variation is shown by the fact that two species of monkeys, both presumably more closely related to humans than rats, chickens, swine, or cattle, had vast differences in their ability to resist selenium deficiency (21). This difference was thought to be related to the distribution of selenium and GSH-Px activity in the blood. The pattern of selenium and selenoenzyme distribution in human blood was found to be closer to that of the resistant (rhesus) than to the nonresistant (squirrel) species of monkey. Any possible implications of these findings for human selenium requirements await further research.

Metabolic-balance studies represent a second approach that has been used by nutritionists to estimate human requirements for many different minerals. One study has shown that healthy North American men need ~80 μg of dietary selenium per day to stay in selenium balance, whereas women need only 57 μg/day (231). This sex difference in the amount of selenium needed to maintain balance was thought to be due to differences in body size between men and women. In fact, when the balance data were expressed on a body weight basis, both sexes needed ~1 μg selenium per kilogram body weight per day in order to achieve balance. On the other hand, selenium intakes as low as 24 and 9 μg/day were sufficient to allow balance in New Zealand women and Chinese men, respectively (405,451). These results demonstrate the profound influence of historical dietary selenium intakes on the amount of selenium needed to maintain balance in people. Thus, at intakes above minimum requirements, selenium balance only reflects equilibrium input and output and is not particularly helpful in determining true human requirements.

The third approach used to estimate human requirements for selenium was to compare dietary selenium intakes in those areas with and without deficiency disease. With the identification of Keshan disease as a selenium-responsive condition in humans, it became possible to compare dietary selenium intakes in those areas of China where the disease was present versus those areas where the disease was absent. Dietary selenium intakes were 7.7 and 6.6 μg/day in endemic and 19.4 and 14.1 μg/day in nonendemic Keshan disease areas for male and female adults, respectively (476). These figures represent what could be considered minimum daily requirements (MDRs) for selenium.

The fourth approach, that of depletion–repletion, was used in an attempt to estimate selenium requirements for North Americans, but their selenium reserves prevented plasma selenium levels from dropping to values typically reported in residents of other countries with lower selenium intakes (Finland, New Zealand), even after a depletion period of almost 7 weeks (235). Yang *et al.* (476) repleted Chinese men of very low selenium status (the typical dietary intake was about 10 μg/day) with graded doses of selenomethionine and followed increases in plasma selenium levels and GSH-Px activities. The plasma GSH-Px activity plateaued at similar values in all the groups given ≥30 μg of supplemental selenium daily. On this basis, these workers suggested what they called a physiological selenium requirement of ~40 μg/day for Chinese adult males. Presumably, this figure might have to be adjusted to take into account differences in the body weight of various other population groups.

Clearly, considerable progress has been made in delineating human nutritional requirements for selenium. Many questions still need to be addressed, however, including whether selenium intakes at pharmacological levels are of any possible value against certain human degenerative diseases such as cancer or cardiovascular disease (see Section V).

B.   Selenium in Animal Feeds and Forages

The levels of selenium in animal feeds and forages have been extensively studied. These levels vary widely with the plant species and with the selenium status of the soils in which they have grown. The effect of species is most apparent with the variously called accumulator, converter, or indicator plants that occur in seleniferous areas and that carry selenium concentrations frequently lying between 1000 and 3000 μg/g. One sample of *Astragalus racemosus* from Wyoming was reported to contain no less than 14,920 μg/g selenium, and >4000 μg/g on the dry basis was observed in the annual legume *Neptunia amplexicaulis* growing on a selenized soil in Queensland (206). The selenium in these plants is present in organic combinations that are readily water extractable and that play an important part in the incidence of selenosis in grazing stock, as considered in Section VI,A. For all ordinary edible grasses and legumes the primary determinant of selenium concentration is the level of available selenium in the soil. This is apparent from the area studies of forages carried out by Allaway and associates (3,290) and from numerous investigations in New Zealand and elsewhere. Pastures and forages in areas where animals are free from selenium-responsive diseases generally contain ≥0.1 μg/g selenium, while in areas with a variable incidence of such diseases the levels are mostly <0.05 μg/g and sometimes as low as 0.02 μg/g (dry basis). Forages and grains from British Columbia exhibited an exceedingly wide range of selenium concentrations, with a considerable proportion of the samples containing between 0.12 and 0.32 μg/g (271). The mean selenium concentration of all forages was 0.19 μg/g, with no significant differences between grasses and clovers, and of all grains 0.27 μg/g, with wheat significantly higher (0.32 μg/g) than barley and oats (0.20 μg/g). Much lower selenium levels in cereal grains from selenium-deficient areas, down to 0.006 to 0.007 μg/g, have been reported. The richest sources of selenium observed in the Canadian study were salmon and herring meals (1.9 μg/g). The relatively high selenium content of such marine products is also evident from the comprehensive study of feedstuffs carried out by Scott and Thompson (376).

The value of the different feedstuffs as a source of selenium depends on the form in which the element is present, as well as on its concentration, at least for chicks and rats. Some workers have found that plant sources generally have a much higher selenium availability than animal products (52), whereas others have found just the opposite (108). Much of the selenium in wheat and probably other grains (316) and alfalfa (6) is thought to be in the form of selenomethionine. The low bioavailability of selenium in certain fish meals is unexplained. Originally, it was thought that heavy metals, especially mercury, might complex selenium and render it unavailable, but a number of observations now seem inconsistent with such a hypothesis (225).

C. Selenium in Human Foods and Dietaries

The selenium content of human foods is influenced by a number of factors including the class of food itself, the geographic origin of the food, and the extent and type of processing and cooking. Seafoods, organ meats, and muscle meats are generally good sources of selenium (>0.2 $\mu g/g$, see ref. 277). Grains and cereal products contain variable amounts of selenium depending on where they are grown (see below). Fruits and vegetables are mostly poor sources (≤ 0.01 $\mu g/g$ wet weight).

The level of selenium in individual foods of plant origin is highly variable depending mainly on the soil conditions under which they are grown. Perhaps the greatest range in the selenium content of grain products was that reported from the People's Republic of China, where the selenium content of samples from areas of human selenosis was three orders of magnitude higher than that of samples from areas with Keshan disease (474). For example, corn from seleniferous regions averaged 8.1 $\mu g/g$, whereas corn from selenium-deficient regions averaged only 0.005 $\mu g/g$. Less extreme, though still significant, differences have been observed in other countries. In the United States and Canada, samples of hard wheat contained an average of 0.53 μg selenium per gram (range 0.05–1.09), whereas soft wheat contained an average of only 0.06 $\mu g/g$ (range 0.02–0.13) (245). Needless to say, such variability in the selenium content of basic staple foods makes it difficult to select a representative value for inclusion in tables of food composition.

In early work, considerably lower levels of selenium were found in white flour and bread than in whole-wheat flour and bread (277), which suggested that milling grains might have an appreciable effect on the selenium concentration in the final cereal product. The food samples used in those studies, however, were randomly selected from local supermarkets and had unknown histories. In a later study in which samples were collected from the same milling run, the level of selenium was only 14% less in the wheat flour than in the original parent grain (100). Likewise, Lorenz found that milling hard wheat into flour resulted in an average 19% decrement in selenium content (245). Thus, the decreases in selenium content of grain products as a result of milling are smaller than those observed with many other nutritionally essential trace elements (79).

Broiling steaks or chops had little effect on their selenium content (168), but frying caused a 40–70% increase in the selenium content of bacon, probably because of fat loss (110). Certain high-selenium vegetables such as asparagus or mushrooms lost 30–40% of their selenium after boiling, perhaps because of leaching into the cooking water (168).

Total daily dietary selenium intakes by humans will clearly vary greatly with the source of the foods consumed and to some extent with the choice of foods, particularly the extent of the consumption of marine fish. The greatest extremes in selenium intakes have been found in the People's Republic of China (474). In a high-selenium area with reported cases of chronic human selenosis, the daily intake averaged 4990 μg, whereas the intake in a low-selenium area with Keshan disease was only ~11 μg/day. Lesser extremes have been observed in other countries. For example, Finland and New Zealand, with their low-selenium soils, reported intakes of ~30 and 28–32 μg/day, respectively (421,436), whereas Venezuela and South Dakota, United States, both high-selenium areas, have reported intakes >200 μg/day (273,324). Self-selected diets of free-living adults in the U.S. state of Maryland provided an average selenium intake of ~80 μg/day in surveys conducted over a period of several years (231,234,453). National food surveys carried out in the United States ("market basket" studies) showed that the overall adult mean dietary selenium intake from 1974 to mid-1982 was 108 μg/day with annual daily means ranging from 83 to 129 μg (332). Most of the selenium in those composite diets was furnished by grain and cereal products (56%), followed by meats, fish, and poultry (38%) and dairy products (6%). Selenium intakes in Canada were somewhat higher than in the United States, since four composite diets from three Canadian cities, each representing the daily per capita consumption of foods, contained 191, 220, 113, and 150 μg (419). Again, cereals provided the most selenium, followed by meats and dairy products. When individual food items in the United States were ranked according to their contribution to the selenium intake in American diets by multiplying the selenium content of the food item by the amount of the food consumed as shown by national surveys, it was found that the five highest ranked foods (beef, white bread, eggs, chicken, and pork) contributed about half of the total selenium in American diets (137).

Water supplies do not normally constitute a significant human source of selenium. In the United States, >90% of the tap water samples collected contained <1 μg of selenium per liter (134). On the other hand, water from 10 of 44 wells in a seleniferous area of South Dakota contained 50–300 μg/liter (398).

D. Potency of Different Forms of Selenium

Variations exist among different chemical forms of selenium in their capacity to meet the needs of animals and to induce toxic effects. In the early studies of

Schwarz, selenium compounds were divided into three categories with respect to their potency against liver necrosis in rats (370). The first category includes elemental selenium and certain compounds that are practically inactive due to poor absorption. Elemental selenium was later shown to be almost completely unavailable for the prevention of ED in chicks (52). Compounds in the second category, which included most inorganic salts such as selenites and selenates and the selenium analogs of cystine and methionine, were found to be more or less equally protective against liver necrosis. Later studies have revealed significant differences within this group and even in the effectiveness of the same selenium compound in protecting against different manifestations of selenium deficiency. For example, the selenium from selenomethionine is less effective than that from sodium selenite in preventing ED in chicks (302), but the reverse is true for the prevention of pancreatic fibrosis (53). In fact, selenomethionine was four times as effective as either selenite or selenocystine for this latter purpose. The potency of selenomethionine is decreased in diets containing suboptimal levels of methionine (410). Some workers have found that vitamin B_6 deficiency interferes with the utilization of selenomethionine (477), but others have not (411). In short-term studies, both D- and L-selenomethionine had similar biopotencies for restoring hepatic GSH-Px activity when injected into selenium-depleted rats (258).

The third category of selenium compounds consists of organic forms more active per unit of selenium than those of the second group. An organic component of hydrolysates of kidney and other potent sources was shown by Schwarz (367) to be three to four times as potent against liver necrosis as selenite–selenium. This fraction was designated factor 3 and established as organic in nature, strongly bound to proteins, and separable into two factors known as α- and β-factor 3. The isolation and identification of these potent preparations have been handicapped by low yields and chemical instability of the purified fractions, but Schwarz and Fredga (371) demonstrated remarkable differences in the capacity of various aliphatic monoseleno- and diselenodicarboxylic acids to prevent liver necrosis in rats. In the course of further studies on the structure–activity interrelationships of other organic selenium compounds in the prevention of dietary liver necrosis, comparable differences in biopotency were demonstrated by these workers (372). It is evident that potency is critically dependent on the nature of the organic moieties attached to the selenium atom, but the basis of this dependence is not clear.

E. Bioavailability of Selenium in Human Foods

A number of methods have been proposed to determine the nutritional availability of selenium in human foods (225). In animal models, a commonly used technique is the elevation of selenium levels and restoration of GSH-Px activities in animals that have been previously depleted of selenium. When the restoration of hepatic GSH-Px in selenium-depleted rats was used as the criterion of selenium bioavailability, values ranging from 83 to 217% (selenite = 100%) were obtained for the selenium in various wheat products, Brazil nuts, and beef kidney (1,58,90). On the other hand, the selenium in various tuna products was only 22–57% as available as that in selenite (1,90). The selenium in mushrooms was only ~4% as available as selenite (58).

Several factors complicate the interpretation of currently used animal assays of selenium bioavailability. For example, Zhou et al. (481) found a positive relationship between the bioavailability of selenium in a diet to rats and the protein content of a diet. Thus, the protein content of a diet must be controlled in order to obtain valid bioavailability estimates. Combs et al. (71) found that the growth rate of chicks used in selenium bioassays influenced the apparent bioavailability of selenite or selenomethionine. These workers suggested the use of a short-term (12-hr) bioassay based on restoration of GSH-Px activity as a way of eliminating this variable. More research is needed to develop animal bioassays for estimating the nutritional availability of selenium in foods and to establish the applicability of these assays to humans.

Very few data exist concerning the bioavailability to humans of selenium from various sources. A placebo-controlled, double-blind study was carried out in Finland because the low selenium status of the people there would allow demonstration of increases in plasma selenium level and platelet GSH-Px activity after supplementation (238). Three criteria were used to assess selenium bioavailability: (1) short-term changes in plasma selenium level and platelet GSH-Px activity (immediate availability), (2) medium-term plasma selenium levels (tissue retention), and (3) long-term platelet GSH-Px activity after discontinuation of supplements (conversion of retained to biologically active selenium). Each selenium source (wheat, yeast, selenate) produced a characteristic response pattern by the three criteria examined. Wheat provided the most biologically active selenium by all three criteria, and those results indicated that, in agreement with animal studies, selenium in wheat was readily utilized by humans. The complexity of the experimental design used in this study shows the work needed to obtain valid selenium bioavailability data in humans.

F. Selenium Supplementation

The methods available for providing selenium supplements to animals include the use of selenium-containing feed additives, provision of salt mixtures or licks containing small amounts of selenium salts, periodic injections or oral dosing with selenium salts, administration of selenium-containing heavy pellets to ruminants, interregional blending of feeds from selenium-rich areas, and treatment of the soil with selenium compounds (407).

Several of these techniques were evaluated by Whanger *et al.* (463), who found that selenium given to ewes as heavy pellets, in salt, as injections, or in drench (10 mg selenium as sodium selenite) was highly effective in preventing WMD in their lambs. More variable results were obtained by giving a feed (oat) supplement furnishing 0.1 or 0.2 µg selenium per gram in the total diet or by giving 5 mg selenium in the drench. MacPherson and Chalmers compared three methods of selenium supplementation to cattle (subcutaneous injection, intra-ruminal pellet, and addition to water) and a fourth method to lambs (oral supple-mentation with a sodium selenite solution) and concluded that all four methods worked effectively for periods ranging from 4 months to 1 year after treatment (248). The method of choice depends primarily on economics and the conditions of husbandry, and is influenced by the amounts of selenium introduced into the soil–plant–animal–human selenium cycle.

For confined animals, the easiest way to supplement with selenium is to add it directly to the feed. The selenium-containing feed additives allowed in the United States by the Food and Drug Administration are sodium selenite and sodium selenate (433). Selenium levels permitted are 0.1 µg/g in the complete feed for beef and dairy cattle, sheep, chickens, ducks, swine (0.3 µg/g in starter and prestarter rations), and 0.2 µg/g for turkeys (408). Before permission was obtained to use selenium feed additives in the United States in 1973, grains and alfalfa grown in areas where the selenium levels in the soil are naturally high were often obtained and shipped to selenium-low areas. This system of blending feed ingredients, however, imposed the need for controlled analyses of the feeds (6), because the selenium content of feeds grown in the so-called high-selenium parts of the country varies widely.

In the United States, selenium addition to salt–mineral mixtures for free-choice feeding is also permitted (433). Up to 20 or 30 µg selenium can be added per gram to salt–mineral mixtures for beef cattle and sheep, so as not to exceed an intake of 0.23 and 1 mg selenium per head per day, respectively.

Direct subcutaneous selenium injections, usually as sodium selenite, con-stitute the best procedure in terms of the amounts of selenium used and are widely practiced to control a range of selenium-responsive diseases, particularly in cattle. New Zealand experience indicates that doses ranging from 10 mg for calves to 30 mg for adults, given at 3-month intervals, are satisfactory (9). Excessive selenium concentrations in edible animal products from the use of selenium injections or oral doses in the amounts prescribed have not been ob-served (74,88,214,318). With sheep, oral dosing with selenite or selenate in doses from 1 to 5 mg selenium at intervals as described previously is the most common means of preventing selenium-responsive diseases in this species, be-cause the drenching can often be carried out when the animals are yarded for routine management procedures. Both injections and oral dosing have the advantage of providing known amounts of selenium to individual animals but have the disadvantage of requiring individual handling and movement of stock.

The disadvantages attached to selenium dosing or injecting led Kuchel and Buckley (210) to investigate the possibilities of selenium-containing heavy pellets for grazing sheep, to be used in a similar manner to the heavy cobalt pellets. Such pellets, in which the selenium sources were calcium selenate, barium selenate, and elemental selenium, were compared when dosed singly to sheep grazing pastures of normal selenium status. Blood and tissue selenium levels were significantly enhanced by all three types of pellets for periods up to 12 months. In an experiment with pellets composed of finely divided metallic iron and elemental selenium and containing 1.25, 2.5, 5, or 10% selenium, blood selenium concentrations were increased significantly in all cases within 1 week. Within 10 to 12 weeks the extent of the rise was related to the proportion of selenium in the pellets. The pellets were retained in the reticulorumen, and the selenium concentrations in the tissues of the treated sheep after 6 to 12 months were no greater than those reported for untreated lambs in the United States (331) and New Zealand (74). Handreck and Godwin (153) have confirmed these findings with sheep, using dense pellets consisting of elemental iron and selenium (9:1 by weight) labeled with ^{75}Se. During the experimental period of 1 month the pellets released 0.5–1.3 mg selenium per day, of which ~30% was excreted in the urine and ~1% in the expired air. Blood selenium levels leveled out at 0.18 to 0.34 μg/ml, and there was no evidence of toxicity or excessive selenium accumulation in the tissues. The rate of release of selenium from the pellet is a function of the size of the grains of elemental selenium in the pellet, and it was suggested that the optimum size of the selenium grains for both an immediately usable and continuously adequate release of selenium was on the order of 35 to 45 μm (334). Although selenium pellets are a convenient and economical means for supplementing grazing ruminants that are not given other concentrate feeds, this product is not yet approved for use in the United States (408).

Treatment of the soil with various selenium compounds, by spraying of herbage or by application alone or following incorporation of the selenium into normal fertilizers, offers further possibilities for optimizing the selenium levels in pastures and forages. In terms of the amounts of selenium required and introduced into the selenium cycle, this is an expensive and hazardous process because of the inefficiency with which such added selenium is taken up by most plants, especially from acid soils (6). Under experimental conditions Allaway et al. (5) showed that the addition of 2 lb selenium per acre as Na_2SeO_3 increased the selenium content of alfalfa from very low levels to concentrations that protect lambs from WMD. Furthermore, the effects of one application lasted for at least 3 years. A serious difficulty in supplying animal requirements by application to pasture is the high levels that occur immediately following application, whether these are due to foliar or root uptake or to initial surface contamination. The problem is to supply sufficient selenium in a form that will maintain adequate concentrations of the element in the plants for a sufficiently long period to avoid the necessity of frequent application, without inducing potentially toxic levels following treatment.

Watkinson and Davies (450) maintain that "with proper precautions to minimize pasture contamination," 1 oz/acre and possibly 2 oz/acre (as sodium selenite) should present no hazard, "at least for a few years." A further possibility has been proposed by Allaway et al. (4) involving the use of "selenized superphosphate." This material, prepared by incorporating the selenium into the superphosphate during the treatment of the rock phosphate with acid, seems to have promise on some soils, since plants containing adequate selenium levels can be produced at additions equivalent to 16 oz selenium/acre, and five times this amount has not resulted in plants with toxic levels of selenium. On very sandy soils the apparent safe range of application of selenized superphosphate is much narrower.

Gissel-Nielsen (124) found that both foliar application of selenite or use of selenite-enriched calcium ammonium nitrate were efficient and safe means of improving the selenium status of pasture crops to prevent deficiencies in grazing animals. The method of choice would depend on local fertilization practices and the availability of selenium solutions or selenium-enriched nitrogen fertilizer. In New Zealand, Watkinson (449) reported that plant and animal blood selenium levels were within the expected range after annual topdressing of pasture deficient for grazing stock with sodium selenate pellets (1% selenium), either alone or in fertilizer, at a rate not exceeding 10 g/ha.

Several different approaches have now been used to supplement humans with selenium. In the Keshan disease areas of China, children are given tablets containing sodium selenite (202). The dose used is 0.5 mg of sodium selenite per week for children 1–5 years old and 1.0 mg of sodium selenite per week for those 6–9 years old. This level of selenium administration has been shown to be highly effective in preventing Keshan disease (202). The Chinese scientists have also studied the addition of selenium to table salt (10–15 μg/g as sodium selenite) and the spraying of sodium selenite on the leaves of crops during their preflowering stage (61). Although some success was obtained with both techniques, salt is not the best vehicle for reaching the target populations concerned (infants and young children), and economic difficulties were encountered in the practical implementation of foliar spraying under Chinese conditions (61).

Agricultural and public health authorities in Finland have made the decision that the selenium content of Finnish feed and food should be increased by the addition of sodium selenate to fertilizers (207). Thus, Finland is the first country to raise the selenium intake of the general population by the deliberate introduction of selenium into the food chain. This will be done by the addition of selenium as selenate to all the main NPK fertilizers, granular in formulation, for both cereal and grassland crops at a concentration of 16 or 6 μg/g, respectively. The manufacture of these selenized fertilizers began in the summer of 1984 and they will be used at application rates of 10 g selenium per hectare per growing season. It is anticipated that the selenium level of the cereal and grassland crops will be raised to 0.1 and 0.15–0.20 μg/g (dry basis), respectively. These measures are expected to increase the average national selenium intake in Finland by humans to a level of 50 μg/day in those years when no imported grains are required. In those years when it is necessary to import grain (usually higher in selenium than the locally grown grain), the average daily intake is still expected to stay <100 μg.

People who are being fed intravenously for extended periods of time (e.g., persons on home TPN) are at risk for developing low selenium status, because most fluids used for such purposes contain very low levels of selenium (226). A product containing selenium, formulated for TPN use, is commercially available. However, the form of selenium in this material is selenite, which could be reduced to the biologically unavailable elemental selenium if it is mixed with TPN solutions containing certain reducing agents (e.g., vitamin C) (389).

The increasing interest about the role of selenium in human nutrition has led to the growing popularity of so-called health food supplements containing selenium for human use. The need for such products, however, is not evident, especially in countries like the United States where the dietary selenium intakes appear to be more than adequate to prevent any selenium deficiency diseases in humans (104). In fact, an outbreak of human selenium poisoning has been reported in the United States because of the consumption of superpotent over-the-counter selenium preparations (11,163,198). Therefore, the use of such products by persons not suffering any ill effects due to selenium deprivation is discouraged.

V. SELENIUM AND HUMAN DISEASE

Reports have now appeared concerning both naturally occurring human selenium deficiency and toxicity (see Section III,I and VI,D, respectively). In light of these positive findings, the question arises whether marginally depressed or elevated selenium intakes can have any implications for human health aside from Keshan disease or human selenosis. A number of workers have attempted to link differences in selenium status with differences in the occurrence of certain degenerative human diseases, particularly cancer and cardiovascular disease.

A. Cancer

An area of considerable current research interest is the possible chemopreventive role of selenium in carcinogenesis (140). Early ecological comparisons suggested an inverse association between selenium intake and the risk of cancer (362,385,386). Clark *et al.* (66) have since conducted a case–control study and found that patients with skin neoplasms living in the U.S. state of North Carolina had plasma selenium levels of 0.141 µg/g, whereas the plasma selenium level of controls was 0.155 µg/g. Willett *et al.* (467) showed that the prediagnostic serum selenium level of 111 cancer patients in the United States (0.129 ± SEM 0.002 µg/ml) was significantly lower than that of 210 cancer-free matched controls (0.136 ± 0.002 µg/ml). The risk of cancer for subjects in the lowest quintile of serum selenium (<0.115 µg/ml) was twice that of subjects in the highest (>0.154 µg/ml). Salonen *et al.* (356) reported that the mean serum selenium concentration of 128 cancer cases in eastern Finland was 50.5 ± 1.1 (SE) µg/liter, whereas that of their matched controls was 54.3 ± 1.0 µg/liter. In this study, there was a relative risk of cancer of 3.1 with serum selenium concentrations <45 µg/liter. In a subsequent study that analyzed serum samples collected in 1977 [the serum samples in the earlier study (356) were drawn in 1972], Salonen *et al.* (357) found that the mean serum selenium concentrations in 51 cancer cases versus their matched controls were 54 and 61 µg/liter, respectively. The somewhat higher serum selenium levels in the 1977 versus the 1972 study reflect the increased importation of high-selenium grain into Finland during the late 1970s (289).

These epidemiological associations have generated considerable interest. However, the age-adjusted mortality rates for breast and colonic cancer reported in Finland are considerably lower than those reported in the United States (448) despite the well-documented lower dietary selenium intakes in Finland (see Section IV,C). Earlier, Allaway had critcized the reported associations between cancer mortality and the geographical distribution of selenium as lacking strength and consistency (2). Presumably, these statistics suggest that if selenium is playing any role in the etiopathogenesis of human cancer, its effect can be overwhelmed by other factors. It would be of obvious interest to identify the nature of these other factors.

Other evidence concerning a possible role for selenium as an anticarcinogenic agent comes from animal studies. The first report of a possible beneficial effect of selenium against cancer in animals was that of Clayton and Baumann in 1949 (67), who found that only 2 of 9 surviving rats given 5 µg/g dietary selenium as selenite and fed a carcinogenic azo dye developed liver tumors, whereas 4 of 10 survivors not given selenium had such tumors. Since that time, a voluminous literature has appeared that conclusively demonstrates that dietary selenium can have a protective effect against chemically induced (183,184, 252,384), spontaneous (presumably virally induced) (264,361), and transplantable tumors (338) in rats and mice under a wide variety of conditions. Despite the large amount of research effort in this area, however, little information exists at the present time concerning the mechanism by which selenium may exert its antitumorigenic effect. It seems likely that more than one mechanism may be involved. For example, Ip and Sinha (183) found that selenium deficiency increased the incidence and yield of mammary tumors induced by dimethylbenz[a]anthracene (DMBA). This increased tumor incidence due to selenium deficiency, however, was seen only in rats fed diets high in polyunsaturated fat (25% corn oil) and was not observed either in rats fed diets high in saturated fat (24% coconut oil plus 1% corn oil) or in rats fed low-fat diets containing polyunsaturated fat (1–5% corn oil). This effect of selenium deficiency in potentiating the tumorigenicity of DMBA may be related to its antioxidant function as GSH-Px.

On the other hand, Ip has also shown that excess levels of dietary selenium (2.5–5 µg/g) inhibit DMBA-induced mammary tumors in rats and that at these levels the effect of selenium is apparently unrelated to GSH-Px (181). Medina et al. (265) agree that the chemopreventive effects of elevated selenium intakes in rodents cannot be attributed to maintaining high levels of GSH-Px. Rather, these workers suggest that perhaps synthetic activities related to DNA should be investigated as critical sites of selenium action. But Ip and Daniel (182) found no effect of selenium deficiency or excess on the binding of tritiated DMBA to rat liver or mammary gland DNA. Also, dietary selenium level had no quantitative or qualitative effect on the formation of DMBA–DNA adducts in the liver. Moreover, in their system, selenium inhibited mammary carcinogenesis after the carcinogen was administered, so they stressed the need to postulate mechanisms for the chemopreventive effect of excess selenium other than those involving alterations of carcinogen metabolism and DNA binding. Clearly, more research is needed to elucidate the mechanism(s) of the anticarcinogenic action(s) of selenium.

Although the above discussion might convey the impression that selenium protects universally against cancer in animals, certain experiments suggest that such is not the case. For example, Whanger et al. (464) found that the protective effect of selenium against the development of spontaneous mammary tumors in C3H mice was highly dependent on the type of diet fed. They concluded that any possible tumor inhibition by selenium may be limited to specific conditions and that selenium cannot be assumed to be effective under all circumstances. Selenium had no effect on the induction of neoplasia by Rauscher leukemia virus in mice (99) and in fact appeared to facilitate the induction of breast fibroadenomas by adenovirus type 9 in rats (10). Moreover, as pointed out by Underwood (432), neoplasia is not observed among the various lesions attributed to selenium deficiency in animals. Simpson (392) found no relationship between selenium treatment and the prevalence of small intestinal carcinoma in New Zealand sheep.

The concept that selenium might be carcinogenic (as opposed to anticarcinogenic) is based on early studies by Nelson et al. (296), who found liver cell adenoma or low-grade carcinoma in 11 of 43 rats that had developed hepatic cirrhosis after being fed a low-protein diet containing high levels of selenium as seleniferous grain or as a mixed inorganic selenide for 18 to 24 months. Since liver tumors were observed only in rats that developed hepatic cirrhosis, histological evaluation of the tissue was difficult because of possible confusion between neoplasia and regenerative hyperplasia. Later Tscherkes and co-workers (429) studied the effects of selenate with rats on low- and high-protein diets. Although liver tumors were observed in these studies, the experiments were criticized (432) because no controls, without added selenium, were used. However, a later communication from this group (441) stated that not a single case of spontaneous liver cancer was seen in 10,000 other rats used in their laboratory. On the other hand, subsequent experiments failed to demonstrate any tumorigenic effects of selenate in rats (441). Harr et al. (155) fed selenite and selenate at varying levels to rats and evaluated their carcinogenic potential in comparison with the known hepatocarcinogen, N-2-fluorenylacetamide. Hyperplastic liver lesions occurred in the selenium-fed rats that did not regress when the added selenium was withdrawn. Of the 63 neoplasms observed, however, none could be attributed to the addition of selenium. The experiments of Schroeder and Mitchener (363), in which rats were given selenite in the drinking water at 2 μg/ml selenium for a year and then selenite or selenate at 3 μg/ml selenium for the remainder of their life span are equally unsatisfactory from the point of view of demonstrating a carcinogenic effect from selenium. As Scott (374) has pointed out, the incidence of tumors was lower in the selenite-fed than in the control rats, and the higher incidence of tumors in the selenate-fed animals may have been due to the beneficial effects of the selenate on their longevity. In a later experiment by Schroeder and Mitchener (364) with mice given selenite or selenate at 3 μg/ml selenium in the drinking water for life, no significant effect on the incidence of spontaneous tumors was observed.

Although these later studies would tend to refute the results of Nelson et al., no one in fact has duplicated exactly the experimental conditions of these earlier workers (low-protein diet with organic or inorganic selenide forms of selenium). Birt et al. (24) found that feeding a diet containing 2.5 μg selenium per gram as sodium selenite increased the yield of pancreatic ductular carcinoma in male hamsters injected with bis(2-oxopropyl)nitrosamine. Moreover, there are three studies that show carcinogenic effects that are probably related to particular compounds that contain selenium rather than to selenium itself. Adenomatous hepatic hyperplasia and multiple thyroid adenomas were found in rats fed 0.05% bis-4-acetaminophenylselenium dihydroxide for 105 days (382), and an increased incidence of hepatomas, lymphomas, and pulmonary tumors was observed in mice given the maximal tolerated dose of selenium diethyldithiocarbamate (ethyl selenac) for 82 weeks (180). High doses of selenium sulfide, an ingredient of certain antidandruff shampoos, given by gavage resulted in an increased incidence of hepatocellular carcinoma in male and female rats and female mice (294). In this study, the increased incidence of liver tumors was not related to the development of hepatic cirrhosis. The female mice given selenium sulfide also developed an increased number of alveolar–bronchiolar carcinomas, which would not be expected to be related to any liver damage.

B. Cardiovascular Disease

As in the case of cancer, the first suggestion that low selenium status might be associated with increased heart disease came from ecological comparisons (387). Since that time, a number of epidemiological studies have appeared, some of which support this concept and some of which do not. In a study from Finland that showed an unequivocally inverse association between serum selenium level and the risk of death from ischemic heart disease, the mean serum selenium concentration for all cases was 52 μg/liter, whereas that for all controls was 55 μg/liter (355). A serum selenium level of ≤35 μg/liter was associated with a 6.9-fold increased risk of death due to ischemic heart disease compared with persons with a serum selenium of ≥45 μg/liter.

In the United States, Moore *et al.* (274) reported an inverse correlation between the plasma selenium level and the severity of arteriographically defined coronary atherosclerosis. The mean plasma selenium concentration of patients with "zero-vessel" disease (defined as no narrowing as great as 50% of any coronary arterial lumen visible on the arteriogram) was 136 ± 7 (SEM) μg/liter (range 90–179 μg/liter), whereas the mean plasma selenium level of patients with "three-vessel" disease (defined as a narrowing of ≥50% in the three major coronary arteries or their branches) was 105 ± 4 (SEM) μg/liter (range, 45–137 μg/liter).

On the other hand, neither Robinson and Thomson in New Zealand (345) nor Ellis *et al.* in England (96) found any correlation between blood selenium concentrations or GSH-Px activities and the traditional risk factors for cardiovascular disease. Moreover, three studies from Finland found little or no association between selenium status and the risk of death from ischemic heart disease (268,358,440). But Salonen (354) critiqued these Finnish studies and showed that two of the three had relatively poor statistical power, and in one of these the mean serum selenium level was relatively high by Finnish standards (73 μg/liter) because of the increased importation of selenium-rich grain during the late 1970s (289). He concluded that in total all four Finnish studies (268,355,358,440) did in fact lend support to the concept of an increased risk of ischemic heart disease under conditions of low selenium intake (reflected by a mean serum selenium concentration of <60 μg/liter).

A possible link between selenium and heart disease is supported by certain animal experiments. Platelets from selenium-deficient rats exhibit increased aggregability (254,359) and increased production of thromboxane A_2 after stimulation with collagen (359). Furthermore, selenium deficiency impairs the production of prostacyclin-like activity by rat aorta (239,254). Since GSH-Px, the enzyme thought to be responsible for removing lipid hydroperoxides *in vivo*, is depressed in selenium deficiency, this decreased production of prostacyclin could be due to an inhibition of prostacyclin synthetase by elevated levels of lipid peroxides (353). Depressed levels of platelet GSH-Px activity in selenium-deficient rats are readily restored by selenium supplementation (237). Similarly, the activity of GSH-Px in platelets of Finnish men of low selenium status increased almost two-fold after supplementation with selenium (238). Further research is needed to establish whether these biochemical changes have any relevance to cardiovascular health in humans.

C. Other Diseases

A number of attempts have also been made to link various human health problems other than cancer and heart disease with marginally excessive or deficient intakes of selenium, but in general these have met with little success. For example, individuals residing in seleniferous areas of South Dakota were reported to have poor dental health (398), but later epidemiological studies of the relationship between selenium status and dental caries produced conflicting results (48,78,146). In animals, selenium has been shown to have either caries-promoting or caries-inhibiting effects depending on the conditions of the experiment (28,30,47). Moreover, the selenium dose levels used in these experiments (≥ 0.8 μg/ml in the drinking water) were much greater than would be found in water supplies except in the most highly seleniferous areas. The U.S. National Research Council concluded that there seemed to be no reason to suspect that selenium is important to cariogenesis in humans (72).

Kilness and Hochberg (204) noted an unusual cluster of four cases of amyotrophic lateral sclerosis (ALS) in a high-selenium area of South Dakota where selenium poisoning of farm animals is endemic and suggested that excess selenium might be an environmental factor predisposing to this disease. Others, however, pointed out that the association of ALS with selenium is probably spurious (213,304,369).

On the basis of histopathological changes in monkeys, Wallach (447) proposed that selenium deficiency may play an important role in the etiology of cystic fibrosis. Subsequent surveys however, showed that whole-blood or serum selenium levels in patients with cystic fibrosis were above the deficient range and that whole-blood GSH-Px activity was normal (56,243,300). Some workers have found decreased GSH-Px activities in the blood of patients with multiple sclerosis (193,414), whereas others have not (266). Mazzella *et al.* (256) reported decreased erythrocyte GSH-Px activities in multiple sclerosis patients, even though their red cell and plasma selenium levels were similar to controls. These workers concluded that the modified GSH-Px activity observed in erythrocytes from multiple sclerosis patients is probably due to genetic factors rather than dietary selenium intake.

VI. SELENIUM TOXICITY

A. Selenosis in Animals

Based on extensive field experience in seleniferous areas of the Great Plains of North America, Rosenfeld and Beath (348) postulated three discrete types of selenium poisoning in livestock: acute, chronic of the blind staggers type, and chronic of the alkali disease type. Localized seleniferous areas have also been identified in Ireland (101), Israel (341), Australia (206), the Soviet Union (208), Venezuela (186), China (474), and South Africa (34). In these areas toxic selenium intakes by animals arise either from the consumption of selenium accumulator plants, the ingestion of more normal forage species with relatively high selenium concentrations due to the presence of above-normal levels of available selenium in the soils of the affected areas, or from both. Selenium accumulator plants play an important dual role in the incidence of selenosis in grazing stock. They have the ability to absorb selenium from soils in which the selenium is present in forms relatively unavailable to other plant species, and on their death to return selenium to the soil in organic forms that are available to these other species. Hence the alternate name ''converter'' plants. Accumulator or converter plants thus provide a direct source of selenium to animals consuming them, and they convert unavailable forms into available forms, so raising the selenium levels in plants that might otherwise contain safe levels of the element. In some areas this may not be important, but in others the presence of converter plants can intensify the severity of selenosis and extend its incidence.

Animals that ingest sufficient quantities of selenium as highly seleniferous accumulator plants are acutely poisoned. Obvious signs of acute distress are apparent in the animal such as abnormal movement and posture, difficult breathing, prostration, and diarrhea, with death often occurring in a few hours. In the field, acute selenium poisoning is rather infrequent because grazing stock usually do not consume selenium accumulator plants except when other forage is not available. Acute selenium toxicity has also occurred experimentally or accidentally by giving selenium compounds to livestock (72).

Blind staggers occurs in animals that consume a limited amount of selenium accumulators during a period of weeks or months (348). The animals suffer from blindness, abdominal pain, salivation, grating of the teeth, and some degree of paralysis. Respiration is disturbed, and death results from respiratory failure. Death also results from starvation and thirst because, in addition to loss of appetite, the lameness and pain from the condition of the hooves are so severe the animals are unwilling to move about to secure food and water. This situation is comparable to that of animals in fluorosis areas. This syndrome has not been produced in animals given pure selenium compounds, so it is possible that alkaloids or other toxic materials that occur in some seleniferous plants may play a role in blind staggers (246,434).

Animals that consume grains containing 5–40 µg/g selenium over weeks or months develop alkali disease. This chronic selenium poisoning is characterized by dullness and lack of vitality, emaciation and roughness of coat, loss of hair from the mane and tail of horses and the body of pigs, malformations, soreness and sloughing of the hooves, stiffness and lameness due to erosion of the joints of the long bones, atrophy of the heart ("dishrag" heart), cirrhosis of the liver, and anemia. Although some workers have not been able to produce alkali disease in cattle by feeding inorganic selenium compounds (246), several others have shown a causal relationship between the disease and the consumption of seleniferous grains and grasses and were able to reproduce the syndrome by feeding inorganic selenium salts (310). Alkali disease has been studied mainly in cattle, but a similar condition in sheep has been described in the Soviet Union under natural conditions (97).

In the rat and dog, a marked restriction of food intake occurs, together with anemia and severe pathological changes in the liver. This organ becomes necrotic, cirrhotic, and hemorrhagic to varying degrees. Anemia is a common manifestation of selenosis in all species, but in the rat and dog a microcytic, hypochromic anemia of progressive severity usually develops, and animals may die with hemoglobin levels as low as 2 g/dl (283,400).

Growing chicks exhibit a reduction in food intake and growth rate when consuming seleniferous diets, and there is a fall in egg production in hens. Franke and Tully (106) showed that the eggs produced by these hens have low hatchability at selenium concentrations in the feed too low to cause manifest signs of poisoning in other animals. The eggs are fertile, but a proportion produce grossly deformed embryos, characterized by missing eyes and beaks and distorted wings and feet. Such monstrosities can also be produced by injecting various selenium compounds directly into the air sac of fertile chicken eggs (322). Severe reproductive problems have been reported in aquatic birds in the San Joaquin Valley in California due to high levels of selenium in irrigation drainwater ponds (306).

Consumption of seleniferous diets interferes with the normal development of the embryo in rats (347), pigs (442), sheep (347), and cattle (84). This effect is also apparent from the birth of foals and calves with deformed hooves in seleniferous areas. On the other hand, no embryonic malformations were observed in pregnant hamsters injected intravenously with a barely sublethal dose of sodium selenite, and in fact the sodium selenite partially protected against sodium arsenate- or cadmium sulfate-induced teratogenesis (174). The rate of reproduction of cattle on ranches with seleniferous ranges is consistently low, and Olson (309) concluded that a reduction in reproductive performance is the main economic effect of selenium poisoning of the alkali disease type.

B. Factors Affecting Selenium Toxicity in Animals

The toxicity of selenium to animals varies with the amounts and chemical forms of the selenium ingested, the duration and continuity of the selenium intake, the nature of the rest of the diet, and to some extent with the species. Limited evidence suggests that monkeys (28) are much more vulnerable to selenium in the drinking water than rats (144) or hamsters (145). Munsell *et al.* (288) place the minimum dietary selenium levels at which signs of toxicity will ultimately arise at 3 to 4 $\mu g/g$, but this will clearly vary with the extent to which other dietary components with which selenium interacts are present. Selenium levels of 3 to 4 $\mu g/g$ do not adversely affect the growth of chicks or the hatchability of eggs, but at 5 $\mu g/g$ hatchability is slightly reduced and at 10 $\mu g/g$ is reduced to zero (282). For normal growth of chicks <5 $\mu g/g$ is necessary (339). In a later study 2 $\mu g/g$ of added selenium as selenite had no adverse effects on the growth or mortality of chicks or on the production or hatchability of their eggs, but 8 $\mu g/g$ significantly reduced hatchability (14).

In rats and dogs given diets containing 5–10 μg selenium per gram, signs of chronic poisoning arise, and at 20 $\mu g/g$ there is complete refusal of food and death in a short time (288). Young pigs fed seleniferous diets containing 10–15 $\mu g/g$ selenium developed signs of selenosis within 2 to 3 weeks (347). The minimum toxic levels for grazing stock probably lie close to 4 to 5 $\mu g/g$ (72). Edible herbage in seleniferous areas commonly contains 5–20 $\mu g/g$ selenium. At the lower levels, signs of toxicity in cattle would take weeks or even months to appear. Mature sheep fed regular oral doses of sodium selenite, up to 600 μg selenium per kilogram body weight per day, for periods as long as 15 months revealed no pathological changes in the tissues. After 5 to 6 months of treatment, a depression in food consumption and body weight increase was observed at daily selenium intakes of 200 and 300 $\mu g/kg$ per day, with some mortality at 400 $\mu g/kg$ (336). The 200-$\mu g/kg$ daily treatment is equivalent to a dietary selenium intake of ~8 $\mu g/g$.

High-protein diets afford some protection against potentially toxic selenium intakes. Smith (397) found that 10 $\mu g/g$ selenium in a 10% protein diet was highly toxic to rats, whereas with an additional 20% protein as casein, scarcely any adverse effects were evident. Linseed oil meal is superior to casein in protecting rats against selenium poisoning, and protein is not responsible for its superior effect (151). Rather, the protective factors appear to be two newly isolated cyanogenic glycosides (323,396). A curious feature of this protective action of linseed oil meal is that it is accompanied by higher selenium levels in

the liver and kidneys compared with those of rats fed casein-containing seleniferous diets (232). Cyanide itself has been shown to have a partial protective
effect against selenium poisoning in rats (320). The sulfur-containing amino
acids cannot be implicated with any certainty in the protein-protecting effect
(314). On the other hand, methionine has a beneficial effect against selenite
toxicity in rats that appears to depend on the presence of vitamin E or other fat-
soluble antioxidants in the diet (230). Dietary levels of 1.0% sulfate decreased
the chronic toxicity caused by selenate in young rats but had a much smaller
effect in rats fed selenite (115). Since sulfate increased the urinary excretion of
selenium as selenate but had little or no effect on selenium as selenite, the
possibility of a specific antagonism between the sulfate and selenate anions was
suggested. The protective effect of high-protein diets may be related to the
production of endogenous sulfate.

The toxicity of selenium can be greatly modified by the dietary levels of
arsenic, silver, mercury, copper, and cadmium, with each element apparently
exerting its protecting action by its own mechanism. The effect of arsenic was
first demonstrated by Moxon (281), who showed that 5 mg/liter as arsenite in the
drinking water prevented all signs of selenosis in rats. Arsenic has since been
used successfully to alleviate selenium poisoning in pigs, dogs, chicks, and
cattle. Sodium arsenite and arsenate are equally effective, arsenic sulfides are
ineffective, and such organic forms as arsanilic acid and 3-nitro-4-hydroxyphen-
ylarsonic acid provide partial protection (164,214,443). The protection afforded
by arsenic is due at least in part to increased biliary selenium excretion (229).
The main target organ of selenium toxicity, the liver, can apparently rid itself of
excess selenium in this way. Although 8 or 15 μg arsenic per gram as arsenite
afforded some protection against the toxicity to chicks of 8 μg/g selenium as
selenite, the selenium levels in the liver and other tissues were not significantly
reduced by the arsenic treatment (14).

The inclusion of mercuric chloride, cupric sulfate, or cadmium sulfate in the
diet of chicks has been shown by Hill (169) to overcome partially the growth
retardation and mortality induced by feeding 40 μg/g selenium as selenium
dioxide. When the mercury and selenium were both fed in inorganic forms, the
most effective ratio in preventing the selenium toxicity was 1 : 1. At a comparable ratio copper was as effective as mercury in reducing the mortality but did not
prevent the growth inhibition as mercury did. Evidence was obtained suggesting
that mercury, copper, and cadmium exert their protection by reacting with selenium, probably within the intestinal tract, to form relatively innocuous compounds. Jensen (195) similarly found copper to be highly effective in preventing
mortality in chicks fed for 2 weeks 20, 40, or 80 μg/g selenium as selenite, with
a much smaller favorable effect on the growth inhibition. The copper sulfate was
fed in these experiments at the extremely high copper level of 1000 μg/g. A
marked protective effect against both the mortality and growth retardation in-

duced by the same levels of dietary selenium from silver fed as silver nitrate was also demonstrated, again at the abnormally high silver level of 1000 $\mu g/g$. Evidence was obtained that silver modifies the selenium toxicity by interfering with its absorption and causing the accumulation of a nondeleterious selenium compound in the tissues, whereas copper achieves its effect only by the latter process.

A further factor affecting selenium toxicity is adaptation by the animal. The adaptive responses to selenium seen in bacteria (221,390) may have a parallel in animals, since Jaffe and Mondragon (187) showed that young rats born from mothers fed a selenium-containing diet lost selenium from their livers, whereas rats bred on the stock diet accumulated this element under the same conditions. These results were confirmed and extended under low- (0.5 $\mu g/g$) or moderate-selenium (4.5 $\mu g/g$) diets to the mothers and high- (10 $\mu g/g$), moderate- or low-selenium diets to the young, with results pointing to an adaptation to chronic selenium intake (188). Ermakov and Kovalsky (97) presented evidence for a similar adaptive response in farm animals, since the increase in selenium retention in certain tissues after feeding a selenite load was less in sheep with a history of high selenium intake than in sheep with a history of normal intakes. The mechanism(s) whereby animals could adapt to excessive selenium exposure are unknown. Hepatic nonprotein sulfhydryls (mainly GSH) and oxidized glutathione (GSSG) were elevated in rats chronically poisoned with sodium selenite, and it was suggested that the increases in GSH might be an adaptive change initiated in an attempt to maintain a normal intracellular GSSG : GSH ratio (219). However, any postulated adaptive mechanism must take into account the fact that selenium exposure history also affects the toxicity of end products of selenium metabolism such as dimethylselenide and trimethylselenonium ion (327).

C. Prevention and Control of Selenosis in Animals

Three possibilities exist for the prevention or treatment of selenium poisoning in animals: (1) treatment of the soil so that selenium uptake by plants is reduced and maintained at nontoxic levels, (2) treatment of the animal so that selenium absorption is reduced or excretion increased thus preventing toxic accumulations in the tissues, and (3) modifying the diet of the animal by the inclusion of substances that antagonize or inhibit the toxic effects of selenium within the body tissues and fluids.

The addition of sulfur or gypsum to soils in a toxic area in North America has been unsuccessful in reducing the absorption of selenium by cereals (105), probably because these soils are mostly already high in gypsum and carry a high proportion of their selenium in organic combinations that are relatively little affected by changes in the inorganic sulfur–selenium ratio. On the other hand, the addition of sulfur to soils to which selenate has been added can inhibit selenium absorption by plants. Selenium uptake by alfalfa from a seleniferous soil has been strikingly reduced by additions of calcium sulfate and barium chloride (341). The latter salt reduced the selenium levels in plants by 90–100% when applied in quantities that did not affect the plant growth or result in significant concentrations of barium in the tissues. The practical possibilities of these interesting findings appear to be limited.

Urinary loss of selenium from the body can be enhanced by the administration of bromobenzene to rats and dogs fed a seleniferous diet, and to steers on seleniferous range (284), but this form of treatment has obvious practical limitations. On the other hand, dietary modifications including high protein and sulfate intakes and the feeding of arsenic at appropriate levels, or mercury or copper as discussed in the previous section, have a possible potential in alleviating selenium toxicity, where dietary control of the animals can be achieved. Much more information is required on the quantitative interactions of these elements and their own toxic hazards when used for this purpose before such treatments can be considered practical possibilities. The position is even more difficult under range conditions. Early studies by Moxon et al. (285) indicated that 25 μg/g arsenic as sodium arsenite added to the salt of cattle on seleniferous range gave some protection, but observations by ranchers and additional studies by Dinkel and co-workers (83) showed that this method of control is ineffective, probably because the arsenic intake is neither high enough nor regular enough. Various management possibilities exist, following mapping of the land into pastures of high and low selenium contents (309). Furthermore, the more highly seleniferous areas can be used for grain production and the grain sold into normal market channels. In this way it would become so diluted that it should not contribute to a public health problem, and could help in raising selenium intakes by animals in selenium-deficient areas.

D. Human Selenosis

After the discovery that selenium was the cause of alkali disease in livestock raised in certain areas of the Great Plains of the United States, public health authorities became concerned about the possible deleterious effects of dietary selenium overexposure to people. Smith et al. (399) studied the health of 111 rural families living in areas known to have a history of alkali disease but could not detect any problems they considered pathognomonic of human selenium toxicity. The incidence of vague symptoms of ill health such as anorexia, indigestion, and general pallor was thought to be high. Moreover, symptoms indicative of injury to the liver, kidneys, skin, and joints were reported. Bad teeth, yellowish discoloration of the skin, skin eruptions, chronic arthritis, diseased nails, and subcutaneous edema were observed. These workers were surprised that they could find no more substantial evidence of disease, especially in those residents who had higher urinary selenium levels. Therefore, a second, more thorough survey was carried out to investigate the relationship between urinary selenium levels and any possible symptoms of human selenosis. Smith and Westfall (398) concluded that none of the signs or symptoms they observed in 100 subjects chosen on the basis of high urinary selenium levels found in the previous survey (399) could be considered specific for human selenium poisoning. However, they did observe a rather high incidence of gastrointestinal disturbances and skin discoloration and suggested that the latter might be due to liver dysfunction caused by excessive selenium intake.

A field study carried out in a seleniferous area of Venezuela showed that the hemoglobin and hematocrit values of children living there were lower than those reported in Caracas children (186). However, there was no correlation between these hematological indices and either blood or urine selenium levels, and the incidence of intestinal parasitic infestation was greater in the seleniferous zone than in Caracas. Dermatitis, loose hair, and pathological nails were more frequent among the children of the seleniferous zone, but the cause of these signs was considered doubtful due to the lack of difference in any of the biochemical tests performed (189).

An epidemic of endemic human selenium intoxication was reported in the People's Republic of China (474). The most heavily affected villages suffered a morbidity of nearly 50% during the peak prevalence years (1961–1964). Although the most common sign of intoxication was hair and nail loss, dental, nervous, and skin disorders may have occurred in the areas of high incidence. One middle-aged female developed both motor and sensory abnormalities that progressed slowly to hemiplegia. Unfortunately, no food, excreta, or tissue samples from the time of the outbreak of poisoning were available for analysis. However, samples collected some years after the peak prevalence of selenosis had subsided contained very high selenium levels. For example, the daily dietary intake of selenium based on total diet analyses averaged almost 5 mg with a range from 3.2 to 6.7 mg. Blood samples drawn in the selenosis area averaged 3.2 μg selenium per milliliter (range 1.3–7.5), and hair averaged 32.2 μg selenium per gram (range 4.1–100).

Twelve cases of human selenium toxicity were reported to the U.S. FDA and the Centers for Disease Control (CDC) in 1984 because of the ingestion of overly potent selenium tablets meant to be consumed as a "health food" supplement (11,163,198). Analyses revealed that the tablets contained 27–31 mg of selenium each or about 182 times higher than stated on the label. About 25 mg of the selenium was sodium selenite, and the reminder was present as either elemental and/or organic selenium. The most common symptoms reported in these cases were nausea and vomiting, nail changes, hair loss, fatigue, and irritability. Other symptoms were abdominal cramps, watery diarrhea, paresthesias, dryness of hair, and garlicky breath. No abnormalities of blood chemistry were seen in 8 of the 12 victims, and renal and liver function tests were normal. The total estimated doses of selenium consumed by the victims ranged from 27 to 2387 mg. The person who ingested the highest dose took one tablet containing 31 mg selenium per day for 77 days. This person was also taking large doses of vitamin C, which may have minimized her toxicity, since the ascorbic acid could reduce selenite to the poorly absorbed elemental selenium.

E. Mechanisms of Selenium Toxicity

The precise ways in which selenium at toxic intakes interferes with tissue structure and function are not completely understood. Most likely, different chemical forms of selenium (e.g., selenite, selenomethionine) may exert their toxic effect at least partially by independent mechanisms. Moreover, the route of exposure (oral, dermal, pulmonary) may have a bearing on the nature of the signs or symptoms observed. The discussion below describes some of the biochemical hypotheses that have been proposed in an attempt to explain the toxicity of selenium (224).

By analogy with effects of heavy metals, it was postulated that selenium toxicity might be mediated via inhibition of certain sulfhydryl enzymes that are crucial for energy metabolism, such as succinic dehydrogenase (205). Later work, however, indicated that selenite is actually a poor inhibitor of sulfhydryl enzymes, and it was suggested that a selenite-catalyzed oxidation of various lower molecular weight substances such as GSH might play a more important role (430,431). Chung and Maines (65) reported that injecting rats with selenite (7 μmol/kg, three times daily) caused increases in hepatic γ-glutamylcysteine synthetase and GSH reductase activity that were thought to represent cellular responses to selenium-mediated perturbations in the GSH:GSSG levels and ratio. Anundi et al. (12) found an increased oxidation of GSH and loss of NADPH in isolated hepatocytes incubated with 30 to 100 μM selenite and suggested that the cytotoxicity of selenite is mediated via a loss of NADPH. Vernie and colleagues (438,439) showed that selenodiglutathione (GSSeSG), formed by the reaction of selenite with GSH (111), is a potent inhibitor of protein biosynthesis in a cell-free system prepared from rat liver. Selenite is a powerful inhibitor of protein synthesis in rabbit reticulocyte lysates, and it apparently acts by blocking eukaryotic initiation factor 2 (351). Additional research in this area may help explain the biochemical basis of selenium toxicity and clarify the feasibility of using selenite as a cytotoxic agent against tumor cells (437).

Although selenium is generally regarded as an antioxidant because of its role in the peroxide-destroying enzyme GSH-Px, it now appears that under certain conditions selenite–selenium may have pro-oxidant effects that may clarify some of its toxic properties. The pro-oxidant nature of high levels of selenite was pointed out by Dougherty and Hoekstra (89), who observed a large increase in ethane exhalation (an indicator of *in vivo* lipid peroxidation) in rats fed a diet deficient in vitamin E and selenium when they were injected with sodium selenite at a dose of 2 mg selenium per kilogram. This burst of ethane production was not observed when rats fed a diet supplemented with vitamin E were injected with selenite. Nor was increased ethane evolution seen in rats injected with selenium as selenate, regardless of whether the diet fed was supplemented or deficient in vitamin E. The concept that the pro-oxidant activity of selenite may play a role in its toxicity is consistent with the observation that chronic poisoning can be induced with relatively low levels of dietary selenite (1.25 μg/g selenium) in rats severely deficient in vitamin E (469). Moreover, Csallany *et al.* (76) found increased levels of organic solvent-soluble lipofuscin pigment in the livers of mice fed normal dietary levels of vitamin E and given only 0.1 μg/ml selenium as selenite in the drinking water. Since the amount of this pigment was reduced by feeding the selenite-treated mice a diet containing 10 times the normal level of vitamin E, the authors concluded that increased oxidative stress due to the selenite exposure was responsible for pigment formation. In this regard, it should be noted that several years ago Pletnikova (337) observed increases in the concentration of oxidized GSH in the blood of rabbits given very low levels of selenium as sodium selenite in the drinking water (calculated to be approximately equivalent to a dietary selenium level of 0.063 μg/g) for a period of 2 months. After seven months of selenite exposure, hepatic elimination of bromsulfalein was slower and the activity of succinic dehydrogenase in the liver was decreased. More work is needed to clarify the physiological significance of these biochemical alterations due to the long-term low-level exposure to selenite.

Selenomethionine is partially catabolized to selenide but is also incorporated as such into protein (409), so it may have toxic effects independent of inorganic selenium. For example, the resistance to denaturation of the β-galactosidase from a selenium-tolerant substrain of *Escherichia coli* was decreased by substituting about half of the methionines by selenomethionine (179). An understanding of the toxic effects peculiar to selenomethionine is especially important in that this amino acid is thought to account for a large fraction of the selenium in grains and cereals (316).

The production of methylated selenium metabolites has always been considered a means by which organisms can detoxify selenium. However, Parizek and co-workers have shown that dimethylselenide is highly toxic to animals under certain conditions (327). For example, lactating rats or rats acutely exposed to arsenic or mercuric compounds are highly susceptible to the toxic effects of methylated selenium compounds. Moreover, male rats fed a low-selenium diet are very sensitive to an injected dose of dimethylselenide, whereas rats previously exposed to selenium are not. Whether this change in sensitivity plays any role in the phenomenon of adaptation to selenium is not known. The biosynthesis of dimethylselenide requires methyl groups from S-adenosylmethionine (SAM)

(178), and mice acutely exposed to selenite experience a sudden reduction in tissue levels of SAM (173). This reduction is the result of both increased demand for methyl groups as well as inhibition of methionine adenosyltransferase, the enzyme that catalyzes the formation of SAM. Additional research is needed to determine whether depletion of tissue methyl group levels plays any role in acute selenium poisoning.

REFERENCES

1. Alexander, A. R., Whanger, P. D., and Miller, L. T. (1983). *J. Nutr.* **113**, 196.
2. Allaway, W. H. (1972). *Ann. N.Y. Acad. Sci.* **199**, 17.
3. Allaway, W. H., and Hodgson, J. F. (1964). *J. Anim. Sci.* **23**, 271.
4. Allaway, W. H., Cary, E. E., Kubota, J., and Ehlig, C. F. (1964). *Proc. Cornell Nutr. Conf., 1964* p. 9.
5. Allaway, W. H., Moore, D. P., Oldfield, J. E., and Muth, O. H. (1966). *J. Nutr.* **88**, 411.
6. Allaway, W. H., Cary, E. E., and Ehlig, C. F. (1967). *In* "Selenium in Biomedicine" (O. H. Muth, ed.), p. 273. Avi Publ. Co., Westport, Connecticut.
7. Allaway, W. H., Kubota, J., Losee, F., and Roth, M. (1968). *Arch. Environ. Health* **16**, 342.
8. Anderson, P. (1960). *Acta Pathol. Microbiol. Scand., Suppl.* **134.**
9. Andrews, E. D., Hartley, W. J., and Grant, A. B. (1968). *N. Z. Vet. J.* **16**, 3.
10. Ankerst, J., and Sjögren, H. O. (1982). *Int. J. Cancer* **29**, 707.
11. Anonymous (1984). *FDA Bull.* **14**, 19.
12. Anundi, I., Stahl, A., and Hogberg, J. (1984). *Chem.-Biol. Interact.* **50**, 277.
13. Arnold, R. L., Olson, O. E., and Carlson, C. W. (1972). *Poult. Sci.* **51**, 341.
14. Arnold, R. L., Olson, O. E., and Carlson, C. W. (1973). *Poult. Sci.* **52**, 847.
15. Arvilommi, H., Poikonen, K., Jokinen, I., Muukkonen, O., Rasanen, L., Foreman, J., and Huttunen, J. K. (1983). *Infect. Immun.* **41**, 185.
16. Bai, J., Wu, S., Ge, K., Deng, X., and Su, C. (1980). *Acta Acad. Med. Sin.* **2**, 29.
17. Baker, S. S., Lerman, R. H., Krey, S. H., Crocker, K. S., Hirsh, E. F., and Cohen, H. (1983). *Am. J. Clin. Nutr.* **38**, 769.
18. Bass, D. A., DeChatelet, L. R., Burk, R. F., Shirley, P., and Szejda, P. (1977). *Infect. Immun.* **18**, 78.
19. Behne, D., and Hofer-Bosse, T. (1984). *J. Nutr.* **114**, 1289.
20. Behne, D., and Wolters, W. (1983). *J. Nutr.* **113**, 456.
21. Beilstein, M. A., and Whanger, P. D. (1983). *J. Nutr.* **113**, 2138.
22. Bendich, A., Gabriel, E., and Machlin, L. J. (1983). *J. Nutr.* **113**, 1920.
23. Binnerts, W. T., and El Boushy, A. R. (1984). *In* "Trace Element Metabolism in Man and Animals-5 Abstracts," p. 136. Cambridge Univ. Press, London and New York.
24. Birt, D. F., Julius, A. D., and Pour, P. M. (1984). *Proc. Am. Assoc. Cancer Res.* **25**, 133.
25. Bisjberg, B., Jochumsen, P., and Rasbech, N. O. (1970). *Nord. Veterinaermed.* **22**, 532.
26. Blaxter, K. L. (1963). *Br. J. Nutr.* **17**, 105.
27. Blincoe, C., and Dye, W. B. (1958). *J. Anim. Sci.* **17**, 224.
28. Bowen, W. H. (1972). *J. Ir. Dent. Assoc.* **18**, 83.
29. Boyne, R., and Arthur, J. R. (1981). *J. Comp. Pathol.* **91**, 271.
30. Britton, J. L., Shearer, T. R., and DeSart, D. J. (1980). *Arch. Environ. Health* **35**, 74.
31. Broderius, M. A., Whanger, P. D., and Weswig, P. H. (1973). *J. Nutr.* **103**, 336.
32. Brown, D. G., and Burk, R. F. (1973). *J. Nutr.* **103**, 102.

33. Brown, D. G., Burk, R. F., Seely, R. J., and Kiker, K. W. (1972). *Int. J. Vitam. Nutr. Res.* **42,** 588.
34. Brown, J. M. M., and DeWet, P. J. (1967). *Onderstepoort J. Vet. Res.* **34,** 161.
35. Buchanan-Smith, J. G., Nelson, E. C., Osburn, B. I., Wells, M. E., and Tillman, A. D. (1969). *J. Anim. Sci.* **29,** 808.
36. Bull, R. C., and Oldfield, J. E. (1967). *J. Nutr.* **91,** 237.
37. Bunk, M. J., and Combs, G. F. (1981). *Proc. Soc. Exp. Biol. Med.* **167,** 87.
38. Burk, R. F. (1983). *Annu. Rev. Nutr.* **3,** 53.
39. Burk, R. F., and Gregory, P. E. (1982). *Arch. Biochem. Biophys.* **213,** 73.
40. Burk, R. F., Whitney, R., Frank, H., and Pearson, W. N. (1968). *J. Nutr.* **95,** 420.
41. Burk, R. F., Brown, D. G., Seely, R. J., and Scaief, C. C. (1972). *J. Nutr.* **102,** 1049.
42. Burk, R. F., Seely, R. J., and Kiker, K. W. (1973). *Proc. Soc. Exp. Biol. Med.* **142,** 214.
43. Burton, V., Keeler, R. F., Swingle, K. F., and Young, S. (1962). *Am. J. Vet. Res.* **23,** 962.
44. Butler, G. W., and Peterson, P. J. (1961). *N. Z. J. Agric. Res.* **4,** 484.
45. Butler, J. A., Whanger, P. D., Patton, N. M., and Weswig, P. H. (1980). *Fed. Proc., Fed. Am. Soc. Exp. Biol.* **39,** 339.
46. Butler, J. A., Whanger, P. D., Patton, N. M., and Weswig, P. H. (1981). *Fed. Proc., Fed. Am. Soc. Exp. Biol.* **40,** 943.
47. Buttner, W. (1963). *J. Dent. Res.* **42,** 453.
48. Cadell, P. B., and Cousins, F. B. (1960). *Nature (London)* **185,** 863.
49. Calvin, H. I. (1978). *J. Exp. Zool.* **204,** 445.
50. Calvin, H. I., Cooper, G. W., and Wallace, E. (1981). *Gamete Res.* **4,** 139.
51. Cantor, A. H., and Scott, M. L. (1974). *Poult. Sci.* **53,** 1870.
52. Cantor, A. H., Scott, M. L., and Noguchi, T. (1975). *J. Nutr.* **105,** 96.
53. Cantor, A. H., Langevin, M. L., Noguchi, T., and Scott, M. L. (1975). *J. Nutr.* **105,** 106.
54. Cantor, A. H., Moorhead, P. D., and Brown, K. I. (1978). *Poult. Sci.* **57,** 1337.
55. Casey, C. E., Guthrie, B. E., Friend, G. M., and Robinson, M. F. (1982). *Arch. Environ. Health* **37,** 133.
56. Castillo, R., Landon, C., Eckhardt, K., Morris, V., Levander, O., and Lewiston, N. (1981). *J. Pediatr.* **99,** 583.
57. Caygill, C. P. J., and Diplock, A. T. (1973). *FEBS Lett.* **33,** 172.
58. Chansler, M. W., Morris, V. C., and Levander, O. A. (1983). *Fed. Proc., Fed. Am. Soc. Exp. Biol.* **42,** 927.
59. Chen, R. W., Whanger, P. D., and Weswig, P. H. (1975). *Bioinorg. Chem.* **4,** 125.
60. Chen, X., Yang, G., Chen, J., Chen, X., Wen, Z., and Ge, K. (1980). *Biol. Trace Elem. Res.* **2,** 91.
61. Chen, X., Chen, X., Yang, G., Wen, Z., Chen, J., and Ge, K. (1981). *In* "Selenium in Biology and Medicine" (J. E. Spallholz, J. L. Martin, and H. E. Ganther, eds.), p. 171. Avi Publ. Co., Westport, Connecticut.
62. Chow, C. K., and Tappel, A. L. (1974). *J. Nutr.* **104,** 444.
63. Christensen, M. J., Janghorbani, M., Steinke, F. H., Istfan, N., and Young, V. R. (1983). *Br. J. Nutr.* **50,** 43.
64. Chu, C., and Tsui, T. (1978). *Chin. Med. J.* **4,** 309.
65. Chung, A., and Maines, M. D. (1981). *Biochem. Pharmacol.* **30,** 3217.
66. Clark, L. C., Graham, G. F., Crounse, R. G., Grimson, R., Hulka, B., and Shy, C. M. (1984). *Nutr. Cancer* **6,** 13.
67. Clayton, C. C., and Baumann, C. A. (1949). *Cancer Res.* **9,** 575.
68. Combs, G. F., Jr., and Combs, S. B. (1984). *Annu. Rev. Nutr.* **4,** 257.
69. Combs, G. F., Jr., and Scott, M. L. (1974). *J. Nutr.* **104,** 1297.
70. Combs, G. F., Liu, C. H., Lu, Z. H., and Su, Q. (1984). *J. Nutr.* **114,** 964.

71. Combs, G. F., Zhou, Y., Su, Q., and Wu, K. (1984). *Int. Symp. Selenium Biol. Med., 3rd, 1984.* Abstracts, p. 32.
72. Committee on Medical and Biologic Effects of Environmental Pollutants—National Research Council (1976). "Selenium." Natl. Acad. Sci., Washington, D.C.
73. Correia, M. A., and Burk, R. F. (1978). *J. Biol. Chem.* **253,** 6203.
74. Cousins, F. B., and Cairney, I. M. (1961). *Aust. J. Agric. Res.* **12,** 927.
75. Creger, C. R., Mitchell, R. H., Atkinson, R. L., Ferguson, T. M., Reid, B. L., and Couch, J. R. (1960). *Poult. Sci.* **39,** 59.
76. Csallany, A. S., Su, L., and Menken, B. Z. (1984). *J. Nutr.* **114,** 1582.
77. Cummins, L. M., and Martin, J. L. (1967). *Biochemistry* **6,** 3162.
78. Curzon, M. E. J. (1981). *In* "Selenium in Biology and Medicine" (J. E. Spallholz, J. L. Martin, and H. E. Ganther, eds.), p. 379. Avi Publ. Co., Westport, Connecticut.
79. Czerniejewski, C. P., Shank, C. W., Bechtel, W. G., and Bradley, W. B. (1964). *Cereal Chem.* **41,** 65.
80. Dam, H., Nielsen, G. K., Prange, I., and Sondergaard, E. (1958). *Nature (London)* **182,** 802.
81. Davies, T. S. (1982). *Lancet* **2,** 935.
82. Dickson, R. C., and Tomlinson, R. N. (1967). *Clin. Chim. Acta* **16,** 311.
83. Dinkel, C. A., Minyard, J. A., Whitehead, E. I., and Olson, O. E. (1937). *Circ.—S. D., Agric. Exp. Stn.* **135.**
84. Dinkel, C. A., Minyard, J. A., and Ray, D. E. (1963). *J. Anim. Sci.* **22,** 1043.
85. Diplock, A. T. (1976). *CRC Crit. Rev. Toxicol.* **4,** 271.
86. Diplock, A. T. (1979). *Adv. Pharmacol. Ther., Proc. Int. Congr. Pharmacol., 7th, 1978,* Vol. 8, p. 25.
87. Diplock, A. T., Caygill, C. P. J., Jeffrey, E. H., and Thomas, C. (1973). *Biochem. J.* **134,** 283.
88. Dornenbal, H. (1975). *Can. J. Anim. Sci.* **55,** 325.
89. Dougherty, J. J., and Hoekstra, W. G. (1982). *Proc. Soc. Exp. Biol. Med.* **169,** 209.
90. Douglass, J. S., Morris, V. C., Soares, J. H., and Levander, O. A. (1981). *J. Nutr.* **111,** 2180.
91. Drake, C., Grant, A. B., and Hartley, W. J. (1960). *N. Z. Vet. J.* **8,** 4 and 7.
92. Drake, J. R., and Fitch, C. D. (1980). *Am. J. Clin. Nutr.* **33,** 2386.
93. Edmonds, L. J., Veillon, C., Robinson, M. F., Thomson, C. D., Morris, V. C., and Levander, O. A. (1984). *Fed. Proc., Fed. Am. Soc. Exp. Biol.* **43,** 473.
94. Eggert, R. G., Patterson, E., Akers, W. T., and Stokstad, E. L. R. (1957). *J. Anim. Sci.* **16,** 1037.
95. Ehligh, C. F., Hogue, D. E., Allaway, W. H., and Hamm, D. J. (1967). *J. Nutr.* **92,** 121.
96. Ellis, N., Lloyd, B., Lloyd, R. S., and Clayton, B. E. (1984). *J. Clin. Pathol.* **37,** 200.
97. Ermakov, V. V., and Kovalsky, V. V. (1968). *Tr. Biogeokhim. Lab., Akad. Nauk SSSR* **12,** 204.
98. Ewan, R. C., Pope, A. L., and Baumann, C. A. (1967). *J. Nutr.* **91,** 547.
99. Exon, J. H., Koller, L. D., and Elliott, S. C. (1976). *Clin. Toxicol.* **9,** 273.
100. Ferretti, R. J., and Levander, O. A. (1974). *J. Agric. Food Chem.* **22,** 1049.
101. Fleming, G. A., and Walsh, T. (1957). *Proc. R. Ir. Acad., Sect. B* **58,** 151.
102. Flohe, L. (1971). *Klin. Wochenschr.* **49,** 669.
103. Flohe, L., Gunzler, W. A., and Schock, H. H. (1973). *FEBS Lett.* **32,** 132.
104. Food and Nutrition Board—National Research Council (1976). *Nutr. Rev.* **34,** 347.
105. Franke, K. W., and Painter, E. P. (1938). *Cereal Chem.* **15,** 1.
106. Franke, K. W., and Tully, W. C. (1935). *Poult. Sci.* **14,** 273; **15,** 316 (1936).
107. Fuss, C. N., and Godwin, K. O. (1975). *Aust. J. Biol. Sci.* **28,** 239.
108. Gabrielson, B. O., and Opstvedt, J. (1980). *J. Nutr.* **110,** 1096.

109. Gallagher, M. L., Webb, P., Crounse, R., Bray, J., Webb, A., and Settle, E. A. (1984). *Nutr. Res.* **4**, 577.
110. Ganapathy, S. N., Joyner, B. T., Sawyer, D. R., and Hafner, K. M. (1978). *In* "Trace Element Metabolism in Man and Animals-3" (M. Kirchgessner, ed.), p. 322. Tech. Univ. Munich, F. R. G., West Germany.
111. Ganther, H. E. (1968). *Biochemistry* **7**, 2898.
112. Ganther, H. E. (1978). *Environ. Health Perspect.* **25**, 71.
113. Ganther, H. E. (1979). *Adv. Nutr. Res.* **2**, 107.
114. Ganther, H. E. (1980). *Ann. N.Y. Acad. Sci.* **355**, 212.
115. Ganther, H. E., and Baumann, C. A. (1962). *J. Nutr.* **77**, 408.
116. Ganther, H. E., and Corcoran, C. (1969). *Biochemistry* **8**, 2557.
117. Ganther, H. E., and Sunde, M. L. (1974). *J. Food Sci.* **39**, 1.
118. Ganther, H. E., Goudie, C., Sunde, M. L., Kopecky, M. J., Wagner, P., Oh, S., and Hoekstra, W. G. (1972). *Science* **175**, 1122.
119. Ganther, H. E., Hafeman, D. G., Lawrence, R. A., Serfass, R. E., and Hoekstra, W. G. (1976). *In* "Trace Elements in Human Health and Disease" (A. S. Prasad and D. Oberleas, eds.), Vol. 2, p. 165. Academic Press, New York.
120. Gardiner, M. R. (1962). *Aust. Vet. J.* **38**, 387.
121. Gardiner, M. R., and Gorman, P. C. (1966). *Aust. J. Exp. Agric. Anim. Husb.* **3**, 284.
122. Gardner, R. W., and Hogue, D. E. (1967). *J. Nutr.* **93**, 418.
123. Ge, K., Xue, A., Bai, J., and Wang, S. (1983). *Virchows Arch. A: Pathol. Anat. Histol.* **401**, 1.
124. Gissel-Nielsen, G. (1984). *Biol. Trace Elem. Res.* **6**, 281.
125. Glover, J. R. (1976). *In* "Selenium–Tellurium in the Environment," p. 279. Industrial Health Foundation, Pittsburgh, Pennsylvania.
126. Godwin, K. O. (1965). *Nature (London)* **217**, 1275.
127. Godwin, K. O. (1965). *Q. J. Exp. Physiol. Cogn. Med. Sci.* **50**, 282.
128. Godwin, K. O., and Frazer, F. J. (1966). *Q. J. Exp. Physiol. Cogn. Med. Sci.* **51**, 94.
129. Godwin, K. O., and Fuss, C. N. (1972). *Aust. J. Biol. Sci.* **25**, 865.
130. Godwin, K. O., Kuchel, R. E., and Fuss, C. N. (1974). *Aust. J. Biol. Sci.* **27**, 633.
131. Godwin, K. O., Fuss, C. N., and Kuchel, R. E. (1975). *Aust. J. Biol. Sci.* **28**, 251.
132. Grant, A. B., and Wilson, G. F. (1968). *N. Z. J. Agric. Res.* **11**, 733.
133. Grant, C. A., and Thafvelin, B. (1958). *Nord. Veterinaermed.* **10**, 657.
134. Greathouse, D. G., and Craun, G. F. (1979). *In* "Trace Substances in Environmental Health-12" (D. D. Hemphill, ed.), p. 31. Univ. of Missouri Press, Columbia.
135. Green, J., and Bunyan, J. (1969). *Nutr. Abstr. Rev.* **39**, 321.
136. Green, R. C., and O'Brien, P. J. (1970). *Biochim. Biophys. Acta* **197**, 31.
137. Greenland, A. S., Holden, J. M., and Wolf, W. R. (1985). *Fed. Proc., Fed. Am. Soc. Exp. Biol.* **44**, 1510.
138. Greger, J. L., and Marcus, R. E. (1981). *Ann.Nutr. Metabl.* **25**, 97.
139. Gries, C. L., and Scott, M. L. (1972). *J. Nutr.* **102**, 1287.
140. Griffen, A. C. (1982). *In* "Molecular Interrelations of Nutrition and Cancer" (M. S. Arnott, J. van Eys, and Y. M. Wang, eds.), p. 401. Raven Press, New York.
141. Griffiths, N. M., Stewart, R. D. H., and Robinson, M. F. (1976). *Br. J. Nutr.* **35**, 373.
142. Grossman, A., and Wendel, A. (1983). *Eur. J. Biochem.* **135**, 549.
143. Gu, B. Q. (1983). *Chin. Med. J.* **96**, 251.
144. Hadjimarkos, D. M. (1966). *Experientia* **22**, 117.
145. Hadjimarkos, D. M. (1970). *Nutr. Rep. Int.* **1**, 175.
146. Hadjimarkos, D. M., and Bonhorst, C. W. (1961). *J. Pediatr.* **59**, 256.
147. Hadjimarkos, D. M., and Shearer, T. R. (1973). *J. Dent. Res.* **52**, 389.

148. Hafeman, D. G., Sunde, R. A., and Hoekstra, W. G. (1974). *J. Nutr.* **104,** 580.
149. Hakkarainen, J., Lindberg, P., Bengtsson, G., Jonsson, L., and Lannek, N. (1978). *J. Anim. Sci.* **46,** 1001.
150. Halpin, K. M., and Baker, D. H. (1984). *J. Nutr.* **114,** 606.
151. Halverson, A. W., Hendrick, C. M., and Olson, O. E. (1955). *J. Nutr.* **56,** 51.
152. Hamilton, E. I., Minski, M. J., and Cleary, J. J. (1972–1973). *Sci. Total Environ.* **1,** 341.
153. Handreck, K. A., and Godwin, K. O. (1970). *Aust. J. Agric. Res.* **21,** 71.
154. Hansson, E., and Jacobson, S. O. (1966). *Biochim. Biophys. Acta* **115,** 285.
155. Harr, J. R., Bone, J. F., Tinsley, I. J., Weswig, P. H., and Yamamoto, R. S. (1967). *In* "Selenium in Biomedicine" (O. H. Muth, ed.), p. 153. Avi Publ. Co., Westport, Connecticut.
156. Harris, P. L., Ludwig, M. I., and Schwarz, K. (1958). *Proc. Soc. Exp. Biol. Med.* **97,** 686.
157. Hartley, W. J. (1963). *Proc. N.Z. Soc. Anim. Prod.* **23,** 20.
158. Hartley, W. J. (1967). *In* "Selenium in Biomedicine" (O. H. Muth, ed.), p. 79. Avi Publ. Co., Westport, Connecticut.
159. Hartley, W. J., and Grant, A. B. (1961). *Fed. Proc., Fed. Am. Soc. Exp. Biol.* **20,** 679.
160. Hartley, W. J., and Hansen, B. (1975). Private communication.
161. Hawkes, W. C., and Tappel, A. L. (1983). *Biochim. Biophys. Acta* **739,** 225.
162. He, G. Q. (1979). *Chin. Med. J.* **92,** 416.
163. Helzlsouer, K., Jacobs, R., and Morris, S. (1985). *Fed. Proc., Fed. Am. Soc. Exp. Biol.* **44,** 1670.
164. Hendrick, C., Klug, H., and Olson, O. E. (1953). *J. Nutr.* **51,** 131.
165. Herrman, J. L. (1977). *Biochim. Biophys. Acta* **500,** 61.
166. Hidiroglou, M., Carson, R. B., and Brossard, G. A. (1965). *Can. J. Anim. Sci.* **45,** 197.
167. Hidiroglou, M., Jenkins, R. J., Carson, R. B., and Mackay, R. R. (1968). *Can. J. Anim. Sci.* **48,** 335.
168. Higgs, D. J., Morris, V. C., and Levander, O. A. (1972). *J. Agric. Food Chem.* **20,** 678.
169. Hill, C. H. (1974). *J. Nutr.* **104,** 593.
170. Hill, K. E., and Burk, R. F. (1982). *J. Biol. Chem.* **257,** 10668.
171. Hill, K. E., and Burk, R. F. (1985). *Arch. Biochem. Biophys.* **240,** 166.
172. Hoffman, I., Jenkins, K. J., Meranger, J. C., and Pigden, W. J. (1973). *Can. J. Anim. Sci.* **53,** 61.
173. Hoffman, J. L. (1977). *Arch. Biochem. Biophys.* **179,** 136.
174. Holmberg, R. E., and Ferm, V. H. (1969). *Arch. Environ. Health* **18,** 873.
175. Hopkins, L. L., Pope, A. L., and Baumann, C. A. (1966). *J. Nutr.* **88,** 61.
176. Hove, E. L., Fry, G. S., and Schwarz, K. (1958). *Proc. Soc. Exp. Biol. Med.* **98,** 1 and 27.
177. Howe, M. (1979). *Arch. Environ. Health* **34,** 444.
178. Hsieh, H. S., and Ganther, H. E. (1977). *Biochim. Biophys. Acta* **497,** 205.
179. Huber, R. E., and Criddle, R. S. (1967). *Biochim. Biophys. Acta* **141,** 587.
180. Innes, J. R. M., Ulland, B. M., Valerio, M. G., Petrucelli, L., Fishbein, L., Hart, E. R., Pallotta, A. J., Bates, R. R., Falk, H. L., Gart, J. J., Klein, M., Mitchell, I., and Peters, J. (1969). *J. Natl. Cancer Inst. (U.S.)* **42,** 1101.
181. Ip, C. (1983). *Biol. Trace Elem. Res.* **5,** 317.
182. Ip, C., and Daniel, F. B. (1985). *Cancer Res.* **45,** 61.
183. Ip, C., and Sinha, D. (1981). *Cancer Res.* **41,** 31.
184. Jacobs, M. M., Frost, C. F., and Beams, F. A. (1981). *Cancer Res.* **41,** 4458.
185. Jacobson, S. O., and Oksanen, H. E. (1966). *Acta Vet. Scand.* **7,** 66.
186. Jaffe, W. G. (1976). *In* "Selenium–Tellurium in the Environment," p. 188. Industrial Health Foundation, Pittsburgh, Pennsylvania.
187. Jaffe, W. G., and Mondragon, M. C. (1969). *J. Nutr.* **97,** 431.

188. Jaffe, W. G., and Mondragon, M. C. (1975). *Br. J. Nutr.* **33,** 387.
189. Jaffe, W. G., Ruphael, M. D., Mondragon, M. C., and Cuevas, M. A. (1972). *Arch. Latinoam. Nutr.* **22,** 595.
190. Janghorbani, M., Kasper, L. J., and Young, V. R. (1984). *Am. J. Clin. Nutr.* **40,** 208.
191. Jenkins, K. J. (1968). *Can. J. Biochem.* **46,** 1417.
192. Jenkins, K. J., and Hidiroglou, M. (1971). *Can. J. Anim. Sci.* **51,** 389.
193. Jensen, G. E., Gissel-Nielsen, G., and Clausen, J. (1980). *J. Neurol. Sci.* **48,** 61.
194. Jensen, L. S. (1975). *Proc. Soc. Exp. Biol. Med.* **149,** 113.
195. Jensen, L. S. (1975). *J. Nutr.* **105,** 769.
196. Jensen, L. S., and McGinnis, J. (1960). *J. Nutr.* **72,** 23.
197. Jensen, L. S., Peterson, R. P., and Falen, L. (1974). *Poult. Sci.* **53,** 57.
198. Jensen, R., Clossen, W., and Rothenberg, R. (1984). *Morbid. Mortal. Wkly. Rep.* **33,** 157.
199. Jones, G. B., and Godwin, K. O. (1963). *Aust. J. Agric. Res.* **14,** 716.
200. Kaarde, J. (1965). *Wien. Tieraerztl. Monatsschr.* **52,** 391; *Vet. Bull. (London)* **36,** 234 (abstr.) (1966).
201. Kaplowitz, N. (1980). *Am. J. Physiol.* **239,** G439.
202. Keshan Disease Research Group (1979). *Chin. Med. J.* **92,** 471.
203. Keshan Disease Research Group (1979). *Chin. Med. J.* **92,** 477.
204. Kilness, A. W., and Hochberg, F. H. (1977). *JAMA, J. Am. Med. Assoc.* **237,** 2843.
205. Klug, H. L., Moxon, A. L., Petersen, D. F., and Potter, V. R. (1950). *Arch. Biochem. Biophys.* **28,** 253.
206. Knott, S. G., and McCay, C. W. R. (1959). *Aust. Vet. J.* **35,** 161.
207. Koivistoinen, P., and Huttunen, J. K. (1986). *In* "Trace Elements in Man and Animals-5" (in press).
208. Kovalsky, V. (1955). *Nutr. Abstr. Rev.* **25,** 544.
209. Ku, P. K., Ely, W. T., Groce, A. W., and Ullrey, D. E. (1972). *J. Anim. Sci.* **34,** 208.
210. Kuchel, R. E., and Buckley, R. A. (1969). *Aust. J. Agric. Res.* **20,** 1099.
211. Kumpulainen, J., Vuori, E., Kuitunen, P., Makinen, S., and Kara, R. (1983). *Int. J. Vitam. Nutr. Res.* **53,** 420.
212. Kumpulainen, J., Salmenpera, L., Siimes, M. A., Perheentupa, J., and Koivistoinen, P. (1984). *In* "Trace Element Metabolism in Man and Animals-5 Abstracts," p. 41. Cambridge Univ. Press, London and New York.
213. Kurland, L. T. (1977). *JAMA, J. Am. Med. Assoc.* **238,** 2365.
214. Kuttler, K. L., and Marble, D. M. (1961). *Am. J. Vet. Res.* **22,** 422.
215. Ladenstein, R., Epp, O., Bartels, K., Jones, A., Huber, R., and Wendel, A. (1979). *J. Mol. Biol.* **134,** 199.
216. Latshaw, J. D. (1975). *J. Nutr.* **105,** 32.
217. Latshaw, J. D., and Biggert, M. D. (1981). *Poult. Sci.* **60,** 1309.
218. Lawrence, R. A., Parkhill, L. K., and Burk, R. F. (1978). *J. Nutr.* **108,** 981.
219. LeBoeuf, R. A., and Hoekstra, W. G. (1983). *J. Nutr.* **113,** 845.
220. Lee, Y. W., Mirocha, C. J., Schroeder, D. J., and Walser, M. M. (1985). *Appl. Environ. Microbiol.* **50,** 102.
221. Letunova, S. V. (1970). *In* "Trace Element Metabolism in Animals-1" (C. F. Mills, ed.), p. 432. Churchill-Livingstone, Edinburgh and London.
222. Levander, O. A. (1977). *Environ. Health Perspect.* **19,** 159.
223. Levander, O. A. (1979). *Environ. Health Perspect.* **29,** 115.
224. Levander, O. A. (1982). *In* "Clinical, Biochemical, and Nutritional Aspects of Trace Elements" (A. S. Prasad, ed.), p. 345. Alan R. Liss, Inc., New York.
225. Levander, O. A. (1983). *Fed. Proc., Fed. Am. Soc. Exp. Biol.* **42,** 1721.
226. Levander, O. A. (1984). *Bull. N.Y. Acad. Med.* [2] **60,** 144.

227. Levander, O. A. (1985). *Fed. Proc., Fed. Am. Soc. Exp. Biol.* **44**, 2579.
228. Levander, O. A., and Argrett, L. C. (1969). *Toxicol. Appl. Pharmacol.* **14**, 308.
229. Levander, O. A., and Baumann, C. A. (1966). *Toxicol. Appl. Pharmacol.* **9**, 98 and 106.
230. Levander, O. A., and Morris, V. C. (1970). *J. Nutr.* **100**, 1111.
231. Levander, O. A., and Morris, V. C. (1984). *Am. J. Clin. Nutr.* **39**, 809.
232. Levander, O. A., Young, M. L., and Meeks, S. A. (1970). *Toxicol. Appl. Pharmacol.* **16**, 79.
233. Levander, O. A., Morris, V. C., and Higgs, D. J. (1973). *Biochemistry* **12**, 4586.
234. Levander, O. A., Morris, V. C., and Moser, P. B. (1981). *Fed. Proc., Fed. Am. Soc. Exp. Biol.* **40**, 890.
235. Levander, O. A., Sutherland, B., Morris, V. C., and King, J. C. (1981). *Am. J. Clin. Nutr.* **34**, 2662.
236. Levander, O. A., Sutherland, B., Morris, V. C., and King, J. C. (1981). *In* "Selenium in Biology and Medicine" (J. E. Spallholz, J. L. Martin, and H. E. Ganther, eds.), pp. 256. Avi Publ. Co., Westport, Connecticut.
237. Levander, O. A., DeLoach, D. P., Morris, V. C., and Moser, P. B. (1983). *J. Nutr.* **113**, 55.
238. Levander, O. A., Alfthan, G., Arvilommi, H., Gref, C. G., Huttunen, J. K., Kataja, M., Koivistoinen, P., and Pikkarainen, J. (1983). *Am. J. Clin. Nutr.* **37**, 887.
239. Levander, O. A., Morris, V. C., and Mutanen, M. (1985). *Fed. Proc., Fed. Am. Soc. Exp. Biol.* **44**, 1670.
240. Li, C., Huang, J., and Li, C. (1984). *Int. Symp. Selenium Biol. Med., 3rd, 1984.* Abstracts, p. 73.
241. Lindberg, P. (1968). *Acta Vet. Scand., Suppl.* **23.**
242. Liu, C. H., Chen, Y. M., Zhang, J. Z., Huang, M. Y., Su, Q., Lu, Z. H., Yin, R. X., Shao, G. Z., Feng, D., and Zheng, P. L. (1982). *Acta Vet. Zootech. Sin.* **13**, 73 [cited in Combs *et al.*(70)].
243. Lloyd-Still, J. D., and Ganther, H. E. (1980). *Pediatrics* **65**, 1010.
244. Lopez, P. L., Preston, R. L., and Pfander, W. H. (1968). *J. Nutr.* **94**, 219; **97**, 123.
245. Lorenz, K. (1978). *Cereal Chem.* **55**, 287.
246. Maag, D. D., and Glenn, M. W. (1967). *In* "Selenium in Biomedicine" (O. H. Muth, ed.), p. 127. Avi Publ. Co., Westport, Connecticut.
247. Mace, D. L., Tucker, J. A., Bills, C. B., and Ferreira, C. J. (1963). *Calif., Dep. Agric., Bull.* (1), 21.
248. MacPherson, A., and Chalmers, J. S. (1984). *Vet. Rec.* **115**, 544.
249. Mahan, D. C., and Moxon, A. L. (1978). *J. Anim. Sci.* **47**, 456.
250. Mahan, D. C., Moxon, A. L., and Cline, J. H. (1975). *J. Anim. Sci.* **40**, 624.
251. Marsh, J. A., Dietert, R. R., and Combs, G. F. (1981). *Proc. Soc. Exp. Biol. Med.* **166**, 228.
252. Marshall, M. V., Arnott, M. S., Jacobs, M. M., and Griffin, A. C. (1978). *Cancer Lett.* **7**, 331.
253. Martin, J. L., and Hurlbut, J. A. (1976). *Phosphorus Sulfur* **1**, 295.
254. Masakawa, T., Goto, J., and Iwata, H. (1983). *Experientia* **39**, 405.
255. Mathias, M. M., Hogue, D. E., and Loosli, J. K. (1967). *J. Nutr.* **93**, 14.
256. Mazzella, G. L., Sinforiani, E., Savoldi, F., Allegrini, M., Lanzola, E., and Scelsi, R. (1983). *Eur. Neurol.* **22**, 442.
257. McAdam, P. A., Lewis, S. A., Helzlsouer, K., Veillon, C., Patterson, B., and Levander, O. A. (1985). *Fed. Proc., Fed. Am. Soc. Exp. Biol.* **44**, 1671.
258. McAdam, P. A., Wood, L. E., and Levander, O. A. (1985). *Am. J. Clin. Nutr.* **41**, 864.
259. McConnell, K. P., and Roth, D. M. (1962). *Biochim. Biophys. Acta* **62**, 503; *J. Nutr.* **84**, 340 (1964).
260. McConnell, K. P., and Wabnitz, C. H. (1957). *J. Biol. Chem.* **226**, 765.

261. McConnell, K. P., Burton, R. M., Kute, T., and Higgins, P. J. (1979). *Biochim. Biophys. Acta* **588**, 113.
262. McCoy, K. E. M., and Weswig, P. H. (1967). *J. Nutr.* **98**, 383.
263. McLean, J. W., Thompson, G. G., and Claxton, J. H. (1959). *N. Z. Vet. J.* **7**, 47; McLean, J. W., Thompson, G. G., and Lawson, B. M. (1963). *Ibid.* **11**, 59.
264. Medina, D., and Shepherd, F. (1980). *Cancer Lett.* **8**, 241.
265. Medina, D., Lane, H. W., and Tracey, C. M. (1983). *Cancer Res.* **43**, 2460s.
266. Mehlert, A., Metcalfe, R. A., Diplock, A. T., and Hughes, R. A. C. (1982). *Acta Neurol. Scand.* **65**, 376.
267. Meyer, W. R., Mahan, D. C., and Moxon, A. L. (1981). *J. Anim. Sci.* **52**, 302.
268. Miettinen, T. A., Alfthan, G., Huttunen, J. K., Pikkarainen, J., Naukkarinen, V., Mattila, S., and Kumlin, T. (1983). *Br. Med. J.* **287**, 517.
269. Millar, K. R., and Sheppard, A. D. (1972). *N. Z. J. Sci.* **15**, 3.
270. Millar, K. R., Craig, J., and Dawe, L. (1973). *N. Z. J. Agric. Res.* **16**, 301.
271. Miltmore, J. E., van Ryswyck, A. L., Pringle, W. L., Chapman, F. M., and Kalnin, C. M. (1975). *Can. J. Anim. Sci.* **55**, 101.
272. Mo, D. (1984). *Int. Symp. Selenium Biol. Med., 3rd, 1984.* Abstracts, p. 72.
273. Mondragon, M. C., and Jaffe, W. G. (1976). *Arch. Latinoam. Nutr.* **26**, 341.
274. Moore, J. A., Noiva, R., and Wells, I. C. (1984). *Clin. Chem. (Winston-Salem, N.C.)* **30**, 1171.
275. Morris, J. G., Cripe, W. S., Chapman, H. L., Walker, D. F., Armstrong, J. B., Alexander, J. D., Miranda, R., Sanchez, A., Sanchez, B., Blair-West, J. R., and Denton, D. A. (1984). *Science* **223**, 491.
276. Morris, J. S., Stampfer, M. J., and Willett, W. (1983). *Biol. Trace Elem. Res.* **5**, 529.
277. Morris, V. C., and Levander, O. A. (1970). *J. Nutr.* **100**, 1383.
278. Motsenbocker, M. A., and Tappel, A. L. (1982). *Biochim. Biophys Acta* **719**, 147.
279. Motsenbocker, M. A., and Tappel, A. L. (1984). *J. Nutr.* **114**, 279.
280. Moxon, A. L. (1937). *S. D., Agric. Exp. Stn. [Bull.]* **311.**
281. Moxon, A. L. (1938). *Science* **88**, 81.
282. Moxon, A. L., and Poley, W. E. (1938). *Poult. Sci.* **17**, 77.
283. Moxon, A. L., and Rhian, M. (1943). *Physiol. Rev.* **23**, 305.
284. Moxon, A. L., Schaefer, A. E., Lardy, H. A., Dubois, K. P., and Olson, O. E. (1940). *J. Biol. Chem.* **132**, 785.
285. Moxon, A. L., Rhian, M., Anderson, H. D., and Olson, O. E. (1944). *J. Anim. Sci.* **3**, 299.
286. Mudd, A. J., and Mackie, I. L. (1973). *Vet. Rec.* **93**, 197.
287. Mulhern, S. A., Taylor, G. L., Magruder, L. E., and Vessey, A. R. (1985). *Nutr. Res.* **5**, 201.
288. Munsell, H. E., Devaney, G. M., and Kennedy, M. H. (1936). *U.S., Dep. Agric., Tech. Bull.* **534.**
289. Mutanen, M., and Koivistoinen, P. (1983). *Int. J. Vitam. Nutr. Res.* **53**, 102.
290. Muth, O. H., and Allaway, W. H. (1963). *J. Am. Vet. Med. Assoc.* **142**, 1379.
291. Muth, O. H., Weswig, P. H., Whanger, P. D., and Oldfield, J. E. (1971). *Am. J. Vet. Res.* **32**, 1603.
292. Nahapetian, A. T., Janghorbani, M., and Young, V. R. (1983). *J. Nutr.* **113**, 401.
293. Nahapetian, A. T., Young, V. R., and Janghorbani, M. (1984). *Anal. Biochem.* **140**, 56.
294. National Cancer Institute (1980). *DHHS Publ. No. (NIH) 80-1750.* U.S. Dept. of Health and Human Services, Bethesda, Maryland.
295. National Research Council—Committee on Dietary Allowances (1980). "Recommended Dietary Allowances," 9th ed. Natl. Acad. Sci., Washington, D.C.
296. Nelson, A. A., Fitzhugh, O. G., and Calvery, H. O. (1943). *Cancer Res.* **3**, 230.

297. Nesheim, M. C., and Scott, M. L. (1958). *J. Nutr.* **65**, 601.
298. Nesheim, M. C., and Scott, M. L. (1961). *Fed. Proc., Fed. Am. Soc. Exp. Biol.* **20**, 674.
299. Nesterov, A. I. (1964). *Arthritis Rheum.* **7**, 29.
300. Neve, J., van Geffel, R., Hanocq, M., and Molle, L. (1983). *Acta Paediatr. Scand.* **72**, 437.
301. Nockels, C. F. (1979). *Fed. Proc., Fed. Am. Soc. Exp. Biol.* **38**, 2134.
302. Noguchi, T., Cantor, A. H., and Scott, M. L. (1973). *J. Nutr.* **103**, 1502.
303. Noguchi, T., Langevin, M. L., Combs, G. F., Jr., and Scott, M. L. (1973). *J. Nutr.* **103**, 444.
304. Norris, F. H., and Sang, K. (1978). *JAMA, J. Am. Med. Assoc.* **239**, 404.
305. Ochoa-Solano, A., and Gitler, C. (1968). *J.Nutr.* **94**, 243.
306. Ohlendorf, H. M., Hoffman, D. J., Saiki, M. K., and Aldrich, T. W. (1986). *Sci. Total Environ.* (in press).
307. Oksanen, H. E. (1967). *In* "Selenium in Biomedicine" (O. H. Muth, ed.), p. 215. Avi Publ. Co., Westport, Connecticut.
308. Oldfield, J. E., Schubert, J. R., and Muth, O. H. (1963). *J. Agric. Food Chem.* **11**, 388.
309. Olson, O. E. (1969). *Proc. Ga. Nutr. Conf.* p. 68.
310. Olson, O. E. (1978). *In* "Effects of Poisonous Plants on Livestock" (R. F. Keeler, K. R. Van Kampen, and L. F. James, eds.), p. 121. Academic Press, New York.
311. Olson, O. E., and Palmer, I. S. (1976). *Metab., Clin. Exp.* **25**, 299.
312. Olson, O. E., and Palmer, I. S. (1984). *J. Food Sci.* **49**, 446.
313. Olson, O. E., Dinkel, C. A., and Kamstra, L. D. (1954). *S. D. Farm Home Res.* **6**, 12.
314. Olson, O. E., Carlson, C. W., and Leitis, E. (1958). *Tech. Bull.—S. D., Agric. Exp. Stn.* **20.**
315. Olson, O. E., Schulte, B. M., Whitehead, E. I., and Halverson, A. W. (1963). *J. Agric. Food Chem.* **11**, 531.
316. Olson, O. E., Novacek, E. J., Whitehead, E. I., and Palmer, I. S. (1970). *Phytochemistry* **8**, 1161.
317. Omaye, S. T., and Tappel, A. L. (1974). *J. Nutr.* **104**, 747.
318. Orstadius, K., and Aberg, B. (1961). *Acta Vet. Scand.* **2**, 60.
319. Pallini, V., and Bacci, E. (1979). *J. Submicrosc. Cytol.* **11**, 165.
320. Palmer, I. S., and Olson, O. E. (1979). *Biochem. Biophys. Res. Commun.* **90**, 1379.
321. Palmer, I. S., Gunsalus, R. P., Halverson, A. W., and Olson, O. E. (1970). *Biochim. Biophys. Acta* **208**, 260.
322. Palmer, I. S., Arnold, R. L., and Carlson, C. W. (1973). *Poult. Sci.* **52**, 1841.
323. Palmer, I. S., Olson, O. E., Halverson, A. W., Miller, R., and Smith, C. (1980). *J. Nutr.* **110**, 145.
324. Palmer, I. S., Olson, O. E., Ketterling, L. M., and Shank, C. E. (1983). *J. Am. Diet. Assoc.* **82**, 511.
325. Parizek, J., and Ostadalova, I. (1967). *Experientia* **23**, 142.
326. Parizek, J., Ostadalova, I., Kalouskova, J., Babicky, A., and Benes, J. (1971). *In* "Newer Trace Elements in Nutrition" (W. Mertz and W. E. Cornatzer, eds.), p. 85. Dekker, New York.
327. Parizek, J., Kalouskova, J., Benes, J., and Pavlik, L. (1980). *Ann. N.Y. Acad. Sci.* **355**, 347.
328. Pascoe, G. A., Wong, J., Soliven, E., and Correia, M. A. (1983). *Biochem. Pharmacol.* **32**, 3027.
329. Patterson, E. L., Milstrey, R., and Stokstad, E. L. R. (1957). *Proc. Soc. Exp. Biol. Med.* **95**, 617.
330. Paulson, G. D., Pope, A. L., and Baumann, C. A. (1966). *Proc. Soc. Exp. Biol. Med.* **122**, 321.
331. Paulson, G. D., Pope, A. L., and Baumann, C. A. (1968). *J. Anim. Sci.* **27**, 195.
332. Pennington, J. A. T., Wilson, D. B., Newell, R. F., Harland, B. F., Johnson, R. D., and Vanderveen, J. E. (1984). *J. Am. Diet. Assoc.* **84**, 771.

333. Peplowski, M. A., Mahan, D. C., Murray, F. A., Moxon, A. L., Cantor, A. H., and Ekstrom, K. E. (1981). *J. Anim. Sci.* **51**, 344.

334. Peter, D. W., Hunter, R. A., and Hudson, D. R. (1981). *In* "Trace Element Metabolism in Man and Animals-4" (J. McC. Howell, J. M. Gawthorne, and C. L. White, eds.), p. 218. Aust. Acad. Sci., Canberra.

335. Peterson, R. P., and Jensen, L. S. (1975). *Poult. Sci.* **54**, 795.

336. Pierce, A. W., and Jones, G. B. (1968). *Aust. J. Exp. Agric. Anim. Husb.* **8**, 277.

337. Pletnikova, I. P. (1970). *Hyg. Sanit.* **35**, 176.

338. Poirier, K. A., and Milner, J. A. (1983). *J. Nutr.* **113**, 2147.

339. Poley, W. E., Wilson, W. O., Moxon, A. L., and Taylor, J. B. (1941). *Poult. Sci.* **20**, 171.

340. Quarterman, J., Mills, C. F., and Dalgarno, A. C. (1966). *Proc. Nutr. Soc.* **25**, xxiii.

341. Ravikovitch, S., and Margolin, M. (1959). *Emp. J. Exp. Agric.* **27**, 235.

342. Reiter, R., and Wendel, A. (1983). *Biochem. Pharmacol.* **32**, 3063.

343. Reiter, R., and Wendel, A. (1984). *Biochem. Pharmacol.* **33**, 1923.

344. Robinson, J. R., Robinson, M. F., Levander, O. A., and Thomson, C. D. (1985). *Am. J. Clin. Nutr.* **41**, 1023.

345. Robinson, M. F., and Thomson, C. D. (1983). *Nutr. Abstr. Rev. Clin. Nutr., Ser. A* **53**, 3.

346. Rosenfeld, I., and Beath, O. A. (1945). *J. Nutr.* **30**, 443.

347. Rosenfeld, I., and Beath, O. A. (1954). *Proc. Soc. Exp. Biol. Med.* **87**, 295.

348. Rosenfeld, I., and Beath, O. A. (1964). "Selenium," 2nd ed. Academic Press, New York.

349. Rotruck, J. T., Pope, A. L., Ganther, H. E., Swanson, A. B., Hafeman, D. G., and Hoekstra, W. G. (1973). *Science* **179**, 588.

350. Rudert, C. P., and Lewis, A. R. (1978). *Rhod. J. Agric. Res.* **16**, 109.

351. Safer, B., Jagus, R., and Crouch, D. (1980). *J. Biol. Chem.* **255**, 6913.

352. Salisbury, R. M., Edmondson, J., Poole, W. H. S., Bobby, F. C., and Birnie, H. (1962). *World's Poult. Congr., Proc., 12th, 1962* p. 379.

353. Salmon, J. A., Smith, D. R., Flower, R. J., Moncada, S., and Vane, J. R. (1978). *Biochim. Biophys. Acta* **523**, 250.

354. Salonen, J. T. (1985). *In* "Trace Elements in Health and Disease" (H. Bostrom and N. Ljungstedt, eds.), p. 172. Almqvist and Wiksell International, Stockholm.

355. Salonen, J. T., Alfthan, G., Huttunen, J. K., Pikkarainen, J., and Puska, P. (1982). *Lancet* **2**, 175.

356. Salonen, J. T., Alfthan, G., Huttunen, J. K., and Puska, P. (1984). *Am. J. Epidemiol.* **120**, 342.

357. Salonen, J. T., Salonen, R., Lappeteläinen, R., Mäenpää, P., Alfthan, G., and Puska, P. (1985). *Brit. Med. J.* **290**, 417.

358. Salonen, J. T., Salonen, R., Pentillä, I., Herranen, J., Jauhiainen, J., Kantola, M., Lappeteläinen, R., Mäenpää, P., Alfthan, G., and Puska, P. (1985). *Am. J. Cardiol.* **56**, 226.

359. Schoene, N. W., Morris, V. C., and Levander, O. A. (1984). *Fed. Proc., Fed. Am. Soc. Exp. Biol.* **43**, 477.

360. Schrauzer, G. N., and White, D. A. (1978). *Bioinorg. Chem.* **8**, 303.

361. Schrauzer, G. N., White, D. A., and Schneider, C. J. (1976). *Bioinorg. Chem.* **6**, 265.

362. Schrauzer, G. N., White, D. A., and Schneider, C. J. (1977). *Bioinorg. Chem.* **7**, 23.

363. Schroeder, H. A., and Mitchener, M. (1971). *J. Nutr.* **101**, 1531.

364. Schroeder, H. A., and Mitchener, M. (1972). *Arch. Environ. Health.* **24**, 66.

365. Schroeder, H. A., Frost, D. V., and Balassa, J. J. (1970). *J. Chronic Dis.* **23**, 227.

366. Schubert, J. R., Muth, O. H., Oldfield, J. E., and Remmert, L. F. (1961). *Fed. Proc., Fed. Am. Soc. Exp. Biol.* **20**, 689.

367. Schwarz, K. (1961). *Fed. Proc., Fed. Am. Soc. Exp. Biol.* **20**, 666.

368. Schwarz, K. (1962). *Vitam. Horm. (N.Y.)* **20**, 463.

369. Schwarz, K. (1977). *JAMA, J. Am. Med. Assoc.* **238**, 2365.

370. Schwarz, K., and Foltz, C. M. (1957). *J. Am. Chem. Soc.* **79**, 3293; *J. Biol. Chem.* **233**, 245 (1958).
371. Schwarz, K., and Fredga, A. (1969). *J. Biol. Chem.* **244**, 2103.
372. Schwarz, K., and Fredga, A. (1972). *Bioinorg. Chem.* **2**, 47 and 171; **3**, 153 (1974); **4**, 235 (1975).
373. Scott, M. L. (1966). *Ann. N.Y. Acad. Sci.* **138**, 82.
374. Scott, M. L. (1973). *J. Nutr.* **103**, 803.
375. Scott, M. L. (1973). *Proc. Cornell Nutr. Conf.* p. 123.
376. Scott, M. L., and Thompson, J. N. (1971). *Poult. Sci.* **50**, 1742.
377. Scott, M. L., Olson, G., Krook, L., and Brown, W. R. (1967). *J. Nutr.* **91**, 573.
378. Segerson, E. C., and Ganapathy, S. N. (1981). *J. Anim. Sci.* **51**, 386.
379. Segerson, E. C., and Johnson, B. H. (1981). *J. Anim. Sci.* **51**, 395.
380. Segerson, E. C., Murray, F. A., Moxon, A. L., Redman, D. R., and Conrad, H. R. (1977). *J. Dairy Sci.* **60**, 1001.
381. Segerson, E. C., Getz, W. R., and Johnson, B. H. (1981). *J. Anim. Sci.* **53**, 1360.
382. Seifter, J., Ehrich, W. E., Hudyma, G., and Mueller, G. (1946). *Science* **103**, 762.
383. Serfass, R. E., and Ganther, H. E. (1975). *Nature (London)* **255**, 640.
384. Shamberger, R. J. (1970). *J. Natl. Cancer Inst. (U.S.)* **44**, 931.
385. Shamberger, R. J., and Frost, D. V. (1969). *Can. Med. Assoc. J.* **100**, 682.
386. Shamberger, R. J., Tytko, S. A., and Willis, C. E. (1976). *Arch. Environ. Health* **31**, 231.
387. Shamberger, R. J., Willis, C. E., and McCormak, L. J. (1980). *In* "Trace Substances in Environmental Health-13" (D. D. Hemphill, ed.), p. 59. Univ. of Missouri Press, Columbia.
388. Shearer, T. R., and Hadjimarkos, D. M. (1975). *Arch. Environ. Health* **30**, 230.
389. Shils, M. E., and Levander, O. A. (1982). *Am. J. Clin. Nutr.* **35**, 829.
390. Shrift, A., and Kelly, E. (1962). *Nature (London)* **195**, 732.
391. Sies, H., and Moss, K. M. (1978). *Eur. J. Biochem.* **84**, 377.
392. Simpson, B. H. (1972). *N. Z. Vet. J.* **20**, 91.
393. Sirichakwal, P. P., Newcomer, C. E., Young, V. R., and Janghorbani, M. (1984). *J. Nutr.* **114**, 1159.
394. Slen, S. B., Demiruren, A. S., and Smith, A. D. (1961). *Can. J. Anim. Sci.* **41**, 263.
395. Smith, A. M., Picciano, M. F., and Milner, J. A. (1982). *Am. J. Clin. Nutr.* **35**, 521.
396. Smith, C. R., Weisleder, D., Miller, R. W., Palmer, I. S., and Olson, O. E. (1980). *J. Org. Chem.* **45**, 507.
397. Smith, M. I. (1939). *Public Health Rep.* **54**, 1441.
398. Smith, M. I., and Westfall, B. B. (1937). *Public Health Rep.* **52**, 1375.
399. Smith, M. I., Franke, K. W., and Westfall, B. B. (1936). *Public Health Rep.* **51**, 1496.
400. Smith, M. I., Stohlman, E. F., and Lillie, R. D. (1937). *J. Pharmacol. Exp. Ther.* **60**, 449.
401. Smith, M. I., Westfall, B. B., and Stohlman, E. F. (1937). *Public Health Rep.* **52**, 1171; **53**, 1199 (1938).
402. Snook, J. T., Palmquist, D. L., Moxon, A. L., Cantor, A. H., and Vivian, V. M. (1983). *Am. J. Clin. Nutr.* **38**, 620.
403. Spallholz, J. E. (1981). *In* "Diet and Resistance to Disease" (M. Phillips and A. Baetz, eds.), p. 43. Plenum, New York.
404. Stadtman, T. C. (1980). *Annu. Rev. Biochem.* **49**, 93.
405. Stewart, R. D. H., Griffiths, N. M., Thomson, C. D., and Robinson, M. F. (1978). *Br. J. Nutr.* **40**, 45.
406. Su, Y., and Yu, W. H. (1983). *Chin. Med. J.* **96**, 594.
407. Subcommittee on Selenium—Committee on Animal Nutrition (1971). "Selenium in Nutrition." Natl. Acad. Sci., Washington, D.C.
408. Subcommittee on Selenium—Committee on Animal Nutrition (1983). "Selenium in Nutrition—Revised Edition." Natl. Acad. Sci., Washington, D.C.

409. Sunde, R. A. (1984). *J. Am. Oil Chem. Soc.* **61,** 1891.
410. Sunde, R. A., Gutzke, G. E., and Hoekstra, W. G. (1981). *J. Nutr.* **111,** 76.
411. Sunde, R. A., Sonnenburg, W. K., Gutzke, G. E., and Hoekstra, W. G. (1981). *In* "Trace Element Metabolism in Man and Animals-4" (J. McC. Howell, J. M. Gawthorne, and C. L. White, eds.), p. 165. Aust. Acad. Sci., Canberra.
412. Swanson, C. A., Reamer, D. C., Veillon, C., King, J. C., and Levander, O. A. (1983). *Am. J. Clin. Nutr.* **38,** 169.
413. Swanson, C. A., Reamer, D. C., Veillon, C., and Levander, O. A. (1983). *J. Nutr.* **113,** 793.
414. Szeinberg, A., Golan, R., Ben Ezzer, J., Sarova-Pinhas, I., Sadeh, M., and Braham, J. (1979). *Acta Neurol. Scand.* **60,** 265.
415. Tappel, A. L. (1962). *Vitam. Horm. (N.Y.)* **20,** 493.
416. Taussky, H. H., Washington, A., Zubillaga, E., and Milhorat, A. T. (1963). *Nature (London)* **200,** 1211; **206,** 509 (1965).
417. Thompson, J. N., and Scott, M. L. (1969). *J. Nutr.* **97,** 335.
418. Thompson, J. N., and Scott, M. L. (1970). *J. Nutr.* **100,** 797.
419. Thompson, J. N., Erdody, P., and Smith, D. C. (1975). *J. Nutr.* **105,** 274.
420. Thomson, C. D. (1974). *N. Z. Med. J.* **80,** 163.
421. Thomson, C. D., and Robinson, M. F. (1980). *Am. J. Clin. Nutr.* **33,** 303.
422. Thomson, C. D., and Stewart, R. D. H. (1973). *Br. J. Nutr.* **30,** 139.
423. Thomson, C. D., and Stewart, R. D. H. (1974). *Br. J. Nutr.* **32,** 47.
424. Thomson, C. D., Robinson, B. A., Stewart, R. D. H., and Robinson, M. F. (1975). *Br. J. Nutr.* **34,** 501
425. Thomson, C. D., Rea, H. M., Doesburg, V. M., and Robinson, M. F. (1977). *Br. J. Nutr.* **37,** 457.
426. Thomson, C. D., Burton, C. E., and Robinson, M. F. (1978). *Br. J. Nutr.* **39,** 579.
427. Thomson, C. D., Robinson, M. F., Campbell, D. R., and Rea, H. M. (1982). *Am. J. Clin. Nutr.* **36,** 24.
428. Trinder, N., Wordhouse, C. D., and Renton, C. P. (1969). *Vet. Rec.* **85,** 550.
429. Tscherkes, L. A., Volgarev, M. N., and Aptekar, S. G. (1963). *Acta Unio Int. Cancrum* **19,** 632.
430. Tsen, C. C., and Collier, H. B. (1959). *Nature (London)* **183,** 1237.
431. Tsen, C. C., and Tappel, A. L. (1958). *J. Biol. Chem.* **233,** 1230.
432. Underwood, E. J. (1977). "Trace Elements in Human and Animal Nutrition," 4th ed. Academic Press, New York.
433. U.S. Department of Health and Human Services, Food and Drug Administration (1984). 21 CFR Pt. 573. Food additives permitted in the feed and drinking water of animals: Selenium. *Fed. Regist.* **49,** 627.
434. Van Kampen, K. R., and James, L. F. (1978). *In* "Effects of Poisonous Plants on Livestock" (R. F. Keeler, K. R. Van Kampen, and L. F. James, eds.), p. 135. Academic Press, New York.
435. Van Vleet, J. F. (1976). *Am. J. Vet. Res.* **37,** 1415.
436. Varo, P., and Koivistoinen, P. (1980). *Acta Agric. Scand., Suppl.* **22,** 165.
437. Vernie, L. N. (1984). *Biochim. Biophys. Acta* **738,** 203.
438. Vernie, L. N., Ginjarr, H. B., Wilders, I. T., and Bont, W. S. (1978). *Biochim. Biophys. Acta* **518,** 507.
439. Vernie, L. N., Collard, J. G., Eker, A. P. M., DeWildt, A., and Wilders, I. T. (1979). *Biochem. J.* **180,** 213.
440. Virtamo, J., Vakela, E., Alfthan, G., Punsar, S., Huttunen, J. K., and Karvonen, M. (1985). *Am. J. Epidemiol.* **122,** 276.
441. Volgarev, M. N., and Tscherkes, L. A. (1967). *In* "Selenium in Biomedicine" (O. H. Muth, ed.), p. 179. Avi Publ. Co., Westport, Connecticut.

442. Wahlstrom, R. C., and Olson, O. E. (1959). *J. Anim. Sci.* **18,** 141.
443. Wahlstrom, R. C., Kamstra, L. D., and Olson, O. E. (1955). *J. Anim. Sci.* **14,** 105.
444. Wahlstrom, R. C., Goehring, T. B., Johnson, D. D., Libal, G. W., Olson, O. E., Palmer, I. S., and Thaler, R. C. (1984). *Nutr. Rep. Int.* **29,** 143.
445. Wallace, E., Calvin, H. I., and Cooper, G. W. (1983). *Gamete Res.* **4,** 377.
446. Wallace, E., Cooper, G. W., and Calvin, H. I. (1983). *Gamete Res.* **4,** 389.
447. Wallach, J. D., and Garmaise, B. (1980). *In* "Trace Substances in Environmental Health—13" (D. D. Hemphill, ed.), p. 469. Univ. of Missouri Press, Columbia.
448. Waterhouse, J., Muir, C., Correa, P., and Powell, J. (1976). "Cancer Incidence in Five Continents," Vol. III. Int. Agency Res. Cancer, Lyon.
449. Watkinson, J. H. (1984). *Int. Symp. Selenium Biol. Med., 3rd, 1984.* Abstracts, p. 61.
450. Watkinson, J. H., and Davies, E. B. (1967). *N. Z. J. Agric. Res.* **10,** 116 and 122.
451. Wei, H. J., Luo, X. M., Yang, C. L., Xing, J., Liu, X., Liu, J., Qiao, C. H., Feng, Y. M., Liu, Y. X., Wu, Q., Guo, J. S. Stoecker, B. J., Spallholz, J. E., and Yang, S. P. (1984). *Fed. Proc., Fed. Am. Soc. Exp. Biol.* **43,** 473.
452. Welsh, S. O. (1979). *J. Nutr.* **109,** 1673.
453. Welsh, S. O., Holden, J. M., Wolf, W. R., and Levander, O. A. (1981). *J. Am. Diet. Assoc.* **79,** 277.
454. Westermarck, T., Raunu, P., Kirjarinta, M., and Lappalainen, L. (1977). *Acta Pharmacol. Toxicol.* **40,** 465.
455. Westfall, B. B., Stohlman, E. F., and Smith, M. I. (1938). *J. Pharmacol. Exp. Ther.* **64,** 55.
456. Whanger, P. D. (1973). *Biochem. Med.* **7,** 316.
457. Whanger, P. D. (1976). *In* "Selenium-Tellurium in the Environment," p. 234. Industrial Health Foundation, Pittsburgh, Pennsylvania.
458. Whanger, P. D., Weswig, P. H., Muth, O. H., and Oldfield, J. E. (1969). *J. Nutr.* **99,** 331.
459. Whanger, P. D., Pedersen, N. D., and Weswig, P. H. (1973). *Biochem. Biophys. Res. Commun.* **53,** 1031.
460. Whanger, P. D., Pedersen, N. D., Hatfield, J., and Weswig, P. H. (1976). *Proc. Soc. Exp. Biol. Med.* **153,** 295.
461. Whanger, P. D., Weswig, P. H., Oldfield, J. E., Cheeke, P. R., and Schmitz, J. A. (1976). *Nutr. Rep. Int.* **13,** 159.
462. Whanger, P. D., Weswig, P. H., Schmitz, J. A., and Oldfield, J. E. (1977). *J. Nutr.* **107,** 1298.
463. Whanger, P. D., Weswig, P. H., Schmitz, J. A., and Oldfield, J. E. (1978). *J. Anim. Sci.* **47,** 1157.
464. Whanger, P. D., Schmitz, J. A., and Exon, J. H. (1982). *Nutr. Cancer* **3,** 240.
465. Whitacre, M. E., and Combs, G. F. (1983). *J. Nutr.* **113,** 1972.
466. Whitacre, M. E., Combs, G. F., and Parker, R. S. (1983). *Fed. Proc., Fed. Am. Soc. Exp. Biol.* **42,** 928.
467. Willett, W. C., Morris, J. S., Pressel, S., Taylor, J. O., Polk, B. F., Stampfer, M. J., Rosner, B., Schneider, K., and Hames, C. G. (1983). *Lancet* **2,** 130.
468. Williams, M. M. F. (1983). *Proc. Univ. Otago Med. Sch.* **61,** 20.
469. Witting, L. A., and Horwitt, M. K. (1964). *J. Nutr.* **84,** 351.
470. Wright, P. L., and Bell, M. C. (1964). *J. Nutr.* **84,** 49.
471. Wright, P. L., and Bell, M. C. (1966). *Am. J. Physiol.* **211,** 6.
472. Wu, A. S. H., Oldfield, J. E., Shull, L. R., and Cheeke, P. R. (1979). *Biol. Reprod.* **20,** 793.
473. Wu, S. H., Oldfield, J. E., Whanger, P. D., and Weswig, P. H. (1973). *Biol. Reprod.* **8,** 625.
474. Yang, G., Wang, S., Zhou, R., and Sun, S. (1983). *Am. J. Clin. Nutr.* **37,** 872.
475. Yang, G., Zhu, L., Liu, S., and Gu, L. (1984). Private communication.

476. Yang, G., Zhu, L., Liu, S., Gu, L., Qian, P., Huang, J., and Lu, M. (1984). *Int. Symp. Selenium Biol. Med., 3rd, 1984.* Abstracts, p. 40.

477. Yasumoto, K., Iwami, K., and Yoshida, M. (1979). *J. Nutr.* **109,** 760.

478. Young, S., and Keeler, R. F. (1962). *Am. J. Vet. Res.* **23,** 955 and 966.

479. Yu, W. H. (1982). *Jpn. Circ. J.* **46,** 1201.

480. Zakowski, J. J., Forstrom, J. W., Condell, R. A., and Tappel, A. L. (1978). *Biochem. Biophys. Res. Commun.* **84,** 248.

481. Zhou, R., Sun, S., Zhai, F., Man, R., Guo, S., Wang, H., and Yang, G. (1983). *Acta Nutr. Sin.* **5,** 137.

4

Lead

JOHN QUARTERMAN

Department of Inorganic Biochemistry
The Rowett Research Institute
Aberdeen, United Kingdom

I. INTRODUCTION

Since the last edition of this book was published there has been an enormous increase in the number of publications about the biology of lead (Pb), but compared with some other metals the increase in understanding of its metabolism and the means by which it produces its toxic effects has been very disappointing. Many enzymes, membranes, and biochemical processes have been shown to be affected by lead, but none has been shown to be both sensitive and of key importance in explaining the manifestations of toxicity, especially at low levels of the metal. It may be that there is not one or even a small number of points at which lead manifests its toxicity but a large number of metabolic lesions that produce an additive debilitating effect. However, the constant stream of papers claiming to demonstrate harmful effects of the metal at low environmental levels is stimulating much fundamental work and, especially in the field of neurochemistry, we may expect important developments.

II. ESSENTIALITY OF LEAD

Evidence for the essentiality of lead has been produced in two laboratories, and the results are suggestive but fall short of being completely convincing (202). In 1974 Schwarz reported that rats grown in an ultraclean environment and with a diet supplying no more than 0.2 μg lead per gram, grew better when the diet was supplemented with lead at 1.0 or 2.5 $\mu g/g$ (273). Since then Reichlmayr-Lais and Kirchgessner have undertaken a series of studies in which rats were maintained through three generations on a diet containing 18 μg lead per kilogram or the same diet supplemented with lead at 1 $\mu g/g$. In rats of the F_1, but not the F_0 or F_2 generation, lead deprivation significantly depressed growth, decreased liver iron stores, changed the activity of some enzymes including alkaline phosphatase, and produced hypochromic microcytic anemia. The effects were not seen in all rats of the F_1 generation and were transient (246–249). In a later experiment lead depletion was again shown to retard growth and decrease carcass and tissue iron concentrations but, contradictorily, there was an increased positive iron balance (250).

III. LEAD IN FOOD AND WATER

Lead is ubiquitous in the environment and varies widely in concentration, although in most cases the amount reaching humans or animals at risk can be reduced. Research has been concerned with discovering those sources that contribute most lead to the environment and defining the amounts that reach the subjects at risk. However, this is often difficult by analysis of environmental samples, especially for human subjects whose behavior and feeding habits can vary extremely, and estimates of intake have to be made from analyses of the subjects' own tissues, usually blood.

A. Air

The lead content of air varies with location, time, and weather, and has been reported to be 0.04 $\mu g/m^3$ in a rural village and 0.27 $\mu g/m^3$ in an urban area (86).

B. Water

The lead in water may be derived from soils and rocks and from fallout from dusts and vehicular exhausts. In the United States, surface drainage water from urban areas contained 69.5 μg lead per liter, of which 6.3 $\mu g/liter$ was soluble, and that from rural areas 7.4 $\mu g/liter$ of which 2.1 $\mu g/liter$ was soluble. However, most of this lead is removed by treatment, and treated water generally contains <10 $\mu g/liter$ before distribution (87,105). Much of the lead subsequently found in water is derived from pipes, and a survey in Great Britain found that ~20% of "first draw" and 10% of random daytime drinking water samples contained >50 μg lead per liter (79). The figures for Scotland were higher than for other parts of the country. The above figures compare with the WHO recommended limit for lead in drinking water of 100 $\mu g/liter$, the European Economic

Community (EEC) limit of 50 µg/liter (with an "action level" of 100 µg/liter), and the U.S. standard of 50 µg/liter. The dissolution of lead from piping and tanks is influenced by the length of time the water is in contact with these and the chemical properties of the water itself—its pH, hardness, chloride and nitrate concentrations, and temperature (79). Reduction of pH by liming decreases the plumbosolvency of water. Soldered joints in copper piping can contribute significantly to lead in drinking water (168).

Much of the lead in water used for cooking is taken up by the food (185). The lead in water usually has some effect on the blood lead concentration of people drinking it, although of course it will vary with particular circumstances. The blood lead concentration of women in a U.K. city, for example, was correlated significantly with the cube root of the domestic water lead concentrations (184).

C. Human Foods

Lead is widely distributed in foods, but the lead content of staple foods such as bread, vegetables, fruit, and meat is generally small. A survey in 1975 in the United Kingdom found that the average lead content of the national diet was 90 µg/kg based on a Total Diet Study, and the estimated mean weekly intake from all sources is ~0.7 mg (177). These figures compare well with the provisional tolerable weekly intake of 3 mg per person established by the joint FAO/WHO Expert Committee on Food Additives in 1972 (135). In the United Kingdom a general limit of 1 mg lead per kilogram of food came into operation in 1980 (177).

A principal source of lead can be the solder in cans containing food, and this was particularly important in baby foods. In 1972 regulations were introduced in the United Kingdom to regulate this, and lead concentrations in baby foods were (in 1975) 70 µg/kg compared with 240 µg/kg before 1972. Lead dissolution is increased if food is stored in open cans. The mean lead contents of fruit and fruit juices was higher for lacquered cans (1.16 mg/kg) than for plain cans (0.09 mg/kg) (177). However, cans made without solder are being introduced and should decrease human lead intake from that source. In general shellfish contain more lead than other foods, but the widest variation was found in root and green vegetables (176). This variation is an indication of contamination by soil and domestic and industrial fallout (8), and this lead is not all removed from the plant by washing. For example, in London, carefully washed radish roots contained from 0.1 to 1.1 mg lead per kilogram fresh weight, and the lead content was closely related to the ethylenediaminetetraacetate (EDTA)-extractable lead in soil (74). In an area of Wales, where soil was heavily contaminated by spoil from old lead mining, the blood lead concentration in women was 50% higher than that in those living in an uncontaminated area, and those who consumed home-grown vegetables had blood lead levels 28% higher than those who consumed no locally grown vegetables (101).

Meat products are little affected by the lead content of the animals' diet. When the dietary lead of sheep (91,309), pigs (310), or chickens (200) was increased to 500 or 616 mg/kg there was only a slight increase in muscle lead content; that of other tissues rose to ≤4 mg/kg. Only kidneys and bone showed larger increases, to ~20 and 90 mg/kg, respectively.

Concentrations of lead in milk are normally ≤0.02 mg/kg (177). When the supply from one locality is contaminated with lead, dilution with uncontaminated milk may render the final product acceptable. For example, in an area of high soil lead originating from previous mining activities, silage containing up to 300 mg lead per kilogram dry matter was produced and fed to dairy cattle. Some samples of milk were found to contain up to 0.14 mg lead per kilogram, but when this milk was bulked with milk from uncontaminated areas the product supplied to the public contained only 0.02 mg/kg.

Some lead-210 (^{210}Pb) is formed in the atmosphere from the decay of radon. The annual fallout is about 2.0 nCi/m^2, and amounts to ~1.28 pCi in the total human diet. The annual fallout of stable lead is about 20 mg/m^2, which may amount to a contribution to the daily intake of 13 μg, or ~15% of the total dietary lead (177).

The contribution of air and water lead to the total lead intake of humans can vary widely from a few percentage points to over half, depending on local circumstances (177,285).

The blood lead concentrations of children are higher in areas of high soil lead than are those of adults. This may be because the children ingest more soil or because they absorb more of what they ingest or, because of their higher activity and metabolic rate, they simply ingest more food per kilogram of body weight. In different areas where soil lead ranged from means of 518 to 14,000 μg/g, house dust lead levels ranged from 565 to 2582 μg/g and hair lead levels from 7.5 to 20.2 μg/g. The blood lead level of mothers in these areas was ~14 μg/dl and unrelated to soil lead, whereas that of their children varied from 20.9 to 29.0 μg/dl in relation to the soil lead content (18).

Street dust from whatever origin can have very high lead contents. In Manchester, United Kingdom, samples of street dust collected from areas used by children had a mean lead content of 970 μg/g ranging from 90 to 10,200 μg/g. The authors believe that while adults may not be seriously affected children may ingest enough lead from this source to exceed the tolerable limits (77).

Leaded paint can also be an important source of environmental lead contamination (294), but in the United Kingdom and other countries the lead is being limited or replaced by other metal oxide opacifiers as a result of legislation.

D.   Animal Feeds

In a survey of 588 individual animal foodstuffs in Germany, including grains, milling by-products, oilseed meals, and protein concentrates, most were found to contain <1 mg lead per kilogram, although the mean lead contents of some cottonseed and coconut cake meals and meat-and-bone meal exceeded 4 mg/kg (113).

In herbage and other crops there is a seasonal pattern in lead concentration, with a minimum during the period of rapid growth, then a rapid increase during the autumn to reach a peak in midwinter (178,242). The capacity of soils to bind lead depends on pH and soil composition. Only a small amount of lead is usually available to the plant, so that large increases in soil lead are needed to increase the lead content of the plant, especially those parts above ground (292,330). By contrast, changes in aerial lead rapidly produce corresponding changes in lead concentrations in the leaves. This is particularly apparent in herbage and other crops growing near roads. Near busy roads an increase in herbage lead is detectable at 50 to 100 m and within 20 m may be two or three times the concentration in distant herbage (105). Such contaminated herbage is of course usually diluted by ingestion of uncontaminated grazing further from the roadside. The highest lead concentrations due to aerial contamination are found in leafy vegetables (e.g., lettuce), but ~50% of the deposited lead is removed by washing of the outer leaves. Soft fruit and cereals may contain up to 1 mg/kg near main roads. In root crops and tree fruit most of the lead is concentrated on the skin (177).

The application of sewage sludge to agricultural land leads to a large increase in the soil content of a number of metals, including lead, which may be very long-lasting or permanent (220), but as is the case with increases in soil lead for other reasons the effect on the lead concentration of washed crops is small (105,220). Unwashed vegetables may contain higher concentrations of lead. Animals grazing land with a high soil lead content are also at risk, because they necessarily ingest large amounts of soil as well as the grazed plants. From 9 to >80% of the total lead intake may be contributed by soil (260). Detritus lying on the surface of soils is often high in lead content and may be consumed by herbivores. For example, dairy cattle, grazing on or eating conserved fodder prepared from an area over which clay pigeon shooting had taken place regularly developed severe lead poisoning seen as inappetence, poor production, stiff joints, diarrhea, stillbirths, and abortions (100). A serious source of lead for animals grazing hill pastures could be metal-tolerant vegetation. In the Pennines (United Kingdom), mosses have been described that accumulate lead and other metals, especially in the older growth at the base of the moss carpet or as a powdery crust (278). This crust may contain up to 60 g lead per kilogram.

IV. LEAD ABSORPTION

The absorption of lead varies widely and depends on the form in which it occurs, its route of administration, the species, age, and physiological state of the animal, and, if the lead is ingested orally, the nature of the diet that accompanies it.

A.   Absorption from Inhaled Lead

The deposition of lead-containing particles in the lungs depends on the size and shape of the particles and the rate and depth of respiration. Experimental and theoretical models have been made to relate particle size to deposition at different sites in the lung. These models assume uniform regular particles and may not be applicable to real aerosols, since these are not only extremely irregular in size and shape but differ at different locations and may even change shape while they are in the lung (58,160). Young and active animals will inhale air in greater amounts and more deeply than older, less active animals and are therefore likely to acquire more lead from air. Only those particles deposited in the nonciliated regions of the lung, that is mainly the alveoli, are believed to be absorbed completely (160). Those deposited in the ciliated bronchioles are swept up to be swallowed or expectorated. In addition, much lead is deposited in the nasal sinuses. Estimates of the fraction of inhaled lead that may be removed in this way ranged from 30 to 70% in one study (143), although some workers obtained a smaller (59) and others a higher figure (204), indicating almost complete movement of particles >1 μm diameter into the gastrointestinal tract.

Chamberlain *et al.* (59) used ^{203}Pb-labeled car exhaust gases as a source of lead to measure the fate of inhaled lead and, bearing in mind that the above considerations will influence individual cases, they found 17–23, 24, 40, and 68% deposition for particles of mean diameter 0.5, 0.09, 0.04, and 0.02 μm, respectively (respiratory cycle of 4 sec). Forty-eight percent of a general urban aerosol was deposited. Over 80% of deposited lead was removed by 24 hr, and <1% remained in the lungs a week later. ^{203}Pb concentration was greatest after ~30 hr in the blood, and over half of the retained lead was found in the blood at this time. These workers found that there was a mean increment of ~2 μg lead per deciliter of blood for each microgram of lead per cubic meter of air inhaled by the subjects, but there was a wide scatter of results among the small number of volunteers studied.

The contribution of aerial lead compared with other sources of lead to the total-body burden of lead or blood lead is an important subject (with political and economic as well as medical implications) and is under intense scrutiny at this time (80,86,88).

B.   Absorption through the Skin

Very little inorganic lead is absorbed through the skin, but tetraethyl lead is absorbed very well, as it is through lungs and gut (159).

C. Intestinal Absorption—General Factors

The absorption of ingested lead is subjected to a wide variety of influences that can cause large variations in the fraction transported across the gut. Most species of animals have a fractional absorption of the order of 1% (see below), although in humans figures of ~10% are usually given (33,142,238). It is not certain if this difference is a real species effect or if it is a consequence of the more highly refined diets normally consumed by humans. The importance of considering dietary differences in such comparisons is evident from one study indicating that when rats were given a semipurified diet consisting mainly of casein, sucrose, and oil, and paint chips containing 10% lead, the retention of lead in the tissues was from 5 to 50 times higher than in rats given a commercial feed (198).

The variation in absorption between individuals and with the same subject at different times can be very great, ranging in one experiment with adult humans from 10.1 to 98.5% (32).

In rats, there was a greater retention and higher toxicity of lead in males than females (152), and in sheep, castrates and females were less severely affected than intact males (235). Human males had higher blood lead concentrations than females in the same environment (268). In apparent contradiction, rats given a male hormone (metandrostenolone) had some resistance to lead toxicity and decreased tissue lead levels (107).

Short periods of fasting (i.e., ≥16 hr) increased lead absorption several fold in rats (226) and humans (33,238). This period of fasting in rats increased the absorption of a number of metals as well as lead. The length of fast needed to increase absorption depended on the age of the animal, and the stimulation of mucosal uptake occurred mainly in the duodenum (226). Chronic food restriction also increased lead retention in rats (232). Water deprivation and high temperature but not high humidity increased the susceptibility of mice to lead poisoning (13), and the stress of confinement increased blood lead concentrations in the monkey (45).

Adaptation to continued lead ingestion occurs. When rats (183), sheep, (192,235), or cattle (145) were given food or water containing a constant level of lead, tissue levels rose to a maximum after ~6 months and then decreased. In sheep given low levels of lead, blood aminolevulinic acid dehydratase (ALAD) activity decreased to a plateau after ~8 weeks. The amount of enzyme per unit volume of blood had increased, although it was partly inhibited by the lead (123). A large dose of lead, given ip to mice, protected against subsequent acute lead poisoning but not against the effects of chronic low-level lead ingestion (102,268).

The fraction of lead retained is independent of the lead content of the diet in sheep (36) and rats (16,103,232) up to ~1000 μg/g diet, and of lead given to rats ip up to 5000 μg (153). The retention does, however, differ according to the chemical species of the lead, ranging from 44 to 164% of the retention of the acetate (16). Only metallic lead was very much less available. Evidence that absorption of metallic lead is inversely proportional to particle size (mean diameter 6 versus >50 μm) may be questioned on the grounds of increased surface oxidation of smaller particles, which can contain up to 5% lead oxide (17).

D.   Mechanism of Absorption by Gut

The evidence about the mechanism of intestinal absorption of lead is conflicting, when different techniques or even the same technique are used by different workers.

In experiments in which different parts of the gut in rats were isolated *in situ* by ligation, the absorption of lead was found to occur solely from the small intestine and was maximal in the duodenum (70). Uptake of lead by the mucosa was rapid and was followed by slower transport into the tissues reaching a maximum after 2 to 3 hr, the rate of transport increasing with increasing concentration of lead in the lumen.

Some evidence from everted sacs of rat small intestine suggested that uptake of lead by the mucosa and transfer into the serosa was similar in all regions of the intestine. Active transport was negligible, and movement of lead was by passive diffusion linked to water transport (31,116). Other work has provided evidence that lead is absorbed by the duodenal everted sacs actively and against a concentration gradient. Jejunal and ileal sacs did not actively transport lead (20).

From studies of absorption using rats in which the vascular supply to the gut was cannulated, evidence was also obtained that active transport does not play a role in lead absorption, although in this case the jejunum was found to transfer more lead than other parts of the small intestine (104). In sheep ~1.4% of an abomasal dose of ^{203}Pb was absorbed into the lymphatic system (141).

The absorption of lead, as of other metals, is very high (~50%) in suckling animals; it decreases rapidly at weaning and steadily thereafter (98,181,225). In suckling mice some lead is absorbed, at least in the distal jejunum and ileum by pinocytosis. Evidence that lead absorption in adults, in contrast to sucklings, is not inhibited by cortisone (144) suggests that pinocytic absorption is less extensive in the mature rat. The retention of lead in the tissues of young rats is greater than in older rats, but its toxicity is less (136,181,225), and intraperitoneally administered chelating agents released less lead from the young animals (137). Children, up to the age of 8 years at least, absorbed and retained more orally ingested lead than adults, up to 50%, although wide variations were found (4,329). Further, in contrast to rats, the lead absorbed is believed to be more biologically active, since a smaller fraction of the body lead is stored in bone and this fraction is more labile, giving rise to higher soft tissue (including brain) levels (19,67).

Some lead may be converted to alkyl derivatives, particularly Me_4Pb, by intestinal microorganisms (270). Such alkyl derivatives that readily cross lipid membranes and are more efficiently absorbed than inorganic lead may have a greatly enhanced toxicity.

In suckling rats given tracer doses of ^{203}Pb into the stomach, the uptake of label was found to occur in both duodenum and ileum. There were, however, great changes in this uptake with age. The uptake by intestinal tissue was much greater in 10-day-old than in 14- and 24-day-old pups, whereas the uptake by the ileum did not change so markedly with age. Between 2 and 24 hr after the dose of lead, duodenal lead content decreased and ileal lead increased in concentration. The authors assumed that the decrease of duodenal lead was due to transfer to the serosa and concluded that the duodenum was the principal site of absorption (127). It is possible, however, that some lead that is initially adsorbed on the duodenal mucosa may be released into the lumen and adsorbed again lower down in the gut. Such a movement of metal ions is believed to explain some observations with copper and zinc (307) and could be a factor in the last-mentioned work with lead.

The process of intestinal absorption is not well understood for any metal, and that of lead absorption may not be clearer until more is known about the process in general and in particular of the metals such as calcium and iron with which lead interacts in the gut.

E. Influence of Dietary Composition

The influence of dietary composition and food intake on lead absorption can be very great and may in some circumstances be a more important consideration than differences in the content of lead in the food or the environment.

1. Effects of Dietary Calcium, Phosphorus, and Vitamin D

Of the many nutrients that influence the absorption of lead, calcium has the greatest effect. It was first observed in 1940 that lead absorption into the tissues of rats was inversely related to the dietary content of calcium and phosphate (161,276,290). These observations have been amply confirmed. Thus in rats given 96 μg lead per milliliter of drinking water for 10 weeks, reducing dietary calcium from 7 to 1 g/kg increased femur lead from 100 to 708 μg/g wet weight and kidney lead from 6.9 to 629 μg/g wet weight (170,283). Diets low in phosphate also increased lead absorption but to a smaller extent than low-calcium diets. Variations in dietary phosphate and calcium at dietary concentrations below requirements for these elements appear to influence lead absorption independently (229). Increases of dietary calcium and phosphate above requirement reduced lead absorption, but the effects were not additive (5,194,231,234).

Effects of calcium and phosphate on lead absorption have been reported also in humans (34), lambs (191,192), pigs (131), horses (323), and mallards (165).

The rate of loss of lead from tissues and from the carcass is also influenced by dietary calcium and phosphate, but in this case the two minerals have different effects. If dietary phosphate or both calcium and phosphate were reduced below requirement, lead loss was increased. If dietary calcium alone was reduced no lead was lost from the carcass, even though there was considerable skeletal resorption (229,230). Such observations confirm earlier work showing that lead deposition in bone is independent of calcium deposition or removal (290). Dietary calcium above requirements also reduced lead loss, while high dietary phosphate had no effect (234). Thus the fastest loss of lead occurs, at least in the rat, when dietary calcium is at requirement level.

There is evidence that the principal interaction between calcium and phosphate and lead during absorption takes place in the gut. Thus the retention of intraperitoneally administered lead is unaffected by diet (161,229). This view has been contested by workers who found that dietary calcium or phosphate deficiency diminished lead excretion without affecting absorption, although elevated intraluminal calcium or phosphate reduced lead absorption, probably by competing for common mucosal receptors (21,22) or rendering the lead insoluble. However, in their experiments the only evidence of calcium or phosphate deficiency in the rats was a slightly decreased femur weight. Some have suggested, with no direct evidence, that tissue phosphate ions play an important role in the regulation of lead absorption (130), and others from everted-sac experiments that the interaction is only in relation to passive transport (114). In chickens no direct interaction was found between lead and calcium in the intestinal lumen, although low calcium intake did increase lead absorption (195). The details of this interaction are clearly in need of further investigation, in particular to distinguish between the roles of luminal calcium and phosphate and physiological deficiencies of these minerals.

Lead absorption was low in vitamin D-deficient rats and was increased by vitamin D dosing, but dosing of rats of normal vitamin D status caused no further stimulation of absorption (286). Doses in excess of requirement may, however, increase tissue lead levels (289). In children there is a seasonal variation in blood lead levels with a peak in the summer months, but this may not be related to the expected variations in vitamin D status. Although there was a highly significant negative correlation between blood lead concentration and 25-hydroxyvitamin D_3 (39,171,291), the serum 25-hydroxyvitamin D_3 levels were, in those cross-sectional studies, found to be related mainly to vitamin intake and not to season, with high blood lead levels being found in those children with low calcium and vitamin D intakes (258).

The action of vitamin D on lead absorption can be differentiated from that on calcium (196). Cholecalciferol, although accelerating calcium transport across both the intestinal mucosa and basolateral membrane, accelerated the movement of lead across only the basolateral membrane. Dihydroxycholecalciferol (1,25-OH$_2$-D$_3$) also has different effects on the absorption from and secretion into the lumen of both lead and calcium. Absorption of lead was found to occur at a similar rate in all parts of the gut, but the duodenum proved most sensitive, with lead absorption falling in vitamin D deficiency and increasing on repletion with 1,25-OH$_2$-D$_3$. By contrast, DeLuca and others, using everted sacs, found that it was lead absorption by the distal small intestine that responded most to vitamin D stimulation (286).

Wasserman and co-workers have found that there was a close relationship between changes in lead absorption in chicks and the calcium-binding protein (CaBP) content of the intestine (85). Stimulation of lead (and calcium) absorption only occurs if there is the potential to enhance 1,25-OH$_2$-D$_3$ synthesis. Thus dietary calcium was without effect on lead absorption or CaBP levels when chicks had been given 1,25-OH$_2$-D$_3$ as their source of vitamin D. Lead binds to the intestinal CaBP induced by vitamin D in rats (22) and chicks, and in the chick the protein has a greater affinity for lead than for calcium (85).

Other effects of calcium on lead absorption may depend on changes in eating behavior. Lead pica is a puzzling behavioral phenomenon involving the continued voluntary ingestion of the toxic metal. Snowdon has shown that such pica is induced in rats and monkeys (288) by dietary calcium deficiency and to a smaller extent by deficiencies of magnesium and zinc, and ceases as soon as these deficiencies are rectified.

2. Magnesium

Magnesium deficiency has been shown to increase tissue lead concentrations in rats and beagles, and the addition of magnesium to a stock diet already adequate in magnesium reduced lead toxicity in the rat (92,282). Cerklewski gave pregnant female rats diets with 150 or 600 µg magnesium and 200 µg lead per gram during pregnancy and lactation, and found that low dietary magnesium increased tissue lead concentrations in the dams and the offspring (53).

3. Iron

In dietary iron deficiency of rats and mice the absorption of lead is several times that observed in the iron-replete animals (24,118,240,284), and supplementary iron reduced lead absorption in rats and human subjects (94,298). The effect of dietary iron deficiency on lead absorption is potentially as great as that of calcium and, taking into account the prevalence of iron deficiency in humans, must be an important consideration in subjects exposed to lead. When pregnant female rats were given a low-iron diet (30 µg/g), the milk contained more lead and the pups at weaning had higher concentrations of lead in blood and bone than the milk and pups of iron-supplemented dams receiving 150 µg iron per gram of diet (52).

The effects of dietary iron deficiency on lead absorption must be distinguished from the effects of anemia produced by other means. When mice or rats were made anemic by bleeding (23,95,96), or iron absorption in rats was stimulated by hypoxia (190), there was no increase in the absorption of a tracer dose of lead. Two other groups have found increases of lead absorption after bleeding, but it may be relevant that they used larger doses of lead—54 mg/kg or 1 mg per rat (10,219)—than the other experimenters.

Effects of dietary iron restriction on lead absorption are evident before the development of iron-deficiency anemia. Thus rats given an iron-deficient diet for 7 days had no signs of anemia but did absorb more ^{203}Pb from ligated loops (189).

From growing evidence of the complexity of the relationships between iron metabolism and lead absorption, including the possibility that adaptation to low iron intake or to higher dietary lead intakes may develop, it is not surprising that there are difficulties in interpreting the results of epidemiological studies. Thus some groups have found a significant correlation between the severity of anemia and the elevation of blood lead concentrations (29,327), while another found that oral iron dosing of children increased blood lead values (11). Male and female adults may differ in the relationship between serum iron and blood lead concentrations (320). In human studies using radioactive tracers, iron absorption was directly correlated with lead absorption in one report (318), but in another (94) there was no difference in lead absorption among a group of subjects who differed considerably in iron absorption. However, as observed above, more needs to be known about the causes of altered iron absorption in each individual before a correlation with lead absorption can be expected.

There is little evidence for the nature of the mechanism of the interaction of iron and lead in absorption. Lead is bound by ferritin and transferrin (149) and by hemosiderinlike compounds in liver (259), as well as by erythrocyte membranes, hemoglobin, and other components of blood (140). Since in all cases lead binding is competitive with that of iron, a wide variety of sites of interaction between iron and lead appear to be possible. However, when lead and iron absorption are both stimulated by fasting, there is evidence that different transport mechanisms are involved (227).

4. Zinc

Diets low or deficient in zinc increase lead absorption and tissue lead concentrations (55). Pregnant and lactating rats given 12 µg zinc per gram of diet had higher concentrations of lead in maternal tissues, and milk and pup tissues were higher in lead, than when rats and their pups were given 120 μg zinc per gram (51). There is no evidence for the mechanism of this interaction, except that since ip zinc is ineffective it must take place in the gastrointestinal tract. It is unlikely that metallothionein is involved, since lead does not induce (216) or bind to the protein (299). The level of zinc in the diet did not affect the rate of lead excretion. Rats were given 200 μg lead per gram of diet for 3 weeks and then a diet with no lead and either 12 or 200 μg zinc per gram. Tissue lead levels decreased at the same rate in each group (54).

A number of experiments have been described that involve the administration of very large amounts of zinc, and conflicting results have been obtained that may however be due to species differences. A group of storage battery workers were given 60 mg zinc as gluconate and 2 g vitamin C daily for 24 weeks. Their mean blood lead levels dropped from 61.6 to 46.0 μg/dl, and hemoglobin levels rose from 13.4 to 15.3 g/dl (211). Including 6.3 mg zinc per gram in the diet of rats decreased tissue lead levels (304), but 4 mg/g feed given to pigs (131), 40 mg/kg body weight to horses (322), and 5 mg/g feed to rabbits (129) all increased the lead content of some tissues without affecting others. The symptoms of lead poisoning were exacerbated or unaffected by the zinc supplements.

5. Copper

The level of dietary copper has little effect on tissue lead concentration or lead retention. When rats were given food containing lead at 200 μg/g, a reduction of dietary copper from 20 to 1 μg/g decreased lead concentration only in the kidney (57). With a low copper intake, urine aminolevulinic acid (ALA) concentrations were decreased considerably in the group given lead, but hemoglobin and plasma ceruloplasmin concentrations were decreased whether or not there was lead in the food. Other work, however, conflicts with this, in that rats given low dietary copper (0.5 μg/g) had decreased hemoglobin levels only when they were also receiving lead (147). This may be a case where other dietary factors such as iron content and level of food intake and the duration and severity of the copper deficiency have to be taken into account.

6. Selenium

As dietary selenium given to lead-exposed rats was increased from 0.015 to 0.5 μg/g, tissue lead and urinary ALA concentrations decreased, and blood and liver ALAD activity increased. With selenium at 1.0 μg/g diet, tissue lead concentrations increased, as did urinary ALA and the inhibition of ALAD (56). These effects of high selenium have been reported in other work with rats (243) and quail (296), and in all cases dietary lead decreased tissue selenium concentrations.

7. Sulfur

Sulfate supplements (3.0 g sulfur per kilogram as sulfate) given to lead ace-
tate-poisoned sheep decreased tissue lead concentrations, reduced the extent to
which lead reduced hemoglobin levels, and increased survival time (235). This
effect was probably due to the reduction of sulfate to sulfide in the rumen and the
production of less available lead sulfide. Sulfate supplements had no effect on
lead retention in the rat (223).

8. Protein and Amino Acids

While the mineral composition of the diet is probably the most important
factor influencing lead uptake, organic components have a significant effect.
Many early reports claimed that diets low in protein resulted in an increased
absorption of lead, but most did not take into account the reduction in voluntary
food intake that accompanied protein-deficient diets. Such restricted food intake
is now known to increase heavy-metal uptake even without a change in dietary
composition (see above), and when food intake was taken into account a low-
protein diet had no effect on, or even reduced, lead retention (228). In the rat at
least, age may also influence the effects of a low-protein diet on lead absorption
(70). In practice, of course, protein deficiency will usually decrease food intake
and will result in increased efficiency of lead absorption and retention.

The effect on lead absorption of dietary amino acid supplements, usually 5–10
g/kg, differed according to the particular amino acid and the age of the rats used
(71,197,223,224). Most amino acids, including the sulfur amino acids, decreased lead retention in rats >100 g in weight and increased retention in weanling rats (224).

9. Lipids

When dietary fat was increased from ~50 to ~200 g/kg, lead retention in the
tissues of rats and mice was increased from two- to sevenfold (15,78), and toxic
effects, especially hepatic lesions, were severely aggravated (60). Lecithin and
bile salts at 5 g/kg diet doubled lead retention in rats, and choline and cholesterol
also had some effect (233,301). The effect of lecithin was not due to the formation of phospholipid-bound lead, since such a complex is no better absorbed than
lead as acetate (155). The stimulation of lead retention varied with the type of fat;
this was not due to the degree of saturation but may be related to the phospholipid
content of the fat (233). Lead secreted in the bile is more readily absorbed than
inorganic lead given by mouth (64), and when the bile duct was exteriorized
there was no significant absorption of lead (233). Ethanol ingestion caused
increased lead toxicity and tissue lead levels (169).

10. Milk

In 1973 Kello and Kostial found a spectacular 57-fold increase in the absorption of an oral dose of [203]Pb when adult rats were given liquid milk as their only food for 3 days (146). The production of this effect requires that milk be given for at least 2 days before the test dose of lead and is abolished if the milk is supplemented with iron at 40 μg/ml (222). The increased absorption in adult rats can be due in part to the lactose (28,44) and in part to the high content of fat and phospholipid, but some stimulation of lead absorption has been achieved by oral dosing of rats given a solid diet with human (but not bovine) lactoferrin in quantities similar to those expected from the ingestion of liquid milk (221,222). These observations were made with weaned rats given milk as their only or main source of food. It cannot be assumed that milk consumed in addition to another source of food will increase or decrease lead absorption, since the extra supply of some component such as the fat or calcium in milk could have an influence depending on the composition of the rest of the diet and the amount of milk consumed. For example, in humans, a milk supplement decreased the absorption of a dose of [203]Pb (34), probably because of an increase of dietary calcium.

11. Lead-Chelating and Binding Agents

Some chelating agents given by mouth or intraperitoneally increase the absorption of lead from the gut. This is the case with citrate, D-penicillamine, ascorbate, EDTA, 2,3-dimercaptopropanol (BAL), nitrilotriacetic acid, and diethylenetriaminepentaacetic acid (68,103,138). Some of these chelating agents only render the lead water soluble and facilitate its movement through the tissues, ultimately increasing its excretion. BAL (61), CaEDTA (62), and penicillamine (106) are most commonly used therapeutically, but the use of 2,3-dimercaptosuccinic acid has been proposed since it is a powerful chelating agent for lead relatively free of toxic side effects (112). EDTA and BAL are less effective in promoting the excretion of lead in young rats than in adult animals (137), which the authors suggest is explicable from other evidence that a greater fraction of the body lead in young animals is in an unavailable form, probably in bone tissue, and therefore less toxic to the animal.

Phytate, when present in the food as either calcium phytate or soybean meal, reduced lead absorption and toxicity (7,324). The strength of binding of phytate for lead, as for other metals, depends on the concentrations of calcium (325) and possibly magnesium, and when a diet containing 6 g calcium per kilogram was supplemented to provide 12 g calcium and 10 g phytate per kilogram, lead retention in the femur was reduced to one-tenth of the unsupplemented level (255).

When alginate was added at 20 or 100 g/liter to milk, the absorption of lead by rats was approximately halved (48,154), but when it was added to a standard laboratory diet lead absorption was increased (48). In humans, a single dose of 5 g alginate had no effect on the uptake of a dose of ^{203}Pb (125). Pectin was found to bind lead strongly *in vitro* and reduced lead absorption in two studies with rats (201,212) but not in a third (317), although in this last work toxic effects of lead were mitigated. Tannic acid (214) and kaolin reduced the severity of lead toxicity (317).

12. Vitamins

The anemia of lead poisoning is more severe in vitamin E-deficient rats. It is believed that there is a synergistic effect on the resistance to mechanical trauma of red blood cells and that the anemia is potentiated by rendering them more susceptible to splenic sequestration (162). Rats given ascorbic acid (3.5 g/dl drinking water) excreted more lead in the urine and had slightly decreased tissue lead concentrations (109). Beneficial effects of ascorbic acid on lead-induced anemia have been reported in rabbits (328) and humans (208,210). Of the B vitamins, beneficial effects on lead poisoning have been reported for thiamine (40), nicotinic acid (2,139,280), and pyridoxine (218). The conversion of tryptophan to nicotinic acid may be affected by lead poisoning (303).

13. Acidosis and Alkalosis

Metabolic acidosis caused a low fractional excretion in urine of serum ultra-filterable lead. Alkalosis increased the fraction of this pool of lead excreted, probably not by a change in glomerular filtration, but by a change in peritubular membrane transport (312).

V. LEAD METABOLISM

A. Kinetics

The kinetics of lead metabolism in humans has been studied extensively by Rabinowitz *et al.* (239) with stable isotopes. Their technique avoided the necessity of using the radioactive isotopes ^{203}Pb or ^{210}Pb or the use of large doses of lead, which may have disturbed the normal physiology of the subject accustomed to low environmental levels of the metal. Their subjects were given daily 79–204 μg of enriched ^{204}Pb for periods of 1 to 124 days. The data suggested a three-compartment model for lead metabolism. The first compartment was considered to include blood but is 1.5–2.2 times larger than the blood mass; it contains ~2 mg lead, has a mean half-life (mean residence time) of 35 days, and is in direct communication with ingested lead, urinary lead, and the other pools. The second pool is identified largely with the soft tissues, contains ~0.6 mg lead, and has a mean half-life of ~40 days. This pool is believed to account for the lead in hair, sweat, and alimentary secretions. Pool three is ascribed to the skeleton, contains the vast quantity of body lead, and has a very long mean half-life. Bones appeared to differ from each other in their rate of lead turnover.

B. Lead in Tissues

The total mean amount of lead found in the bodies of 150 human accident victims in the United States was 121 mg, ranging from 50 to 205 mg, of which >90% was in the skeleton (272). The lead concentration in several tissues, including aorta, kidneys, liver, and bone, increased with age but differed greatly in different parts of the world. The highest concentrations of lead are usually found in bone, kidneys, liver, aorta, and hair (1–2 μg/g fresh weight), while the lowest are in muscle, adipose tissues, and brain (<100 μg/kg). Of the different parts of the skeleton, lead has been found in highest concentration in the teeth (38 μg/g fresh weight) and then in the long bones (~26 μg/g) (297). The assay of lead in teeth has provided a useful means of surveying lead exposure in human populations, particularly of children (199). In pigs the highest concentration of lead in bone occurred in the primary spongiosa. It increased during primary calcification and greatly decreased in maturing bone tissue (237). An oral dose of tracer [203]Pb is found after 2 days mostly in the spleen, liver, and kidneys, with very little in the heart (14). Lead is found, however, in all compartments of the cells of these tissues. For example, in the kidneys ~42% was in the soluble fraction, 27% in the microsomes, 15% in the nuclei, and 14% in the mitochondria. After 7 days only 30–40% of the tracer remained, having been lost most rapidly from the soluble fraction.

The molecular forms in which most of the lead occurs in the cell have not been clearly identified, except for the intranuclear inclusion bodies. It is known, though, that whether lead is given as the bivalent or tetravalent form it is recovered in the body as divalent lead (302). Most of the lead in the cell is bound, since only 10% can be removed by dialysis and very little lead is lipid soluble (264). In liver cytosol several lead-binding fractions have been separated, but most of the lead was associated with a high molecular weight protein fraction (264). Two smaller proteins of molecular weights 11,500 and 63,000 bound lead in kidneys and brain, but no binding to small proteins was observed in liver or lung cytosol (206). In the microsomes, two-thirds of the lead was associated with the rough and one-third with the smooth membrane (264). Lead is bound particularly strongly by mitochondria, both in the protein portion of the membrane (50) and in the matrix (316), and binding is accompanied by ultrastructural and functional changes that are nonspecific and reversible (72,110).

One of the earliest manifestations of exposure to lead is the formation of intranuclear inclusion bodies. The formation of the appropriate mRNA is initiated, and microscopically visible deposits of a sulfhydryl-rich protein are formed in nuclei mainly in the epithelium of the proximal tubules (111). The protein contains up to 68.5 μg lead per milligram, and its formation has been observed in animal and human kidney and in other tissues such as osteoclasts (132). These bodies are shed, along with the tubular epithelium, into the urine as the lining cells die, and they may serve the function of concentrating and excreting the lead.

In erythrocytes a low molecular weight (10,000) protein that binds lead appears in response to low levels of lead exposure (241), and while lead normally binds to hemoglobin, the hemoglobin of fetal origin showed a much greater affinity for lead (205).

C. Biochemical Effects

Lead has a multiplicity of biochemical effects arising from its ability to form bonds with a variety of anionic ligands, particularly sulfhydryl groups, sulfur atoms in cysteine residues, imidazole groups in histidine residues, and carboxyls and phosphates (306). The usual result of lead binding to enzymes is inhibition, as for example with glutathione (GSH) S-transferase (245) and several glycolytic (293) and mitochondrial (72,110) enzymes, although lead can substitute *in vitro* for other metals with the retention of some activity, as in the case with the zinc-containing enzyme alkaline phosphatase (263). In practice, the most serious enzymic effects of lead poisoning are seen in the cytochrome P_{450}-linked mixed-function oxidase system and in the enzymes involved in heme synthesis. A liver microsomal preparation from rats given 50 μmol lead per kilogram body weight iv metabolized aminoantipyrine and other drugs ~50% less rapidly than controls given no lead (274). The toxicity of carbon tetrachloride, which depends on the production of CCl_3 radicals, was reduced in lead-poisoned rats (209). A number of human subjects admitted to hospital with symptoms of lead exposure had low rates of elimination of phenazon, a drug that is excreted after metabolism by the P_{450} system. The rate of elimination of the drug increased after chelation therapy (175).

The most common, though not universal, consequence of lead poisoning is anemia, and this is produced by interference with heme synthesis, globin synthesis, and the induction of erythrocyte membrane defects. Cytochrome production is also affected, and reduced cytochrome content may be responsible for the respiratory abnormality observed in kidney mitochondria (251), and for impairment of the mixed-function oxidase system. There are at least five enzymes involved in the production of heme from succinyl-CoA and glycine, and claims have been made that lead affects all of them, but two are strongly inhibited and produce easily measurable biochemical effects. The second enzyme in the sequence, ALAD or porphobilinogen synthase, is found in the cytosol of mature erythrocytes and is very sensitive to lead. Although its role in heme synthesis has become redundant in the mature mammalian erythrocyte, its activity remains inversely correlated with blood lead within the range of values found in the general population (252). While erythrocyte ALAD activity is a sensitive indicator of blood lead concentrations, a low activity does not necessarily indicate failure of other tissues to synthesize heme. Dogs that had been given lead and had almost no erythrocyte ALAD activity were bled and subsequently found to synthesize hemoglobin as rapidly as dogs with high ALAD activity (173). ALAD is a sulfhydryl enzyme; GSH is required for its activation and restores its activity after inhibition by lead. In lead-poisoned rats there was an increased concentration of GSH in erythrocytes and other tissues that may represent a compensatory mechanism (133). The ALAD molecule also contains zinc, and the activity of erythrocyte ALAD was increased with an increase in dietary zinc in humans (1) and in rats given lead in the food (269). When ALAD activity is reduced by lead ingestion, the urinary excretion of its substrate, ALA, is not so closely related to lead exposure as is erythrocyte ALAD activity (3), possibly because ALA synthesis may also be increased by a positive-feedback mechanism in response to the level of heme (156).

A second enzyme strongly inhibited by lead is heme synthetase (ferrochelatase), which catalyzes the insertion of Fe^{3+} into protoporphyrin IX. The effects of lead on its activity are seen principally in the bone marrow. The iron-free protoporphyrin (EP) produced binds zinc in place of iron, and the resulting compound fluoresces strongly and is easily estimated. Ferrochelatase responds less rapidly than ALAD to lead exposure. Women and children have a greater increase in EP production than men in response to lead ingestion (252). Iron deficiency anemia also results in an elevation of EP (157), and the role of iron and other metals and sulfhydryl compounds in heme synthesis has been reviewed (172). Lead feeding of rats depressed ferrochelatase in heart mitochondria and was associated with ultrastructural changes in the myocardium (182).

Very low levels of lead *in vitro* decrease the activity of erythrocyte membrane Na^+,K^+-ATPase and specifically increase potassium permeability (213). Erythrocyte membrane stability may decrease. Thus, blood lead levels of 60 to 80 μg/dl in lead workers were associated with a decreased life span of erythrocytes (128). Schoolchildren with blood lead levels of 20 to 40 μg/dl had lower erythryocte Na^+,K^+-ATPase activities and GSH levels and higher activities of enzymes associated with an immature erythrocyte population than children with blood lead levels of <20 μg/dl (9).

Pb^{2+} is extremely efficient at depolymerizing RNA, and site-specific cleavages of tRNA can be brought about by the action of Pb^{2+} ions. Yeast tRNA[Phe] in 1 mM lead acetate at pH 7.5 shows sugar–phosphate strand scission between residues 17 and 18 (41). The authors believe that the action of Pb^{2+} is catalytic, not stoichiometric, and that the fact that the reaction, unlike that with most other metals, takes place at physiological pH is important. Lead-binding sites *in vivo* are not well documented compared with those of other metals, and it may exert its toxicity *in vivo* by its ability to cleave RNA rapidly rather than by binding to reactive groups of proteins as is one common assumption. As a complement to this work, another group of workers (81) gave rats lead acetate and [1-^{14}C]leucine ip. Liver RNA concentration increased, the protein–RNA ratio decreased, RNA : DNA increased, and [^{14}C]leucine incorporation decreased. The increased cellular concentration of RNA-reacting material might result from the degradation of polymeric RNA. Some years before this in 1972 another group had observed a defect in the production of the protein of hemoglobin in the reticulocytes of lead-poisoned children, manifested by an excessive production of β-chains over α-chains (319). The authors believe that lead was affecting hemoglobin protein synthesis directly as well as through its known effects on heme synthesis. The plasma concentrations of individual amino acids altered and the albumin–globulin ratio decreased with complex changes in the γ-globulin fraction (117,151). This last may be related to the defects on immunity in lead poisoning, which are discussed elsewhere. A further biochemical disorder of blood induced by lead is impaired glycosylation of hemoglobin (84). The decreased collagen synthesis observed in mouse fibroblasts is believed to be due to the inhibition of proline hydroxylation. The hydroxylation is iron dependent, and kinetic analysis of the lead–iron interaction suggests that the mechanism is competitive (315).

Lead has a strong affinity for mitochondria, and many of its pathological effects may be due to its ultrastructural and functional changes. Swelling and distortion of mitochondrial cristae have been observed in liver, kidney, and heart cells in animals and humans (72); there is also interference with respiration, ATP hydrolysis, and calcium transport in mitochondria in these tissues and in the brain (42).

The uptake of iodine by the thyroid gland was decreased in patients poisoned with lead through drinking "moonshine," which is a liquor often produced in lead-contaminated stills (266). Female rats took up less iodine than male rats into the thyroid after receiving 1000 μg lead per liter in the drinking water for 8 months (265).

When rats were given lead either by suckling lead-fed dams or by iv injection there was an increased water intake and urine volume and increased excretion of sodium, potassium, calcium, and water, especially after stress such as extracellular expansion (134,193). There was a poor ability to decrease sodium excretion in response to a low-sodium diet; plasma renin activity was increased, but there was no change in renin substrate concentration (97). However, not all species may respond to lead poisoning in this way. In lead-poisoned sheep, urine production and sodium and potassium excretion decreased (236), while in humans there was renal insufficiency with raised serum urea concentrations and hyperuricemia (47), low plasma renin activity (267), and low urinary kallikrein excretion (38).

The conversion of cholecalciferol to $1,25\text{-}OH_2\text{-}D_3$ was impaired in chicks given a low-calcium diet with 1000 μg lead per gram. Production of the dihydroxy compound was reduced by about a third and when dietary lead was increased to 6000 μg/g by about four-fifths (85). A similar reduction in $1,25\text{-}OH_2\text{-}D_3$ production in children with high blood lead levels has been reported (257). In rats, lead decreased $1,25\text{-}OH_2\text{-}D_3$ plasma levels with diets low in calcium or phosphorus but not if the diets were supplemented with these minerals. Lead blocked the action of $1,25\text{-}OH_2\text{-}D_3$ on intestinal calcium transport but not the mobilization of calcium from bone or the mineralization of rachitic bone (287).

An iv injection of lead in rats caused a rapid rise in plasma nondialyzable calcium, but chronic oral administration of lead induced hyperplasia of parathyroid C cells and increased the concentration of calcitonin in blood and thyroid tissue (203,215,256).

In rats adequate or deficient in selenium or vitamin E, lead increased the respiratory output of ethane, which was taken to be evidence of increased lipid peroxidation (279).

Lead in turn affects the metabolism of other minerals. After lead dosing of rats the uptake of ip ^{45}Ca and $^{32}PO_4$ by bone was reduced (65,326). Lead increased passive, but not active, intestinal transport of ^{45}Ca by everted sacs (114) but decreased absorption by intact rats (115). There are also reports that lead affects absorption or distribution of iron (83), copper (126), zinc (27,311), selenium (243,296), and mercury (69).

These last two sections on absorption and metabolism have described some well-established effects of lead but have also revealed many more where the evidence is fragmentary or contradictory. Often where the gross effect of lead is clear as in anemia or kidney accumulation, there are many details known about the molecular chemistry involved, such as the inhibition of the heme synthesizing enzymes or the formation of intranuclear inclusion bodies. Other metal-binding proteins have not been identified, and the starting point for many studies of lead transport and activity is lacking. Part of the problem is that lead tends to associate with very high molecular weight proteins, but newer techniques for fractionating these proteins may yield valuable information.

VI. TOXICITY

In this section are discussed the toxic effects of lead on particular organs or systems. In each case failures of function are described that can be attributed to lead ingestion. However, as is the case with the biochemical effects discussed previously, few of these effects are specific to lead toxicity and can be used as a specific diagnosis. These few include blood ALAD activity, urine ALA concentration, and the occurrence of renal intranuclear inclusion bodies, but at low levels of lead intake these are not necessarily related to functional disorders. The diagnosis and assessment of severity of lead poisoning in any particular case depends on observing one or more of the disorders that have been ascribed to lead ingestion along with evidence of elevated tissue lead concentrations.

A. Kidney

Pathological changes in the kidney are very frequent consequences of acute or chronic lead poisoning. The changes are principally observed in the epithelia of the proximal convoluted tubules, most characteristically in the form of intranuclear inclusion bodies, but also by mitochondrial damage, tubular swelling, atrophy, and fibrosis, a Fanconi-like syndrome of impaired tubular reabsorption, and hyperuricemia.

These and other aspects of lead toxicity in the kidney have been examined after chronic low-level exposure of rats (0.5–250 μg lead per gram body weight per day) for periods up to 9 months. Renal lead levels of 5 μg/g were associated with a blood lead concentration of 11 μg/dl and with cytomegaly and karyomegaly in renal proximal tubule cells (99).

A survey in Britain of households with >100 μg lead per liter of drinking water found that the concentration of lead in the water was correlated with that in blood and with renal insufficiency of the householders. Evidence of renal insufficiency included elevated serum urea and uric acid concentrations, although there was no evidence of clinical disease (47). Gout often occurs as a result of chronic lead exposure, and, when accompanied by elevated blood lead concentrations, there was evidence of renal impairment (25). Production of gout by lead poisoning may be due to the inhibition of guanine aminohydrolase (89).

B. Vascular System

Symptoms of cardiac disease are frequently observed as part of the syndrome of lead intoxication. In both chronic and acute poisoning there is impaired cardiac performance, and evidence of ultrastructural damage, ECG abnormalities, defects in the formation of tropomyosin and other proteins, an exaggerated response to norepinephrine, and central and peripheral nervous system defects have all been reported. In their review (321) Williams *et al.* point out that neonatal animals are particularly susceptible in these respects to small doses of lead. Even older rats (200 g) given ≤1 μg lead per milliliter of drinking water for 25 weeks showed ultrastructural changes in the cardiac mitochondria, namely loss of the regular spacing and orientation of the cristae (187).

Hypertension has been reported in rat pups given lead through gestation and lactation and was associated with defects in the renin–angiotensin system (313). Other workers have reported lead-induced elevation of plasma lipoproteins and cholesterol esters in rats (301). In human male hypertensives in Glasgow (United Kingdom), there was a statistically significant excess of cases with high blood lead levels (26), but in a study of >7000 middle-aged men in the United Kingdom no relation whatever was found between blood lead concentration and renal function or blood pressure (217).

C. Reproduction

Lead is transferred readily across the placenta in humans (262), rats (52,82), and goats (174), and the blood lead levels of the fetus are usually >50% of and closely correlated with those of the mother.

When pregnant rats were given drinking water containing lead at 50 mg/liter (i.e., 1000 times the present drinking water standard), there were increased fetal losses but no abnormalities (82). The possibility of toxic effects of lead during pregnancy in the human population has arisen usually when there is a high level of exposure to the metal (253). At these high levels of lead intake, dietary iron (52) and zinc (82) supplements reduced fetal lead and gave some protection against lead toxicity. Maternal lead intake had the effect of decreasing calcium absorption in young rats (305).

At low levels of maternal lead intake, some lead is still transferred to the fetus (186) but there is little evidence of toxic effects. A small fraction of maternal lead is found in milk. In rats <1% of ^{203}Pb injected ip into the dams, or about half the absorbed fraction of an oral dose, was found in the tissues of suckling rats (179,180).

D. Disease and Infection

Reduced antibody synthesis is responsible for the increased mortality from bacterial and viral diseases in animals that are exposed to lead.

Chronic exposure to lead in mice given as little lead as 13.75 μg/g diet produced a significant decrease in antibody synthesis, particularly IgG, indicating that the memory cell is involved (150). It is the macrophage-dependent immune response that is suppressed but not the macrophage-independent antigen synthesis. These effects were reversed by 2-mercaptoethanol and occurred at levels of lead intake below those that affected weight gain and water intake (35). Antibody production was decreased in rabbits, but in rats, while hepatic phagocytic activity was impaired, there was no loss of immune response.

E. Nervous System

The neurological effects of lead were among the first to be recognized and are, of all the toxic reactions of lead, the subject of most concern and active research at present. In adults acutely exposed to lead, symptoms include optical atrophy, tremors, and wrist drop, with structural and functional changes in peripheral nerves including demyelination and slow conductance velocity. In children, advanced lead intoxication is manifest as encephalopathy, with vascular damage and neuronal degeneration in the brain. Any child with a blood lead content of >80 μg/100 ml is considered to be at risk of encephalopathy and requires immediate hospitalization. In animals, numerous effects of acute lead poisoning have been described earlier, but there is also a wealth of evidence demonstrating neurological effects of low levels of lead in young animals and in *in vitro* preparations (73,121,261,277). In general, in experiments with suckling rats and mice given chronic low levels of lead, dopamine levels were unchanged or decreased in brain while norepinephrine levels, the concentrations of catecholamine metabolites, and monoamine oxidase activity were increased. Acetylcholine metabolism was influenced by lead in parts of the brain, and lead inhibited synaptic transmission. In the peripheral nervous system there was reduced motor nerve conduction velocity and impairment of fine motor coordination. Calcium pretreatment *in vivo* or administration *in vitro* is able to protect against the inhibition by lead of acetylcholine release from ganglia or of neuromuscular functions. Despite this work, the causal relationships between lead ingestion and behavioral dysfunction are not clear.

VII. TOXICITY IN ANIMALS AND HUMANS

Much information has been given about lead metabolism in various animals elsewhere in this chapter, and a number of dietary and other factors have been described that have to be taken into account when considering what effect a particular intake of lead will have, and what level of lead intake can be considered to be toxic. Thus, grazing animals are liable to ingest more lead from soil than other animals and, for ruminants but not other species, the sulfate content of the diet may be important. Otherwise, similar considerations apply to dietary interactions in all animals.

There are, however, considerable differences in the susceptibility of different species to lead ingestion, but the reasons for these differences have not been investigated. The pig, for example (see below), can tolerate high levels of lead in the food and in its tissues, but the reason is not known. It may be that an investigation of the pig's tolerance to lead would provide a better understanding of the nature of lead toxicity and of individual variation in susceptibility to lead poisoning.

A. Cattle

In cattle the most common signs of acute lead poisoning are blindness, excessive salivation, muscle twitching, sudden death, hyperirritability, depression, convulsions, grinding teeth, and slow reflexes (63,207). However, death some-times occurs with few or no clinical signs (164). The amount of ingested lead required to produce these signs is not known and blood lead concentrations varied widely in cases of acute poisoning, but kidney and liver lead concentra-tions were ~10 μg/g fresh weight or less. Chronic lead poisoning probably begins to occur after prolonged ingestion of >6 mg of lead per kilogram of body weight per day (12,122,145,166,167). Below this level of intake erythrocyte ALAD is severely decreased, but there is little or no effect on hemoglobin concentrations or growth. At about this rate of lead ingestion (~300 μg/kg feed), toxic signs, particularly anemia and growth retardation, may appear. These may appear rapidly (within 2 months) (12) or not at all (5,164), depending on the physiological state of the animal (e.g., age, health, pregnancy) and the composition of the diet. The determination of blood lead concentration has limited value in the diagnosis of lead poisoning, since when cattle were given a constant intake of lead there were wide fluctuations in blood lead levels, reaching a maximum after 3 to 5 months and a minimum after 25 months (5,275). Calves given a milk-replacer diet showed signs of lead poisoning within 1 to 3 weeks with 2.7 mg of lead per kilogram of body weight per day (331).

B. Sheep

Chronic lead poisoning in sheep in the field has been reported from several areas of Britain (46,66,295). Only lambs are affected, and clinical signs take the form of ill thrift, stiffness of the limbs associated with osteoporosis, and hydro-nephrosis. Clinical signs do not appear every year on the affected farms nor in all the animals exposed to lead. In experimental chronic lead poisoning there were no clinical effects with daily lead intakes of 3 or 4.5 mg/kg body weight or 200 or 1000 μg/kg feed (6,49,90,91). However, these or lower levels of lead pro-duced transitory sterility and abortions (5,275), and severe toxicity and deaths occurred when dietary calcium, phosphate, or sulfur was inadequate (192,235). Male sheep were more susceptible than castrates or nonpregnant ewes (235). Impaired responses to auditory stimuli were found in sheep that had been given 100 mg of lead per kilogram of body weight per day for 9 weeks (308).

C. Pigs, Goats, and Rabbits

Pigs (158,163), goats (76,108), and rabbits (254) appear to be more resistant to lead. When pigs were given 60 mg lead/kg feed there was only a small effect on hemoglobin concentrations after 15 weeks and no effect on weight gain for 4 weeks despite considerable lead absorption producing concentrations of 130 μg/dl ml blood and ~210 μg/g in liver and kidney (163). Other workers found behavioral changes in pigs given lead at 1 g/kg feed within 1 or 2 weeks, although these effects were dependent on the dietary content of calcium, phosphate, and zinc (131). There was no anemia or reduced growth in rabbits given 25 mg/kg per day.

D. Horses

The first sign of lead poisoning in horses is the noisy respiratory efforts known as "roaring" due to pharyngeal and laryngeal paralysis. This was produced in horses weighing 450 kg that had been receiving an estimated intake of 6.4 mg of lead per kilogram of body weight per day for a few months when there were no signs of any hematological changes except some stippled erythrocytes (148), while 240-kg horses eating hay providing 7.4 mg of lead and 0.19 mg of cadmium per kilogram body weight per day died within 100 days, exhibiting a variety of toxicological signs (43). In another report there was clinical lead poisoning in horses receiving 2.4 mg of lead per kilogram of body weight per day from contaminated hay, but this may not have been the only source of lead (122). As is the case with other species, young horses are more sensitive to lead than older animals (271). Interactions of lead with dietary calcium, phosphate, and zinc in horses have been mentioned elsewhere (322,323).

E. Birds

Two-week-old chickens given water that provided 12.5 mg of lead per kilogram of body weight per day for 71 days showed no hematological or other signs of lead poisoning except a decrease in ALAD (124), and quail tolerated 100 μg per gram of food but not 500 μg/g (188). In contrast to this, birds are very sensitive to metallic lead in the gizzard. The ingestion of one lead shot was lethal to 19% of mallards within 20 days (165), and 5 of 9 doves given four lead shot died within 9 days (314). Female birds had more lead in tissues and eggs than the tissues in males, and the tissue lead levels were correlated with the number of eggs laid after dosage (93). In the United Kingdom there is a heavy mortality among wild birds, especially swans, in some areas. In one study, of 320 mute swan carcasses examined, 224 were deemed to have died of lead poisoning, 170 had lead fishing weights and 4 had gunshot in the gizzard (281). Kidney lead levels were found to be the best indicator of lead poisoning, and there were characteristic renal intranuclear inclusions and liver and erythrocyte abnormalities (30).

F. Dogs and Cats

When dogs were given a diet containing 350–450 g fat <1 g calcium per kilogram, lead absorption increased threefold and they were predisposed to encephalopathy compared with dogs given a commercial food (119). Dogs receiving this diet and 5 mg of lead per kilogram of body weight per day developed nervous signs and brain lesions within 24 days (120). In another study it was found that dogs tended to have higher blood lead levels than cats in the same area, and the blood levels of lead in dogs differed between town and country whereas that of cats did not (37).

G. Fish

The maximum tolerable concentrations of lead in water for trout were found to be 18–37 μg/liter hard water and 4-7 μg/liter soft water at the eyed-egg stage, and 7–14 μg/liter after hatching (75). The effectiveness of lead in causing black tails in rainbow trout depended on the size of the fish and their growth rate when they encountered the lead (130a).

H. Humans

Many manifestations of lead poisoning in humans have been described above under various headings, including toxicity to kidney, heart, and vascular system, and there is great interest in the possible effects of low intakes on the nervous system, particularly in children where claims have been made that mental retardation may be correlated with drinking water lead concentrations (153). There is considerable literature on this subject, which it is not appropriate to detail here but which has been reviewed exhaustively by Ratcliffe (244), among others. The matter is hotly debated, but there is no doubt that in experimental animals and in *in vitro* systems neurological effects can be produced by very low levels of lead. In humans the problems of proving such effects of lead are those of estimating the intake and body burden of lead on the one hand, and of isolating and measuring any particular effect on intelligence, hyperactivity, or other mental function on the other. Then it must be shown that the effect is not related to any other environmental, hereditary, or physiological parameter. Lead concentrations in blood, teeth, and hair, and blood ALAD and FEP and urine ALA have all been used to estimate lead intake, and the limitations of these have been discussed. However, while nearly every study claiming to associate lead ingestion and nervous disorders has been criticized in one way or another, the number and variety of such studies showing a positive association is now so large that the probability that low levels of lead ingestion can damage the central nervous system must be taken seriously.

REFERENCES

1. Abdulla, M., and Svensson, S. (1981). *In* "Trace Element Metabolism in Man and Animals-4" (J. McC. Howell, J. M. Gawthorne, and C. L. White, eds.), pp. 584–587. Aust. Acad. Sci., Canberra.

2. Acocella, G. (1966). *Acta Vitaminol.* **20,** 195–202.
3. Alessio, L., Bertazzi, P. A., Monelli, D., and Foa, V. (1976). *Int. Arch. Occup. Environ. Health* **37,** 89–105.
4. Alexander, F. W., Delves, H. T., and Clayton, B. E. (1973). *Comm. Eur. Communities [Rep.] EUR* **EUR-5004.**
5. Allcroft, R. (1951). *Vet. Rec.* **63,** 583–590.
6. Allcroft, R., and Blaxter, K. L. (1950). *J. Comp. Pathol.* **60,** 209–218.
7. Anders, E., Bagnall, C. R., Krigman, M. R., and Mushak, P. (1982). *Bull. Environ. Contam. Toxicol.* **28,** 61–67.
8. Anderson, R. J., and Davies, B. E. (1980). *J. Geol. Soc., London* **137,** 547–558.
9. Angle, C. R., and McIntire, M. S. (1977). *Int. Conf. Heavy Met. Environ. [Symp. Proc.], 1st, 1975,* pp. 13–15.
10. Angle, C. R., McIntire, M. S., and Brunk, G. (1977). *J. Toxicol. Environ. Health* **3,** 557–563.
11. Angle, C. R., and Stelmak, K. L. (1976). *In* "Trace Substances in Environmental Health-9" (D. D. Hemphill, ed.), pp. 377–386. Univ. of Missouri Press, Columbia.
12. Aronson, A. L. (1972). *Am. J. Vet. Res.* **33,** 627–629.
13. Baetjer, A. M., Joardar, S. N. D., and McQuarry, W. A. (1960). *Arch. Environ. Health* **1,** 463–477.
14. Barltrop, D., Barrett, A. J., and Dingle, J. T. (1971). *J. Lab. Clin. Med.* **77,** 705–712.
15. Barltrop, D., and Khoo, H. E. (1976). *Sci. Total Environ.* **6,** 265–273.
16. Barltrop, D., and Meek, F. (1975). *Postgrad. Med. J.* **51,** 805–809.
17. Barltrop, D., and Meek, F. (1979). *Arch. Environ. Health* **33,** 280–285.
18. Barltrop, D., Strehlow, C. D., Thornton, I., and Webb, J. S. (1975). *Postgrad. Med. J.* **51,** 59–62.
19. Barry, P. S., and Mossman, D. B. (1970). *Br. J. Ind. Med.* **27,** 339–351.
20. Barton, J. C. (1984). *Am. J. Physiol.* **247,** G193–G198.
21. Barton, J. C., and Conrad, M. E. (1981). *Am. J. Clin. Nutr.* **34,** 2192–2198.
22. Barton, J. C., Conrad, M. E., Harrison, L., and Nuby, S. (1978). *J. Lab. Clin. Med.* **91,** 366–376.
23. Barton, J. C., Conrad, M. E., and Holland, R. (1981). *Proc. Soc. Exp. Biol. Med.* **166,** 64–69.
24. Barton, J. C., Conrad, M. E., Nuby, S., and Harrison, L. (1978). *J. Lab. Clin. Med.* **92,** 526–547.
25. Batuman, Y., Maesaka, J. K., Haddad, B., Tepper, E., Landy, E., and Wedeen, R. P. (1981). *N. Engl. J. Med.* **304,** 520–523.
26. Beevers, D. G. (1976). *Lancet* **1,** 1–3.
27. Behari, J. R. (1981). *Chemosphere* **10,** 1067–1072.
28. Bell, R. R., and Spicket, J. T. (1981). *Food Cosmet. Toxicol.* **19,** 429–436.
29. Betts, P. R., Astley, R., and Raine, D. N. (1973). *Br. Med. J.* **1,** 402–406.
30. Birkhead, M. E. (1982). *Tissue Cell* **14,** 691–701.
31. Blair, J. A., Coleman, I. P. L., and Hilburn, M. E. (1979). *J. Physiol. (London)* **286,** 343–350.
32. Blake, K. C. H. (1976). *Environ. Res.* **11,** 1–4.
33. Blake, K. C. H., Barbezat, G. O., and Mann, M. (1983). *Environ. Res.* **30,** 182–187.
34. Blake, K. C. H., and Mann, M. (1983). *Environ. Res.* **30,** 188–194.
35. Blakley, B. R., and Archer, D. L. (1981). *Toxicol. Appl. Pharmacol.* **61,** 18–26.
36. Blaxter, K. L. (1950). *J. Comp. Pathol.* **60,** 140–159.
37. Bloom, H., Noller, B. N., and Shenman, G. (1976). *Aust. Vet. J.* **52,** 312–316.
38. Boscolo, P., Salimei, E., Adams, A., and Porcelli, G. (1977). *Life Sci.* **20,** 1715–1722.

39. Box, V., Cherry, N., Waldron, H. A., Dattani, J., Griffiths, K. D., and Hill, F. G. H. (1981). *Lancet* **2,** 373.
40. Bratton, G. R., Zmudzki, J., Bell, M. C., and Warnock, L. G. (1981). *Toxicol. Appl. Pharmacol.* **59,** 164–172.
41. Brown, R. S., Hingerty, B. E., Dewan, J. C., and Klug, A. (1983). *Nature (London)* **303,** 543–546.
42. Bull, R. J. (1977). *In* "Biochemical Effects of Environmental Pollutants" (S. D. Lee, ed.). Ann Arbor Sci. Publ., Ann Arbor, Michigan.
43. Burrows, G. E., and Borchard, R. E. (1982). *Am. J. Vet. Res.* **43,** 2129–2133.
44. Bushnell, P. J., and De Luca, H. F. (1983). *J. Nutr.* **113,** 365–378.
45. Bushnell, P. J., Shelton, S. E., and Bowman, R. E. (1979). *Bull. Environ. Contam. Toxicol.* **22,** 819–826.
46. Butler, E. J., Nisbet, D. I., and Robertson, J. M. (1957). *J. Comp. Pathol.* **67,** 378–396.
47. Campbell, B. C., Beattie, A. D., Moore, M. R., Goldberg, A., and Reid, A. G. (1977). *Br. Med. J.* **1,** 482–485.
48. Carr, T. E. F., Nolan, J., and Durakovic, A. (1969). *Nature (London)* **224,** 1115.
49. Carson, T. L., van Gelder, G. A., Buck, W. B., and Hoffman, L. J. (1973). *Clin. Toxicol.* **6,** 389–403.
50. Castellino, N., and Aloj, S. (1969). *Br. J. Ind. Med.* **26,** 139–143.
51. Cerklewski, F. L. (1979). *J. Nutr.* **109,** 1703–1709.
52. Cerklewski, F. L. (1980). *J. Nutr.* **110,** 1453–1457.
53. Cerklewski, F. L. (1983). *J. Nutr.* **113,** 1443–1447.
54. Cerklewski, F. L. (1984). *J. Nutr.* **114,** 550–554.
55. Cerklewski, F. L., and Forbes, R. M. (1976a). *J. Nutr.* **106,** 689–696.
56. Cerklewski, F. L., and Forbes, R. M. (1976b). *J. Nutr.* **106,** 778–783.
57. Cerklewski, F. L., and Forbes, R. M. (1977). *J. Nutr.* **107,** 143–146.
58. Chamberlain, A. C., Clough, W. S., Heard, M. J., Newton, D., Stott, A. N. B., and Wells, A. C. (1975). *Proc. R. Soc. London, Ser. B* **192,** 77–110.
59. Chamberlain, A. C., Heard, M. J., Little, P., Newton, D., Wells, A., and Wiffen, R. D. (1978). *U.K. At. Energy Res. Establ., Rep.* **AERE-R9198.**
60. Chiodi, H., and Cardeza, A. F. (1949). *Arch. Pathol.* **48,** 395–404.
61. Chisholm, J. J. (1968). *J. Pediatr.* **73,** 1–38.
62. Chisholm, J. J. (1970). *Pediatr. Clin. North Am.* **17,** 591–595.
63. Christian, R. G., and Tryphonas, L. (1971). *Am. J. Vet. Res.* **32,** 203–216.
64. Cikrt, M., and Tichy, M. (1975). *Experientia* **31,** 1320–1321.
65. Cimasoni, G., and Collet, R. A. (1964). *Helv. Odontol. Acta* **8,** 142–147.
66. Clegg, F. G., and Rylands, J. M. (1966). *J. Comp. Pathol.* **76,** 15–22.
67. Cohen, N., Kneip, T. J., Goldstein, D. H., and Muchmore, E. A. S. (1972). *J. Med. Primatol.* **1,** 142–155.
68. Coleman, I. P. L., Blair, J. A., and Hilburn, M. E. (1982). *Int. J. Environ. Stud.* **18,** 187–191.
69. Congiu, L., Corongiu, F. R., Dore, M., Montaldo, C., Vargiolu, S., Casula, D., and Spiga, G. (1978). *Toxicol. Appl. Pharmacol.* **51,** 363–366.
70. Conrad, M. E., and Barton, J. C. (1978). *Gastroenterology* **74,** 731–740.
71. Cook, J. A., and DiLuzio, N. R. (1973). *Exp. Mol. Pathol.* **19,** 127–138.
72. Cramer, K., Goyer, R. A., Jagenburg, R., and Wilson, M. H. (1974). *Br. J. Ind. Med.* **31,** 113–127.
73. Damstra, T. (1977). *Environ. Health Perspect.* **19,** 297–307.
74. Davies, B. E., Conway, D., and Holt, S. (1979). *J. Agric. Sci.* **93,** 749–752.
75. Davies, P. H., Grettl, J. P., Sinley, J. R., and Smith, N. F. (1976). *Water Res.* **10,** 198–206.

76. Davis, J. N., Libke, K. G., Watson, D. F., and Bibb, T. L. (1976). *Cornell Vet.* **66,** 490–497.

77. Day, J. P., Hart, M., and Robinson, M. S. (1975). *Nature (London)* **253,** 343–345.

78. DeLuca, J., Hardy, C. A., Burright, R. G., Donovick, P. J., and Tuggy, R. L. (1982). *J. Toxicol. Environ. Health* **10,** 441–447.

79. Department of the Environment, U.K. (1977). "Lead in Drinking Water," Pollut. Pap. No. 12 H.M. Stationery Office, London.

80. Department of Health and Social Security (1980). "Lead and Health." H.M. Stationery Office, London.

81. Dhar, A., and Banerjee, P. K. (1983). *Int. J. Vitam. Nutr. Res.* **53,** 349–354.

82. Dilts, P. V., and Ahokas, R. A. (1979). *Am. J. Obstet. Gynecol.* **135,** 940–946.

83. Dobbins, A., Johnson, D. R., and Nathan, P. (1978). *J. Toxicol. Environ. Health* **4,** 541–550.

84. Dobryszycka, W., Sawicki, G., and Gasiorowski, K. (1984). *Bromatol. Chem. Toksykol.* **17,** 17–21.

85. Edelstein, S., Fullmer, C. S., and Wasserman, R. H. (1984). *J. Nutr.* **114,** 692–700.

86. Elwood, P. C., Gallacher, J. E. J., Phillips, K. M., Davies, B. E., and Toothill, C. (1984). *Nature (London)* **310,** 138–140.

87. Environmental Protection Agency (1975). *In* "Chemical Analysis of Interstate Carries Water Supply Systems," Publ. No. EPA 430/9-75-005. USEPA, Washington, D.C.

88. Facchetti, S., and Garibaldi, G. (1983). *Riv. Soc. Ital. Sci. Aliment.* **12,** 519–522.

89. Farkas, W. R., Stanawitz, T., and Schneider, M. (1978). *Science* **199,** 786–787.

90. Fassbender, C. P., and Rang, H. (1975). *Zentralbl. Veterinaermed., Reihe A* **22,** 533–548.

91. Fick, K. R., Ammerman, C. B., Miller, S. M., Simpson, C. F., and Loggins, P. E. (1976). *J. Anim. Sci.* **42,** 515–523.

92. Fine, B. P., Barth, A., Sheffet, A., and Lavenhar, M. (1976). *Environ. Res.* **12,** 224–227.

93. Finley, M. T., Dieter, M. P., and Locke, N. N. (1976). *Bull. Environ. Contam. Toxicol.* **16,** 261–269.

94. Flanagan, P. R., Chamberlain, M. J., and Valberg, L. S. (1982). *Am. J. Clin. Nutr.* **36,** 823–829.

95. Flanagan, P. R., Haist, J., and Valberg, L. S. (1980). *J. Nutr.* **110,** 1754–1763.

96. Flanagan, P. R., Hamilton, D. L., Haist, J., and Valberg, L. S. (1979). *Gastroenterology* **77,** 1074–1081.

97. Fleischer, N., Mouw, D. R., and Vander, A. J. (1980). *J. Lab. Clin. Med.* **95,** 759–770.

98. Forbes, G. B., and Reino, J. C. (1972). *J. Nutr.* **102,** 647–652.

99. Fowler, B. A., Kimmel, C. A., Woods, J. S., McConnell, E. E., and Grant, L. D. (1980). *Toxicol. Appl. Pharmacol.* **56,** 59–77.

100. Frape, D. L., and Pringle, J. D. (1984). *Vet. Rec.* **114,** 615–616.

101. Gallacher, J. E. J., Elwood, P. C., Phillips, K. M., Davies, B. E., Ginnever, R. C., Toothill, C., and Jones, D. T. (1984). *J. Epidemiol. Commun. Health* **38,** 173–176.

102. Garber, B. T., and Wei, E. (1972). *Am. Ind. Hyg. Assoc. J.* **33,** 756–760.

103. Garber, B. T., and Wei, E. (1974). *Toxicol. Appl. Pharmacol.* **27,** 685–691.

104. Gerber, G. B., and Deroo, J. (1974). *Environ. Physiol. Biochem.* **5,** 314–318.

105. Getz, L. L., Haney, A. W., Larimore, R. W., McNurney, J. W., Leland, H. V., Price, P. W., Rolfe, G. L., Wartman, R. L., Hudson, L. J., Solomon, R. L., and Reinbold, K. A. (1977). *In* "Lead in the Environment" (W. R. Bogess and B. G. Wixson, eds.), N.S.F./R.A.-770214. U.S. National Science Foundation, Washington, D.C.

106. Gibbs, K., and Walshe, J. M. (1966). *Lancet* **1,** 175–179.

107. Goldberg, N. W., and Gestin, I. (1972). *Rev. Roum. Med., Med. Int.* **9,** 75–78.

108. Gonda, I. M., Youssef, S. A. H., Abdel Aziz, S. A., and Soliman, M. M. (1983). *Vet. Med. J.* **31,** 369–379.

109. Goyer, R. A., and Cherian, M. G. (1979). *Life Sci.* **24**, 433–438.
110. Goyer, R. A., and Rhyne, B. C. (1975). *In* "Pathobiology of Cell Membranes" (B. F. Trump and A. V. Arstila, eds.), Vol. 1, pp. 383–426. Academic Press, New York.
111. Goyer, R. A., and Rhyne, B. C. (1973). *Int. Rev. Exp. Pathol.* **12**, 1–77.
112. Graziano, J. H. (1978). *In* "Trace Element Metabolism in Man and Animals-3" (M. Kirchgessner, ed.), pp. 608–610. Tech. Univ. Munich, West Germany.
113. Grossmann, G., and Egels, W. (1975). *Dtsch. Tieräertzl. Wochenschr.* **82**, 273–275.
114. Gruden, N. (1975). *Toxicology* **5**, 163–166.
115. Gruden, N., and Buben, M. (1979). *Environ. Res.* **18**, 270–275.
116. Gruden, N., and Stanić, M. (1975). *Sci. Total Environ.* **3**, 288–292.
117. Halaceva, L., Boyadziev, V., and Nikolova, P. (1968). *Scr. Sci. Med.* **6**, 93–98.
118. Hamilton, D. L. (1978). *Toxicol. Appl. Pharmacol.* **46**, 651–661.
119. Hamir, A. N., Sullivan, N. D., and Handson, P. D. (1982). *Aust. Vet. J.* **58**, 266–268.
120. Hamir, A. N., Sullivan, N. D., Handson, P. D., Wilkinson, J. S., and Leveille, R. B. (1981). *Aust. Vet. J.* **57**, 401–406.
121. Hammond, P. B. (1977). *Annu. Rev. Pharmacol. Toxicol.* **17**, 197–214.
122. Hammond, P. B., and Aronson, A. L. (1964). *Vet. Toxicol.* **111**, 595–611.
123. Hapke, H. (1973). *Comm. Eur. Communities [Rep.] EUR* **EUR-5004**, 239–248.
124. Hapke, H., and Frers, E. (1981). *Dtsch. Tieräertztl. Wochenschr.* **88**, 220–222.
125. Harrison, G. E., Carr, T. E. F., Sutton, A., and Humphreys, E. R. (1969). *Nature (London)* **224**, 1115–1116.
126. Hemingway, R. G., Inglis, J. S. S., and Brown, N. A. (1964). *Res. Vet. Sci.* **5**, 7–16.
127. Hennig, S. J., and Leeper, L. L. (1984). *Biol. Neonate* **46**, 27–35.
128. Hernberg, S. (1967). *Work Environ. Health* **3**, Suppl. 1.
129. Hietanen, E., Aitio, A., Koivusaari, V., Kilpio, J., Nevalainen, T., Närhi, N., Savolainen, H., and Vainio, M. (1982). *Toxicology* **25**, 113–127.
130. Hilburn, M. E., Coleman, I. P. L., and Blair, J. A. (1980). *Environ. Res.* **23**, 301–308.
130a. Hodson, P. V., Dixon, D. G., Spry, D. J., Whittle, D. M., and Sprague, J. B. (1982). *Can. J. Fish. Aquat. Sci.* **39**, 1243–1251.
131. Hsu, F. S., Krook, L., Pond, W. G., and Duncan, J. R. (1975). *J. Nutr.* **105**, 112–118.
132. Hsu, F. S., Krook, L., Shively, J. N., Duncan, J. R., and Pond, W. G. (1973). *Science* **181**, 447–448.
133. Hsu, J. M. (1981). *J. Nutr.* **111**, 26–33.
134. Johnson, D. R., and Kleinman, L. I. (1979). *Toxicol. Appl. Pharmacol.* **48**, 361–367.
135. Joint FAO/WHO Expert Committee on Food Additives (1972). "Evaluation of Certain Food Additives," FAO Nutr. Meet. Rep. Ser. No. 51. FAO/UN, Rome.
136. Jugo, S. (1977). *Environ. Res.* **13**, 36–46.
137. Jugo, S., Maljković, T., and Kostial, K. (1975). *Environ. Res.* **10**, 271–279.
138. Jugo, S., Maljković, T., and Kostial, K. (1975). *Toxicol. Appl. Pharmacol.* **34**, 259–263.
139. Kao, L. C., and Forbes, R. M. (1973). *Arch. Environ. Health* **27**, 31–35.
140. Kaplan, M. L., Jones, A. G., Davis, M. A., and Kopoito, L. (1975). *Life Sci.* **16**, 1545–1554.
141. Kay, R. N. B., and Quarterman, J. (1981). *Proc. Physiol. Soc., London* **322**, 27P.
142. Kehoe, R. A. (1961a). *R. Inst. Public Health Hyg. J.* **24**, 81–100.
143. Kehoe, R. A. (1961b). *Pure Appl. Chem.* **3**, 129–144.
144. Keller, C. A., and Doherty, R. A. (1980). *Am. J. Physiol.* **239**, G114–G122.
145. Kelliher, D. J., Hilliard, E. P., Poole, D. B. R., and Collins, J. D. (1973). *Ir. J. Agric. Res.* **12**, 61–69.
146. Kello, D., and Kostial, K. (1973). *Environ. Res.* **6**, 355–360.
147. Klauder, D. S., and Petering, H. G. (1977). *J. Nutr.* **107**, 1779–1785.
148. Knight, H. D., and Burau, R. G. (1973). *J. Am. Vet. Med. Assoc.* **162**, 781–786.

149. Kochen, J., and Greener, Y. (1975). *Pediatr. Res.* **9**, 323.
150. Koller, L. D., and Kovacic, S. (1974). *Nature (London)* **250**, 148–150.
151. Kośmider, S., and Petocka, Z. (1967). *Zentralbl. Arbeitsmed. Arbeitsschutz* **17**, 170–174.
152. Kostial, K., Maljković, T., and Jugo, S. (1974). *Arch. Toxikol.* **31**, 265–269.
153. Kostial, K., and Momćilović, B. (1974). *Arch. Environ. Health* **29**, 28–30.
154. Kostial, K., Simonović, I., and Pisović, M. (1971). *Environ. Res.* **4**, 360–363.
155. Ku, Y., Alvarez, G. H., and Mahaffey, K. R. (1978). *Bul. Environ. Contam. Toxicol.* **20**, 561–567.
156. Kusell, M., O'Cheskey, S., and Gerschenson, L. E. (1978). *J. Toxicol. Environ. Health* **4**, 503–513.
157. Lamola, A. A., and Yamane, T. (1974). *Science* **186**, 936–938.
158. Lassen, E. D., and Buck, W. B. (1979). *Am. J. Vet. Res.* **40**, 1359–1364.
159. Laug, E. P., and Kunze, F. M. (1948). *J. Ind. Hyg. Toxicol.* **30**, 256–259.
160. Lawther, P. J., Commins, B. T., Ellison, J. McK., and Biles, B. (1973). *Comm. Eur. Communities [Rep.] EUR* **EUR-5004**, 373–389.
161. Lederer, L. G., and Bing, F. C. (1940). *JAMA, J. Am. Med. Assoc.* **114**, 2457–2461.
162. Levander, O. A., Morris, V. C., Higgs, D. J., and Ferretti, R. J. (1975). *J. Nutr.* **105**, 1481–1485.
163. Link, R. P., and Pensinger, R. R. (1966). *Am. J. Vet. Res.* **27**, 759–763.
164. Logner, K. R., Neathery, M. M. W., Miller, W. J., Gentry, R. P., Blackmon, D. M., and White, F. D. (1984). *J. Dairy Sci.* **67**, 1007–1013.
165. Longcore, J. R., Andrews, R., Locke, L. N., Bagley, G. E., and Young, L. T. (1974). "Toxicity of Lead and Proposed Substitute Shot to Mallards," Spec. Sci. Rep.—Wildl. No. 183. U.S. Govt. Printing Office, Washington, D.C.
166. Lynch, G. P., Jackson, E. D., Kiddy, C. A., and Smith, D. F. (1976). *J. Dairy Sci.* **59**, 1490–1494.
167. Lynch, G. P., Smith, D. F., Fisher, M., Pike, T. L., and Weinland, B. T. (1976). *J. Anim. Sci.* **42**, 410–421.
168. Lyon, T. D. B., and Lenihan, J. M. A. (1976). *Br. Corros. J.* **12**, 41–45.
169. Mahaffey, K. R., and Goyer, R. A. (1974). *Arch. Environ. Health* **28**, 217–222.
170. Mahaffey, K. R., Goyer, R. A., and Haseman, J. K. (1973). *J. Lab. Clin. Med.* **82**, 92–100.
171. Mahaffey, K. R., Rosen, J. F., Chesney, R. W., Peeler, J. T., Smith, C. M., and DeLuca, H. F. (1982). *Am. J. Clin. Nutr.* **35**, 1327–1331.
172. Maines, M. D., and Kappas, A. (1977). *Science* **188**, 1215–1221.
173. Maxfield, M. E., Stopps, G. J., Barnes, J. R., D'Snee, R., and Agar, A. (1972). *Am. Ind. Hyg. Assoc. J.* **33**, 326–327.
174. McLellan, J. S., von Smolinski, A. W., Bederka, J. P., and Boulos, R. M. (1974). *Fed. Proc., Fed. Am. Soc. Exp. Biol.* **33**, 288 (Abstr. 479).
175. Meredith, P. A., Campbell, B. C., Moore, M. R., and Goldberg, A. (1977). *Eur. J. Clin. Pharmacol.* **12**, 235–239.
176. Ministry of Agriculture, Fisheries and Food (1972). "Survey of Lead in Food." H.M. Stationery Office, London.
177. Ministry of Agriculture, Fisheries and Food (1982). "Survey of Lead in Food: Second Supplementary Report." H.M. Stationery Office, London.
178. Mitchell, R. L., and Reith, J. W. S. (1966). *J. Sci. Food Agric.* **17**, 437–440.
179. Momčilović, B. (1978). *Arch. Environ. Health* **33**, 115–117.
180. Momčilović, B. (1979). *Experientia* **35**, 517–518.
181. Momčilović, B., and Kostial, K. (1974). *Environ. Res.* **8**, 214–220.
182. Moore, M. R. (1975). *Postgrad. Med. J.* **51**, 760–764.
183. Moore, M. R., Goldberg, A., Carr, K., Toner, P., and Lawrie, T. D. V. (1974). *Scott Med. J.* **19**, 155–156.

184. Moore, M. R., Goldberg, A., Pocock, S. J., Meredith, A., Stewart, I. M., MacAnespie, H., Lees, R., and Low, A. (1982). *Scott Med. J.* **27**, 113–121.
185. Moore, M. R., Hughes, M. A., and Goldberg, D. J. (1979). *Int. Arch. Occup. Environ. Health* **44**, 81–90.
186. Moore, M. R., Meredith, P. A., and Goldberg, A. (1977). *Lancet* **1**, 717–719.
187. Moore, M. R., Meredith, P. A., Goldberg, A., Carr, K. E., Toner, P. G., and Lawrie, T. D. V. (1975). *Clin. Sci. Mol. Med.* **49**, 337–341.
188. Morgan, G. W., Edens, F. W., Thaxton, P., and Parkhurst, C. R. (1975). *Poult Sci.* **54**, 1636–1642.
189. Morrison, J. N., and Quarterman, J. (1981). *Proc. Nutr. Soc.* **41**, 21A.
190. Morrison, J. N., and Quarterman, J. (1985). *In* "Trace Elements in Man and Animals-5" (C. F. Mills, P. J. Aggett, I. Bremner, and J. K. Chesters, eds.), pp. 519–521. Cambridge Univ. Press, London and New York.
191. Morrison, J. N., Quarterman, J., and Humphries, W. R. (1974). *Proc. Nutr. Soc.* **33**, 88A.
192. Morrison, J. N., Quarterman, J., and Humphries, W. R. (1977). *J. Comp. Pathol.* **87**, 417–429.
193. Mouw, D. R., Vander, A. J., Cox, J., and Fleischer, N. (1978). *Toxicol. Appl. Pharmacol.* **46**, 435–447.
194. Mouw, D. R., Wagner, J. D., Kalitis, K., Vander, A. J., and Mayor, G. H. (1978). *Environ. Res.* **15**, 20–27.
195. Mykkänen, H. M., and Wasserman, R. H. (1981). *J. Nutr.* **111**, 1757–1765.
196. Mykkänen, H. M., and Wasserman, R. H. (1982). *J. Nutr.* **112**, 520–527.
197. Mylroie, A. A., Moore, L., and Erogbogbo, V. (1977). *Toxicol. Appl. Pharmacol.* **41**, 361–367.
198. Mylroie, A. A., Moore, L., Olyai, B., and Anderson, M. (1978). *Environ. Res.* **15**, 57–64.
199. Needleman, H. L., Tuncay, O. C., and Shapiro, I. M. (1972). *Nature (London)* **235**, 111–112.
200. Negel, K. (1980). Müehle Mischfutterlich. **117**, 30–32.
201. Niculescu, T., Rafaila, E., Eremia, R., and Balasa, E. (1968). *Igiena* **17**, 421–428.
202. Nielsen, F. H. (1984). *Annu. Rev. Nutr.* **4**, 21–41.
203. Norimatsu, H., and Talmage, R. V. (1979). *Proc. Soc. Exp. Biol. Med.* **161**, 94–98.
204. O'Neill, I. K., Harrison, R. M., and Williams, C. R. (1982). *Trans.—Inst. Min. Metall., Sect. C* **91**, C84–C90.
205. Ong, C. N., and Lee, W. R. (1980). *Br. J. Ind. Med.* **37**, 292–298.
206. Oskarsson, A., Squibb, K. S., and Fowler, B. A. (1982). *Biochem. Biophys. Res. Commun.* **104**, 290–298.
207. Osweiler, G. D., Buck, W. B., and Lloyd, W. E. (1973). *Clin. Toxicol.* **6**, 367–376.
208. Pal, D. R., Chatterjee, J., and Chatterjee, G. C. (1975). *Int. J. Vitam. Nutr. Res.* **45**, 429–437.
209. Pani, P., Corongiu, F. P., Sanna, A., and Congiu, L. (1975). *Drug Metab. Dispos.* **3**, 148–154.
210. Papaioannou, R., and Sohler, A. (1978). *Fed. Proc., Fed. Am. Soc. Exp. Biol.* **37** (Abstr. 1018).
211. Papaioannou, R., Sohler, A., and Pfeiffer, C. C. (1978). *J. Orthomol. Psychiatry* **7**, 1–13.
212. Paskins-Hurlbert, A. J., Tanaka, Y., Skoryna, S. C., Moore, W., and Stara, J. F. (1977). *Environ. Res.* **14**, 128–140.
213. Passow, H., Rothstein, A., and Clarkson, T. W. (1961). *Pharmacol. Rev.* **13**, 185–224.
214. Peaslee, M. H., and Einhellig, F. A. (1977). *Experientia* **33**, 1206.
215. Peng, T.-C., Gitelman, H. J., and Garner, S. C. (1979). *Proc. Soc. Exp. Biol. Med.* **160**, 114–117.
216. Piotrowski, J. K., and Symanśka, J. A. (1976). *J. Toxicol. Environ. Health* **1**, 991–1002.

217. Pocock, S. J., Shaper, A. G., Ashby, D., Delves, T., and Whitehead, T. P. (1984). *Br. Med. J.* **289,** 872–874.
218. Pokotilenko, G. M. (1964). *Farmakol. Toksikol. (Moscow)* **27,** 88–89.
219. Pollack, S., George, J. N., Reba, R. C., Kaufman, R. M., and Crosby, W. H. (1965). *J. Clin. Invest.* **44,** 1470–1473.
220. Purves, D. (1972). *Environ. Pollut.* **3,** 17–24.
221. Quarterman, J. (1983). *Proc. Nutr. Soc.* **42,** 45A.
222. Quarterman, J. (1983). *In* "Spurenelement-Symposium 4" (M. Anke, ed.), pp. 187–193. Friedrich-Schiller Universität, Jena.
223. Quarterman, J., Humphries, W. R., and Morrison, J. N. (1976). *Proc. Nutr. Soc.* **35,** 33A.
224. Quarterman, J., Humphries, W. R., Morrison, J. N., and Morrison, E. (1980). *Environ. Res.* **23,** 54–67.
225. Quarterman, J., and Morrison, E. (1978). *Environ. Res.* **17,** 78–83.
226. Quarterman, J., and Morrison, E. (1981). *Br. J. Nutr.* **46,** 277–287.
227. Quarterman, J., and Morrison, E. (1981). *In* "Industrial and Environmental Xenobiotics" (I. Gut, M. Cikrt, and G. L. Plaa, eds.), pp. 37–44. Springer-Verlag, Berlin and New York.
228. Quarterman, J., Morrison, E., Morrison, J. N., and Humphries, W. R. (1978). *Environ. Res.* **17,** 66–77.
229. Quarterman, J., and Morrison, J. N. (1975). *Br. J. Nutr.* **34,** 351–362.
230. Quarterman, J., Morrison, J. N., and Carey, L. F. (1974). *In* "Trace Substances in Environmental Health-7" (D. D. Hemphill, ed.), pp. 289–294. Univ. of Missouri Press, Columbia.
231. Quarterman, J., Morrison, J. N., and Humphries, W. R. (1975). *Proc. Nutr. Soc.* **34,** 89A.
232. Quarterman, J., Morrison, J. N., and Humphries, W. R. (1976). *Environ. Res.* **12,** 180–187.
233. Quarterman, J., Morrison, J. N., and Humphries, W. R. (1977). *Proc. Nutr. Soc.* **36,** 103A.
234. Quarterman, J., Morrison, J. N., and Humphries, W. R. (1978). *Environ. Res.* **17,** 60–67.
235. Quarterman, J., Morrison, J. N., Humphries, W. R., and Mills, C. F. (1977). *J. Comp. Pathol.* **87,** 405–416.
236. Quarterman, J., Morrison, J. N., and Morrison, E. (1977). *Proc. Nutr. Soc.* **36,** 102A.
237. Quint, P., Gessler, M., Althoff, J., and Hoehling, H. J. (1984). *Fresenius' Z. Anal. Chem.* **317,** 653–655.
238. Rabinowitz, M. B., Kopple, J. D., and Wetherill, G. W. (1980). *Am. J. Clin. Nutr.* **33,** 1784–1788.
239. Rabinowitz, M. B., Wetherill, G. W., and Kopple, J. D. (1976). *J. Clin. Invest.* **58,** 260–270.
240. Ragan, H. A. (1977). *J. Lab. Clin. Med.* **90,** 700–706.
241. Raghavan, S. R. V., Culver, B. D., and Gonick, H. C. (1980). *Environ. Res.* **22,** 264–270.
242. Rains, D. W. (1971). *Nature (London)* **233,** 210–211.
243. Rastogi, S. C., Clausen, J., and Srivastava, K. C. (1976). *Toxicology* **6,** 377–388.
244. Ratcliffe, J. M. (1981). "Lead in Man and the Environment." Ellis Harwood Ltds., Chichester, U.K.
245. Reddy, C. C., Scholz, R. W., and Massaro, E. J. (1981). *Toxicol. Appl. Pharmacol.* **61,** 460–468.
246. Reichlmayr-Lais, A. M., and Kirchgessner, M. (1981). *Ann. Nutr. Metab.* **25,** 281–288.
247. Reichlmayr-Lais, A. M., and Kirchgessner, M. (1981). *Z. Tierphysiol., Tierernaehr. Futtermittelkd.* **46,** 1–8.
248. Reichlmayr-Lais, A. M., and Kirchgessner, M. (1981). *Z. Tierphysiol., Tierernaehr. Futtermittelkd.* **46,** 8–14.
249. Reichlmayr-Lais, A. M., and Kirchgessner, M. (1981). *Z. Tierphysiol., Tierernaehr. Futtermittelkd.* **46,** 145–150.
250. Reichlmayr-Lais, M. M., and Kirchgessner, M. (1985). *In* "Trace Elements in Man and

Animals-5'' (C. F. Mills, P. J. Aggett, I. Bremner, and J. K. Chesters, eds.), pp. 283–286. Cambridge Univ. Press, London and New York.

251. Rhyne, B. C., and Goyer, R. A. (1971). *Exp. Mol. Pathol.* **14,** 386–391.

252. Roels, H., Buchet, J. P., Lauwerys, R., Hubermont, G., Bruaux, P., Claeys-Thoreau, F., Lafontaine, A., and van Overschelde, J. (1976). *Arch. Environ. Health* **31,** 310–316.

253. Rom, W. N. (1976). *Mt. Sinai J. Med.* **43,** 542–552.

254. Roscoe, D. E., Nielsen, S. W., Eaton, H. D., and Rousseau, J. E. (1975). *Am. J. Vet. Res.* **36,** 1225–1230.

255. Rose, H. E., and Quarterman, J. (1984). *Environ. Res.* **35,** 482–489.

256. Rosen, J. F. (1972). *Clin. Res.* **4,** 755.

257. Rosen, J. F., Chesney, R. W., Hamstra, A., DeLuca, H. F., and Mahaffey, K. R. (1980). *N. Engl. J. Med.* **302,** 1128–1131.

258. Rosen, J. F., and Roginsky, M. (1973). *Pediatr. Res.* **7,** 393.

259. Rüssell, H. A. (1970). *Bull. Environ. Contam. Toxicol.* **5,** 115–124.

260. Russell, K., Brebner, J., Thornton, I., and Suttle, N. (1985). *In* "Trace Elements in Man and Animals-5'' (C. F. Mills, P. J. Aggett, I. Bremner, and J. K. Chesters, eds.). Cambridge Univ. Press, London and New York.

261. Rutter, M., and Jones, R. R. (1983). "Lead versus Health." Wiley, Chichester, U.K.

262. Ryu, J. E., Ziegler, E. E., and Fomon, S. J. (1978). *J. Pediatr.* **93,** 476–478.

263. Sabbioni, E., Girardi, F., and Marafante, E. (1976). *Biochemistry* **15,** 271–276.

264. Sabbioni, E., and Marafante, E. (1976). *Chem-Biol. Interact.* **15,** 1–20.

265. Sandstead, H. H. (1967). *Proc. Soc. Exp. Biol. Med.* **124,** 18–20.

266. Sandstead, H. H. (1973). *In* "Trace Substances in Environmental Health-6'' (D. D. Hemphill, ed.), pp. 223–236. Univ. of Missouri Press, Columbia.

267. Sandstead, H. H., Michaelis, A. M., and Temple, T. E. (1970). *Arch. Environ. Health* **20,** 356–363.

268. Sania, G. H., Hasegawa, T., and Yoshikawa, H. (1972). *J. Occup. Med.* **14,** 301–305.

269. Schenkel, H., Lantzsch, H.-J., and Scheuermann, S. (1981). *In* "Trace Element Metabolism in Man and Animals-4'' (J. McC. Howell, J. M. Gawthorne, and C. L. White, eds.), pp. 575–577. Aust. Acad. Sci., Canberra.

270. Schmidt, V., and Huber, F. (1976). *Nature (London)* **259,** 157–158.

271. Schmitt, N., Devlin, E. L., Larsen, A. A., McCausland, E. D., and Saville, J. M. (1971). *Arch. Environ. Health* **23,** 185–195.

272. Schroeder, H. A., and Tipton, J. H. (1968). *Arch. Environ. Health* **17,** 965–978.

273. Schwarz, K. (1974). *In* "Trace Element Metabolism in Animals-2'' (W. G. Hoekstra, J. W. Suttie, H. E. Ganther, and W. Mertz, eds.), pp. 355–380. University Park Press, Baltimore, Maryland.

274. Scoppa, P., Roumengous, M., and Penning, W. (1973). *Experientia* **28,** 970–972.

275. Sharma, R. M., and Buck, W. B. (1976). *Vet. Toxicol.* **18,** 186–188.

276. Shields, J. B., and Mitchell, H. H. (1940). *J. Nutr.* **21,** 541–552.

277. Shih, T.-M., and Hanin, I. (1978). *Life Sci.* **23,** 877–888.

278. Shimwell, D. W., and Laurie, A. E. (1972). *Environ. Pollut.* **3,** 291–301.

279. Sifri, M., and Hoekstra, W. G. (1978). *Fed. Proc., Fed. Am. Soc. Exp. Biol.* **37,** 757.

280. Silvestroni, A., and Balletta, A. (1964). *Folia Med. (Naples)* **47,** 1121–1129.

281. Simpson, V. R., Hunt, A. E., and French, M. C. (1979). *Environ. Pollut.* **18,** 187–202.

282. Singh, N. P., Thind, I. S., Vitale, L. F., and Pawlow, M. (1979). *Arch. Environ. Health* **34,** 168–173.

283. Six, K. M., and Goyer, R. A. (1970). *J. Lab. Clin. Med.* **76,** 933–942.

284. Six, K. M., and Goyer, R. A. (1972). *J. Lab. Clin. Med.* **79,** 128–136.

285. Smart, G. A., Warrington, M., and Evans, W. H. (1981). *J. Sci. Food Agric.* **32,** 129–133.

286. Smith, C. M., DeLuca, H. F., Tanaka, Y., and Mahaffey, K. R. (1978). *J. Nutr.* **108,** 843–847.
287. Smith, C. M., DeLuca, H. F., Tanaka, Y., and Mahaffey, K. R. (1981). *J. Nutr.* **111,** 1321–1329.
288. Snowdon, C. T. (1977). *Physiol. Behav.* **18,** 885–893.
289. Sobel, A. E., and Burger, M. (1955). *J. Biol. Chem.* **212,** 105–110.
290. Sobel, A. E., Yuska, H., Peters, D. D., and Kramer, B. (1940). *J. Biol. Chem.* **132,** 239–265.
291. Sorrell, M., Rosen, J. F., and Roginsky, M. (1977). *Arch. Environ. Health* **32,** 160–164.
292. Spittler, T. M., and Feder, W. A. (1979). *Commun. Soil Sci. Plant Anal.* **10,** 1195–1210.
293. Sporn, A., Dinu, I., Boghianu, L., Ozeranschi, L., and Botescu, E. (1971). *Nahrung* **15,** 373–380.
294. Stark, A. D., Quah, R. F., Meigs, J. W., and DeLouise, E. R. (1982). *Environ. Res.* **27,** 372–383.
295. Stewart, W. L., and Allcroft, R. (1956). *Vet. Rec.* **68,** 723–728.
296. Stone, C. L., and Soares, J. H. (1976). *Poult. Sci.* **55,** 341–349.
297. Strehlow, C. D., and Kneip, T. J. (1969). *Am. Ind. Hyg. Assoc. J.* **30,** 372–379.
298. Suzuki, T., and Yoshida, A. (1979). *J. Nutr.* **109,** 982–988.
299. Suzuki, Y., and Yoshikawa, H. (1976). *Ind. Health* **14,** 25–31.
301. Tarugi, P., Calandra, S., Borella, P., and Vivoli, G. F. (1982). *Atherosclerosis* **45,** 221–234.
302. Taylor, M. M., and Bednekoff, A. G. (1971). *Bios (Madison, N.J.)* **42,** 124–138.
303. Tenconi, L. T., and Acocella, G. (1966). *Acta Vitaminol.* **20,** 189–194.
304. Thawley, D. G. (1975). Dissertation, University of Guelph, Ontario, Canada.
305. Toraason, M. A., Barbe, J. S., and Knecht, E. A. (1981). *Toxicol. Appl. Pharmacol.* **60,** 52–65.
306. Vallee, B. L., and Ulmer, D. D. (1972). *Annu. Rev. Biochem.* **41,** 91–128.
307. Van Barneveld, A. A., and Van den Hamer, C. J. A. (1984). *Nutr. Rep. Int.* **29,** 173–182.
308. Van Gelder, G. A., Carson, T. L., Smith, R. M., and Buck, W. B. (1973). *Clin. Toxicol.* **6,** 405–418.
309. Vemmer, H., and Oslage, H. J. (1976). *Landbauforsch. Voelkenrode* **26,** 17–22.
310. Vemmer, H., and Petersen, V. (1980). *Landwirtsch. Forsch.* **33,** 424–425.
311. Victery, W., Thomas, D., Schoeps, P., and Vander, A. J. (1981). *Biol. Trace Elem. Res.* **4,** 211–219.
312. Victery, W., Vander, A. J., and Mouw, D. R. (1978). *Fed. Proc., Fed. Am. Soc. Exp. Biol.* **37,** Abs. 3727.
313. Victery, W., Vander, A. J., Shulak, J. M., Schoeps, P., and Julius, S. (1982). *J. Lab. Clin. Med.* **99,** 354–362.
314. Viet, H. P., Kendall, R. J., and Scanlon, P. F. (1983). *Poult. Sci.* **62,** 952–956.
315. Vistica, D. T., Ahrens, F. A., and Ellison, W. R. (1977). *Arch. Biochem. Biophys.* **179,** 15–23.
316. Walton, J. R. (1973). *Nature (London)* **243,** 100–101.
317. Wapnir, R. A., Moak, S. A., and Lifshitz, F. (1980). *Am. J. Clin. Nutr.* **33,** 2303–2310.
318. Watson, W. S., Hume, R., and Moore, M. R. (1980). *Lancet* **2,** 236–237.
319. White, J. M., and Harvey, D. R. (1972). *Nature (London)* **236,** 71–73.
320. Wibowo, A. A. E., Del Castillo, P., Herber, R. F. M., and Zielhuis, R. L. (1977). *Int. Arch. Occup. Environ. Health* **39,** 113–120.
321. Williams, B. J., Hejtmancik, M. R., and Abreu, M. (1983). *Fed. Proc., Fed. Am. Soc. Exp. Biol.* **42,** 2989–2993.
322. Willoughby, R. A., MacDonald, E., McSherry, B. J., and Brown, G. (1972). *Can. J. Comp. Med.* **36,** 348–359.

323. Willoughby, R. A., Thirapatsukan, T., and McSherry, B. J. (1972). *Am. J. Vet. Res.* **33,** 1165–1173.
324. Wise, A. (1982). *Bull. Environ. Contam. Toxicol.* **29,** 550–553.
325. Wise, A., and Gilburt, D. G.(1981). *Toxicol. Lett.* **9,** 45–50.
326. Yamomoto, T., Yamaguchi, M., and Sukita, Y. (1974). *Toxicol. Appl. Pharmacol.* **27,** 204–205.
327. Yip, R., Norris, T. N., and Anderson, A. S. (1981). *J. Pediatr.* **98,** 922–925.
328. Yun, H. C. (1976). *Yakhak Hoe Chi* **19,** 21–29; *Chem Abstr.* **84,** 13115 (1976).
329. Ziegler, E. E., Edwards, B. B., Jensen, R. L., Mahaffey, K. R., and Fomon, S. J. (1978). *Pediatr. Res.* **12,** 29–33.
330. Zimdahl, R. L., and Koeppe, W. E. (1977). *In* "Lead in the Environment" (W. R. Burgess, and B. G. Wixson, eds.), N.S.F./R.A.-770124. U.S. National Science Foundation, Washington, D.C.
331. Zmudski, J., Bratton, G. R., Womac, C., and Rowe, L. (1983). *Bull. Environ. Contam. Toxicol.* **30,** 435–441.

5

Cadmium

KRISTA KOSTIAL

Institute for Medical Research
Zagreb, Yugoslavia

I. CADMIUM IN ANIMAL TISSUES AND FLUIDS

Cadmium (Cd) is virtually absent from the human body at birth and accumulates with age up to ~50 years (251). At this age the total-body burden of a "standard" nonexposed middle-aged person varies from about 5 to 20 mg. Great regional variations in the cadmium body burden are observed (44,119,250,252). Values in most countries (Europe and United States) are ~10 mg (5–7 mg for nonsmokers and 8–13 mg for smokers), while concentrations in the Japanese are considerably higher—20 mg, regardless of smoking history (109), or more (159). These values reflect differences in cadmium intakes, mainly from the food but also from smoking, which significantly contributes to the body burden. Smokers have on the average about twice the tissue cadmium concentration of nonsmokers (44,190,252).

In long-term, low-level exposure about half the body burden of cadmium is localized in the kidneys and liver and approximately one-third in the kidneys. The cadmium concentration in the kidneys of "normal" people is about 10 to 15 times higher than in the liver (94,109). A cadmium gradient is present in the human kidney, with the concentrations in the outer cortex being twice those of the inner medulla (134). Values from 15 to 70 μg/g wet weight of cadmium were recorded in the kidney cortex of persons aged 40–50 years. Levels of 11 to 14 μg/g were obtained in nonsmokers, 23 to 28 μg/g in smokers in most countries, and higher levels of 54 to 100 μg/g were found in Japan. Liver values ranged from 0.5 to 1.0 μg/g in nonsmokers, 1.0 to 3.2 μg/g in smokers, and in Japan from 5 to 10 µg/g wet weight. Women often have higher renal concentrations than men (e.g.,78,109,250).

These data were mainly obtained from autopsy studies. Recent results of *in vivo* neutron activation measurements of kidney and liver concentrations (49,203) are similar, but this method is still not sufficiently sensitive to measure tissue levels in a "normal population." In general, all these measurements observed substantial individual variations in cadmium concentrations, even in people from the same area. The individual concentrations in kidney, liver, and other organs followed a log normal distribution (44,119,249,250,255).

Cadmium concentrations increase with age (4,9,24,44,119,180,209). Very little cadmium was detectable in the kidney and liver of human infants, but thereafter concentrations increased 200-fold during the first 3 years of life (85). The total increase in renal cadmium was even greater than is shown by the concentration data because of increasing weights of the kidneys and liver between birth and 3 years of age (140). Autopsy data demonstrate that there is a tendency to accrete cadmium until ~50 years of age. Beyond that age the levels of renal cadmium remain essentially constant or decrease (109).

In other tissues cadmium levels are much lower. Organs that accumulate cadmium include testes, lungs, pancreas, spleen, and various endocrine organs. In contrast, the concentration in bone, brain, fat, and muscle tissue is very low (83,234). Hamiltion and Minski (80) reported the following mean values (in micrograms per gram wet weight) for adult human tissues in England: whole kidney 13.9 ± 0.7, kidney cortex 14.3 ± 2.9, kidney medulla 12.3 ± 2.8, liver 4.3 ± 1.0, lung 2.3 ± 0.8, testes 0.3 ± 0.09, ovary 0.10 ± 0.03, and muscle 0.03 ± 0.01.

The particular concentration of cadmium in the liver and kidneys, particularly in the kidney cortex, is apparent from many studies in several animal species. The levels of cadmium in the liver and kidneys and to a lesser extent those in other tissues reflect dietary cadmium intakes over a wide range.

Data on cadmium concentrations in liver, kidney, and muscle of animals fed ≤15 µg/g dietary cadmium were summarized by the Subcommittee on Mineral Toxicity in Animals (233). These data include results obtained by various authors in several animal species, that is, cattle (221), sheep (152,221), swine (81,221), and chickens (128,221). The concentrations in livers and kidneys of control animals were mostly <1 µg/g wet weight. As the cadmium exposure level and/or time of exposure increased, the concentrations of cadmium in liver, kidney, and muscle increased.

In adult male Wistar rats given 50 µg/ml cadmium in drinking water, cadmium accumulated in tissues linearly with time for 12 weeks (8). Suzuki (239) reported a study in which male rats were injected subcutaneously with cadmium, 0.5 mg/kg body weight, 6 days a week. A linear accumulation of cadmium was observed in the first 5–7 weeks. The author described this first stage as "nontoxic" accumulation. There is considerable evidence that when the cadmium concentration in the renal cortex reaches ~200 μg/g wet weight, damage to the proximal tubule occurs (47,110,157), at which time renal cadmium concentration may decrease, since cadmium is being excreted in urine.

The retention of cadmium in the liver and kidneys is related to its selective storage or sequestration in the protein metallothionein discovered by Margoshes and Vallee (141) in equine renal cortex. Great advances have since then been made in the field of metallothionein research (87,98). Metallothioneins are a class of low molecular weight (6000), cysteine-rich metal-binding proteins found ubiquitously in nature (31,98,115,131,172). The protein is known to bind various metal ions such as cadmium, zinc, copper, and mercury (97), and its biosynthesis is closely regulated by the levels of exposure of an organism to salts of these metals (27,226,253). For these reasons it is widely accepted that metallothionein functions as a detoxifying agent by sequestering toxic metals (260), but it has also been suggested that the protein may function in the regulation and/or metabolism of essential metals (15,18,33,95). The binding of cadmium by metallothionein and deposition or synthesis of the metal–protein complex in the kidney and other soft tissues apparently accounts for its very long half-life in the body.

Normal human blood is low in cadmium. In nonoccupationally exposed persons the mean blood level is usually <1 μg/liter. Smith et al. (224) reported values of 0.60 μg/liter in persons without exposure, aged 2 months to 31 years. Blood levels in the newborn are correlated with the maternal levels (124). Smokers are reported as having cadmium levels 50% greater than nonsmokers (265).

Bernard et al. (12) studied blood levels of cadmium in rats receiving 2, 20, and 200 μg/liter of cadmium. The blood concentration increased to plateau values after ~3 months and was proportional to the concentration in drinking water. Blood is therefore considered to reflect recent exposure. This is in agreement with results obtained in humans (45,125).

The urinary cadmium excretion of individuals with no known abnormal exposure to cadmium is low, with mean concentrations ranging from <0.5 to 2.0 μg/liter. Several studies indicate that cadmium excretion in urine increases with age (45,119,240,250). Urinary excretion in smokers is higher than in nonsmokers (15,119). Cadmium exposure from whatever source tends to increase the daily urinary output of the element. The significant correlation between cadmium levels in the renal cortex and in urine found in animals (12,25,45) indicates that urinary cadmium is mainly a reflection of body (renal) burden, but this correlation is valid only before renal kidney damage occurs.

Mother's milk generally contains <1 μg/liter in New Zealand and Sweden

(24,122), but German data indicate higher levels (96,215,216). The cadmium concentration of cow's milk is also low (~5 µg/liter) and varies among individuals and different locations (139).

The cadmium concentration in hair ranges from 0.5 to 3.5 µg/g (71). Hair of infants has been shown to have high levels of cadmium that decline thereafter throughout life (78,119). Kowal *et al.* (119) reported negative correlation between hair cadmium and cadmium levels in urine and blood. Data for calf hair show similar concentrations. The cadmium concentration of wool ranged from 0.55 to 1.2 µg/g and was not significantly increased by dietary supplementation (41).

II. CADMIUM METABOLISM

The salient features of cadmium metabolism are (1) lack of an effective homeostatic control mechanism, (2) retention in the body with an unusually long half-life, sufficiently long that active accumulation of the element occurs over most of the lifetime, (3) accumulation in soft tissues, chiefly in kidney and liver, and (4) powerful interactions with other divalent metals, both at the absorptive level and in the tissues (25).

Cadmium metabolism is difficult to study because of the long-term aspects of cadmium's movement within the body and of the low quantities of cadmium in the diet, tissues, and body fluids (65). The literature on tissue distribution of cadmium and factors that affect cadmium metabolism has been summarized in several comprehensive reviews (e.g.,6,16,60,65,98,188,204,233,262).

Cadmium may be introduced into the body from a number of sources including air, food, and water.

A. Inhalation

The absorption of airborne cadmium compounds varies greatly. The relative amount of cadmium inhaled and absorbed via the pulmonary tract depends on the physicochemical form of airborne cadmium as well as the subsequent fate of cadmium deposited in the respiratory tract. The extent of deposition in the pulmonary tract is a function of the particle size and solubility (83). Friberg *et al.* (71) and Elinder *et al.* (44) calculated that ~50% of cadmium inhaled via cigarette smoke could be absorbed. In various acute and chronic animal experiments 10–40% of inhaled cadmium was absorbed (71).

B. Gastrointestinal Absorption

The absorption of ingested cadmium differs by animal species and by type of compounds. There appears to be no homeostatic control mechanism to limit absorption and retention below a nontoxic threshold (233). Gastrointestinal absorption is influenced by a number of dietary and physiological factors.

While ingestion constitutes the major part of human cadmium intake, only a small proportion (i.e., ~6%) of dietary cadmium is absorbed and the rest passes into feces (55,147,198). Various animal experiments have given lower absorption figures of ≤2%, depending on the species used (71,104,155).

Large amounts of cadmium are taken up by the intestinal tract but are thereafter lost via the feces. The cadmium retained by the gastrointestinal tract appears to represent primarily the fraction that is most rapidly cleared from the body. This phase usually takes 4–12 days in rats (117,154), goats (151), and cows (150). Cadmium tends to concentrate in the wall of the small intestine, presumably in the mucosal cells. Hamilton and Valberg (79) found that cadmium absorption into mucosal cells and transfer to plasma were dependent on the oral dose administered. Cadmium uptake into the walls of intact strips of rat duodenum–jejunum has also been studied (205). Fox et al. (62,63) showed that a marginally adequate level of dietary zinc markedly increased the retention of cadmium by the duodenum and jejunum–ileum in young Japanese quail as compared with the control level. A severe depression by cadmium of ^{64}Cu uptake from ligated segments of rat intestine has also been demonstrated (254). A combined supplement of zinc, manganese, and copper markedly decreased tissue concentrations of cadmium of young Japanese quail (92). A significant reduction in ^{59}Fe absorption was similarly found in chicks fed 75 μg/g of dietary cadmium (67), and in mice fed a low-iron diet and given cadmium in the drinking water or as intragastric cadmium dose (79). These findings have been confirmed in human beings (55). Women with low body iron stores had on average twice as high a gastrointestinal absorption rate as a control group of women. Diets low in calcium are associated with significantly higher levels of absorption and deposition of cadmium in the tissues (71,105,113,236,256). Diets deficient in vitamin D also lead to increased cadmium absorption (238,261). Gontzea and Popescu (77) have demonstrated that the quantity and quality of dietary protein significantly affect cadmium uptake and toxicity.

Competition for binding sites on ligands functioning in intestinal absorption provide a reasonable explanation for the mutual antagonism between cadmium and essential elements such as zinc and iron when the amount of cadmium intake approaches that of the essential elements. It is less apparent whether and how essential elements affect the absorption of very low levels of cadmium (64,65). The complex relationship between cadmium and various metals and nutrients has been reviewed (17,66,233).

Since interactions with dietary factors influence not only the absorption but also the toxicity of cadmium, more data on these interactions are given in Section IV and in the chapters on copper, zinc, and iron in relation to cadmium interactions with those elements.

Although the influence of age on cadmium absorption is well established in animals, very few data in humans are available. Results reported by Alexander (1) indicate that infants absorb up to 55% of ingested cadmium as compared to much lower values in adults (~6%). Experiments on mice (144) show a 10 times higher absorption of cadmium in young than in adult mice. Kello and Kostial (104), Sasser and Jarboe (208), and Engström and Nordberg (51) also demonstrated high absorption in neonatal rats and mice. Some human (234) and animal data (105) indicate that cadmium absorption might be higher in females than in males.

C. Distribution and Transport

Data from animal experiments show that cadmium is taken up from the blood into the liver, where incorporation into metallothionein occurs (19). Cadmium is then slowly released from the liver into the blood for transport to other organs, especially the kidneys.

In humans the largest stores of cadmium are in the liver and kidneys, with the renal cortex showing the highest concentrations. With increased exposure a greater proportion of the body burden will be found in the liver (71,72). In spite of low concentrations of cadmium in muscles, bone, and skin, these tissues might represent a significant contribution to the body burden due to their mass.

The placenta and mammary gland effectively limit cadmium transport into fetus and milk (137); thus, the concentration in organs of an embryo, fetus (30), or a newborn baby (85) is lower by three orders of magnitude than in an adult woman.

In animals numerous investigators have demonstrated the rapid concentration of cadmium in the liver and kidneys and the gradual shift of cadmium from other tissues to the kidneys. After a single exposure the highest organ burden of cadmium was initially found in the liver. However, kidney levels of cadmium increased at later intervals after exposure and sometimes exceeded the liver level. For example, Miller et al. (151) found that 14 days after giving a single oral dose of $^{109}CdCl_2$ to young goats, the tissue concentrations of cadmium-109 (^{109}Cd) were highest in the kidneys, followed by liver, duodenum, and abomasum.

D. Excretion

The continuous synthesis of cadmium-binding metalloprotein in the liver and kidneys causes cadmium to be trapped there; thus, elimination is very slow. It has been estimated that <0.01% of the body burden is excreted daily, to a large extent via urine (71), but also via bile, the gastrointestinal tract, saliva, the skin, sweat, and so on (45,76,121). Urinary output is seen to increase slowly with age

(15,109). As the cadmium body burden of humans or animals decreases in the presence of renal dysfunction, urinary levels of cadmium may increase markedly (9,71,123,168,223,238). Biliary excretion occurs in animals but has not yet been demonstrated to be significant in humans. There are no quantitative data available showing the degree of net gastrointestinal excretion in humans (46). Since only 6% of cadmium is absorbed from the gastrointestinal tract, the bulk of cadmium in the intestinal tract is lost in the feces. Fecal excretion therefore appears to reflect the dietary intake closely (114,248). Animal studies summarized by Friberg *et al.* (71) show that a small amount of an injected dose will be excreted in the feces within a few days after injection. The mechanism for such excretion probably involves a transfer of cadmium via the intestinal mucosa, but biliary excretion may also be involved. The biliary excretion within 24 hr after an injection is dependent on the dose (35,111). Both during and after parenteral exposure to cadmium, the total gastrointestinal excretion is considerably higher than the urinary excretion (166).

Cadmium is also eliminated through hair (e.g., 26,167), skin (153), and breast milk (201), but these routes are of limited importance.

E. Biological Half-Life

All evidence indicates that the half-life of cadmium in the whole body is very long. For humans values of 15 to 30 (71), 18 to 38 (83), and 10 to 30 (73) years have been reported.

Ellis *et al.* (48), by using *in vitro* neutron activation analysis, estimated the biological half-life at 10 to 33 years. Rahola *et al.* (198), using radioisotopes of calcium, stated that it was not possible to determine accurately the biological half-life of cadmium but suggested a range from 130 days to infinity. A half-life of 26 years has been estimated, based on a 2-year observation period of a human subject given a single dose of radioactive cadmium (219).

In order to estimate half-life more accurately, other approaches have been used. One of these involves comparing the total daily excretion with the total-body burden, applying a one-compartment model to the body as a whole (e.g., 71,245). An elaborate model including separate compartments and incorporating relevant variations in daily intake, tissue weights, and renal function has been subsequently developed (107). From this model the half-lives in the liver and kidneys were estimated at 7.5 and 12 years, respectively.

Experimental studies, whatever the conditions may be, produce much shorter estimates of half-life of cadmium in animals than in humans, ranging from several weeks in mice to 2 years in monkeys. Variations in exposure time, animal species, and interactions between cadmium and other exposure factors may explain the wide differences (71). Later studies indicate that the biological half-life of cadmium might be a function of the dose (51).

III. SOURCES OF CADMIUM

Cadmium occurs widely in nature, in close association with zinc; the cadmium–zinc ratio is generally between 1 : 100 and 1 : 1000. Substantial amounts of cadmium are continuously added to soil, water, and air as a consequence of anthropogenic activities. The total production of cadmium in the world was about 50 metric tons in 1910 and 17,000 metric tons in 1969 (246), but since then the level of production has not increased (232).

Cadmium and its compounds have many uses. Hutton (88) quantified the major sources of cadmium in the European Economic Community and assessed the relative significance of such inputs to the environmental compartments: air, land, and water. The steel industry, waste incineration, volcanic action, and zinc production, in that order, are estimated to account for the largest part of emissions of atmospheric cadmium in the region. Waste disposal results in the single largest input of cadmium to land. The quantity of cadmium associated with this source is greater than the total from the four other major sources: coal combustion, iron and steel production, phosphate fertilizers, and zinc production. The characterization of cadmium inputs to aquatic systems is incomplete, but of the sources considered, the manufacture of cadmium-containing products accounts for the largest discharge, followed by phosphate fertilizer manufacture and zinc production.

Since little or no recycling of cadmium occurs, the amount entering the environment roughly equals the amount produced or used. This steady accumulation in the environment is of considerable concern when estimating cadmium exposure.

In the absence of cadmium-emitting factories such as zinc refineries, the cadmium levels in air approximate 0.001 $\mu g/m^3$, which leads to an average inhaled amount of approximately 0.02–0.03 $\mu g/day$ for adults. The highest concentrations are found in industrialized cities (up to 0.06 $\mu g/m^3$) and in the vicinity of smelting operations (22,56).

The amount inhaled from the air is in most circumstances insignificant compared with that ingested with food, with the exception of heavy smokers who have a cadmium intake of 3 to 5 $\mu g/day$ or more from this source alone (70,132,133,190). Moreover, such inhaled cadmium is absorbed much more efficiently than ingested cadmium, as mentioned in the previous section.

Drinking water also contributes relatively little to the average daily intake. A survey of U.S. community water supplies (146) revealed an average cadmium concentration of 1.3 $\mu g/liter$. On the basis of an average adult consumption rate of 2 liters/day, drinking water contributes not more than 3–4 $\mu g/day$ to the average total cadmium intake (204), and food is the major source of cadmium for animals and nonsmoking humans.

A. Cadmium in Human Food and Dietaries

Estimates of daily dietary cadmium intakes by humans are extremely variable. Estimates derived from new data are generally lower than earlier figures, most probably due to better analytical techniques and quality control. Data published by the U.S. Food and Drug Administration (FDA) on market basket surveys over 7 years showed that the mean cadmium intake of 15- to 20-year-old males in the United States, including cadmium from water, was 38 ± 12 μg/day. Adjustment of the 38-μg/day intake rate, by the recommended daily calorie intake for various age groups, results in an average daily cadmium intake from birth to age 50 of 33 μg/day for men and 26 μg/day for women (204).

An alternate procedure for determining cadmium intake is to analyze the cadmium concentration in feces and then calculate the intake from the absorption factor. Data based on fecal excretion suggest intakes of 18 to 21 μg/day for teenage males in the United States (109,119).

Human dietary cadmium intake shows regional differences. While cadmium intake in some European countries, New Zealand, and the United States is on the average 15–20 μg/day, it is several times higher in Japan. In polluted areas in Japan the daily intake may amount to ≥200 μg. There are no data on dietary intakes in other Asian countries (190). Analysis of rice from several countries indicates that the daily intake of cadmium might be high in Taiwan, Singapore, and Java (142,237).

In addition to regional differences, cadmium intake by humans also varies considerably with the amounts and types of food consumed. Cadmium is selectively concentrated by certain food crop classes. Based on the 1973 FDA Total Diet Study, plant food classes supplied the following percentages of the total daily cadmium intake: grain and cereals 23, fruits 18, potatoes 18, leafy vegetables 6, garden foods 3, root vegetables 1.5, and legumes 0.8 (139).

It is well established that increasing the cadmium content of soil increases the cadmium content of plants grown in those soils. Cadmium uptake by plants varies with soil characteristics (particularly pH) and the type of plant. In general, leaves, roots (including tubers), and seeds accumulate the largest amount of cadmium. An extensive report on the uptake of cadmium by plants has been published (244). The hazard of increasing cadmium in plants through crop fertilization, particularly with sludge, is under comprehensive evaluation (14,81,204,264). When domestic animals consume high-cadmium plants, the liver and kidneys accumulate the element with only relatively minor increases in muscle. The cadmium content also depends on the age of the animals at slaughter. Aquatic food species including fish, crabs, oysters, and shrimp bioconcentrate cadmium. Bioaccumulation of cadmium in marine intervertebrates depends largely on the concentration of cadmium leached from the sediment into the water. However, the bioavailability is relatively low as a result of formation of stable complexes (235).

The body burden of cadmium also depends on the bioavailability of cadmium from foods. No studies on cadmium bioavailability from intrinsically labeled foods have been carried out yet in humans (66). Based on animal experiments measuring jejunal and ileal cadmium concentrations in young Japanese quail, Fox *et al.* (61) obtained relative biological values of 38 and 48% for oysters, 62% for scallops, 40% for liver, and 32% for spinach, as compared to cadmium chloride. Rats fed a commercial rat diet retained markedly smaller doses of ^{109}Cd than rats receiving milk (104) or meat, bread, and milk (196). Similar effects were observed in adult male mice fed mouse pellets or milk, each with or without additional cadmium chloride (50). Other interacting nutrients include zinc, iron, copper, selenium, calcium, pyridoxine, ascorbic acid, and protein. In general, intakes of the interacting nutrients in amounts greater than the requirement decrease, and deficiencies increase, the absorption and effects of cadmium (65,197).

B. Cadmium in Animal Feeds and Roughages

Data on the cadmium content in animal feeds have been reviewed (233). The seeds of some plants such as corn contain less cadmium than the leaves, whereas for others, such as wheat or soybeans, the quantities are similar. Most forages and plant materials fed to animals contain levels of cadmium well below 0.5 µg/g on a dry-weight basis (10,25,233). Use of high-cadmium sludges for fertilizing feedcrop lands has been shown to increase substantially the cadmium content of the crop (37). Phosphate fertilizers contain some cadmium but not nearly as much as urban sewage sludge applied at equivalent levels for fertilization. The cadmium in plants grown on sludge-fertilized soil was shown to be available for absorption by guinea pigs (74), mice (26), and sheep (84).

IV. HEALTH EFFECTS

A. General

Data suggesting that cadmium might be an essential element are very limited and not yet adequately analyzed (181,217).

Cadmium is toxic to virtually every system in the animal body, whether ingested, injected, or inhaled. Extensive information exists on the acute and chronic effects of cadmium in humans and experimental animals. Many of the data in animals are derived from studies using relatively high parenteral doses. Of much greater importance are studies that investigate adverse health effects with chronic exposure at lower levels as they may be encountered in the environment.

There are large differences between the effects of high, single exposures and of chronic exposures to smaller doses of cadmium. Present data suggest that the toxicity of cadmium is possibly determined by the capacity of the tissues to synthesize metallothionein.

The signs reported after single injections of high-cadmium doses into animals relate to reproductive organs and the nervous system, but a number of lesions in other organs may also occur. At sufficiently high doses necrosis of the sensory ganglia (76), and testicular or placental necrosis was observed [reviewed by Parizek (179)]. Testicular necrosis can be induced by relatively low doses that do not damage other organs. However, in acute toxicity experiments where only mortality is recorded the effects of cadmium on the liver may be the most important. Such effects have been most consistently demonstrated after single intraperitoneal or subcutaneous injections of various cadmium salts at concentrations exceeding 0.5–1.0 mg/kg body weight (e.g., 204). A vast literature on various aspects of these effects has been published, but it is of limited value for an evaluation of human or animal health effects from environmental exposure.

In animals, lung damage is predominant after short-term inhalation, and at high inhalation exposures lethal edema might occur. In humans, the cadmium-induced acute pulmonary disorder is a chemical pneumonitis or sometimes a pulmonary edema. Approximately 5 mg/m^3 inhaled over a period of 8 hr may be lethal, and ~1 mg/m^3 inhaled over the same time period gives rise to clinically evident symptoms in sensitive individuals (71,86).

After short-term oral exposure the type of damage is to some extent dependent on animal species. Rats, for example, may tolerate large concentrations without gastrointestinal reactions. Therefore, both liver necrosis and other lesions may be observed after such exposure, whereas the gastrointestinal reactions are dominant in humans. The symptoms are nausea, vomiting, abdominal cramps, and headache. In more severe cases shock may develop. The symptoms appear within minutes after ingestion (169). The concentration of cadmium in water that gives rise to vomiting is about 15 mg/liter, but when administered with protein-containing foods, higher concentrations are probably required. Long-term exposure to cadmium gives rise primarily to renal tubular proteinuria, regardless of whether exposure is oral, by inhalation or by means of repeated long-term injection. When lower concentrations are inhaled during longer periods of time, lung damage is less prominent and renal tubular proteinuria dominates. With higher concentrations of cadmium in the diet a wide range of adverse effects can occur in animals, including depressed growth, enteropathy, anemia, poor bone mineralization, severe kidney damage, cardiac enlargement, hypertension, and fetal malformation. These changes were reviewed by a Subcommittee on Mineral Toxicity in Animals. These data include the effects of graded levels of cadmium exposure in various animal species including cattle (138,194,221,263), sheep (41,42,152,263), goats (5), swine (38,81,221), chickens (5,67,128,195,221),

Japanese quail (59,90,91,143,199,200), dogs (7,53), rats (53,100,181), and mice (53,213). The mechanism by which cadmium exposure induces changes in specific organs is discussed later.

Friberg (69) found in animal experiments that exposure to cadmium causes a special type of proteinuria, similar to the one he demonstrated in exposed workers. Cadmium causes primarily renal tubular lesions, but there may also be glomerular lesions. These effects are generally seen at average renal cortex concentrations of 200–300 μg/g wet weight, but some effects have also been reported at considerably lower concentrations. Such results were obtained in different animal species such as rabbits (159,160,231), rats (11), swine (38), and monkeys (162).

The effects of cadmium at low-level oral exposure include damage to the absorptive cells of the intestinal villi (143), changes in blood pressure, cardiac hypertrophy, and kidney damage (177,181,212). Other studies also show induction of lung fibrosis and cardiotoxicity by similar low-level chronic cadmium exposure (185). Observations in experimental animals indicating biochemical effects of cadmium on the activity of certain enzymes, its binding to certain amino acids such as histidine as well as to DNA, and its adverse effects on oxidative phosphorylation, cytochrome P_{450} protein synthesis, and cell replication suggest that even a small degree of cadmium accumulation might be undesirable (93,207).

In conditions of oral cadmium exposure it is claimed that 5 μg/kg are usually required to produce physiological effects (66). However, minimum toxic levels or maximum safe dietary cadmium levels cannot be given with any precision, because cadmium metabolism is so strongly influenced by dietary interactions, notably with zinc, copper, and iron. The most important interactions are between cadmium and zinc. Accumulation of cadmium in liver and kidneys will also increase zinc levels in these organs by binding to metallothionein. Cadmium and zinc are intimately connected, and an adequate zinc supply might reduce some of the effects of cadmium. Maternal exposure to cadmium leads to greatly altered iron and copper metabolism in neonates (16,185).

Some results indicate that in part the oral toxicity of cadmium compounds may result from disturbances of the metabolism of zinc, copper, and iron, especially when administered at low doses. This has been found to be the case when dietary intakes of these essential metals were optimal or even excessive according to accepted dietary standards (185).

The biochemical basis of these physiological perturbations in conditions of low-level exposure is still unknown. New data from Petering *et al.* (184) indicate that exposure of kidneys to low levels of cadmium can reduce the number of binding sites in metallothionein for zinc and copper, thereby inhibiting the normal function of metallothionein in the kidney. It is assumed that under this condition new net synthesis of metallothionein does not occur. If these results are sustained, then one must look carefully at the effects of cadmium that may occur in the kidneys and other tissues below threshold for induction of new thionein, for in the present view it is only the induction of thionein and its irreversible steady-state binding of cadmium that protects cells from this metal.

In humans the two target organs after long-term exposure to cadmium are the lungs and the kidneys. Lung impairment has been described only in workers exposed to cadmium by inhalation, but even in those one can detect signs of renal dysfunction before the occurrence of respiratory impairment (230). The kidneys are undoubtedly the organs that exhibit the first adverse effect following long-term moderate to excessive exposure by inhalation or ingestion (171), although this type of cadmium-induced damage does not frequently occur in the general population. Tubular proteinuria (the first sign of chronic adverse effects) so far has been demonstrated only in exposed workers and in exposed Japanese people. Some studies on the health effect of cadmium in the general population in other countries are now under way in England and Belgium (23,82,126,202).

Kidney dysfunction was first discovered in workers occupationally exposed to cadmium by Friberg (68). The main feature of renal dysfunction caused by cadmium is an increased excretion of low molecular weight proteins (i.e., β_2-microglobulins) (69). In addition, a reduction in glomerular filtration rate was observed in heavily exposed workers. Aminoaciduria, glycosuria, and phosphaturia in addition to proteinuria could also occur in cases of chronic cadmium poisoning (102,186).

There is general agreement that the kidney lesions found in Japan among aged women who have ingested contaminated food and who sometimes also exhibited signs of osteomalacia (*itai-itai* disease) are mainly due to cadmium.

However, there are large differences between men and women with regard to secondary effects caused by renal dysfunction. Renal stones have been common among Swedish male workers, whereas osteomalacia has been found in women with cadmium-induced renal damage (190). The Swedish male workers had high calcium intakes, whereas the Japanese women had low calcium intake and lost large amounts of calcium during pregnancy and lactation. Imbalances of other nutrients such as vitamin D, proteins, and iron might also contribute to osteomalacia in Japanese women (201).

Early stages of renal dysfunction are not associated with any serious clinical complaints. In England several studies (e.g., 23) have been conducted in the village of Shipman, Sommerset, where extremely high levels of cadmium in soil due to an old zinc mine have been found. Although Harvey *et al.* (82) found that average liver concentrations of cadmium, by *in vivo* neutron activation analysis, were about five times higher than in the control region, the number of people living in this area is too small to allow definite conclusions. It was also found that the body burden of cadmium is higher in women from Liège (where nonferrous smelters have been active since the end of the last century), compared to other cities in Belgium. A higher total protein excretion was also found, but it was not accompanied by a higher excretion of β_2-microglobulin (126,202).

While adverse health effects resulting from heavy past occupational and environmental exposure to cadmium have been amply documented, the relationship between long-term, low-level exposure to cadmium and early effects is less clear (103). The extent to which a slight renal tubular proteinuria constitutes a health hazard is also one of the controversial issues (190). The increased excretion of β_2-microglobulin has in itself no health significance, but it indicates that the kidney is saturated with cadmium and that other cadmium-induced dysfunctions are possible.

B. Renal Effects

Data on the mechanism of cadmium-induced effects in the kidneys in relation to metallothionein have been reviewed by several authors (73,127,176,192,229). Most researchers in the field now agree that cadmium damage is typically located in the proximal tubules, resulting in decreased reabsorption of proteins (73). The reabsorption of cadmium metallothionein is almost complete at low levels of cadmium in plasma, but may be less effective at higher levels (58,161). The reabsorbed cadmium metallothionein is catabolized in the tubules, and cadmium is split from the metallothionein and bound to newly formed metallothionein in tubular cells. It is believed (71,164,175) that kidney damage is prevented as long as the kidney can produce enough metallothionein. Beyond that stage the non-metallothionein-bound cadmium ions become very toxic. It is obvious that because of differences in transport, the metabolism of individual forms of cadmium is entirely different (174). It also depends on the rate of administration. For example, when a single dose of metallothionein-bound cadmium is given intravenously, an almost immediate and complete uptake occurs in the renal tubule. It is conceivable that under such conditions the resynthesis and rebinding processes are insufficient to sequester cadmium from the sensitive tissue receptors, and renal damage occurs at total tissue concentrations much lower than when renal cadmium concentrations rise slowly. This explains the wide range of cadmium concentrations (10–200 μg/g wet weight) in the renal cortex associated with renal tubular dysfunction in experimental animals (176).

The mechanism by which renal toxicity develops during cadmium intoxication is still unclear. It is assumed, as mentioned earlier, that cadmium is transported directly to the renal tubular cells in the form of cadmium–thionein (27,220), and studies with injected cadmium–metallothionein have shown that this small cadmium-containing protein induces damage similar to ionic cadmium, by specifically affecting proximal tubular cells (32,170,227). Cadmium–metallothionein might damage the proximal tubular cell membrane in the process of its reabsorption from the tubular lumen (32). However, it might also be that the toxic species is Cd^{2+} released from the protein moiety, following degradation of the protein within the proximal tubular cell (20,228,257). Results of Squibb and Fowler (229) indicate that the renal toxicity of cadmium is due to its rapid uptake and degradation by renal proximal tubular cells. Effects on lysosomal enzyme activities, RNA synthesis, and possibly membrane fusion processes appear to be due to the rapid release of toxic Cd^{2+} ions within the cell.

Tubular dysfunction, once established, is irreversible (73,127). This is in agreement with data showing a buildup of cadmium levels in the kidneys long after cessation of the exposure. According to Piscator (192), it is conceivable that the initial effect of cadmium is at the site in the proximal tubule where metallothionein is initially reabsorbed. High local concentrations of metallothionein and cadmium in the tubular cells will eventually influence the reabsorption of β_2-microglobulin. Cadmium will spread upwards or downwards in the tubules, and eventually the reabsorption sites for more anionic proteins like albumin or cationic proteins like lysozymes will be affected. This means that there is a multistage process, depending on at what stage the local critical concentrations are reached in different segments of the proximal tubule. It is also possible that, depending on the intensity of exposure, the metallothionein delivered to the renal tubules may vary in charge, which may lead to small changes in absorption rates. It is obvious that more knowledge about metallothionein metabolism and cadmium transport is still needed to validate the various speculations mentioned in this section.

No effective therapy of the renal tubular dysfunction is available at this time. Efforts made in the 1940s and 1950s to find a safe drug without side effects for this purpose have been disappointing. New results from animals are slightly more promising (173).

Chelation therapy of cadmium intoxication is extremely effective when administered immediately after cadmium exposure. The decreasing effectiveness of chelators with time might be due to the inability of the chelating agents to react with cadmium already bound to metallothionein in tissues (34,52), but it might be also due to the intracellular distribution of cadmium as opposed to the extracellular distribution of chelators (polyamino-polycarbohydrates, especially diethylenetriaminepentaacetate, DTPA) (78). While chelators are much less effective when not administered immediately after exposure, new data by Klaassen et al. (112) indicate that repeated administration of chelators does enhance the urinary excretion of cadmium and provide hope that long-term therapy may decrease body burden and toxicity of cadmium. Data by Andersen (2) indicate

that during chronic cadmium exposure 2,3-dimercaptopropanol (BAL) significantly increases bilary excretion of cadmium and most of the mobilized cadmium
is chelated out of metallothionein.

C. Effect on Reproduction and Development

The effect of cadmium and other metals on reproduction and development has
been reviewed (75).

Acute effects after a single parenteral administration of cadmium salts have
already been mentioned. They include acute hemorrhagic necrosis in testes,
hemorrhages and necrosis in nonovulating ovaries, and destruction of the placenta during the last third of pregnancy (179). However, testicular necrosis is not
produced when cadmium is bound to metallothionein (165). This could explain
why testicular changes are not observed in chronic exposure. Metallothionein
binding has also been found to protect against the acute effect of cadmium in the
placenta. Zinc and selenium can also protect against the toxic reproductive
effects of cadmium. Webb and Samarawickrama (258) have demonstrated
that acute exposure to cadmium reduces placental transport of zinc. Thus,
acute exposure to cadmium alters placental viability, structure, and function.
Acute cadmium dosing also blocks embryonic implantation in sexually mature
rats.

Cadmium is taken up in reproductive tissues such as gonads and uterus
(40,145). The animal and human placenta accumulates cadmium, but cadmium
transport into the conceptus is low (148). Chronic treatment with cadmium causes
thickening of the wall of the small blood vessels in the uterus and ovaries of rats
and may also lead to ovarian atrophy (156).

Gastrointestinal uptake of cadmium in the newborn is higher (1) (up to
55%) than in adults and remains high in the suckling period. This is probably related to retention of cadmium by intestinal mucosa, breast, and other
tissues (28,118,149,207). Cadmium levels in mother's milk are low, and exposure of the infant by this route is assumed to be of limited significance
(191).

Damage to reproductive tissues is considered to be the critical effect of cadmium after acute parenteral doses.

Although there are some indications that low-level cadmium exposure can
affect the placental blood vessels, present evidence is insufficient to consider this
as the critical effect of cadmium exposure, especially since it has been observed
only in animals but not in humans.

Cadmium has not been identified as a human teratogen either in clinial case
reports or in epidemiological studies (156). In one report, occupationally exposed women had offspring with lower birth weights than in controls (39).

D. Hypertension

Much evidence exists linking cadmium exposure to hypertension in animals, but the mechanism of chronic cadmium hypertension is still poorly understood. It develops at dose levels insufficient to induce other signs of cadmium toxicity; higher dose levels judged to be toxic on chronic administration have no or opposite effects (116,183,210). Not all investigators were able to demonstrate cadmium-induced hypertension in oral studies in rats and dogs (e.g., 43,183,259). Perry *et al.* (183) discussed some possible causes of this discrepancy. The experimental conditions may have played a predominant role, since it has been observed that selenium, copper, and zinc may counteract the hypertensive action of cadmium.

Some studies have suggested a possible relationship of cadmium exposure with hypertension in humans (130,182,211,212,214), although others have not (109,178).

Two epidemiological studies have been reported linking cadmium in drinking water to blood pressure (13,21). Significant differences were found in mean systolic and diastolic blood pressure between adults drinking water containing cadmium at 3 µg/liter and those drinking water with 1 µg/liter. However, the accuracy of water analytical data has been questioned, and several other elements were also different between the towns (57).

A number of mechanisms have been postulated to explain the effects of cadmium on the cardiovascular system, including interference with catecholamine metabolism, direct action on vascular walls, changes in peripheral compliance or modification of cardiac performance, and involvement of the renin–angiotensin–aldosterone system, possibly triggered by changes in sodium reabsorption.

It will remain unclear whether the effect of cadmium on hypertension observed in animals is relevant to human health until the mechanism of action and interacting factors are better understood.

E. Carcinogenicity

Although carcinogenicity of cadmium is still an unresolved issue, some results obtained in experimental animals indicate evidence for carcinogenicity under some experimental circumstances. Lung carcinomas in rats exposed to cadmium chloride aerosols for 18 months by inhalation (12.4, 25, and 50 μ/m^3) provide sufficient evidence for carcinogenicity of cadmium under these conditions (242). Injection site and testicular tumors in mice and rats given cadmium metal or cadmium salts also indicate carcinogenicity of certain cadmium compounds. Intratracheal instillation of cadmium oxide has produced an increase in mammary tumors and an increase in tumors at multiple sites among male rats (206). Injection of cadmium chloride into the prostate has induced tumors of that tissue

(218). One study suggests that the incidence of pancreatic islet cell tumors may be increased by administration of cadmium chloride by this route (193).

However, no carcinogenic response has been observed with ingested cadmium, and its potency via the oral route is at least 200 times less than that via inhalation in experimental animals (83). Loser (136) reported a 2-year cadmium feeding study in rats, with cadmium doses up to 50 $\mu g/g$ as cadmium chloride. At this level some inhibition of weight gain was noted in males, but mortality was not increased over that observed in control animals, and no suggestion of dose-dependent increases of tumor formation at any site was noted.

Epidemiological studies do not provide sufficient evidence of a risk of prostate cancer from exposure to cadmium. These data have been reviewed (189). However, evidence from the same studies seems to provide better evidence of a lung cancer risk from exposure to cadmium. This applies to the Thun *et al.* (247) study of an excess risk of lung cancer seen in cadmium smelter workers—an updated and enlarged version of the earlier Lemen *et al.* (129) study. Sorahan and Waterhouse (225) also noted an increased risk of lung cancer in their population study. Holden (86) reported an excessive risk of lung cancer in "vicinity" workers that he attributed to the presence of other metals such as arsenic. Andersson *et al.* (3), in their update of the Kjellström *et al.* (108) study, noticed a slight, insignificant lung cancer risk in alkaline battery factory workers.

The interpretation of these epidemiological studies is difficult because the numbers are very small. Yet, one might tentatively conclude on the basis of limited evidence that airborne concentrations of certain cadmium compounds might be carcinogenic in humans.

F. Effect on Bone

Renal dysfunction can cause mineral disturbances that eventually may cause renal stones or osteomalacia. In Japan much attention has been paid to the role of cadmium in inducing bone disease. A special meeting was held in 1979 on cadmium-induced osteopathy (222). Most results indicate that in certain experimental conditions cadmium can induce osteomalacia or osteoporosis in animals (190). Positive findings are all from studies in rats (89,101,158,241). Negative results have been observed from studies in mice, rabbits, and monkeys (99,106,163). One of the reasons for decreased calcium absorption could be the inhibition by cadmium of vitamin D hydroxylation in the renal cortex (54). The formation of the metabolite 1,25-dihydroxycholecalciferol can be almost totally suppressed by high dietary cadmium exposure in rats (135). The concentration of the calcium-binding protein in intestinal mucosa is also decreased by cadmium exposure. In humans the disease is certainly the result of long-term exposure to cadmium and severe deficiency of several essential nutrients, as mentioned earlier.

G. Other Effects

Slight anemia has been found in exposed workers but is not common in industries today (190). Animal studies have shown that it is an iron deficiency anemia (187), and a decreased gastrointestinal absorption of iron due to cadmium might be one of the mechanisms. Hemolytic anemia was also reported by Axelsson and Piscator (9). Hemolysis could mean that metallothionein containing cadmium is released from red cells, thus causing a more rapid increase in renal concentration of cadmium (190).

Liver has an enormous capacity to synthesize metallothionein, which could explain the absence of major changes in liver function in exposed workers (76). However, animal experiments have indicated that morphological and enzymatic changes may occur in the liver (29,231,243).

Cook et al. (36) have shown that cadmium salts increase the susceptibility of rats to infections with Gram-negative bacteria containing endotoxin. A number of studies have since then been performed in order to clarify the mechanism of action of cadmium on the immune system. From these studies it is clear that cadmium has the potency to interfere with the immune system (83,120). Significant immune-suppressive effects have been observed with oral exposure of mice and rabbits to cadmium chloride in drinking water (83).

REFERENCES

1. Alexander, F. W., Clayton, B. E., and Delves, H. T. (1974). Q. J. Med. 43, 89.
2. Andersen, O. (1984). Environ. Health Perspect. 54, 249.
3. Andersson, K., Elinder, C. G., Hogsteadt, C., and Kjellström, T. (1982). "Mortality among Cadmium Workers in a Swedish Battery Factory." Report to the Swedish Work Environment Fund. Stockholm, Sweden.
4. Anke, M., and Schneider, H. J. (1971). Arch. Exp. Veterinaermed. 25, 805.
5. Anke, M., Henning, H. J., Schneider, H., Ludke, H., von Gargen, W., and Schlegel, H. (1970). In "Trace Element Metabolism in Animals-1" (C. F. Mills, ed.), p. 317. Churchill-Livingstone, Edinburgh and London.
6. Anke, M., Henning, A., Schneider, H. J., Groppel, B., Grun, M., Pritschefeld, M., and Ludke, H. (1976). Math. Naturwiss. Unterr. 25, 241.
7. Anwar, R. A., Langham, R. F., Hoppert, C. A., Alfredson, B. V., and Byerrum, R. U. (1961). Arch. Environ. Health 3, 92.
8. Aughy, E., Fell, G. S., Scott, R., and Black, M. (1984). Environ. Health Perspect. 54, 153.
9. Axelsson, B., and Piscator, M. (1966). Arch. Environ. Health 12, 360.
10. Baker, D. E., Amacher, M. C., and Leach, R. M. (1979). Environ. Health Perspect. 28, 45.
11. Bernard, A., Lauwerys, R., and Gengoux, P. (1981). Toxicology 20, 345.
12. Bernard, A., Goret, A., Buchet, J. P., Roels, H., and Lauwerys, R. (1980). J. Toxicol. Environ. Health 6, 175.
13. Bierenbaum, M. L., Fleischmana, M. L., and Dunn, A. L. (1975). Lancet 1, 1008.
14. Bingham, F. T. (1979). Environ. Health Perspect. 28, 39.
15. Bremer, H. (1983). In "Health Evaluation of Heavy Metals in Infant Formula and Junior

Food'' (E. H. F. Schmidt and A. G. Hildebrandt, eds.), p. 175. Springer-Verlag, Berlin and New York.

16. Bremner, I., and Campbell, J. K. (1978). *Environ. Health Perspect.* **25**, 125.
17. Bremner, I. (1979). *In* "The Chemistry, Biochemistry and Biology of Cadmium" (M. Webb, ed.). Elsevier/North-Holland, Amsterdam.
18. Bremner, I., and Mills, C. F. (1981). *Phil Trans. R. Soc. Lond.* **294**, 75.
19. Cain, K., and Webb, M. (1983). *In* "Health Evaluation of Heavy Metals in Infant Formula and Junior Food" (E. H. F. Schmidt and A. G. Hildebrandt, eds.), p. 105. Springer-Verlag, Berlin and New York.
20. Cain, K., and Holt, D. E. (1983). *Chem.-Biol. Interact.* **43**, 223.
21. Calabrese, E. J., and Tuthill, R. W. (1978). *J. Environ. Sci. Health, Part A* **A13**, 781.
22. Caroll, R. E. (1966). *JAMA, J. Am. Med. Assoc.* **198**, 177.
23. Carruthers, M., and Smith, B. (1979). *Lancet* **1**, 845.
24. Casey, C. E. (1977). *N. Z. Med. J.* **85**, 275.
25. Chaney, R. L., Hundemann, P. T., Palmer, W. T., Small, R. J., White, M. C., and Decker, A. M. (1978). *In* "Proceedings of the 1977 National Conference on Compositing of Municipal Residues and Sludges," p. 86. Information Transfer, Inc., Rockville, Maryland.
26. Chaney, R. L., Stoewsand, G. S., Bache, C. A., and Lisk, D. J. (1978). *J. Agric. Food Chem.* **26**, 992.
27. Chang, C. C., Vandermallie, R. J., and Garvey, J. S. (1980). *Toxicol. Appl. Pharmacol.* **55**, 94.
28. Chang, L. W., Wade, P., Pounds, J., and Ruehl, K. (1980). *Adv. Pharmacol. Chemother.* **17**, 195.
29. Chapatwala, K. D., Hobsone, M., Desaiah, D., and Rajanna, B. (1982). *Toxicol. Lett.* **12**, 27.
30. Chaube, S., Nishimura, H., and Swinyard, C. A. (1973). *Arch. Environ. Health* **26**, 237.
31. Chen, K. W., Whanger, P. D., and Weswig, P. H. (1975). *Bioinorg. Chem.* **4**, 125.
32. Cherian, M. G., Goyer, R. A., and Delaquerriere-Richardson, L. (1976). *Toxicol. Appl. Pharmacol.* **38**, 399.
33. Cherian, M. G., and Goyer, R. A. (1978). *Life Sci.* **10**, 1.
34. Cherian, M. G. (1984). *Environ. Health Perspect.* **54**, 243.
35. Cikrt, M., and Tichy, M. (1974). *Br. J. Ind. Med.* **31**, 134.
36. Cook, J. A., Hoffman, E. D., and DiLuzio, N. R. (1975). *Proc. Soc. Exp. Biol. Med.* **150**, 741.
37. Council for Agricultural Science and Technology (1976). "Application of Sewage Sludge to Cropland: Appraisal of Potential Hazards of Heavy Metals to Plants and Animals," EPA No. MCD-33, p. 29. USEPA, Washington, D.C.
38. Cousins, R. J., Barber, A. K., and Trout, J. R. (1973). *J. Nutr.* **103**, 964.
39. Cvetkova, R. P. (1970). *Gig. Tr. Prof. Zabol.* **14**, 31.
40. Dencker, L. (1975). *J. Reprod. Fertil.* **44**, 461.
41. Doyle, J. J., Pfander, W. H., Grebing, S. E., and Pierce, J. O. (1974). *J. Nutr.* **104**, 160.
42. Doyle, J. J., and Pfander, W. H. (1975). *J. Nutr.* **105**, 599.
43. Eakin, D. J., Schroeder, L. A., Whanger, P. D., and Weswis, P. H. (1980). *Am. J. Physiol.* **238**, E53.
44. Elinder, C. G., Kjellström, T., Lind, B., and Linman, R. (1976). *Arch. Environ. Health* **31**, 292.
45. Elinder, C. G., Kjellström, T., Linnman, L., and Pershagen, G. (1978). *Environ. Res.* **15**, 473.
46. Elinder, C. G., and Pannone, M. (1979). *Environ. Health Perspect.* **28**, 123.
47. Elinder, C. G., Nordberg, M., Palm, B., and Piscator, M. (1981). *Environ. Res.* **26**, 22.

48. Ellis, K. J., Vartsky, D., Zanzi, I., Cohn, S. H., and Yasumura, S. (1979). *Science* **205,** 323.
49. Ellis, K. J., Morgan, W. D., Zanzi, I., Yasumura, S., Vartsky, D. D., and Cohn, S. H. (1981). *J. Toxicol. Environ. Health* **7,** 691.
50. Engström, B., and Nordberg, G. F. (1978). *Toxicology* **9,** 195.
51. Engström, B., and Nordberg, G. F. (1979). *Toxicology,* 13, 215.
52. Engström, B. (1984). *Environ. Health Perspect.* **54,** 219.
53. Fairchild, E. J., Lewis, R. J., and Tatkin, R. L. (1977). "Registry of Toxic Effects of Chemical Substances," DHEW Publ. No. (NIOSH) 78-104-B, Vol. 2, p. 526. VSDHEW, Washington, D.C.
54. Feldman, S. L., and Cousins, R. J. (1973). *Nutr. Rep. Int.* **8,** 251.
55. Flanagan, P. R., McLellan, J. S., Haist, J., Cherian, M. F., Chamveralin, M. J., and Valberg, L. S. (1978). *Gastroenterology* **74,** 841.
56. Fleischer, M., Sarofim, A. F., Fassett, D. W., Hammond, P., Schacklette, Nisbet, I. C. T., and Epstein, S. (1974). *EHP, Environ. Health Perspect., Exp. Issue* No. 253.
57. Folsom, A. R., and Prineas, R. J. (1982). *Am. J. Epidemiol.* **115,** 818.
58. Foulkes, E. C. (1982). *In* "Biological Roles of Metallothionein" (E. C. Foulkes, ed.), p. 131. Elsevier/North-Holland, New York.
59. Fox, M. R. S., Fry, B. E., Harland, B. F., Schertel, M. E., and Weeks, C. E. (1971). *J. Nutr.* **101,** 1295.
60. Fox, M. R. S. (1976). *In* "Trace Elements in Human Health and Disease" (A. S. Prasad and D. Oberleas, eds.), Vol. 2, p. 401. Academic Press, New York.
61. Fox, M. R. S., Jacobs, R. M., Jones, A. O. L., Fry, B. E., and Hamilton, R. P., Jr. (1978). *In* "Trace Element Metabolism in Man and Animals-3" (M. Kirchgessner, ed), p. 327. Tech. Univ. Munich, F.R.G., West Germany.
62. Fox, M. R. S. (1979). *Environ. Health Perspect.* **29,** 95.
63. Fox, M. R. S., Jacobs, R. M., Jones, A. O. L., and Fry, B. E., Jr. (1979). *Environ. Health Perspect.* **28,** 107.
64. Fox, M. R. S., Jacobs, R. M., Jones, A. O. L., Fry, B. E., Jr., and Stone, C. L. (1980). *Ann. N.Y. Acad. Sci.* **355,** 249.
65. Fox, M. R. S. (1982). *In* "Clinical, Biochemical, and Nutritional Aspects of Trace Elements" (A. S. Prasad, ed.), *Fed. Am. Soc. Exp. Biol.* p. 537. Alan R. Liss, Inc., New York.
66. Fox, M. R. S. (1983). *Fed. Proc., Fed. Am. Soc. Exp. Biol.* **42,** 1726.
67. Freeland, J. H., and Cousins, R. J. (1973). *Nutr. Rep. Int.* **8,** 337.
68. Friberg, L. (1948). *J. Ind. Hyg. Toxicol.* **30,** 32.
69. Friberg, L. (1950). *Acta Med. Scand., Suppl.* **138,** 240.
70. Friberg, L., Piscator, M., and Nordberg, G. F. (1971). "Cadmium in the Environment" CRC Press, Cleveland, Ohio.
71. Friberg, L., Piscator, M., Nordberg, G. F., and Kjellström, T. (1974). "Cadmium in the Environment," 2nd ed. Chem. Rubber Co. Press, Cleveland, Ohio.
72. Friberg, L., Kjellström, T. Nordberg, G. F., and Piscator, M. (1975). "Cadmium in the Environment III: A Toxicological and Epidemiological Appraisal." Office of Research and Development, USEPA, Washington, D.C.
73. Friberg, L. (1984). *Environ. Health Perspect.* **54,** 1.
74. Furr, A. K., Stoewsand, G. J., Bache, C. A., and Lisk, D. J. (1976). *Arch. Environ. Health* **31,** 87.
75. Furst, A., Mehlman, M., and Vostal, J. (1983). *In* "Reproductive and Developmental Toxicity of Metals" (T. W. Clarkson, G. F. Nordberg, and P. R. Sagar, eds.), p. 3. Plenum, New York.
76. Gabbiani, G., Badonnel, M. C., Mathewson, S. N., and Ryan, G. B. (1974). *Lab. Invest.* **30,** 686.

77. Gontzea, I., and Popescu, F. (1978). *J. Ind. Med.* **35,** 154.
78. Gross, S. B., Yeager, D. W., and Middendof, M. S. (1976). *J. Toxicol. Environ. Health* **2,** 153.
79. Hamilton, D. L., and Valberg, L. S. (1974). *Am. J. Physiol.* **227,** 1033.
80. Hamilton, E. I., and Minski, M. J. (1972–1973). *Sci. Total Environ.* **1,** 375.
81. Hansen, L. G., and Hinesly, T. D. (1979). *Environ. Health Perspect.* **28,** 51.
82. Harvey, T. C., Chettle, D. R., Fremlin, J. H., AlHaddad, I. K., and Downey, S. P. M. J. (1979). *Lancet* **1,** 551.
83. "Health Assessment Document for Cadmium," EPA 600/8-81-023. USEPA, Research Triangle Park, North Carolina.
84. Heffron, C. L., Reid, J. T., Elfving, D. C., Stoewsand, G. S., Hascher, W. M., Telford, J. N., Furr, A. K., Parkinson, T. F., Bache, C. A., Wszolek, P. C., and Lisk, D. J. (1980). *J. Agric. Food Chem.* **28,** 58.
85. Henke, G., Sachs, H. W., and Bohn, G. (1970). *Arch. Toxicol.* **26,** 8.
86. Holden, H. (1980). *Lancet* **1,** 1137.
87. Hunt, C. T., Boulanger, Y., Fesik, S. W., and Armitage, I. M. (1984). *Environ. Health Perspect.* **54,** 135.
88. Hutton, M., ed. (1982). "Cadmium in the European Community, a Prospective Assessment of Sources, Human Exposure and Environmental Impact," MARC Rep. No. 26. Chelsea College, University of London.
89. Itokawa, Y., Nishino, K., Takashima, M., Nakata, T., Kaito, H., Okamoto, E., Daijo, K., and Kawamura, J. (1978). *Environ. Res.* **15,** 206.
90. Jacobs, R. M., Fox, M. R. S., and Aldridge, M. H. (1969). *J. Nutr.* **99,** 119.
91. Jacobs, R. M., Jones, A. O. L., Fox, M. R. S., and Fry, B. E. (1978). *J. Nutr.* **108,** 22.
92. Jacobs, R. M., Jones, A. O. L., Fry, B. E., and Fox, M. R. S. (1978). *J. Nutr.* **108,** 901.
93. Jacobson, K. B., and Turner, J. E. (1980). *Toxicology* **16,** 1.
94. Johnson, D. E., Prevost, R. J., Tillery, J. B., and Thomas, R. E. (1977). Final Rep. Contract No. 68-02-1725. Prepared by Southwest Research Institute, San Antonio, Texas, for USEPA, Washington, D.C.
95. Johnson, D. R., and Foulkes, E. C. (1980). *Environ. Res.* **21,** 360.
96. Kaferstein, F. K.,and Muller, J. (1981). *ZEBS-Ber.* No. 1.
97. Kagi, J. H. R., and Vallee, B. L. (1960). *J. Biol. Chem.* **235,** 3460.
98. Kagi, J. H. R., and Nordberg, M. (1979). "Metalloghionein." Birkhaueser, Boston, Massachusetts.
99. Kajikawa, K., Nakanishi, I., and Kuroda, K. (1981). *Exp. Mol. Pathol.* **34,** 9.
100. Kanisawa, M., and Schroeder, H. A. (1969). *Cancer Res.* **29,** 892.
101. Kawamura, J., Yoshida, O., Nishino, K., and Itokawa, Y. (1978). *Nephron* **20,** 101.
102. Kazantis, G., Flynn, F. V., Spowage, J. S., and Trott, D. G. (1963). *Q. J. Med.* **32,** 165.
103. Kazantis, G., and Armstrong, B. G. (1984). *Environ. Health Perspect.* **54,** 193.
104. Kello, D., and Kostial, K. (1977). *Toxicol. Appl. Pharmacol.* **40,** 277.
105. Kello, D., Dekanic, D., and Kostial, K. (1979). *Arch. Environ. Health* **34,** 30.
106. Kitamura, S. (1982). *Proc. Int. Cadmium Conf., 3rd, 1981* p. 171.
107. Kjellström, T., and Nordberg, G. F. (1978). *Environ. Res.* **16,** 248.
108. Kjellström, T., Friberg, L., and Rahnster, D. (1979). *Environ. Health Perspect.* **28,** 199.
109. Kjellström, T. (1979). *Environ. Health Perspect.* **28,** 169.
110. Kjellström, T., Siroishi, K., and Ervin, P. E. (1977). *Environ. Res.* **13,** 318.
111. Klaassen, C. D., and Kotsonis, F. N. (1977). *Toxicol. Appl. Pharmacol.* **41,** 101.
112. Klaassen, C. D., Waalkes, M. P., and Cantilena, L. R. (1984). *Environ. Health Perspect.* **54,** 233.

113. Kobayashi, J. (1974). *In* "Trace Substances Environmental Health-7" (D. D. Hempshill, ed.), Univ. of Missouri Press, Columbia.

114. Kojima, S., Haga, Y., Kurihara, T., Yamawaki, T., and Kjellström, T. (1977). *Environ. Res.* **14**, 436.

115. Kojima, Y., and Kagi, J. H. R. (1978). *Trends Biochem. Sci.* **3**, 90.

116. Kopp, S. J., Gloner, T., Erlanger, M., Perry, E. F., Perry, H. M., and Barany, M. (1980). *J. Environ. Pathol. Toxicol.* **4**, 205.

117. Kostial, K., Rabar, I., Blanusa, M., and Landeka, M. (1979). *Proc. Nutr. Soc.* **38**, 251.

118. Kostial, K. (1983). *In* "Reproductive and Developmental Toxicity of Metals" (T. W. Clarkson, G. F. Nordberg, and P. R. Sager, eds.), p. 727. Plenum, New York.

119. Kowal, D. E., Johnson, D. E., Kraemer, D. F., and Pahren, H. R. (1979). *J. Toxicol. Environ. Health* **5**, 995.

120. Kranjc, E. I., Vos, J. G., and van Logtent, M. J. (1983). *In* "Health Evaluation of Heavy Metals in Infant Formula and Junior Food" (E. H. F. Schmidt and A. G. Hildebrandt, eds.), p. 112. Springer-Verlag, Berlin and New York.

121. Langmyhr, F. J., Eyde, B., and Jonsen, J. (1979). *Anal. Chim. Acta* **107**, 211.

122. Larsson, B., Slorach, S. A., Hagman, U., and Hofvander, Y. (1981). *Acta Paediatr. Scand.* **70**, 281.

123. Lauwerys, R. R., Buchet, J. P., Roels, H. A., Brouwers, J., and Stanescu, D. (1974). *Arch. Environ. Health* **28**, 145.

124. Lauwerys, R. R., Buchet, J. P., Roels, H. A., and Hubermont, G. (1978). *Environ. Res.* **15**, 278.

125. Lauwerys, R. R., Roels, H. A., Buchet, J. P., Bernard, A., and Stanescu, D. (1979). *Environ. Health Perspect.* **28**, 137.

126. Lauwerys, R. R., Roels, H. A., Bernard, A., and Buchet, J. P. (1980). *Int. Arch. Occup. Environ. Health* **45**, 271.

127. Lauwerys, R. R., Bernard, A., Roels, H. A., Buchet, J. P., and Vian, C. (1984). *Environ. Health Perspect.* **54**, 147.

128. Leach, R. M., Jr., Wang, K. W. L., and Baker, D. E. (1979). *J. Nutr.* **109**, 437.

129. Lemen, R. A., Lee, J. S., Wagoner, J. K., and Blejer, H. P. (1976). *Ann. N.Y. Acad. Sci.* **271**, 273.

130. Lener, J., and Birbr, B. (1971). *Lancet* **1**, 970.

131. Lerch, K. (1980). *Nature (London)* **284**, 368.

132. Lewis, G. P., Coughlin, L. L., Jusko, W. J., and Hartz, S. (1972). *Lancet* **2**, 291.

133. Lewis, G. P., Jusko, W. J., Coughlin, L. L., and Hartz, S. (1972). *J. Chronic Dis.* **25**, 717.

134. Livingstone, H. D. (1972). *Clin. Chem. (Winston-Salem, N.C.)* **18**, 67.

135. Lorentzon, R., and Larsson, S. E. (1977). *Clin. Sci. Mol. Med.* **53**, 439.

136. Loser, E. (1980). *Cancer Lett.* **9**, 191.

137. Lucis, O. J., Lucis, R., and Shaikh, Z. A. (1972). *Arch. Environ. Health* **25**, 14.

138. Lynch, G. P., Smith, D. F., Fisher, M., Pike, T. L., and Weinland, B. T. (1976). *J. Anim. Sci.* **42**, 410.

139. Mahaffey, K. R., Corneliussen, P. E., Jelinek, P. E., and Fiorino, J. A. (1975). *Environ. Health Perspect.* **12**, 63.

140. Mahaffey, K. R. (1983). *In* "Reproductive and Developmental Toxicity of Metals" (T. W. Clarkson, G. F. Nordberg, and P. R. Sager, eds.), p. 777. Plenum, New York.

141. Margoshes, M., and Vallee, B. L. (1957). *J. Am. Chem. Soc.* **79**, 4813.

142. Masironi, R., Koirtyohann, S. R., and Pierce, J. O. (1977). *Sci. Total Environ.* **7**, 27.

143. Mason, K. E., Richardson, M. E., and Fox, M. R. S. (1977). *Fed. Proc., Fed. Am. Soc. Exp. Biol.* **36**, 1152.

144. Matsusaka, N., Tanaka, M., Nishimura, Y., Yuyama. A., and Kobayashi, H. (1972). *Med. Biol.* **85**, 275.

145. Mattison, D. R., Gates, A. H., Leonard, A., Wide, M., Hemminki, K., and Copius-Peereboom-Stegeman, J. H. J. (1983). *In* ''Reproductive and Developmental Toxicity of Metals'' (T. W. Clarkson, G. F. Nordberg, and P. R. Sager, eds.), p. 41. Plenum, New York.

146. McCabe, L. J., Symons, J. M., Lee, R. D., and Robeck, G. G. (1970). *J. Am. Water Works Assoc.* **62**, 670.

147. McLellan, J. S., Flanagan, P. R., Chamberlain, M. J., and Valberg, L. S. (1978). *J. Toxicol. Environ. Health.* **4**, 131.

148. Miller, R. K., and Shaikh, Z. A. (1983). *In* ''Reproductive and Developmental Toxicity of Metals'' (T. W. Clarkson, G. F. Nordberg, and P. R. Sager, eds.), p. 151. Plenum, New York.

149. Miller, W. J., Lampp, G. W., Powell, G. W., Salotti, C. S., and Blackmon, D. M. (1967). *J. Dairy Sci.* **50**, 1404.

150. Miller, W. J., Blackmon, D. M., and Martin, Y. G. (1968). *J. Dairy Sci.* **51**, 1836.

151. Miller, W. J., Blackmon, D. M., Gentry, R. P., and Pate, F. M. (1969). *J. Dairy Sci.* **52**, 2029.

152. Mills, C. F., and Delgarno, A. C. (1972). *Nature (London)* **239**, 171.

153. Molin, L., and Wester, P. O. (1976). *Scand. J. Clin. Lab. Invest.* **36**, 679.

154. Moore, W., Jr., Stara, J. F., and Crocker, W. C. (1973a). *Environ. Res.* **6**, 159.

155. Moore, W., Jr., Stara, J. F., Crocker, W. C., Malachuk, M., and Iltis, R. (1973b). *Environ. Res.* **6**, 473.

156. Mottet, N. K., and Ferm, V. H. (1983). *In* ''Reproductive and Developmental Toxicity of Metals'' (T. W. Clarkson, G. F. Nordberg, and P. R. Sager, eds.), p. 93. Plenum, New York.

157. Nogawa, K., Ishizaki, A., and Kawano, S. (1978). *Environ. Res.* **15**, 185.

158. Nogawa, K., Kobayashi, E., and Konishi, F. (1981). *Environ. Res.* **24**, 233.

159. Nomiyama, K. (1974). *Kankyo Hoken Repoto* **31**, 53.

160. Nomiyama, K. (1975). *In* ''Progress in Water Technology'' (P. A. Krenkel, ed.), p. 15. Pergamon, Oxford.

161. Nomiyama, K., and Foulkes, E. C. (1977). *Proc. Soc. Exp. Biol. Med.* **156**, 97.

162. Nomiyama, K., Nomiyama, H. H., Nomura, Y., Taguchi, T., Matsui, K., Yotoriyama, M., Akahori, F., Iwao, S., Koizumi, N., Masaoka, T., Kitamura, S., Tsuchiya, K., Suzuki, T., and Kobayashi, K. (1979). *Environ. Health Perspect.* **28**, 223.

163. Nomiyama, K. (1980). *Sci. Total Environ.* **14**, 199.

164. Homiyama, K., and Nomiyama, H. H. (1982). *In* ''Biological Roles of Metallothionein'' (E. C. Foulkes, ed.), p. 47. Elsevier/North-Holland, New York.

165. Nordberg, G. F., Piscator, M., and Lind, B. (1971). *Acta Pharmacol. Toxicol.* **29**, 456.

166. Nordberg, G. F. (1972). *Environ. Physiol. Biochem.* **2**, 7.

167. Nordberg, G. F., and Nishiyama, K. (1972). *Arch. Environ. Health* **24**, 209.

168. Nordberg, G. F., and Piscator, M. (1972). *Environ. Physiol. Biochem.* **2**, 37.

169. Nordberg, G. F., Slorach, S., and Stenstrom, T. (1973). *Lakartid* **70**, 601.

170. Nordberg, G. F., Goyer, R. A., and Nordberg, M. (1975). *Arch. Pathol.* **99**, 192.

171. Nordberg, G. F., ed. (1976). ''Effects and Dose-Response Relationships of Heavy Metals.'' Elsevier/North-Holland, Amsterdam.

172. Nordberg, G. F., and Kojima, Y. (1979). *In* ''Metallothionein'' (J. H. R. Kagi and M. Nordberg, eds.), p. 41. Birkhaeuser, Basel.

173. Nordberg, G. F. (1984). *Environ. Health Perspect.* **54**, 213.

174. Nordberg, M., and Nordberg, G. F. (1975). *Environ. Health Perspect.* **12**, 103.

175. Nordberg, M. (1978). *Environ. Res.* **15**, 381.

176. Nordberg, M. (1984). *Environ. Health Perspect.* **54,** 13.
177. Ohanian, E. V., Iwai, J., Leitl, G., and Tuthill, R. (1978). *Am. J. Physiol.* **235,** H385.
178. Ostergaard, K. (1977). *Lancet* **1,** 677.
179. Parizek, J. (1983). *In* "Reproductive and Developmental Toxicity of Metals" (T. W. Clarkson, G. F. Nordberg, and P. R. Sager, eds.), p. 301. Plenum, New York.
180. Perry, H. M., Jr., Tipton, I. H., Schroeder, H. A., Steiner, R. L., and Cook, M. J. (1961). *J. Chronic Dis.* **14,** 259.
181. Perry, H. M., Jr., Erlanger, M., and Perry, E. F. (1977). *Am. J. Physiol.* **232,** H114.
182. Perry, H. M., Jr., and Perry, E. F. (1974). *Prev. Med.* **3,** 344.
183. Perry, H. M., Jr., Perry, E. F., and Erlanger, M. W. (1980). *J. Environ. Pathol. Toxicol.* **4,** 195.
184. Petering, D. H., Loffsgaarden, J., Schneider, J., and Fowler, B. (1984). *Environ. Health Perspect.* **54,** 73.
185. Petering, H. G., Choudhury, H., and Semmer, K. L. (1979). *Environ. Health Perspect.* **28,** 97.
186. Piscator, M. (1966). *Arch. Environ. Health* **12,** 335.
187. Piscator, M. (1974). *In* "Cadmium in the Environment" (L. Friberg, M. Piscator, G. F. Nordberg, and T. Kjellström, eds.), 2nd ed., p. 71. CRC Press, Cleveland, Ohio.
188. Piscator, M. (1979). "Management and Control of Heavy Metals in the Environment." CEP Consultants, Edinburgh, United Kingdom.
189. Piscator, M. (1981). *Environ. Health Perspect.* **40,** 107.
190. Piscator, M. (1982). *In* "Clinical, Biochemical and Nutritional Aspects of Trace Elements" (A. S. Prasad, ed.), p. 521. Alan R. Liss, Inc., New York.
191. Piscator, M. (1983). *In* "Health Evaluation of Heavy Metals in Infant Formula and Junior Food" (E. H. F. Schmidt and A. G. Hildebrandt, eds.), p. 120. Springer-Verlag, Berlin and New York.
192. Piscator, M. (1984). *Environ. Health Perspect.* **54,** 175.
193. Poirier, L. A., Kasprzak, K. S., Hoover, K. L., and Wenk, M. L. (1983). *Cancer Res.* **43,** 4575.
194. Powell, G. W., Miller, W. J., and Clifton, C. M. (1964). *J. Dairy Sci.* **47,** 1017.
195. Pritzl, M. C., Lie, Y. H., Kienholz, E. W., and Whiteman, C. E. (1974). *Poult. Sci.* **53,** 2026.
196. Rabar, I., and Kostial, K. (1981). *Arch. Toxicol.* **47,** 63.
197. Ragan, H. A. (1983). *Sci. Total Environ.* **28,** 317.
198. Rahola, T., Aaran, R. K., and Miettinen, J. K. (1972). *In* "Assessment of Radioactive Contamination in Man," p. 553. IAEA, Vienna.
199. Richardson, M. E., and Fox, M. R. S. (1974). *Lab. Invest.* **31,** 722.
200. Richardson, M. E., Fox, M. R. S., and Fry, B. E., Jr. (1974). *J. Nutr.* **104,** 323.
201. Roels, H. A., and Lauwerys, R. R. (1983). *In* "Health Evaluation of Heavy Metals in Infant Formula and Junior Food" (E. H. F. Schmidt and A. G. Hildebrandt, eds.), p. 126. Springer-Verlag, Berlin and New York.
202. Roels, H. A., Lauwerys, R. R., Buchet, J. P., and Bernard, A. (1981). *Environ. Res.* **24,** 117.
203. Roels, H. A., Lauwerys, R. R., Buchet, J. P., Bernard, A., Chettle, D. R., Harvey, T. C., and AlHaddad, J. K. (1981). *Environ. Res.* **26,** 271.
204. Ryan, J. A., Pahren, H. R., and Lucas, J. B. (1982). *Environ. Res.* **28,** 251.
205. Sahagian, B. M., Harding-Barlow, I., and Perry, H. M., Jr. (1966). *J. Nutr.* **90,** 259; **93,** 291 (1967).
206. Sanders, C. L., and Mahaffey, J. A. (1984). *Environ. Res.* **33,** 227.
207. Sandstead, H. H., Doherty, R. A., and Mahaffey, K. A. (1983). *In* "Reproductive and

Development Toxicity of Metals'' (T. W. Clarkson, G. F. Nordberg, and P. R. Sager, eds.), p. 205. Plenum, New York.

208. Sasser, L. B., and Jarboe, G. E. (1977). *Toxicol. Appl. Pharmacol.* **41**, 423.
209. Schroeder, H. A.,and Balassa, J. J. (1961). *J. Chronic Dis.* **14**, 236.
210. Schroeder, H. A., and Vinton, W. H., Jr. (1962). *Am. J. Physiol.* **202**, 515.
211. Schroeder, H. A. (1965). *J. Chronic Dis.* **18**, 647.
212. Schroeder, H. A., Balassa, J. J., and Vinton, W. H., Jr. (1965). *J. Nutr.* **86**, 51.
213. Schroeder, H. A., and Mitchener, M. (1971). *Arch. Environ. Health* **23**, 102.
214. Schroeder, H. A. (1974). *Med. Clin. North Am.* **58**, 381.
215. Schulte-Löbbert, F. J., and Bohn, G. (1977). *Arch. Toxicol.* **37**, 155.
216. Schulte-Löbbert, F. J., Bohn, G., and Acker, L. (1978). *Beitr. Ber. Med.* **36**, 491.
217. Schwarz, K., and Spallholz, J. E. (1978). *Ed. Proc.—Int. Cadmium Conf., 1st, 1977*, p. 105.
218. Scott, R., and Aughey, E. (1979). *Br. J. Urol.* **50**, 25.
219. Shaikh, Z. A., and Smith, J. C. (1980). *In* ''Mechanisms of Toxicity and Hazard Evaluation'' (B. Holmstedt, R. Lauwerys, M. Mercer, and M. Roberfroid, eds.), p. 569. Elsevier, Amsterdam.
220. Shaikh, Z. A. (1982). *In* ''Biological Roles of Metallothionein'' (E. C. Foulkes, ed.), p. 69. Elsevier/North-Holland, New York.
221. Sharma, R. P., Street, J. C., Verma, M. P., and Shupe, J. L. (1979). *Environ. Health Perspect.* **28**, 59.
222. Shigematsu, I., and Nomiyama, U. (eds.) (1979). ''Proceedings from the Conference on Cadmium Induced Osteopathy.'' Japan Public Health Association, Tokyo.
223. Singerman, A. (1976). *In* ''Effects and Dose-Response Relationships of Toxic Metals'' (G. F. Nordberg, ed.), p. 207. Elsevier/North-Holland, Amsterdam.
224. Smith, T. J., Temple, A. R., and Reading, J. C. (1976). *Clin. Toxicol.* **9**, 75.
225. Sorahan, T., and Waterhouse, J. A. H. (1983). *Br. J. Ind. Med.* **40**, 293.
226. Squibb, K. S., Cousins, R. J., and Feldman, S. L. (1977). *Biochem. J.* **164**, 223.
227. Squibb, K. S., Ridlington, J. W., Charmichael, and Fowler, B. A. (1979). *Environ. Health Perspect.* **28**, 287.
228. Squibb, K. S., Pritchard, J. B., and Fowler, B. A. (1982). *In* ''Biological Roles of Metallothionein'' (E. C. Foulkes, ed.), p. 181. Elsevier/North-Holland, New York.
229. Squibb, K. S., and Fowler, B. A. (1984). *Environ. Health Perspect.* **54**, 31.
230. Stanescu, D., Veriter, C., Frans, A., Goncette, L. Roels, H. A., Lauwerys, R. R., and Brasseur, L. (1977). *Scand. J. Respir. Dis.* **58**, 289.
231. Stowe, H. D., Wilson, M., and Goyer, R. A. (1972). *Arch. Pathol.* **94**, 389.
232. Stubbs, R. L. (1982). *Proc. Int. Cadmium Conf. 3rd, 1981*, p. 3. Cadmium Association, London, England.
233. Subcommittee on Mineral Toxicity in Animals (1980). ''Mineral Tolerance of Domestic Animals,'' p. 93. Natl. Acad. Sci., Washington, D.C.
234. Sumino, K., Hayakawa, K., Shibata, T., and Kitamura, S. (1975). *Arch. Environ. Health* **30**, 487.
235. Sunda, W. G., Engel, D. W., and Thuotte, R. M. (1978). *Environ. Sci. Technol.* **12**, 409.
236. Suzuki, S., Taguchi, T., and Yokohashi, G. (1969). *Ind. Health* **7**, 155.
237. Suzuki, S., Djuangshi, N., Hyodo, K., and Soemarwoto, O. (1980). *Arch. Environ. Contam. Toxicol.* **9**, 437.
238. Suzuki, T. (1974). *Proc. Jpn. Assoc. Ind. Hyg.*, p. 125.
239. Suzuki, Y. (1980). *J. Toxicol. Environ. Health* **6**, 469.
240. Szadkowski, D., Schultze, H., Schaller, K. H., and Lehnert, G. (1969). *Arch. Hyg. Bakteriol.* **153**, 1.
241. Takashima, M., Moriwaki, S., and Hokawa, Y. (1980). *Toxicol. Appl. Pharmacol.* **54**, 223.

242. Takenaka, S., Oldiges, H., Konig, H., Hochrainer, D., and Oberdoerster, G. (1983). *JNCI, J. Natl. Cancer Inst.* **70**, 367.

243. Tarasenko, N., Yu, Vorobjeva, R. S., Sabalina, L. P., and Cvetkova, R. P. (1975). *Gig. Sanit.* **9**, 22.

244. Task Force (1980). Report No. 83. "Effect of Sewage Sludge on the Cadmium and Zinc Content of Crops." Council for Agricultural Science and Technology, Ames, Iowa.

245. Task Group on Metal Accumulation (1973). *Environ. Physiol. Biochem.* **3**, 65.

246. Teworte, W. (1974). *VDI-Ber.* **203**, 5.

247. Thun, M. J., Schnorr, T. M., Smith, A. B., and Halpern, W. E. (1984). "Mortality Among a Cohort of U.S. Cadmium Production Workers: An Update." Natl. Inst. Occup. Safety & Health, Washington, D.C. (unpublished).

248. Tipton, I. H., and Stewart, P. L. (1970). *Annu. Prog. Rep.—Oak Ridge Natl. Lab., Health Phys. Div.* **ORNL-4446.**

249. Tsuchiya, J., and Iwao, S. (1978). *Environ. Health Perspect.* **25**, 119.

250. Tsuchiya, K., Seki, Y., and Sugita, M. (1976). *Keio J. Med.* **25**, 83.

251. Underwood, E. J. (1977). "Trace Elements in Human and Animal Nutrition," 4th ed. Academic Press, New York.

252. Vahter, M. (ed.) (1982). "Assessment of Human Exposure to Lead and Cadmium through Biological Monitoring." Report prepared for the United Nations Environment Programme and World Health Organization by National Swedish Institute of Environmental Hygiene, Karolinska Institute, Sweden.

253. Vallee, B. L. (1979). *In* "Metallothionein" (J. H. R. Kagi and M. Nordberg, eds.), p. 19. Birkhaeuser, Basel.

254. Van Campen, D. R. (1966). *J. Nutr.* **88**, 125.

255. Vuori, E., Huunan-Seppala, A., Kilpio, J. O., and Salmela, S. S. (1979). *Scand. J. Work Environ. Health* **5**, 16.

256. Washko, P. W., and Cousins, R. J. (1976). *J. Toxicol. Environ. Health* **1**, 1055.

257. Webb, M., and Etienne, A. T. (1977). *Biochem. Pharmacol.* **26**, 25.

258. Webb, M., and Samarawickrama, G. P. (1983). *J. Appl. Toxicol.* **1**, 270.

259. Whanger, P. D. (1979). *Environ. Health Perspect.* **28**, 115.

260. Winge, D. R., Premakumar, R., and Rajagopalan, K. V. (1975). *Arch. Biochem. Biophys.* **170**, 242.

261. Worker, N. A., and Migicovsky, B. B. (1961). *J. Nutr.* **75**, 222.

262. World Health Organization Task Group (1979). "Environmental Health Criteria for Cadmium." World Health Organization Report EHE/EHC/79.20. WHO, Geneva, Switzerland.

263. Wright, F. C., Palmer, J. S., Riner, J. C., Hanfler, M., Miller, J. A., and McBeth, C. A. (1977). *J. Agric. Food Chem.* **25**, 293.

264. Yost, K. J. (1979). *Environ. Health Perspect.* **28**, 5.

265. Zielhuis, R. L., Stuik, E. J., Herber, R. F. M., Salle, H. J. A., Verberk, M. M., Posma, F. D., and Jager, J. H. (1977). *Int. Arch. Environ. Health* **39**, 53.

6

Arsenic

MANFRED ANKE

Karl-Marx-Universität Leipzig
Sektion Tierproduktion und Veterinärmedizin
G.D.R. 6900 Tierernährungschemie
Jena, German Democratic Republic

I. ARSENIC IN ANIMAL TISSUES AND FLUIDS

The arsenic (As) content of tissues and fluids is significantly influenced by arsenic intake, animal species, organ, and possibly age as well. Extreme differences in the arsenic exposure via air, water, soil, animal feeds, and human foods account for the wide range of reported concentrations. In a study of adult human tissues, Smith (243) reported mean arsenic concentrations of most tissues to be between 0.04 and 0.09 μg/g on a dry-weight basis. The variability was extremely high, and the skin, nails, and hair were consistently richer in arsenic than other tissues, with 0.12, 0.36, and 0.65 μg/g, respectively. Hamilton *et al.* (113) obtained much higher mean concentrations for all tissues examined other than muscles. Levander *et al.* (162) summarized the arsenic content of human organs of normal and arsenic-exposed persons. The lungs of normal versus arsenic-exposed persons contained 0.08–0.17 and 2.3–2.6 μg/g fresh weight, respectively (99,107,141). In the liver were found 0.09–0.30 and 4.4–6.9 μg/g fresh weight for these groups (38,99,107,141). The arsenic content of human hair and nails has commanded considerable interest because of its value in the diagnosis of arsenic poisoning. Normal hair always contains arsenic in small amounts that are greatly increased by excessive intakes of the element in certain forms. The levels remain high after cessation of intake and rapidly return to normal (67,243). This statement is equally true for the hair of mammalian animals and the feathers of birds. Hens stored 3.5% of the applied arsenic-76

(^{76}As) in feathers 12 hr after the oral dose. This percentage decreased to 0.4% by 24 hr and at the end of the experiment was at 0.2%. The arsenic incorporation in the hair of goats was with the maximum ^{76}As incorporation after 48 hr (15,120). The median arsenic content in 1000 samples of human hair was 0.51 μg/g, and median concentrations for males and females 0.62 and 0.37 μg/g, respectively. Schroeder and Balassa (229) also found more arsenic in the hair of males than in that of females. Values >3 μg/g indicate possible poisoning (242). Gabor et al. (97) found 10–31 μg/g arsenic in the hair of workers from plants with different arsenic exposure. Hair levels correlated with arsenic in air and in the urine of examined persons. Workers in a copper-processing plant in Czechoslovakia showed a mean arsenic content of 178 μg/g in the hair when they were exposed to air containing arsenic trioxide at 1.0 to 5.1 mg/m^3 but only 57 μg/g when they were exposed to air containing 0.08–0.18 mg/m^3. A nonexposed group had 0.15 μg/g (216). A high arsenic content of drinking water is also reflected in the arsenic content of human hair (nonexposed individuals 0.06, exposed individuals 1.2 μg/g) (203). In such occupational surveys it is important to distinguish between exogenous arsenic from atmospheric pollution and cosmetics on one hand and that from ingestion on the other. The use of a modified Gutzeit apparatus makes it possible to establish or rule out external contamination (214).

The normal arsenic content of nails is 0.4–1.1 μg/g (131) The arsenic exposure during the growing period of the nail can be determined by analysis of different segments of finger- and toenails (117,233). Normal values of 0.001 to 0.008 μ /g arsenic were found in the enamel of human teeth (218), and 0.06 μg/g in the whole tooth (35). The teeth of workers from a copper-processing plant with arsenic emission contained 2.5 μg arsenic/g those of citizens from a nearby town 0.6 μg/g (218).

The level of arsenic reported for human blood varies widely among individuals and between investigations using different analytical methods. Undoubtedly, the variation reflects differences of environmental exposure; in addition, differences due to analytical methods cannot be ruled out. Vallee et al. (265) quote values ranging from 0.01 to 0.64 μg arsenic per gram in while fresh blood, similar to those reported by Iwataki and Horiuchi (133) and ranging from zero to 0.37 μg/g. In a later study of blood samples from healthy human adults in England, Hamilton et al. (113) obtained a mean of 0.2 ± 0.02 μg arsenic per gram wet weight. Wagner and Weswig (267) found 0.01–0.13 μg/g wet weight in control persons without arsenic exposure and 0.03–0.27 in workers exposed to cacodylic acid (dimethylarsenic acid) used as a forest herbicide. Foa et al. (84) found much lower arsenic concentrations in the blood of nonexposed men (5.1 ± 6.9 μg/liter). The arsenic content of the blood resulting from arsenic intoxication and acute renal failure can be significantly reduced by hemodialysis (266).

Arsenic entering the bloodstream of rats is bound mainly to hemoglobin in the red cells (126,128). Two days after the injection of ^{74}As, dogs, chicks, guinea pigs, and rabbits stored $< 0.27\%$ of the injected ^{74}As in all tissues, whereas rats and cats accumulated 79 and 5.6% of the arsenic in blood. None of the ^{74}As was bound to the plasma proteins, a trace in cellular ghosts, and most of it in hemoglobin from which it could not be removed by dialysis (159). In attempts to find another small animal model representative of humans, Peoples (207) studied the distribution of arsenic in rats, guinea pigs, rabbits, and hamsters, and Marafante et al. (173) studied the distribution in rats and rabbits. Hamsters and rabbits are suited as model animals.

The milk of Greek and Indian women contained 0.6–6 and 0.2–1.1 µg arsenic per liter, respectively. There were no differences between colostrum and mature milk with regard to the arsenic content (66,108). On the other hand, normal milk of cows and goats contains 20–60 µg arsenic per liter (113,126,134,255). The feeding of arsenic to cows at 0.05 to 1.25 mg/kg of body weight for 8 weeks did not increase the arsenic content of milk (207). The same results were obtained after feeding methane arsonic acid to cows (208), after arsenic exposure (0.1 mg/kg body weight for 1 month) in the form of sodium arsenate (255), after feeding lead arsenate to heifers during 126 days (0.285 µg/kg body weight) (174), and after doses of 0.34 of arsenic trioxide were fed daily for 3 days to a heifer (83). These results support the conclusion that there is a blood–mammary barrier to arsenic. The most likely explanation is that an active transport mechanism is saturated at normal plasma concentrations. Aliphatic organic and inorganic arsenic compounds apparently use the same transport mechanism (162). Toxic doses of arsenic break down the barrier to arsenic. In experiments with lactating dairy cows, significantly higher levels of arsenic in milk were observed for cows fed either 3.2 or 4.8 mg of arsenic per kilogram of body weight from arsanilic acid or 3-nitro-4-hydroxyphenylarsonic acid. After feeding arsenic at 4.8 mg/kg body weight over 2 weeks the arsenic levels of milk plateaued and remained constant (46).

II. ARSENIC METABOLISM

For the assessment of toxicity and bioavailability of arsenic, knowledge of the extent of absorption, retention, and excretion of the element is important. The chemical form of arsenic as well as the species of animal influence arsenic metabolism to a large extent.

A. Biotransformation

In the nineteenth century, a number of lethal poisoning incidents occurred in people who were living in rooms with wallpaper containing relatively small amounts of arsenical pigments. It was postulated that these deaths had resulted from the inhalation of minute amounts of trimethylarsenic, a volatile, neurotoxic compound, which had been produced by the mold *Scopulariopsis brevicaulis* growing on moist areas of the wallpaper (49). On the other hand, seafood, especially crustaceans, contains quite high arsenic concentrations, often as much as 100 mg/kg. Fortunately, almost all of this arsenic is in the form of organoarsenic compounds, which are nontoxic and are not metabolized to toxic forms in the human body.

1. Reductive Biomethylation of Arsenic

Gosio (105) first observed that a mold, *Penicillium brevicaule* (*Scopulariopsis brevicaulis*), produces the garlic odor typical of wallpapers with arsenic-containing pigments (Scheele's green, Schweinfurt green).

Challenger *et al.* (50) identified the volatile substance as trimethylarsine. Several strains of fungi isolated from soil (254) (*Aspergillus, Fusarium, Penicillium*) or from sewage (61) can produce trimethylarsine from different arsenic compounds (254). This process takes place under aerobic conditions. It is of interest that some fungi can methylate arsonic acids with aromatic substituents to organoarsines (177). Such aromatic arsonic acids are in widespread use as food additives (247). These methylarsines are not usually formed in uncontaminated environments at concentrations that can be detected with the rather sensitive methods available today (7).

The biomethylation of arsenic under anoxic conditions by methanogenic bacteria proceeds in two methylation steps to dimethylarsinic acid as in the aerobic system, but then reduction to dimethylarsine occurs rather than further methylation. Dimethylarsine is a highly reactive substance. This explains its tendency to be rapidly transformed into nonvolatile materials in sewage sludge (177). The inclusion of arsonic acids in swine diets increased the decomposition of dry matter of swine waste in anaerobic storage systems while increasing the retention of nitrogen (247).

2. Biosynthesis of Structurally Complex Organoarsenic Compounds

Algae in the low-phosphate environment of the oceans absorb arsenate and must detoxify it immediately in order to survive. Each of the intermediates of arsenic detoxification is itself toxic (Fig. 1) and cannot be allowed to accumulate.

The major arsenical metabolites of all phytoplankton were identified by Edmonds and Francesconi (75) as 5-dimethylarsenosoribosides of glycerol and glycerol sulfate. These strange arsenicals accumulate in all aquatic plants (197). Another major arsenical, the arsenolipid, was characterized as a glycerolphosphatide (58). Its arsenical moiety, first considered to be a trimethylarsonium

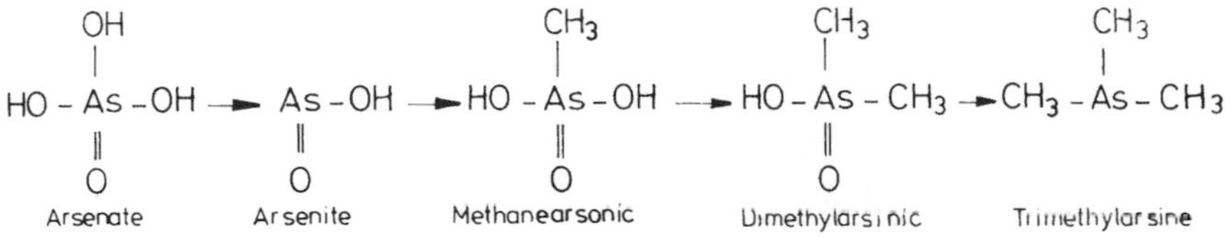

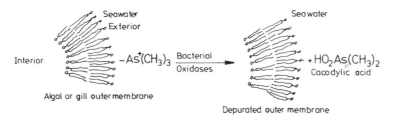

Fig. 1.   Intermediates of arsenate detoxification (32).

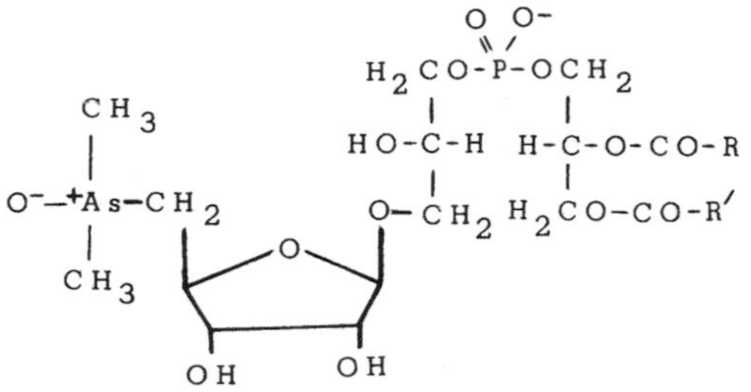

Fig. 2.   Structural formula for 5-dimethylarsenosoriboside of phosphatidylglycerol.

Fig. 3.   Mechanism for arsenic disposal by marine algae and invertebrates (32).

derivative, has been identified as 5-dimethylarsenosoriboside of phosphatidylglycerol (75) (Fig. 2).

The arsenophospholipid 3-O-5'-dimethylarsenoso-5'-deoxyribosylphosphatidylglycerol cacodylate, a seawater component and product of algal arsenate metabolism, is one of the excretion products. Its release from membrane-associated arsenophospholipid by oxidation is a plausible process (Fig. 3).

Many species of algae accumulate considerable concentrations of arsenolipids. How this compound and the arsenosugars are transformed into arsenobetaine in marine animals, still needs clarification (72,73,151). In one diatom, *Chaetoceros gracilis,* a novel water-soluble arsenical has been described in addition to those previously recognized in aquatic plants. The novel compound "D" is conceivably a metabolic precursor of arsenobetaine (31a) (Fig. 4).

In contrast to all observations with phosphate-depleted algal cultures or with

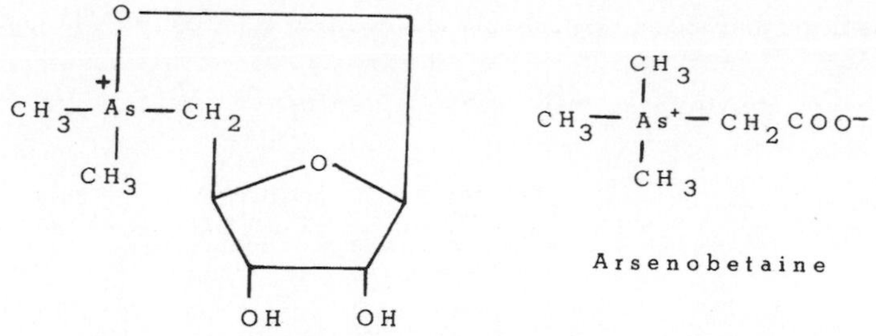

$$CH_3 - \overset{+}{\underset{\underset{CH_3}{|}}{As}} - CH_2$$

$$CH_3 - \overset{\overset{CH_3}{|}}{\underset{\underset{CH_3}{|}}{As^+}} - CH_2COO^-$$

Arsenobetaine

OH   OH

Fig. 4. Proposed structure of arsenical product ''D'' (31b).

tropical oceanic algae, no arsenic appeared in the phytoplankton extracts grown in phosphate-rich waters. The fixed [74]As was unextractable and presumed to be protein bound. It appears that algae exposed to high phosphate concentrations may lack the capability for arsenic detoxification and thus succumb to pollution levels that are tolerated in a low-phosphate environment, as it occurs in the tropical oceans (31).

Regardless of species, the marine fauna of oceans regularly contains arsenic as arsenobetaine (47,74,154,170,200,202,235–237). The intake of arsenobetaine-rich fauna by humans and animals is not dangerous, despite the high arsenic content, because in that form the element is rapidly excreted via the kidneys. The biotransformation of arsenic makes life in water and the consumption of marine fauna by terrestrial beings possible. Terrestrial plants do not have the capacity to transform arsenic into arsonium phospholipid, but freshwater plants are able to do so. The difference indicates that freshwater plants but not terrestrial plants have developed mechanisms for the rapid detoxification of toxic arsenic species (197).

B.   Absorption, Retention, and Excretion

Absorption, retention, and excretion of arsenic are influenced by amounts and chemical forms of arsenic and the species of animals. Arsenic in the form of arsenobetaine is quickly absorbed and rapidly excreted by humans and mammals, regardless of whether arsenobetaine is isolated from marine organisms or synthetically produced. Chapman (51) showed that humans excreted 74% of 25 mg arsenic ingested from lobster within 48 hr, whereas Coulson *et al.* (60) found that arsenic in shrimp was completely excreted by humans within 4 days after consumption. Luten *et al.* (170) reported similar results with arsenobetaine from plaice (*Pleuronectes platessa*). Young men excreted 69–85% of the ingested arsenic via urine within 5 days. Young monkeys excreted 67% of the fish arsenic via urine and 10% via feces within 5 days (52); whereas pigs excreted 68 and

23% via urine and feces, respectively (186). Mice also tolerated intraperitoneal injections of 500 mg/kg arsenobetaine without symptoms of illness and excreted it rapidly (47). Synthesized [73]As-labeled arsenobetaine was completely absorbed by mice and was the only labeled arsenic compound in urine and tissues, indicating that no biotransformation occurred (263).

Compounds that enter the gastrointestinal tract are subjected to the action of bacteria and enzymes, and after absorption they must pass through the liver before reaching the general circulation. This process could alter the chemical form, especially of inorganic arsenic compounds.

Cows and dogs excrete a considerable percentage of trivalent or pentavalent inorganic arsenic given orally as methylated arsenic in the urine (155). Investigations with germ-free mice showed that the methylation of inorganic arsenic cannot be the work of intestinal bacteria only, because germ-free and conventional mice excreted the same percentage (81.8 and 82.3%) of dimethylarsinic acid in the urine after oral doses of arsenic(V) (262).

Six hours after the injection of inorganic arsenic into dogs, >70% of the dose appeared in blood plasma and red blood cells as dimethylarsinic acid (54). The *in vivo* methylation of arsenic may be carried out enzymatically in the liver, whereas *in vitro* methylation might result from microbiological contamination (250). Shortly after arsenic infusion the urinary arsenic was mainly inorganic, but on the following day dimethylarsinic acid was the main excretion product (257). Arsenic(V) must be reduced to arsenic(III) before methylation (261).

Pentavalent organic arsenic compounds such as carbasone, tryparsamide, and cacodylic acid were absorbed directly in proportion to concentration over a 100-fold range by the small intestine of rats. Their absorption half-times were as follows: carbasone, 87 min; tryparsamide, 184 min; and cacodylic acid, 201 min, respectively (130). Hens absorbed more [76]As from As_2O_3 than goats. Absorption, incorporation, and excretion are slowed down by the interposition of rumen and rumen flora (15). Arsenic-deficient goats absorbed more than control animals (15). Arsenic is bound by the protein sulfur of hair. In this way, bezoares (hairs in the rumen) protect against arsenic poisoning (31).

The absorbed arsenic is incorporated into different tissues at different rates. In hens skeleton muscles stored most of the retained [76]As at the measuring points from 45 min to 96 hr after oral intake, followed by skeleton, liver, blood, lungs, kidneys, feathers, and ovaries. The amount of arsenic incorporated into the egg was extremely low but increased with time. The [74]As incorporation into the downy feathers of hens reached its peak 12 hrs after oral intake; that into goat hair took 48 hr to reach its peak (15,120). In mice and golden hamsters Lindgren and Dencker (166) found [76]As, 4 or 30 days after injection, in the epididymis, thyroid, skin, and lens. In the liver of marmoset monkeys 50% of the [74]As was strongly bound to the rough microsomal membranes. Rats accumulate less arsenic from arsenobetaine than from cacodylic acid (239). Both arsenic(III) and arsenic(V) are transferred into the embryo in mice and monkeys (98,165), arsenite at a slower rate than arsenate (165). Teratogenicity of arsenic has been described for hamsters, rats, and mice (22,23,81,123).

Urinary arsenic excretion rises with increasing arsenic intake, and the total urinary arsenic excretion provides a useful index of exposure (42,164,175). The renal excretion rate of arsenic in humans depends on the form. After 4 days, the amount of arsenic excreted in urine represented 46, 78, and 75% of the single oral dose of 500 μg arsenic in form of sodium arsenite, monomethylarsonate, or cacodylate, respectively. Cacodylate was excreted unchanged, monomethylarsonate was slightly (13%) methylated, while roughly 75% of the arsenic excreted after the ingestion of sodium arsenite was methylated (44). Similar results were reported by Tam *et al.* (252). After oral intake of arsenic(III) they found 51% dimethylarsinic acid, 21% monomethylarsenic compound, and 27% inorganic arsenic in urine. The excretion of inorganic [74]As in humans is best represented by a three-component exponential function. Sixty-six percent of the arsenic had a half-time of 2.1 days, 30% of 9.5 days, and 4% of 38 days (215). Arsenite from fish is excreted unchanged in human urine, with a half-life of ~ 1 day (62). Only 0.3% of the fish arsenic was found in the feces (251). The renal excretion of arsenobetaine is not accompanied by increases of inorganic or methylated arsenic (84).

Dogs also excrete >95% of the arsenic via the urine. The half-life of the first component (85% of the dose) was 5.9 hr; that of the second (14% of the dose) was 2.4 days (121). Fifty minutes after dosing, dimethylarsinic acid was detectable in plasma and red blood cells. After 6 hr, >90% of the arsenic in the plasma, red blood cells, and urine was present as dimethylarsinic acid (54).

Methylation of inorganic arsenic to mono- and dimethylated arsenic was demonstrated in rats, mice, hamsters, rabbits, cats, and dogs. The biological methylation of arsenic is a general phenomenon in mammals (173,201).

Arsenic excretion via bile is influenced by the route of intake and chemical form. It is very low after oral intake, but increases to >10% of the dose after intravenous injection in rats (55,145,245). This is important when arsenic-containing drugs (e.g., melarsoprol) are administered intravenously. The arsenic of this drug is excreted via the gallbladder (64), whereas in case of other drugs (acetarsol, tryparsamide) the excreted arsenic was mainly found in the urine (63).

The arsenic of organic compounds such as arsanilic acid is also well absorbed but disappears rapidly from the tissues, mostly into the feces (205). When these forms of arsenic are fed as growth stimulants in the recommended amounts, the element does not accumulate in the tissues to excessive concentrations (96). By 5 to 7 days after the end of arsenic supplementation the arsenic concentration of muscle and adipose tissue decreased to 0.5 mg/kg of body weight, while that in the liver and kidneys was 2.0 mg/kg (82). Monosodium methanearsonate, an herbicide, is accumulated in crawfish, but this arsenic concentration is not dangerous for humans (2).

III. ARSENIC AS AN ESSENTIAL ELEMENT

When discussing arsenic in nutritional terms, one must distinguish between arsenic-deficient rations of <50 ng/g, a normal arsenic supply of 350 to 500 ng/g, and therapeutic doses of 3.5 to 5 μg/g) (13). This differentiation is necessary because therapeutic effects of arsenic in control animals may influence the interpretation of results in arsenic deficiency experiments. Safe and adequate intakes for arsenic are not well defined (178).

The first arsenic deficiency experiments with rats (60,126,230,231,241) were not successful or had confusing results compared to our present state of knowledge, because the arsenic content of "arsenic deficiency rations" was too high or because the arsenic content of control rations was within the therapeutic range and thus had nonnutritional, pharmacodynamic effects. An additional complication arose from the relatively small effect of arsenic on growth rates, even in well-designed experiments.

The essentiality of arsenic in growing, pregnant, and lactating goats, and in minipigs and their offspring has been systematically investigated since 1973, with <10 ng arsenic per gram of semisynthetic rations to induce deficiency and with 350 ng/g in the control rations (11–14,16–18), as it has been in rats (189,194,232,258) and chicks (195,259,260). The influence of arsenic on growth was identified after intrauterine arsenic depletion in minipigs, goats, and rats (16,189,194). On the average, arsenic-deficient goats developed 20% less body weight than control animals (Table I). Intrauterine arsenic depletion led to a significantly reduced birth weight in kids and piglets. The difference between control and arsenic-deficient offspring decreased during the suckling period—

Table I

Growth of Control and Arsenic-Deficient Goats on the Average of the First 167 Experimental Days[a,b]

n	Control goats		Arsenic-deficient goats		p	%[c]
	s	$\bar{x}$	$\bar{x}$	s		
54; 55	29	95	76	43	<.01	80

[a]Ref. (18).
[b]Grams per goat and day.
[c]Control goats ≐ 100%; arsenic-deficient goats ≐ x%.

this is also true for pups (189)—because the arsenic content of the milk of mothers even when fed arsenic-deficient rations is higher than that of semi-synthetic rations. When kids and piglets were fed the semisynthetic ration, the differences again increased significantly (12,18). In growing, pregnant, and lactating goats, arsenic deficiency led to a significantly reduced feed consumption that, on the average, correlated with poor growth rates and milk production.

Arsenic-depleted female goats and minipigs had normal signs of estrus, but the success of the first insemination remained significantly worse than in controls (Table II), and significantly more inseminations were necessary to produce pregnancy. A major effect of arsenic deficiency in goats is a high abortion rate (18); for minipigs and rats, an increased perinatal mortality was reported (12,14, 189,194,258).

Arsenic deficiency had a small but significant influence on the milk yield of fat and protein in goats (16,18), with a reduction of 25 to 30% in both milk components. The significantly decreased milk fat content may be related to the lowered content of triglycerides in the blood serum of arsenic-deficient goats (11).

There were cases of sudden death in lactating goats, without any previous symptoms (11–14,16,18). Mortality was highest between days 17 and 35 of the second lactation. Only one goat with a defective udder and a very low milk production survived the third lactation. Nonpregnant, arsenic-deficient goats, however, can survive to >5 years of age. About one-third of the arsenic-deficient kids died between days 8 and 91 of life. Death frequently occurred between days 7 and 42 of life and after day 84. As a rule, kids died without prodromal symptoms. The high mortality of arsenic-depleted adult goats is consistent with the increased mortality in arsenic-depleted rats (194), but there was no increased mortality in chicks (259,260).

Histological changes in the cardiac muscle of arsenic-deficient kids occur

Table II

Influence of Arsenic Deficiency on Reproduction and Mortality of Goats[a]

Parameter	Control	Arsenic deficiency	p
Success of first insemination (%)	69	47	<.001
Conception rate (%)	84	67	<.05
Pairings per gravidity	1.4	2.1	<.001
Abortion rate (%)	1	33	<.001
Kids per goat	1.5	1.4	>.05
Sex ratio (female = 1)	1:2.0	1.8	>.05
Died, kids (%)	3	32	<.001
Died, mothers (%)	25	40	<.001

[a]Ref. (18).

early (13,14). An atrophy of both types of skeletal and heart muscle fibers could be detected. Ultrastructurally, there was a remarkably increased formation of mitochondrial membrane contours, and a fine granular and electron-dense material in the membrane could be demonstrated. At a more advanced stage this material accumulated at one pole of the mitochondrion, followed by rupture of the mitochondrial membrane and leakage into the cytoplasm. These changes were found in skeletal muscle, myocardium, and liver, with the pathological changes in skeletal muscles and myocardium being more pronounced than those in liver (226, 227). The electron-dense material might contain calcium phosphate. Analysis of the brain did not show any ultrastructural changes. Marked alterations of hepatic mitochondrial structures, albeit of a different nature, are known to occur in arsenic-exposed rats (89,90).

Biochemical changes due to arsenic deficiency have been described in all species of animals examined, but the biological importance of these changes is still unknown. In some investigations a potential pharmacodynamic effect of arsenic supplements that exceeded the requirements must be taken into consideration. This is true above all for the analysis of the effects of arsenic on blood cells and hemoglobin. In the experiments using a dietary arsenic content of 2 and 350 ng/g, no influence of arsenic on hemoglobin, hematocrit, blood cells, or iron metabolism was found in minipigs or in goats, or their offspring. Arsenic deficiency did not affect the iron content of the whole body, of milk, urine, feces, or of 10 organs in three experiments with minipigs and 12 experiments with goats (11–13). In contrast, Nielsen (189), using diets containing 30 and 4500 ng arsenic per gram, and Uthus et al. (258) with diets of 15 and 2000 ng/g, found depressed hematocrit and hemoglobin content of blood in growing arsenic-deficient chicks as compared with controls receiving 2000 ng/g (259,260). The reason for these different results might be the pharmacodynamic influence of high doses of arsenic on the blood picture. Arsenic supplements of 1000 to 5000 ng/g promote the formation of erythrocytes and reduce that of leukocytes (82,118).

Arsenic deficiency caused changes in the mineral composition of 12 examined organs of goats and pigs, of their newborns, and of their milk. A reduced ash content of the skeleton was found in deficient goats but not in minipigs (11,12). The copper concentration of the organs of arsenic-deficient goats and minipigs was significantly increased but remained within physiological limits (12,13). The strongest effect of arsenic deficiency was an increased manganese content in the organs of kids and piglets and in milk (11,13). An increased manganese content was also found in the liver of arsenic-deficient rats (258).

Arsenic-deficient chicks had reduced plasma uric acid concentrations and increased arginase activity of the kidneys (259). Arsenic strongly influences arginine metabolism (195), but manganese and zinc may modify the role of arsenic (196,258). In arsenic-deficient goats a significantly reduced triglyceride content was registered (11). Arsenic might play a role related to lipid phosphorus in biological systems. A novel phospholipid, O-phosphatidyltrimethylarsonium lactic acid, has been identified in algae (32,58). There was a short-term, transient increase of the lipid content of the liver in arsenic-exposed mice (29).

In older rats arsenic deprivation significantly depressed the specific activity of phosphatidylethanolamine methyltransferase, and total liver microsomal activities of phosphatidylethanolamine methyltransferase, phosphatidyldimethylethanolamine methyltransferase, and choline phosphotransferase (59). This suggests that arsenic may not directly affect these three enzymes. The lower total microsomal liver protein found in arsenic-deprived animals rather points to a more general role that may affect amino acid metabolism or protein synthesis.

The pharmacological or therapeutic effects of arsenic and their historical roots have been described comprehensively by Most (184). There was a widespread intake of relatively high amounts of inorganic arsenic by humans and horses, particularly in the Steiermark region of the European Alps, but also in Thuringia and other parts of Germany. The people accustomed to arsenic believed that this element protected against various diseases and that it improved the general state of health and increased virility. Arsenic eaters were described as particularly healthy (184) despite regular daily intakes of 0.5 g arsenic. In unexposed persons 0.1 g arsenic can be lethal (31).

Organic arsenic formulations have been used as feed additives for disease control and improvement of weight gain in swine and poultry since the mid-1940s. These compounds are phenylarsonic acids—arsanilic acid, 3-nitro-4-hydroxyphenylarsonic acid, 4-nitrophenylarsonic acid, and 4-ureidophenylarsonic acid—and their salts (162). The most widely used compounds, generally at Marsenic levels of 50 $\mu g/g$, are arsanilic acid and 3-nitro-4-hydroxyphenylarsonic acid (91,96,225). Poultry responded to therapeutic arsenic doses better than pigs (118); and hens supplemented with 50 μg arsanilic acid or 3-nitro-4-hydroxyphenylarsonic acid per gram of feed produced more egg weight and fewer infertile eggs (8–10). Yet, the use of organic arsenic formulations as feed additives appears to be decreasing. Supplementation with arsenicals must be discontinued 5–7 days before slaughter, which reduces the arsenic concentrations in muscle and liver or kidneys to <0.5 and 2.0 mg/kg, the upper limit of acceptable arsenic concentrations (82).

IV. SOURCES OF ARSENIC

The abundance of arsenic in the continental crust of the earth is generally stated as 1.5 to 2 $\mu g/g$. Despite this low concentration, arsenic is a major constituent of no fewer than 245 mineral species, of which arsenopyrite is by far the most common.

Igneous rocks contain 1.5–3 μg arsenic per gram; lime and sandstones contain similar amounts. Shales, clays (14 μg/g), phosphate rocks (22 μg/g), sedimentary iron ores (400 μg/g), and coal (13 μg/g) are notably rich in arsenic (162).

Arsenic is present in all soils, but the geological history of a particular soil determines its quantity (106,219). The natural arsenic content in virgin soils varies from 0.1 to 40 μg/g. The average is about 5 or 6 μg/g, but it varies considerably among geographical regions (56,125). Apart from the geological origin, the arsenic content of soils is above all determined by arsenical pesticides, smelters, coal-fired power plants, and erosion caused by intensive land use. Notably high levels were found in lake and estuarine sediments (127,157) and alluvial soils (269).

The soil can also be enriched with arsenic by aerosols and floating dust. In a restricted territory an emitter can send off 25–30 metric tons of arsenic into the air every year and a power station 1 ton of arsenic every day (20,24). This fallout affects humans and animals mainly through its incorporation into foods and feedstuffs. Arsenic occurring in house dust comes from soils or herbicides (146).

The arsenic content of seawater amounts to 2 to 5 ng/g. Wide ranges of arsenic concentration from 0.1 to 800 ng/g were found in rivers and lakes and in drinking water (162,246), and extremely high values in the groundwater from territories with thermal activity (109) and with arsenic-rich rocks (37,101). These waters can contain 800–276,000 ng arsenic per gram, but up to 80% of the element can be bound by the iron-rich sediments of spring waters (182). In 1962 the maximally permissible arsenic concentration in drinking water amounted to 50 ng/g (162). Waters from Chile (800 ng/g) (37), the U.S. state of Oregon (0–2000 ng/g) (101,274), and Taiwan (800 ng/g) (79) exceed this limit and pose the risk of chronic arsenic poisoning.

Bacteria of very different species take up arsenic. For example, they are able to convert the herbicide monosodium methanearsonate and other organic arsenicals into inorganic products. Peak rates of demethylation were found after 29 months (1). The rumen microflora are sensitive to arsenic. Arsenite inhibits the fermentative activity and growth of some rumen bacteria more than arsenate, and this inhibition is observed with amounts that are not toxic for the host ruminant (85). Some bacterial species have varients that show resistance to arsenite and arsenate. *Escherichia coli* strains isolated from a Tokyo hospital and a polluted river showed a high frequency of AsO_4^{3-} resistance (240).

The arsenic content of plants is determined by the arsenic exposure via soil, water, air, fertilizers and other chemicals, the geological origin of the soil, and the species, part, and age of plants. Concentrations vary from 0.01 to approximately 5 μg/g (162), but plants growing on arsenic-enriched soils can accumulate extremely high levels (217). Bermuda grass can accumulate up to 45 μg/g arsenic. As a rule, stems contain more arsenic than leaves (217). After fertilizing with arsenic-rich flue ash, in most cases only the first cutting of crops was arsenic enriched (176). In the absence of rain 50% of the monosodium meth-anearsonate used as herbicide was absorbed by Johnson grass after 6 hr and 90% after 155 hr (176). All arsenic-containing herbicides increase the arsenic content of herbage, fruits, and vegetables. This is also true for the fruits of previously treated cultivations (3,6,270). On the other hand, vegetables from old orchard soils treated with arsenic-containing insecticides were not richer in arsenic than control plants (142). Marine plants, particularly algae and seaweed, may have extremely high arsenic contents. For example, the arsenic content of Japanese seaweed varied between 19 and 172 $\mu g/g$; the mean was 112 $\mu g/g$ (268).

Most human foods and animal feeds contain <0.3 $\mu g/g$ and rarely exceed 1 $\mu g/g$ on a dry basis (162). The following ranges of arsenic have been reported in feeds and foods: forage crops 0.1–1 $\mu g/g$, cereals 0.05–0.4 $\mu g/g$, vegetables 0.05–0.8 $\mu g/g$, fruits 0.03–1 $\mu g/g$ dry weight, meat 0.005–0.1 $\mu g/g$ fresh weight, milk 0.01–0.05 $\mu g/g$, eggs 0.01–0.1 $\mu g/g$ fresh weight (69,122,148, 162). Foods of marine origin are much richer in arsenic than other foods. Fish contains 2–80 $\mu g/g$, oysters 3–10 $\mu g/g$, and mussels as much as 10,120 $\mu g/g$ (36,77,129,132,158,169,273). The arsenic content of fish meals used in animal feeding ranged from 2.6 to 19 $\mu g/g$ with a mean of 6 $\mu g/g$ (168).

The total amount of arsenic ingested daily by humans is strongly influenced by the amounts of seafood included in the diet. Early estimates of dietary arsenic intake in humans (400–900 $\mu g/day$) (229) were probably too high. Later sur-veys, which indicate an arsenic intake of 70 to 170 $\mu g/day$ for Japanese indi-viduals (187) and 140 or 130 $\mu g/day$ obtained in the U.S. market basket survey (71) and in Japan (132) are probably more realistic. During the first 12 months of life, infants in Austria ingested 0.03–0.15 mg of arsenic per month (276). Other investigations in Belgium, the United States, and Japan established daily intake ranges of 10 to 130 μg arsenic (45,124,136,178). A range of 100 to 150 μg might be a reasonable estimate for the daily intake, unless the diets contain substantial amounts of seafoods. The arsenic requirement of goats, minipigs, chicks, and rats is <50 $\mu g/kg$ dry matter, and probably <25 μg, as the experi-ments with minipigs and goats prove. Extrapolated from animal data, a sug-gested dietary arsenic requirement of adult humans would be near 6 μg per 1000 kcal or 12–25 μg daily (18,191,193).

V. ARSENIC TOXICITY

The knowledge of the toxicity of arsenic compounds goes back to the fourth centry B.C. Aristotle mentioned the toxicity of sandarac (i.e., of realgar): "San-darac kills horses as well as any kind of draught-cattle. It is put into water and strained through" (43). The symptoms of acute arsenic poisoning in humans by the oral route—nausea, vomiting, diarrhea, burning of mouth and throat, and severe abdominal pains—have frequently been described.

Chronic exposure to smaller toxic doses results in weakness, prostration, muscular aching with few gastrointestinal symptoms, and hearing losses in children at frequencies of 125,250, and 8000 Hz, especially at the lowest frequency range (27). Skin and mucosal changes usually develop together with peripheral neuropathy and linear pigmentations in the fingernails (30). Headache, drowsiness, confusion, and convulsions are found in both acute and chronic arsenic intoxication. The biochemical basis for these disturbances is probably an inhibition of a wide range of enzyme systems by arsenic (26,137,223,224). Enzymes containing active thiol groups are effectively inhibited through combination of arsenic with these groups (5,21,70,90,138,149,150).

Injection of arsenic induces the synthesis of a stress protein (41,144). The maximum long-term arsenic intake compatible with health and well-being in humans cannot be given with any precision because of variations in individual susceptibility and because the chemical form of arsenic greatly affects its toxicity. The fatal dose of ingested arsenic trioxide for humans has been reported to range from 70 to 180 mg (265). Mizuta et al. (181) examined >200 patients who had been poisoned by a soy sauce contaminated with arsenic. This resulted in an average estimated daily ingestion of 3 mg of arsenic for 2 to 3 weeks. Hamamoto (112) reported the poisoning of Japanese infants, who ingested an average of approximate 3.5 mg arsenic per day over a period of 33 days. Orchard workers have been found to ingest as much as 6.8 mg arsenic per day without any signs of intoxication (138).

Arsenic contamination of well water in Taiwan, Minnesota (United States), Canada, and public water supplies in Argentina (19), Chile (37), and Mexico has caused signs and symptoms of arsenic poisoning (102). Negative findings have been reported from California (102), Oregon (183), and Alaska (114,152) in the United States. The reason for these different findings might be the possible presence of other etiological agents in the water supply.

According to Sabbioni et al. (222), the arsenic exposure of humans from the air will increase. The coal-fired power plants of European communities are estimated to be emitting 3340 metric tons of arsenic into the environment by the year 1990. Men in the vicinity of coal-fired power plants, zinc and copper smelters, and workers exposed to arsenic-based weed preservatives and of lead-acid battery manufacturing plants are particularly exposed (28,80,115,156,249). In the vicinity of Swedish smelters the birth weight of children was significantly reduced. This need not be due to arsenic exposure, but might also be caused by lead or cadmium (198). The same is true in conjunction with the significantly increased abortion rate (199).

Napoleon's illness and death on St. Helena Island have been the subject of controversy. His symptoms have been compared with those of arsenic poisoning (86), and arsenic has been found in samples of his hair (87,161,244). Forshufvud and Weider (88) have argued that he was the victim of a poisoning conspiracy, but arsenic was widely used during the nineteenth century. One possible source is paint or wallpaper. Jones and Ledingham (139) examined the wallpaper of Napoleon's residence on St. Helena and found enough arsenic to cause illness, but probably not death.

Body weight gain in turkey poults was significantly decreased by 400 µg/g of arsanilic acid in the diet, about four times the dose required for growth stimulation (4). At twice these levels the mortality of young turkeys up to 28 days of age was 55.6%. Rabbits apparently tolerate much less arsanilic acid than poultry and pigs. They fell ill with <56 µg/g of arsanilic acid in the ration (57). In dogs, doses of 0.7 mg sodium arsenate per kilogram of body weight led to histological changes consisting of mild degeneration and vacuolization of the renal tubular epithelium. A dose of 7.3 mg/kg body weight resulted in alterations detected by urine analysis, but did not markedly affect other clinical pathological measurements (257). Methanearsonate herbicide was significantly more toxic to juvenile toads (*Scaphiopus couchi*) than to adults. Eighty-six percent of a group of juvenile toads was killed by a concentration only one-eighth that recommended for agricultural spraying (140).

Rats have been fed diets supplemented with 50 µg/g arsenic as As_2O_3 without any toxic effects, whereas a significant growth depression was observed at 200 µg/g (172,234).

Bees are particularly sensitive to arsenic. The amount of arsenic recorded in bees during an outbreak of heavy losses (1.1 µg/g) was about 8.5 to 21 times as high as the natural arsenic content (0.048–0.1 µg/g) in healthy bees (253).

Sodium arsenite fed in the drinking water at an arsenic level of 5 µg/ml to mice and rats from weaning to natural death had no effect on growth, health, and longevity despite some arsenic accumulation in the tissues, especially in the aorta and red blood cells. There was no evidence of tumorigenicity or carcinogenicity (230,231). Arsenic in the form of arsenite at a level of 10 µg/g in the drinking water over a period of 15 months has been shown to reduce mammary tumor incidence in female virgin C3H/St mice to 27%, compared to 82% spontaneous tumors in untreated controls. However, arsenic also enhanced the growth rate of mammary tumors once they were established (228). Arsenic levels of 416 µg/g as sodium arsenate were toxic, but there were no differences either in tumor incidence or in the times at which tumors were detected (153). Other experiments with animals also failed to produce evidence for the carcinogenicity of arsenic (25,143). Only Shirachi et al. (238) suggested that sodium arsenite increased the incidence of diethylnitrosamine-initiated renal tumors by acting as a promoter carcinogen.

It is difficult to establish unambiguously that arsenic is a carcinogen in humans, since, apart from arsenic, there are other carcinogenic heavy metals in the environment of examined populations. Smoking may also influence the data. Furthermore, the length of exposure is of importance (40,65,92, 93,95,210,211,213). Skin cancer has occurred in association with exposure to inorganic arsenic compounds in a variety of populations, including patients treated with Fowler's solution, Taiwanese exposed to arsenic in artesian-well water, workers engaged in the manufacture of pesticides, and vintners using arsenic as a pesticide (135,167). Lung cancer has been observed to be associated with inhalation exposure to arsenic in copper smelters (34,39,76,160,209,272), workers in pesticide-manufacturing plants (171,204), and Moselle vintners (167). Hepatic angiosarcoma or hemangioendothelial sarcoma is a rare tumor, which is strongly associated with different kinds of exposure. Long-term arsenic exposure is assumed to promote this form of tumor (78,220). Further investigations must clarify the risk of cancer due to arsenic exposure in humans (39,275).

The total daily urinary arsenic excretion (103,156,180,188) and the levels of arsenic in hair (119) and nails (115) are suited to identify arsenic exposure. Kidneys and liver at autopsy also reflect arsenic exposure (33,48). Precise diagnostic criteria of potentially harmful arsenic intakes or definitions of "safe" long-term dietary arsenic intakes are not available.

Several therapeutic approaches for acute arsenic poisoning have been tested. Thioctic acid alone or in combination with 2,3-dimercapto-1-propanol (BAL), but not BAL alone, may be beneficial in cattle (116).

D-Penicillamine was effective in arsenic-poisoned children (212). BAL (104), 2,3-dimercapto-1-propanesulfonate (DMPS) (147,248), and 2,3-dimercaptosuccinic acid (DMSA) are successfully used to treat arsenic poisoning in humans.

On the other hand, arsenic can give partial protection against chronic selenosis (68,163,185,206), with the vitamin E level influencing the selenium incorporation of tissues. Arsenic also protected against selenium toxicity in suspension cultures of mice fibroblasts (221).

REFERENCES

1. Abdelghani, A. A., Anderson, A. C., Englande, A. J., Mason, J. W., and de Kernion, P. (1977). In "Trace Substances in Environmental Health-10" (D. D. Hemphill, ed.), pp. 419–426. Univ. of Missouri, Press, Columbia.
2. Abdelghani, A. A., Anderson, A. C., Hughes, J., and Englande, A. J. (1980). In "Arsenic, 3. Spurenelement-Symposium" (M. Anke, H.-J. Schneider, and C. Brückner, eds.), pp. 147–153. Wise. Publ. Friedrich-Schiller-Univ., Jena, G.D.R.
3. Abdelghani, A. A., Anderson, A. C., and Mason, J. W. (1979). Bull. Environ. Contam. Toxicol. 23, 797–799.
4. Al-Timimi, A. A., and Sullivan, T. W. (1972). Poult. Sci. 51, 111–116.

5. Anca, Z., and Gabor, S. (1980). *In* "Arsenic, 3. Spurenelement-Symposium" (M. Anke, H.-J. Schneider, and C. Brückner, eds.), pp. 167–171. Wiss. Publ. Friedrich-Schiller-Univ., Jena, G.D.R.

6. Anderson, A. C., Abdelghani, A. A., and Metzger, R. L. (1983). *Sci. Total Environ.* **29**, 113–120.

7. Andreae, M. O. (1980). *In* "Arsenic, 3. Spurenelement-Symposium" (M. Anke, H.-J. Schneider, and C. Brückner, eds.), pp. 131–137. Wiss. Publ. Friedrich-Schiller-Univ., Jena, G.D.R.

8. Andrews, D. K., Bird, H. R., and Sunde, M. L. (1966). *Poult. Sci.* **45**, 838–847.

9. Andrews, D. K., Bird, H. R., and Sunde, M. L. (1966). *Poult. Sci.* **45**, 1305–1313.

10. Anke, M., and Anke, I. (1972). *Jahrb. Tierernaehr. Fuetterung.* **8**, 365–370.

11. Anke, M., Groppel, B., Grün, M., Hennig, A., and Meissner, D. (1980). *In* "Arsenic, 3. Spurenelement-Symposium" (M. Anke, H.-J. Schneider, and C. Brückner, eds.), pp. 25–32. Wiss. Publ. Friedrich-Schiller-Univ., Jena, G.D.R.

12. Anke, M., Grün, M., and Partschefeld, M. (1976). *In* "Trace Substances in Environmental Health-10" (D. D. Hemphill, ed.), pp. 403–409. University of Missouri, Columbia.

13. Anke, M., Grün, M., Partschefeld, M., Groppel, B., and Hennig, A. (1977). *In* "Trace Element Metabolism in Man and Animals-3" (M. Kirchgessner, ed.), pp. 248–252. Tech. Univ. Munich, F.R.G.

14. Anke, M., Hennig, A., Grün, M., Partschefeld, M., Groppel, B., and Lüdke, H. (1976). *Arch. Tierernaehr.* **26**, 742–743.

15. Anke, M., Hoffmann, G., Grün, M., Groppel, B., and Riedel, E. (1982). *IAEA-TEC DOC* **267**, 135–146.

16. Anke, M., Schmidt, A., Groppel, B., and Kronemann, H. (1983). *In* "Lithium-4. Spurenelement-Symposium" (M. Anke, W. Baumann, H. Bräunlich, and C. Brückner, eds.), pp. 87–104. Wiss. Publ. Friedrich-Schiller-Univ., Jena, G.D.R.

17. Anke, M., Schmidt, A., Groppel, B., and Kronemann, H. (1984). *In* "New Results in the Research of Hardly Known Trace Elements" (I. Pais, ed.), pp. 61–71. Budapest, Hungary.

18. Anke, M., Schmidt, A., Kronemann, H., Krause, U., and Gruhn, K. (1984). *In* "Trace Element Metabolism in Man and Animals" (C. F. Mills, ed.). Aberdeen, Scotland (in press).

19. Astolfi, E. (1971). *Prensa Med. Argent.* **58**, 1342–1343.

20. Auermann, E., and Meyer, R. (1980). *In* "Arsenic, 3. Spurenelement-Symposium" (M. Anke, H.-J. Schneider, and C. Brückner, eds.), pp. 103–107. Wiss. Publ. Friedrich-Schiller-Univ., Jena, G.D.R.

21. Baron, D., Kunick, I., Frischmuth, I., and Petres, J. (1975). *Arch. Dermatol. Res.* **253**, 15–22.

22. Baxley, M. N., Hood, R. D., Vedel, G. C., Harrison, W. P., and Szczech, G. M. (1981). *Bull. Environ. Contam. Toxicol.* **26**, 749–756.

23. Beaudoin, A. R. (1974). *Teratology* **10**, 153–156.

24. Bencko, V. (1970). *Wiss. Z. Humboldt-Univ. Berlin, Math.- Naturwiss. Reihe* **14**, 499–501.

25. Bencko, V. (1977). *Environ. Health Perspect.* **19**, 179–182.

26. Bencko, V., and Nemeckova, H. (1971). *J. Hyg., Epidemiol., Milcrobiol., Immunol.* **15**, 104–110.

27. Bencko, V., and Symon, K. (1977). *Environ. Health Perspect.* **19**, 95–101.

28. Bencko, V., and Symon, K. (1977). *Environ. Res.* **13**, 378–385.

29. Benes, B., and Bencko, V. (1981). *J. Hyg., Epidemiol., Microbiol., Immunol.* **25**, 384–392.

30. Bennett, I. L., and Heyman, A. (1966). *In* "Principles of Internal Medicine" (T. R. Harrison *et al.*, eds.), 5th ed., p. 1405. McGraw-Hill, New York.

31. Benson, A. A. (1980). *Naturwiss. Rundsch.* **33**, 114–115.

31a. Benson, A. A. (1984). *In* "Lecture Notes on Coastal and Estuarine Studies, 8. Marine Phytoplankton and Productivity" (O. Holm-Hansen, L. Bolis, and R. Gilles, eds.), pp. 55–59. Springer-Verlag, Berlin and New York.

31b. Benson, A. A. (1985). Personal communication.

32. Benson, A. A., Cooney, R. V., and Summons, R. E. (1980). *In* "Arsenic, 3. Spurenelement-Symposium" (M. Anke, H.-J. Schneider, and C. Brückner, eds.), pp. 139–145. Wiss. Publ. Friedrich-Schiller-Univ., Jena, G.D.R.

33. Bernardini, P., Boscolo, G., Carrelli, G., Innaccone, A., and Sacchettoni-Logroscino, G. (1980). *In* "Arsenic, 3. Spurenelement-Symposium" (M. Anke, H.-J. Schneider, and C. Brückner, eds.), pp. 199–204. Wiss. Publ. Friedrich-Schiller-Univ., Jena, G.D.R.

34. Blejer, H. P., and Wagner, W. (1976). *Ann. N. Y. Acad. Sci.* **271**, 179–186.

35. Bodor, E., and Ghelberg, N. Q. (1980). *In* "Arsenic, 3. Spurenelement-Symposium" (M. Anke, H.-J. Schneider, and C. Brückner, eds.), pp. 287–290. Wiss. Publ. Friedrich-Schiller-Univ., Jena, G.D.R.

36. Bohn, A. (1975). *Mar. Pollut. Bull.* **6**, 87–89.

37. Borgono, J. M., Vincent, P., Venturino, H., and Infante, A. (1977). *Environ. Health Perspect.* **19**, 103–105.

38. Boylen, G. W., Jr., and Hardy, H. L. (1967). *Am. Ind. Hyg. Assoc. J.* **28**, 148–150.

39. Brown, C. C., and Chu, K. C. (1983). *Environ. Health Perspect.* **50**, 293–308.

40. Brown, C. C., and Chu, K. C. (1983). *JNCI, J. Natl. Cancer Inst.* **70**, 455–463.

41. Brown, I. R., and Rush, S. J. (1984). *Biochem. Biophys. Res. Commun.* **120**, 150–155.

42. Browning, E. (1961). "Toxicology of Industrial Metals." Butterworth, London.

43. Brückner, C., and Dietze, B. (1980). *In* "Arsenic. 3. Spurenelement-Symposium" (M. Anke, H.-J. Schneider, and C. Brückner, eds.), pp. 5–10. Wiss. Publ., Friedrich-Schiller-Univ., Jena, G.D.R.

44. Buchet, J. P., Lauwerys, R., and Roels, H. (1981). *Int. Arch. Occup. Environ. Health* **48**, 71–79.

45. Buchet, J. P., Lauwerys, R., Vandevoorde, A., and Pycke, J. M. (1983). *Food Chem. Toxicol.* **21**, 19–24.

46. Calvert, C. C., and Smith, L. W. (1981). *J. Anim. Sci.* **51**, 414–421.

47. Cannon, J. R., Saunders, J. B., and Toia, R. F. (1983). *Sci. Total Environ.* **31**, 181–185.

48. Carmignani, M., Carelli, G., Sacchettoni-Logroscino, G., Gioia, A., and Boscolo, P. (1983). *In* "Lithium, 4. Spurenelement-Symposium" (M. Anke, W. Baumann, H. Bräunlich, and C. Brückner, eds.), pp. 119–126. Wiss. Publ., Friedrich-Schiller-Univ., Jena, G.D.R.

49. Challenger, F. (1945). *Chem. Rev.* **36**, 315–361.

50. Challenger, F., Higginbottom, C., and Ellis, L. (1933). *J. Chem. Soc.*, pp. 95–101.

51. Chapman, A. C. (1926). *Analyst* **51**, 548–563.

52. Charbonneau, S. M., Spencer, K., Bryce, F., and Sandi, E. (1978). *Bull. Environ. Contam. Toxicol.* **20**, 470–477.

53. Charbonneau, S. M., Tam, G. K. H., Bryce, F., and Collins, B. (1978). *In* "Trace Substances in Environmental Health-12" (D. D. Hemphill, ed.), pp. 276–283. Univ. of Missouri Press, Columbia.

54. Charbonneau, S. M., Tam, G. K. H., Bryce, F., Zawidzka, Z., and Sandi, E. (1979). *Toxicol. Lett.* **3**, 107–113.

55. Cikrt, M., and Bencko, V. (1974). *J. Hyg. Epidemiol.,Microbiol., Immunol.* **18**, 129–136.

56. Colbourn, P., Allaway, B. J., and Thornton, I. (1975). *Sci. Total Environ.* **4**, 359–363.

57. Confer, A. W., Ward, B. C., and Hines, F. A. (1980). *Lab. Anim. Sci.* **30**, 234–236.

58. Cooney, R. V., Mumma, R. O., and Benson, A. A. (1978). *Proc. Natl. Acad. Sci. U.S.A.* **75**, 4262–4264.

59. Cornatzer, W. E., Uthus, E. O., Haning, J. A., and Nielsen, F. H. (1983). *Nutr. Rep. Int.* **27**, 821–829.
60. Coulson, E. J., Remington, R. E., and Lynch, K. M. (1935). *J. Nutr.* **10**, 255–270.
61. Cox, D. P., and Alexander, M. (1973). *Bull. Environ. Contam. Toxicol.* **9**, 84–88.
62. Crecelius, E. A. (1977). *Environ. Health Perspect.* **19**, 147–150.
63. Christau, B., Chabas, E., and Placidi, M. (1975). *Ann. Pharm. Fr.* **33**, 577–589.
64. Cristau, B., Placidi, M., and Legait, J.-P. (1975). *Med. Trop.* **35**, 389–401.
65. Cuzick, J., Evans, S., Gillman, M., and Price Evans, D. A. (1982). *Br. J. Cancer* **45**, 904–911.
66. Dang, H. S., Jaiswal, D. D., and Somasundaram, S. (1983). *Sci. Total Environ.* **29**, 171–175.
67. Dewar, W. A., and Lenihan, J. M. (1956). *Scott. Med. J.* **1**, 236–241.
68. Diplock, A. T., and Mehlert, A. (1980). *In* "Arsenic. 3. Spurenelement-Symposium" (M. Anke, H.-J. Schneider, and C. Brückner, eds.), pp. 75–81. Wiss. Publ., Friedrich-Schiller-Univ., Jena, G.D.R.
69. Doyle, J. J., and Spaulding, J. E. (1978). *J. Anim. Sci.* **47**, 398–419.
70. Drummond, G. I. (1981). *Arch. Biochem. Biophys.* **211**, 30–38.
71. Duggan, R. E., and Lipscomb, G. Q. (1969). *Pestic. Monit. J.* **2**, 153–159.
72. Edmonds, J. S., and Francesconi, A. (1977). *Tetrahedron Lett.* **18**, 1543–1546.
73. Edmonds, J. S., and Francesconi, K. A. (1981). *Chemosphere* **10**, 1041–1044.
74. Edmonds, J. S., and Francesconi, K. A. (1981). *Mar. Pollut. Bull.* **12**, 92–96.
75. Edmonds, J. S., and Francesconi, K. A. (1981). *Nature (London)* **289**, 602–604.
76. Enterline, P. E., and Marsh, G. M. (1982). *Am. J. Epidemiol.* **116**, 895–911.
77. Falconer, C., R., Shepherd, R. J., Pirie, J. M., and Topping, G. (1983). *J. Exp. Mar. Biol. Ecol.* **71**, 193–203.
78. Falk, H., Caldwell, G. G., Ishak, K. G., Thomas, L. B., and Popper, H. (1981). *Am. J. Ind. Med.* **2**, 43–50.
79. Fan, C.-J., and Yang, W.-F. (1969). *Kuo Li Tai-wan Ta Hsueh Kung Ch'eng Hsueh K'an* **13**, 95–112.
80. Feldman, R. G., Niles, C. A., Kelly-Hayes, M., Sax, D. S., Dixon, W. J., Thompson, D. J., and Landau, E. (1979). *Neurology* **29**, 939–944.
81. Ferm, V. H., and Kilham, L. (1977). *Environ. Res.* **14**, 483–486.
82. Ferslew, K. E., and Edds, G. T. (1979). *Am. J. Vet. Res.* **40**, 1365–1369.
83. Fitch, L. W. N., Grimmett, R. E. R., and Wall, E. M. (1939). *N. Z. J. Sci. Technol. Sect. A* **21**, 146A–149A.
84. Foa, V., Colombi, A., Maroni, M., Buratti, M., and Calzaferri, G. (1984). *Sci. Total Environ.* **34**, 241–259.
85. Forsberg, C. W. (1978). *Can. J. Microbiol.* **24**, 36–44.
86. Forshufvud, S. (1962). "Who Killed Napoleon?" Hutchinson, London.
87. Forshufvud, S., Smith, H., and Wassen, A. (1961). *Nature (London)* **192**, 103–105.
88. Forshufvud, S., and Weider, B. (1978). "Assassination at St. Helena." Mitchell, Vancouver.
89. Fowler, B. A., Woods, J. S., and Schiller, C. M. (1979). *Lab. Invest.* **41**, 313–320.
90. Fowler, B. A., Woods, J. S., Squibb, K. S., and Davidian, N. M. (1982). *Exp. Mol. Pathol.* **37**, 351–357.
91. Frost, D. V. (1967). *Fed. Proc., Fed. Am. Soc. Exp. Biol.* **26**, 194–208.
92. Frost, D. V. (1980). *In* "Arsenic. 3. Spurenelement-Symposium" (M. Anke, H.-J. Schneider, and C. Brückner, eds.), pp. 17–23. Wiss. Publ., Friedrich-Schiller-Univ., Jena, G.D.R.
93. Frost, D. V. (1983). *In* "Lithium, 4. Spurenelement Symposium." (M. Anke, W. Baumann,

H. Bräunlich, and C. Brückner, eds.), pp. 89–96. Wiss. Publ., Friedrich-Schiller-Univ., Jena, G.D.R.

94. Frost, D. V. (1983). *In* "Biological Availability of Trace Metals" (R. E. Wilding and E. A. Jenne, eds.), pp. 455–466. Elsevier, Amsterdam.

95. Frost, D. V. (1984). *Sci. Total Environ.* **38,** 1–6.

96. Frost, D. V., Overby, L. R., and Spruth, H. C. (1955). *J. Agric. Food Chem.* **3,** 235–243.

97. Gabor, S., Coldea, V., and Ossian, A. (1980). *In* "Arsenic. 3. Spurenelement-Symposium" (M. Anke, H.-J. Schneider, and C. Brückner, eds.), pp. 283–286. Wiss. Publ., Friedrich-Schiller-Univ., Jena, G.D.R.

98. Gerber, G. B., Maes, J., and Eykens, B. (1982). *Arch. Toxicol.* **49,** 159–168.

99. Gerin, C., and de Zorzi, C. (1961). *Zacchia* **24,** 1–19.

100. Gibson, R. S., and Gage, L.-A. (1982). *Sci. Total Environ.* **26,** 33–40.

101. Goldblatt, E. L., van Denburgh, A. S., and Marsland, R. A. (1963). *Oreg. Dep. Health* **II,** 1–24.

102. Goldsmith, J. R. (1972). *Water Res.* **6,** 1133–1136.

103. Gollop, B. R., and Glass, W. I. (1979). *N. Z. Med. J.* **89,** 10–11.

104. Goodman, L. S., and Gilman, A., eds. (1975). "The Pharamcological Basis of Therapeutics," 5th ed., p. 912. Macmillan, New York.

105. Gosio, B. (1893). *Arch. Ital. Biol.* **18,** 253–265.

106. Greaves, J. E. (1913). *Biochem. Bull.* **2,** 519–523.

107. Grigg, F. J. T. (1929). *Analyst* **54,** 659–660.

108. Grimanis, A. P., Vassilaki-Grimani, M., Alexiou, D., and Papadatos, C. (1979). *Nucl. Act. Tech. Life Sci., Proc. Symp., 1978,* p. 241.

109. Grimmett, R. E. R., and McIntosh, I. G. (1939). *N. Z. J. Sci. Techol., Sect. A* **21,** 137A–145A.

110. Gutenmann, W. H., Pakhala, I. S., Churey, D. J., Kelly, W. C., and Lisk, D. J. (1979). *J. Agric. Chem.* **27,** 1393–1395.

111. Halevy, O., and Sklan, D. (1984). *Life Sci.* **34,** 1945–1951.

112. Hamamoto, E. (1955). *Nihon Iji Shimpo* **1649,** 3–12.

113. Hamilton, E. J., Minski, M. J., and Cleary, J. J. (1972-1973). *Sci. Total Environ.* **1,** 341–349.

114. Harrington, J. (1978). *Am. J. Epidemiol.* **108,** 377–385.

115. Hartwell, T. D., Handy, R. W., Harris, B. S., Williams, S. R., and Gehlbach, S. H. (1983). *Arch. Environ. Health* **38,** 284–295.

116. Hatch, R. C., Clark, J. D., and Jain, A. V. (1978). *Am. J. Vet. Res.* **39,** 1411–1414.

117. Henke, G., Nucci, A., and Queiroz, L. S. (1982). *Arch. Toxicol.* **50,** 125–131.

118. Hennig, A. (1972). "Mineralstoffe, Vitamine, Ergotropika." VEB Deutscher Landwirtschaftsverlag, Berlin, G.D.R.

119. Hindmarsh, J. T., McLetchie, O. R., Heffernan, L. P. M., Hayne, O. A., Ellenberger, H. A., McCurdy, R. F., and Thiebaux, H. J. (1977). *J. Anal. Toxicol.* **1,** 270–276.

120. Hoffmann, G., Anke, M., Grün, M., Groppel, B., and Riedel, E. (1980). *In* "Arsenic. 3. Spurenelement-Symposium" (M. Anke, H.-J. Schneider, and C. Brückner, eds.), pp. 41–48. Wiss. Publ., Friedrich-Schiller-Univ., Jena, G.D.R.

121. Hollins, J. G., Charbonneau, S. M., Bryce, F., Ridgeway, J. M., Tam, G. K. H., and Willes, R. F. (1979). *Toxicol. Lett.* **4,** 7–13.

122. Holm, J. (1979). *Fleischwirtschaft* **59** (9), 1345–1349.

123. Hood, R. D., and Harrison, W. P. (1982). *Bull. Environ. Contam. Toxicol.* **29,** 671–678.

124. Horiguchi, S., Teramoto, K., Kurono, T., and Ninomiya, K. (1978). *Osaka City Med. J.* **24,** 131–136.

125. Horvath, A., and Möller, F. (1980). *In* "Arsenic. 3. Spurenelement-Symposium" (M. Anke, H.-J. Schneider, and C. Brückner, eds.), pp. 95–102, Wiss. Publ., Friedrich-Schiller-Univ., Jena, G.D.R.
126. Hove, E., Elvehjem, C. A., and Hart, E. B. (1938). *Am. J. Physiol.* **124**, 205–212.
127. Huang, P. M., and Liaw, W. K. (1978). *Int. Rev. Gesamten Hydrobiol.* **63**, 533–543.
128. Hunter, F. T., Kip, A. F., and Irvine, J. W., Jr. (1942). *J. Pharmacol. Exp. Ther.* **76**, 207–220.
129. Hunter, R. G., Carroll, J. H., and Butler, J. S. (1981). *J. Freshwater Ecol.* **1**, 121–127.
130. Hwang, S. W., and Schanker, L. S. (1973). *Xenobiotics* **3**, 351–355.
131. Ioanid, N. Bors, G., and Popa, I. (1961). *Dtsch. Z. Gesamte Gerichtl. Med.* **52**, 90–94.
132. Ishizaki, M. (1979). *Japn. J. Hyg.* **34**, 605–611.
133. Iwataki, N., and Horiuchi, K. (1959). *Osaka City Med. J.* **5**, 209–211.
134. Iyengar, G. V. (1982). *IAEA-TEC DOC* **269**, 23–25.
135. Jackson, R., and Grainge, J. W. (1975). *CMA J.* **113**, 396–399.
136. Jelinek, C. F., and Corneliussen, P. E. (1977). *Environ. Health Perspect.* **19**, 83–89.
137. Johnson, J. L., and Rajagopalan, K. V. (1978). *Bioorg. Chem.* **8**, 439–444.
138. Johnstone, R. M. (1963). *In* "Metabolic Inhibitors: A Comprehensive Treatise" (R. M. Hochster and J. H. Quastel, eds.), Vol. 2, pp. 99–118. Academic Press, New York.
139. Jones, D. E. H., and Ledingham, K. W. D. (1982). *Nature (London)* **299**, 626–627.
140. Judd, F. W. (1977). *Herpetologica* **33**, 44–46.
141. Katsura, K. (1957). *Shikoku Acta Med.* **11**, 439–444.
142. Kenyon, D. J., Elfving, D. C., Pakkala, I. S., Bache, C. A., and Lisk, D. J. (1979). *Bull. Environ. Contam. Toxicol.* **22**, 221–223.
143. Kerkvliet, N. I., and Steppan, L. B. (1980). *J. Environ. Pathol. Toxicol.* **4**, 65–79.
144. Kim, Y.-J., Shuman, I., Sette, M., and Przybyla, A. (1983). *J. Cell Biol.* **96**, 393–400.
145. Klaassen, C. D. (1974). *Toxicol. Appl. Pharmacol.* **29**, 447–457.
146. Klemmer, H. W., Leitis, E., and Pfenninger, K. (1975). *Bull. Environ. Contam. Toxicol.* **14**, 449–452.
147. Klimova, L. K. (1958). *Farmakol. Toksikol. (Moscow)* **21**, 53–59.
148. Knöppler, H.-O. (1975). *Z. Lebensm-Unters. -Forsch.* **157**, 277–280.
149. Knowless, F. C., and Benson, A. A. (1983). *In* "Lithium, 4. Spurenelement-Symposium" (M. Anke, W. Baumann, H. Bräunlich, and C. Brückner, eds.), pp. 111–114, Wiss. Publ., Friedrich-Schiller-Univ., Jena, G.D.R.
150. Knowless, F. C., and Benson, A. A. (1983). *In* "Lithium, 4. Spurenelement-Symposium" (M. Anke, W. Baumann, H. Bräunlich, and C. Brückner, eds.), pp. 115–118. Wiss. Publ. Friedrich-Schiller-Univ., Jena, G.D.R.
151. Knowless, F. C., and Benson, A. A. (1983). *Trends Biochem. Sci.* **8**, 178–180.
152. Kreiss, K., Zack, M. M., Feldman, R. G., Niles, C. A., Chirico-Post, J., Sax, D. S., Landrigan, P. J., Boyd, M. H., and Cox, D. H. (1983). *Arch. Environ. Health* **38**, 116–121.
153. Kroes, R., van Logten, M. J., Berkvens, J. M., de Vries, T., and van Esch, G. J. (1974). *Food Cosmet. Toxicol.* **12**, 671–679.
154. Kurosawa, S., Yasuda, K., Taguchi, M., Yamazaki, S. Toda, S., Morita, M., Uehiro, T., and Fuwa, K. (1980). *Agric. Biol. Chem.* **44**, 1993–1994.
155. Lakso, J., and Peoples, S. A. (1975). *Agric. Food Chem.* **23**, 674–676.
156. Landrigan, P. J., Costello, R. J., and Stringer, W. T. (1982). *Scand J. Work, Environ. Health* **8**, 169–177.
157. Langston, W. J. (1980). *J. Mar. Biol. Assoc. U. K.* **60**, 869–881.
158. La Touche, Y. D., and Mix, M. C. (1982). *Bull. Environ. Contam. Toxicol.* **29**, 665–670.
159. Lanz, H., Jr., Wallace, P. W., and Hamilton, J. G. (1950). *Univ. Calif., Berkeley, Publ. Pharmacol.* **2**, 263–282.

160. Lee-Feldstein, A. (1983). *JNCI, J. Natl. Cancer Inst.* **70**, 601–609.
161. Leslie, A. C. D., and Smith, H. (1978). *Arch. Toxicol.* **41**, 163–167.
162. Levander, O. A. *et al.* (1977). Natl. Acad. Sci., Washington, D. C.
163. Levander, O. A., and Morris, V. C. (1980). *In* "Arsenic, 3. Spurenelement-Symposium" (M. Anke, H.-J. Schneider, and C. Brückner, eds.), pp. 69–74. Wiss. Publ., Friedrich-Schiller-Univ., Jena, G.D.R.
164. Lilis, R., Valciukes, J. A., Weber, J.-P., Fischbein, A., Nicholson, W. J., Campbell, C., Malkin, J., and Selikoff, I. J. (1984). *Environ. Res.* **22**, 76–95.
165. Lindgren, A., Danielsson, B. R. G., Dencker, L., and Vahter, M. (1984). *Acta Pharmacol. Toxicol.* **54**, 311–320.
166. Lindgren, A., and Dencker, L. (1980). *In* "Arsenic. 3. Spurenelement-Symposium" (M. Anke, H.-J. Schneider, and C. Brückner, eds.), pp. 57–63, Wiss. Publ., Friedrich-Schiller-Univ. Jena, G.D.R.
167. Lüchtrath, H. (1983). *J. Cancer Res. Clin. Oncol.* **105**, 173–182.
168. Lunde, G. (1968). *J. Sci. Food Agric.* **19**, 432–434.
169. Lunde, G. (1970). *J. Sci. Food Agric.* **21**, 242–247.
170. Luten, J. B., Riekwel-Booy, G., and Rauchbaar, A. (1982). *Environ. Health Perspect.* **15**, 165–170.
171. Mabuchi, K., Lilienfeld, A. M., and Snell, L. A. (1979). *Arch. Environ. Health* **34**, 312–320.
172. Mahaffey, K. R., Capar, S. G., Gladen, B. C., and Fowler, B. A. (1981). *J. Lab. Clin. Med.* **98**, 463–481.
173. Marafante, E., Rade, J., Pietra, R., Sabbioni, E., and Bertolero, F. (1980). *In* "Arsenic. 3. Spurenelement-Symposium" (M. Anke, H.-J. Schneider, and C. Brückner, eds.), pp. 49–55. Wiss. Publ., Friedrich-Schiller-Univ., Jena, G.D.R.
174. Marshall, S. P., Hayward, F. W., and Meagher, W. R. (1963). *J. Dairy Sci.* **46**, 580–581.
175. Martinez, G., Cebrian, M., Chamorro, G., and Jauge, P. (1983). *Proc. West. Pharmacol. Soc.* **26**, 171–174.
176. Mason, J. W., Anderson, A. C., Smith, P. M., Abdelghani, A. A., and Englande, A. J., Jr., (1979). *Bull. Environ. Contam. Toxicol.* **22**, 612–616.
177. McBride, B. C., Merrilees, H., Cullen, W. R., and Pickett, W. (1978). *In* "Organometals and Organometalloids. Occurrence and Fate in the Environment" (F. E. Brinkman and J. M. Bellama, eds.), ACS Symp. Ser. No. 82. p.94. Am. Chem. Soc., Washington, D.C.
178. Mertz, W. (1980). *In* "Arsenic. 3. Spurenelement-Symposium" (M. Anke, H.-J. Schneider, and C. Brückner, eds.), pp. 11–15. Wiss. Publ., Friedrich-Schiller-Univ., Jena, G.D.R.
179. Mertz, W. (1981). *Science* **213**, 1332–1338.
180. Milham, S., Jr., and Strong, T. (1974). *Environ. Res.* **7**, 176–182.
181. Mizuta, N., Mizuta, M., Ito, F., Ito, T., Uchida, H., Watanabe, Y., Akama, H., Murakami, T., Hayashi, F., Nakamura, K., Yamaguchi, T., Mizuia, W., Oishi, S., and Matsumura, H. (1956). *Bull. Yamaguchi Med. Sch.* **4**, 131–150.
182. Moenke, H. (1956). *Chem. Erde* **18**, 89–91.
183. Morton, W., Starr, G., Pohl, D., Stoner, J., Wagner, S., and Weswig, P. (1976). *Cancer* **37**, 2523–2532.
184. Most, K.-H. (1939). Dissertation, Univ. Graz, Austria.
185. Moxon, A. L. (1938). *Science* **88**, 81.
186. Munro, I. C. (1976). *Clin. Toxicol.* **9**, 647–663.
187. Nakao, M. (1960). *Osaka-shiritsu Daikaku Igaku Zasshi* **9**, 541–571.
188. Nelson, K. W., and Dungey, C. E. (1980). *In* "Arsenic. 3. Spurenelement-Symposium" (M. Anke, H.-J. Schneider, and C. Brückner, eds.), pp. 173–178. Wiss. Publ., Friedrich-Schiller-Univ., Jena, G.D.R.

189. Nielsen, F. H. (1980). *Adv. Nutr. Res.* **3**, 157–172.
190. Nielsen, F. H. (1982). *In* "Clinical, Biochemical, and Nutritional Aspects of Trace Elements" (R. L. Liss, ed.). 150 Fifth Avenue, New York.
191. Nielsen, F. H. (1984). *Bull. N. Y. Acad. Med.* **60**, 177–195.
192. Nielsen, F. H., Givand, S. H., and Myron, D. R. (1975). *Fed. Proc., Fed. Am. Soc. Exp. Biol.* **34**, 923.
193. Nielsen, F. H., and Mertz, W. (1984). *In* "Present Knowledge in Nutrition" (R. E. Olson *et al.*, eds.), 5th ed., pp. 607–618. Nutr. Found., Inc., Washington, D.C.
194. Nielsen, F. H., Myron, D. R., and Uthus, E. O. (1977). *In* "Trace Element Metabolism in Man and Animals-3" (M. Kirchgessner, ed.), pp. 244–247. Tech. Univ. Munich, F.R.G.
195. Nielsen, F. H., and Shuler, T. R. (1978). *Fed. Proc., Fed. Am. Soc. Exp. Biol.* **37**, 893.
196. Nielsen, F. H., Uthus, E. O., and Cornatzer, W. E. (1983). *Biol. Trace Elem. Res.* **5**, 389–397.
197. Nissen, P., and Benson, A. A. (1982). *Physiol. Plant.* **54**, 446–450.
198. Nordström, S., Beckman, L., and Nordenson, I. (1978). *Hereditas* **88**, 43–46.
199. Nordström, S., Beckman, L., and Nordenson, I. (1978). *Hereditas* **88**, 51–54.
200. Norin, H., and Christakopoulos, A. (1982). *Chemosphere* **11**, 287–298.
201. Odanaka, Y., Matano, O., and Goto, S. (1980). *Bull. Environ. Contam. Toxicol.* **24**, 452–459.
202. Oladimeji, A. A., Quadri, S. U., Tam, G. K. H., and de Freitas, A. S. W. (1979). *Ecotoxicol. Environ. Saf.* **3**, 394–400.
203. Olguin, A., Jauge, P., Cebrian, M., and Albores, A. (1983). *Proc. West. Pharmacol. Soc.* **26**, 175–177.
204. Ott, M. G., Holder, B. B., and Gordon, H. L. (1974). *Arch. Environ. Health* **29**, 250–255.
205. Overby, L. R., and Frost, D. V. (1960). *J. Anim. Sci.* **19**, 140–144.
206. Palmer, I. S., Thiex, N., and Olson, O. E. (1983). *Nutr. Rep. Int.* **27**, 249–257.
207. Peoples, S. A. (1964). *Ann. N. Y. Acad. Sci.* **111**, 644–649.
208. Peoples, S. A. (1969). *Fed. Proc., Fed. Am. Soc. Exp. Biol.* **28**, 359.
209. Pershagen, G. (1980). *In* "Arsenic. 3. Spurenelement-Symposium" (M. Anke, H.-J. Schneider, and C. Brückner, eds.), pp. 179–183, Wiss. Publ., Friedrich-Schiller-Univ., Jena, G.D.R.
210. Pershagen, G. (1981). *Environ. Health Perspect.* **40**, 93–100.
211. Pershagen, G., Wall, S., Taube, A., and Limman, L. (1981). *Scand. J. Work, Environ. Health* **7**, 302–309.
212. Peterson, R. G., and Rumack, B. H. (1977). *J. Pediatr.* **91**, 661–666.
213. Pinto, S. S, Henderson, V., and Enterline. P. E. (1978). *Arch. Environ. Health* **33**, 325–331.
214. Pirl, J. G., Townsend, G. F., Valaitis, A. K., Grohlich, D., and Spikes, J. J. (1983). *J. Anal. Toxicol.* **7**, 216–219.
215. Pomroy, C., Charbonneau, S. M., McCullough, R. S., and Tam, G. K. H. (1980). *Toxicol. Appl. Pharmacol.* **53**, 550–556.
216. Parásik, J., Legáth, V., Puchá, K., and Kratochvil, I. (1966). *Prac. Lek.* **18**, 352–356.
217. Porter, E. K., and Peterson, P. J. (1975). *Sci. Total Environ.* **4**, 365.
218. Rasmussen, E. G. (1974). *Scand. J. Dent. Res.* **82**, 562–565.
219. Risch, M. A., Atadschanev, P., and Teschabaev, C. (1980). *In* "Arsenic, 3. Spurenelement-Symposium" (M. Anke, H.-J. Schneider, and C. Brückner, eds.), pp. 91–93. Wiss. Publ. Friedrich-Schiller-Univ., Jena, G.D.R.
220. Roat, J. W., Wald, A., Mendelow, H., and Pataki, K. I. (1982). *Am. J. Med.* **73**, 933–936.
221. Rössner, P., Bencko, V., and Havrankova, H. (1977). *Environ. Health Perspect.* **19**, 235–237.

222. Sabbioni, E., Goetz, L., Springer, A., and Pietra, R. (1983). *Sci. Total Environ.* **29**, 213–227.

223. Sardana, M. K., Drummond, G. S., Sassa, S., and Kappas, A. (1981). *Pharmacology* **23**, 247–253.

224. Schiller, C. M., Fowler, B. A., and Woods, J. S. (1978). *Chem.-Biol. Interact.* **22**, 25–33.

225. Schmid, A. (1983). *Dtsch. Tieraerztl. Wochenschr.* **90**, 10–13.

226. Schmidt, A., Anke, M., Groppel, B., and Kronemann, H. (1983). *In* "Mengen- und Spurenelemente" (M. Anke, C. Brückner, H. Gürtler, and M. Grün, eds.), Vol. 3, pp. 424–425. Karl-Marx-Univ., Leipzig, G.D.R.

227. Schmidt, A., Anke, M., Groppel, B., and Kronemann, H. (1984). *Exp. Pathol.* **25**, 195–197.

228. Schrauzer, G. N., and Ishmael, D. (1974). *Ann. Clin. Lab. Sci.* **4**, 441–450.

229. Schroeder, H. A., and Balassa, J. J. (1966). *J. Chronic Dis.* **19**, 85–106.

230. Schroeder, H. A., and Balassa, J. J. (1967). *J. Nutr.* **92**, 245–252.

231. Schroeder, H. A., Kanisawa, M., Frost, D. V., and Mitchener, M. (1968). *J. Nutr.* **96**, 37–45.

232. Schwarz, K. (1977). *In* "Clinical Chemistry and Chemical Toxicology of Metals" (S. S. Brown ed.), pp. 3–22. Elsevier/North-Holland Biomedical Press, Amsterdam.

233. Shapiro, H. A. (1967). *J. Forensic Med.* **14**, 65–71.

234. Sharpless, G. R., and Metzger, M. (1941). *J. Nutr.* **21**, 341–346.

235. Shiomi, K., Shimagawa, A., Azuma, M., Yamanaka, H., and Kikuchi, T. (1980). *Comp. Biochem. Physiol.* **74C**, 393–396.

236. Shiomi, K., Shinagawa, A., Igarashi, T., Hirota, K., Yamanaka, H., and Kikuchi, T. (1984). *Bull. Jpn. Soc. Sci. Fish.* **50**, 293–297.

237. Shiomi, K., Shinagawa, A., Yamanaka, H., and Kikuchi, T. (1983). *Bull. Jpn. Soc. Sci. Fish.* **49**, 79–83.

238. Shirachi, D. Y., Johansen, M. G., McGowan, J. P., and Tu, S.-H. (1983). *Proc. West. Pharmacol. Soc.* **26**, 413–415.

239. Siewicki, T. C. (1981). *J. Nutr.* **111**, 602.–609.

240. Silver, S., and Nakahara, H. (1983). *Ind. Biomed. Environ. Perspect.* **19**, 190–199.

241. Skinner, J. T., and McHargue, J. S. (1946). *Am. J. Physiol.* **145**, 500–506.

242. Smith, H. S. (1964). *J. Forensic Sci. Soc.* **4**, 192–199.

243. Smith, H. S. (1967). *J. Forensic Sci. Soc.* **7**, 97–102.

244. Smith, H. S., Forshufvud, S., and Wassen, A. (1962). *Nature (London)* **194**, 725–726.

245. Stevens, J. T., Hall, L. L., Farmer, J. D., DePasquale, L. C., Chernoff, N., and Durham, W. F. (1977). *Environ. Health Perspect.* **19**, 151–157.

246. Strain, W. H., Varnes, A. W., Matisoff, G., and Khourey, C. J. (1980). *In* "Arsenic. 3. Spurenelement-Symposium" (M. Anke, H.-J. Schneider, and C. Brückner, eds.), pp. 83–89. Wiss. Publ., Friedrich-Schiller-Univ., Jena, G.D.R.

247. Sutton, A. L., and Brumm, M. C. (1980). *In* "Arsenic. 3. Spurenelement-Symposium" (M. Anke, H.-J. Schneider, and C. Brückner, eds.), pp. 115–121. Wiss. Publ. Friedrich-Schiller-Univ., Jena, G.D.R.

248. Tadlock, C. H., and Aposhian, V. (1980). *Biochem. Biophys. Res. Commun.* **94**, 501–507.

249. Takahashi, W., Pfenninger, K., and Wong, L. (1983). *Arch. Environ. Health* **38**, 209–214.

250. Tam, G. K. H., Charbonneau, S. M. Bryce, F., Pomroy, C., and Sandi, E. (1979). *Toxicol. Appl. Pharmacol.* **50**, 319–322.

251. Tam, G. K. H., Charbonneau, S. M., Bryce, F., and Sandi, E. (1982). *Bull. Environ. Contam. Toxicol.* **28**, 669–673.

252. Tam, G. K. H., Charbonneau, S. M., Lacroix, G., and Bryce, F. (1979). *Bull. Environ. Contam. Toxicol.* **22**, 69–71.

253. Terzić, L. J., Terzić, V., Krunić, M., and Brajković, M. (1984). *Acta Vet.* **34,** 57–62.
254. Thom, C., and Raper, K. B. (1932). *Science* **76,** 548–550.
255. Tölgyesi, G., and Bokori, J. (1980). *In* "Arsenic. 3. Spurenelement-Symposium" (M. Anke, H.-J. Schneider, and C. Brückner, eds.), pp. 65–68. Wiss. Publ., Friedrich-Schiller-Univ., Jena, G.D.R.
256. Tsukamoto, H., Parker, H. R., Gribble, D. H., Mariassy, A., and Peoples, S. A. (1983). *Am. J. Vet. Res.* **44,** 2324–2330.
257. Tsukamoto, H., Parker, H. R., and Peoples, S. A. (1983). *Am. J. Vet. Res.* **44,** 2331–2335.
258. Uthus, E. O., Cornatzer, W. E., and Nielsen, F. H. (1983). *In* "Arsenic; Industrial, Biomedical, Environmental Perspectives" (W. H. Lederer and R. J. Fensterheim, eds.), pp. 173–189. Van Nostrand-Reinhold, Princeton, New Jersey.
259. Uthus, E. O., and Nielsen, F. H. (1980). *In* "Arsenic. 3. Spurenelement-Symposium" (M. Anke, H.-J. Schneider, and C. Brückner, eds.), pp. 33–39. Wiss. Publ., Friedrich-Schiller-Univ., Jena, G.D.R.
260. Uthus, E. O., and Nielsen, F. H. (1983). *In* " Lithium. 4. Spurenelement-Symposium" (M. Anke, W. Baumann, H. Bräunlich, and C. Brückner, eds.), pp. 105–110. Wiss. Publ. Friedrich-Schiller-Univ., Jena, G.D.R.
261. Vahter, M., and Envall, J. (1983). *Environ. Res.* **32,** 14–24.
262. Vahter, M., and Gustafsson, B. (1980). *In* "Arsenic. 3. Spurenelement-Symposium" (M. Anke, H.-J. Schneider, and C. Brückner, eds.), pp. 121–129. Wiss. Publ., Friedrich-Schiller-Univ., Jena, G.D.R.
263. Vahter, M., Marafante, E., and Dencker, L. (1983). *Sci. Total Environ.* **30,** 197–211.
264. Valentine, J. L., Hang, H., Reisbord, L., and Schluchter, M. (1984). *In* "Trace-5," Element Metabolism in Man and Animals p. 268. Aberdeen, Scotland.
265. Vallee, B. L., Ulmer, D. D., and Wacker, W. E. C. (1960). *AMA Arch. Ind. Health* **21,** 132–151.
266. Vaziri, N. D., Upham, T., and Barton, C. H. (1980). *Clin. Toxicol.* **17,** 451–456.
267. Wagner, S. L., and Weswig, P. (1974). *Arch. Environ. Health* **28,** 77–79.
268. Watanabe, T., Hirayama, T., Takahashi, T., Kokutbo, T., and Ikeda, M. (1979). *Toxicology* **14,** 1–22.
269. Wauchope, R. D. (1975). *J. Environ. Qual.* **4,** 355–358.
270. Wauchope, R. D., and McWhorter, C. G. (1977). *Bull. Environ. Contam. Toxicol.* **17,** (2).
271. Weaver, R. W., Melton, J. R., Wang, D., and Duble, R. L. (1984). *Environ. Pollut.* **33,** 133–142.
272. Welch, K., Higgins, J., Oh, M., and Burchfiel, C. (1982). *Arch. Environ. Health* **37,** 325–335.
273. Westöö, G., and Rydalv, M. (1972). *Var. Föeda* **24,** 21–23.
274. Whanger, P. D., Weswig, P. H., and Stoner, J. C. (1977). *Environ. Health Perspect.* **19,** 139–143.
275. Wildenberg, J. (1978). *Environ. Res.* **16,** 139–152.
276. Woidich, H., and Pfannhause, W. (1980). *Z. Lebensm.-Unters. -Forsch.* **170.** 95–98.

7

Silicon

EDITH MURIEL CARLISLE

School of Public Health
University of California
Los Angeles, California

I. SILICON IN ANIMAL TISSUES AND FLUIDS

Earlier data on the distribution of silicon (Si) in animal tissues have varied greatly, and in general, reported values were considerably higher before the advent of plastic laboratory ware, stricter precautions to minimize dust contamination, and the development of suitable methods. Even with later methods, most of which are modifications of the ammonium molybdate colorimetric reaction, large differences in silicon levels in tissues are often observed (78), probably due to limitations of the colorimetric determination by this technique. It is most important, therefore, to compare the results of the colorimetric assays with an alternative analytical procedure. Some years ago the normal range for silicon in human fetal tissues was given as 18 to 180 µg/g (dry basis) compared with 23 to 460 µg/g for adult human tissues (57). The results of two more later studies for a range of tissues in adult humans (42), rats, and rhesus monkeys (60) are presented in Table I. The very high levels in the human lymph nodes were shown to be associated with the presence of clusters and grains of quartz (42). Comparable levels of silicon in rat tissues to those given in Table I have been reported by McGavack *et al.* (63) and later by Carlisle, who has shown that the consistent low concentrations of silicon in most organs do not appear to vary appreciably during life. Parenchymal tissues such as heart and muscle, for example, range in silicon level from 2 to 10 µg/g dry weight (16). The lungs are an exception.

Normal whole-blood levels in humans (63), monkeys (60), and rats (26,63) average ~ 1 µg/ml, except for ovines, which are higher (5 µg/ml). Normal

Table I
Silicon Concentration in Animal Tissues

Tissue	Silicon concentration (μg/g wet weight)		
	Adult man[a]	Adult rat[b]	Rhesus monkey[b]
Brain	23 ± 4.4	0.8 ± 0.9	1.4 ± 0.7
Kidneys	40 ± 11	0.5 ± 0.7	1.6 ± 1.5
Liver	33.6 ± 13.8	1.6 ± 1.5	1.2 ± 1.2
Lungs	57.4 ± 10.7	1.6 ± 1.4	194 ± 183.2
Muscle	41 ± 0.9	0.9 ± 0.7	1.2 ± 0.5
Testis	3.1 ± 1.6	1.1 ± 1.1	2.0 ± 1.2
Lymph nodes	489 ± 215	4.1 ± 5.5	21.9 ± 10.6

[a]From Hamilton *et al.* (42).
[b]From LeVier (60).

human serum has a narrow range of silicon concentration averaging 0.50 μg/ml (26); the range is similar to that found for most of the other well-recognized trace elements in human nutrition. The silicon is present almost entirely as free soluble monosilicic acid (7), and the concentration of silicon in other examined body fluids has been found to be similar to that of normal serum (34), indicating that silicon is freely diffusible throughout tissue fluid. No correlations of age, sex, occupation, or pulmonary condition with blood silicon concentrations were found as a result of measurements on hundreds of people, although the level increased when silicon compounds were specifically administered (89). Moderate increases have been obtained in rats' blood after feeding sodium metasilicate. Much higher levels have been reached, however, after feeding organic silicates (26).

The highest silicon concentrations generally occur in connective tissues such as aorta, trachea, tendon, bone, skin, and its appendages, as shown by studies in several animal species (16). In the rat, for example, the aorta, trachea, and tendon are four to five times richer in silicon than liver, heart, and muscle. The concentrations in these tissues decline with age, whereas most other tissues display no such age change. For example, Carlisle (16) compared the tissue silicon levels in rabbits at 12 weeks and at 18 to 24 months of age; the silicon levels in the heart, liver, and muscle remained between 5 and 15 μg/g (dry basis) at the two ages, while those of the aorta and skin declined from 80 to 15, and from 46 to 9 μg/g, respectively. Similarly, the mean silicon level in fetal pig skin was reported as 95 μg/g, compared with 10 μg/g in mature pig skin. A comparable decrease in the silicon content of rat skin with age has also been reported (59). The silicon content of the normal human aorta has been found to decrease considerably with age, and the level in the arterial wall has been shown to decrease with the development of atherosclerosis (31,61).

Among the human tissues, epidermis and hair have been reported to contain unusually large amounts of localized silicon. The element accumulates in the cornified epidermis on the surface of skin, and in the epicuticle of hair, as well as the wool and feathers of other animals, in an alkali-insoluble component constituting only 0.4–1.7% of the total tissue weight. It has been suggested (37) that this small alkali-insoluble component with its high silicon content may contribute to the solidity and great chemical resistance of keratinous tissues and may also play a role as a barrier of absorption. High levels (mean 243 ± 18, range 100–450 μg/g dry weight) of silicon have also been reported in human dental enamel (62) and in the head of the monkey femur containing the epiphyses (456.3 ± 71.0 μg/g dry weight) (60).

The high silicon content of connective tissues appears to arise mainly from its presence as an integral component of the glycosaminoglycans and their protein complexes, which contribute to the structural framework of this tissue. The unique properties of silicon atoms in respect to bonding and macromolecular structures are well known (66). Using extraction and fractionation procedures on connective tissues such as bone, cartilage, and skin, Carlisle (18) obtained glycosamino–protein complexes of high silicon content. Silicon was also found as a component of glycosaminoglycans isolated from these complexes. Silicon has also been reported to be a bound component of isolated glycosaminoglycans by Schwarz (77). Strong alkali and acid hydrolysis was found to free the silicon–polysaccharide bond, giving free, dialyzable silicate; however, enzymatic hydrolysis of hyaluronic acid or pectin did not liberate silicic acid but led to products of low molecular weight still containing silicon in bound form. Carlisle (16) found that disaccharides enzymatically derived from chondroitin sulfate A contained considerably more silicon than disaccharides obtained from chondroitin sulfate C. The finding that silicon is a constituent of disaccharide units led Carlisle to suggest that it may be added at the state of formation of the polysaccharide chain from smaller units. Schwarz (77) concluded that "Si is present as silanolate, i.e., an ether (or esterlike) derivative of silicic acid, and that R_1-O-Si-O-R_2 or R_1-O-Si-O-Si-O-R_2 bridges play a role in the structural organization of glycosaminoglycans and polyuronides." He later reported, however, that many of his earlier observations on the occurrence of bound silicon in glycosaminoglycans were in error because they were based partially on results obtained with materials contaminated by silica or polysilicic acid, and the hypothesis that silicon acts generally as a cross-linking agent may have to be modified.

In order to gain information about silicon's function in soft tissues, the subcellular distribution of silicon in whole rat liver was determined in terms of percentage of homogenate total. The element was found to be equally distributed in the supernatant, mitochondria, and nuclei-debris. Little silicon was associated with the microsomal fraction (60).

Few data are available on the silicon content of milk in which acceptable modern methods of analysis have been employed and care taken to avoid contamination from glass. It seems that with cow's milk there is considerable individual variation, a marked decrease from the levels in colostrum to those in milk (58), and little influence of dietary silicon intakes. Thus, Archibald and Fenner (2) reported that the milk of six cows alternately fed a control ration and one containing added sodium silicate at the rate of 1 g/day (230 mg silicon) averaged 1.4 μg/ml, irrespective of treatment.

Little information is available on the silicon content of eggs. The total silicon content of a hen's egg is found to be about 50 μg. The silicon concentration (dry weight) in the yolk is about 3 μg/g and in the white 10 μg/g. However, about two-thirds of the total silicon content of the egg is contributed by the yolk because of its greater weight (29).

II. SILICON METABOLISM

It has been estimated that humans assimilate 9–14 mg of silicon daily. This figure correlates well with the report of Goldwater (40) that humans excrete 9 mg of silicon in urine daily. In a balance study (54) the silicon intake of men on a high-fiber diet was about double that on a low-fiber diet, and although urinary excretions ranged between 12 and 16 mg/day for the low- and high-fiber diets, the differences are not statistically significant. These values appear to be in the same range as those estimated earlier.

A. Absorption

Little is known of the extent or mechanism of silicon absorption from the products entering the alimentary tract derived from food sources such as silica, monosilicic acid, and silicon found in organic combination, such as pectin and mucopolysaccharides. Balance trials in animals indicate that almost all ingested silicon is unabsorbed, passing through the digestive tract to be lost in the feces. Moreover, most of the small proportion that is absorbed is excreted in the urine. The proportion of absorbed silicon actually retained in the body is not known.

In humans silicic acid in foods and beverages has been reported to the readily absorbed across the intestinal wall and rapidly excreted in the urine (7). In guinea pigs absorption is found to occur mainly as monosilicic acid (73), some of which comes from the silica of the plant materials that is partly dissolved by the fluids of the gastrointestinal tract. In sheep the extent of absorption of silica as monosilicic acid varies with the silica content of the diet. Jones and Handreck (47) found that the amounts exreted in the urine increased with increasing silica content of the diet from 0.10 to 2.84%, but reached a maximum of 205 mg SiO_2 per day, this amount representing <4% of the total intake.

The form of dietary silicon has been shown to be an important factor affecting its absorption, appearing to correlate with the rate of production of soluble or absorbable silicon in the gastrointestinal tract (9). Other factors have been reported to influence silicon absorption: the dietary fiber content of the diet has been implicated in studies with humans (54), and changes in silicon absorption and resulting levels of silicon in the blood and intestinal tissues of rats have been found to be related to age, sex, and the activity of various endocrine glands (30).

As is the case for many other elements, silicon availability is also probably affected by excess amounts of certain other mineral elements in the diet, which may result in a diminution of silicon absorption through a reduction in the production of soluble silicon. Molybdenum is an example of such an element; an interrelationship between silicon and molybdenum has been established (20). A marked interaction was demonstrated with both semisynthetic and amino acid diets. Plasma silicon levels were strongly and inversely affected by molybdenum intake; silicon-supplemented chicks on a liver-based diet (molybdenum 3 ppm) had a 348% lower plasma silicon level than chicks on a casein diet (molybdenum 1 ppm). Molybdenum supplementation also reduced silicon levels in those tissues examined. Conversely, plasma molybdenum levels are also markedly and inversely affected by the inorganic silicon intake. Silicon supplementation of an amino acid diet reduced the plasma molybdenum levels of the silicon-supplemented chicks by 280% and the red blood cell molybdenum by 425% compared to the low-silicon group. Reduction in molybdenum tissue retention by silicon also occurred.

Several other elements also suggest themselves as likely candidates on the basis of their ability to depress greatly the amounts of silicic acid yielded in solution from various forms of silica suspended in body fluids (ascitic fluid and serum), especially aluminum and including Fe_2O_3, $Ca(OH)_2$, MgO, and SrO (56). The solubility depressor effect is believed to be a simple precipitation of the insoluble silicate by the metal. On the other hand, manganese availability appears to be influenced by silicon. In poultry rations differences in manganese availability among various inorganic manganese sources have been demonstrated, but only two, the carbonate and silicate (38,76), are relatively unavailable. Furthermore, silicon has been used as a therapeutic agent to alleviate manganese toxicity in plants (e.g., wheat, oats, and rice; 70).

Microscopic particles of silica (phytoliths) have been demonstrated in the lymph nodes and urinary calculi of sheep. Phytoliths are minute bodies of amorphous silica (opal) that are formed within a great variety of plants. Many of the smaller phytoliths are able to pass the intestinal barrier of animals, including humans, and enter the lymphatic and blood vascular systems, thus to be distributed throughout the body. In addition to possibly producing pathological effects, the presence of phytoliths throughout many tissues, especially lungs, liver, lymph nodes, and spleen, makes it difficult to determine the amount of silicon in animal tissues that is actually contributing to physiological mechanisms. The gut wall of humans is also permeable to particles the size of diatoms. Volkenheimer (86) showed that diatomaceous earth particles are absorbed through the intact intestinal mucosa, pass through the lymphatic and circulatory systems, and reach other tissues supplied by arterial blood via the alveolar region of the lung. Examination of human organs has revealed diatoms in lungs, liver, and kidneys, as a consequence of their presence in atmospheric dust and their movement from the respiratory tract. The capacity of these particles to travel in the blood and to penetrate membranes, including the placenta, is illustrated further by their presence in the organs of stillborn and premature infants (39).

B. Intermediary Metabolism

Silicon is found to be freely diffusible throughout tissue fluid. In early studies Baumann (7) showed that monomeric orthosilicic acid penetrates all body liquids and tissues at concentrations less than its solubility (0.01%) and is readily excreted. These findings are supported by later studies where the concentration of silicon in those body fluids examined was found to be similar to that of normal human serum, except for urine (34). This indicates that silicon is freely diffusible throughout tissue fluid; the higher levels and wider range encountered in urine suggesting that the kidney is the main excretory organ. Policard *et al.* (72) have suggested that polymeric molecules of silicic acid containing up to four to five silicon–oxygen units characterize the transport form of silicic acid in blood.

C. Excretion

Increased urinary silicon output with increasing intake, up to fairly well-defined limits has been demonstrated in humans (45), rats (51), guinea pigs (73), and cows (3). In sheep, Nottle (69) found urinary excretion of silica to increase with rising dietary silica intakes up to an intake of 83 mg SiO_2 per day. Thereafter, urinary excretion leveled off at 200 to 250 mg/day. The upper limits of urinary silicon excretion do not seem to be set by the ability of the kidneys to excrete more, because much greater urinary excretion can occur after peritoneal injections (73). These limits are determined by the rate and extent of silicon absorption from the gastrointestinal tract into the blood. In the ruminant this is influenced by the solubility of silica in the rumen fluid (47). Once silicon has entered the bloodstream, it must pass rapidly into the urine and tissues, because even at widely divergent intakes, the silicon level in the blood remains relatively constant. In the experiments of Jones and Handreck (47) cited earlier in this section, it was reported that in sheep the sum of the amounts appearing in the urine and feces was within 1% of the amounts ingested, indicating that body retention was small. Urinary excretion was also low. In three separate experiments with sheep the proportion excreted in the urine, but not the amounts, decreased progressively from 3.3 to 0.55% as the silicon intake increased from 0.4 to 14 g/day (35,47,67). The amounts excreted in the urine represented only small proportions of the amounts ingested, while in the only known silicon balance study in humans (54) urinary excretion represented a large portion of the silicon ingested and more silicon was excreted than was taken in. The recovery of ingested silica in steers and sheep has also been found to often exceed the calculated intake.

III. SILICON DEFICIENCY AND FUNCTIONS

Silicon is essential for growth and skeletal development in rats and chicks, and a mechanism and site of action have been identified.

A. Growth

By means of specially purified diets and a plastic isolator environment, Schwarz and Milne (80) were able to demonstrate significant increases in the growth rate of rats from the addition of 50 mg silicon per 100 g of diet (500 ppm) as sodium metasilicate ($Na_2SiO_3 \cdot 9H_2O$) in aqueous solution. The increases in weight of the weanling rats over that of the controls were 33.8% on one basal diet and 25.5% on another, each over a 26-day period (Table II). Lower levels of supplementary silicon gave statistically insignificant responses. The unsupplemented animals also exhibited an impaired incisor pigmentation that was significantly improved but not prevented by silicon supplements. Similar responses in the growth of chicks from supplementation of silicon supplied as sodium metasilicate at 500 ppm to purified diets were independently demonstrated by Carlisle (14). The results of three studies, in which 30, 30, and 49.8% increases in average daily weight gain over a 23-day period were obtained, are given in Table II. The deficient chicks were smaller but in proportion, with all organs appearing relatively atrophied, and with the legs and comb particularly pale. The deficient chicks had no wattles, and the combs were severely attenuated. Skeletal development was significantly retarded, as discussed in the next section.

B. Calcification

The first indications of physiological role for silicon came from the *in vitro* electron microprobe studies of Carlisle (11) showing silicon to be uniquely

Table II

Growth Effects of Dietary Silicon in Rats and Chicks Maintained in a Trace Element-Controlled
Environment on Low-Silicon Diets

	Number of animals	Average daily weight gain (g)	Percentage increase	p
Rat studies[a]				
Basal diet A				
Control	15	1.51 ± 0.11		
50 mg/dl Si	11	1.02 ± 0.08	33.8	<.005
Basal diet B				
Control	12	1.19 ± 0.06		
50 mg/dl Si	11	1.49 ± 0.06	25.2	<.005
Chick studies[b]				
Study no. 1				
Control	36	2.37 ± 0.11		
Si-supplemented	36	3.10 ± 0.10	30	<.01
Study no. 2				
Control	30	3.25 ± 0.09		
Si-supplemented	30	4.20 ± 0.09	30	<.02
Study no. 3				
Control	48	2.57 ± 0.09		
Si-supplemented	48	3.85 ± 0.11	49.8	<.01

[a]From Schwarz and Milne (80).
[b]From Carlisle (14).

localized in active growth areas in young bone of mice and rats. The amount present in specific very small regions within the active growth areas appeared to be uniquely related to "maturity" of the bone mineral. In the earliest stages of calcification in these regions both the silicon and calcium contents of the osteoid tissue were found to be very low, but as mineralization progressed the silicon and calcium contents rose congruently. In a more advanced stage the amount of silicon fell markedly, so that as calcium approached the proportion present in bone apatite, the silicon was present only at the detection limit. In other words, the more "mature" the bone mineral, the smaller the amount of measurable silicon. Further studies of the calcium–phosphorus ratio in silicon-rich regions gave values <1.0 compared with a calcium–phosphorus ratio of 1.67 in mature bone apatite. These findings suggested strongly that silicon is involved with phosphorus in an organic phase during the series of events leading to calcification.

Subsequent *in vivo* experiments showed that silicon has a demonstrable effect on *in vivo* calcification (12,15). In studies with weanling rats on a low-silicon (<5 ppm) diet or on supplements of 10, 25, or 250 ppm silicon, silicon was found to hasten the rate of bone mineralization (12). Tibia of the rats on the 250-ppm supplement reached a higher degree of mineralization in a shorter time than tibia from the low- and medium-silicon diets. Calcium content of the bone also increased with increased dietary silicon, substantiating the theory of a relationship between mineralization and silicon intake. The tendency of silicon to accelerate mineralization was also demonstrated by its effect on bone maturity, as indicated by the calcium–phosphorus ratio. The concept of an agent that affects the speed of chemical maturity of bone is not new. Muller *et al.* (64) found that the chemical maturity of vitamin D-deficient bone (measured by the amount of heat-produced pyrophosphate), although inferior to control bone during the period of maximum growth, approaches the control level at the end of the experiment.

C. Bone Formation

The earliest studies suggesting a role for silicon in bone formation were those mentioned above. Most significant, however, was the establishment of a silicon deficiency state incompatible with normal skeletal development. In the chick, this is evidenced by reduced circumference, thinner cortex, and less flexibility of leg bones, as well as by smaller, abnormally shaped skulls with the cranial bones appearing flatter (14). Silicon deficiency in rats was also shown to result in skull deformations (80). Subsequent *in vivo* studies with chicks demonstrated a relationship between silicon, magnesium, and fluorine in growing bone (13).

Later studies also emphasize the importance of silicon in bone formation. Skull abnormalities associated with reduced collagen content have been produced in silicon-deficient chicks under conditions promoting optimal growth using a semisynthetic diet containing a natural protein in place of the crystalline amino acid diets used in earlier studies (21). Additionally, a striking difference between the silicon-deficient and silicon-supplemented chicks was observed in the appearance of the skull matrix, the matrix of the deficient chicks totally lacking the normal striated trabecular pattern of the control chicks. The deficient chicks showed a nodular pattern of bone arrangement, indicative of a primitive type of bone.

Using the same conditions as above, and by introducing three different levels of vitamin D, it has also been shown that the effect exerted by silicon on bone formation is substantially independent of the action of vitamin D (24). All chicks on silicon-deficient diets, regardless of the level of dietary vitamin D, had gross abnormalities of skull architecture, and furthermore, the silicon-deficient skulls showed considerably less collagen at each vitamin D level. As in the previous study, the bone matrix of the silicon-deficient chicks totally lacked the normal striated trabecular pattern of the control chicks. In the rachitic groups of chicks, the appearance of the bone matrix was quite different from the groups receiving adequate vitamin D, being considerably less calcified and more transparent, enabling the cells and underlying structure to be seen more easily. The deficient chicks appeared to have a marked reduction in the number of osteoblasts compared to the controls. In these two studies, the major effect of silicon appears to be on the collagen content of the connective tissue matrix, and this is independent of vitamin D.

D. Cartilage and Connective Tissue Formation

In addition to bone, silicon-deficiency is manifested by abnormalities involving articular cartilage and connective tissue (18). Chicks in the silicon-deficient group had thinner legs and smaller combs in proportion to their size. Long-bone tibial joints were markedly smaller and the bones contained 34–35% less water than those of silicon-supplemented chicks. The deficient chicks also revealed a significantly lower hexosamine content in their articular cartilage. In cock's comb also, a smaller amount of connective tissue, a lower total percentage of hexosamines and a lower silicon content were found in the silicon-deficient group. These findings point clearly to an involvement of silicon in glycosaminoglycan formation in cartilage and connective tissue.

Long-bone abnormalities similar to those reported above have been produced in silicon-deficient chicks using a semisynthetic diet containing a natural protein in place of crystalline amino acids used in the earlier studies, again demonstrating (22) a requirement for silicon in articular cartilage formation. Tibia from silicon-deficient chicks had significantly less glycosaminoglycans and collagen, the difference being greater for glycosaminoglycans than collagen. Tibia from silicon-deficient chicks also showed rather marked pathological changes; the most profound being demonstrated in epiphyseal cartilage. The disturbed epiphyseal cartilage sequences resulted in defective endochondral bone growth, indicating that silicon is involved in a metabolic chain of events required for normal growth of bone.

E. Connective Tissue Matrix

Silicon's primary effect in bone and cartilage appears to be on formation of the matrix, although silicon may participate in the mineralization process itself. The above *in vivo* studies have shown silicon to be involved in both collagen and glycosaminoglycan formation. These *in vivo* findings have been corroborated in studies of bone (19) and cartilage (23) in tissue culture, where it was demonstrated that silicon has a marked effect on growth, which appeared to be mainly due to an increase in collagen content. Silicon was also shown to be required for formation of glycosaminoglycans, the other major polymeric molecule of the matrix.

An interaction between silicon and ascorbate has also been shown in cartilage matrix (28). Silicon's greatest effect was on hexosamine content in the presence of ascorbate. Furthermore, silicon and ascorbate interacted to give maximal production of hexosamines. Silicon also appeared to increase hydroxyproline, total protein, and noncollagenous protein beyond the effects of ascorbate.

An effect of silicon on formation of extracellular cartilage matrix components by connective tissue cells has also been demonstrated (27), in chondrocytes isolated from chick epiphyses cultured under silicon-low and silicon-supplemented conditions. The major effect of silicon appeared to be on collagen. Silicon also had a pronounced stimulatory effect on matrix polysaccharides. Silicon's effect on collagen and glycosaminoglycan formation was not due to cellular proliferation but to some system in the cell that participates in their formation.

Additional support for silicon's metabolic role in connective tissue at the cellular level is provided by evidence of its presence in connective tissue cells (17). X-Ray microanalysis of active growth areas in young bone and isolated osteoblasts show silicon to a be a major ion of osteogenic cells, the amounts of silicon being in the same range as that of calcium, phosphorus, and magnesium. Moreover, silicon appeared to be especially high in the metabolically active state of the cell, the osteoblast. Clear evidence that silicon occurs in the osteoblast and is localized in the mitochondria adds strong support to the proposition that silicon is required for connective tissue matrix formation.

F. Enzyme Activity

A dependence on silicon for maximal prolyl hydroxylase activity has been demonstrated (25). Prolyl hydroxylase obtained from frontal bones of 14-day-old chick embryos incubated for 4 or 8 days under low-silicon conditions with 0, 0.2, 0.5, or 2.0 mM silicon added to the media show lower enzyme activity in low-silicon bones with increasing enzyme activity in 0.2, 0.5, and 2.0 mM cultures. The results support the *in vivo* and *in vitro* findings of a requirement for silicon in collagen biosynthesis, the activity of prolyl hydroxylase being a measure of the rate of collagen biosynthesis.

IV. SILICON REQUIREMENTS AND SOURCES

The minimum dietary silicon requirements compatible with satisfactory growth and health are largely unknown, although there is limited evidence obtained from experiments with rats and chicks. The silicon requirements for growth and satisfactory skeletal development in the experiments reported with rats using a basal ration containing ~7 μg/g of silicon approximated 50 μg/g of

dry diet (80), relatively large amounts of silicon provided as water-soluble sodium metasilicate appearing to be needed to produce an optimum response. In a
later report (78) it was mentioned that other silicon compounds were found to be
more effective; however, this work was never published. Whether the silicon
compounds occurring in natural materials would be more available or less available and the silicon requirement therefore lower or higher than the 500 μg/g
tentatively given must await the results of further research. The basal ration used
in the experiments with chicks (14,21) contained ~1 μg/g silicon, and a significant effect on skull formation appeared at a level of 250 μg/g of silicon as water-
soluble sodium metasilicate. Evidence that the dietary silicon requirement is
relatively high relative to those of some of the other trace elements is further
apparent from the experiments of Carlisle (12) on the rate of bone mineralization
in chicks. Silicon supplementation of a low-silicon diet at the rate of 250 μg/g
increased the ash content of the tibia significantly more at 2 weeks than did either
25 or 10 μg/g, although the differences had largely disappeared at 5 weeks.
Unfortunately, a supplementation level of 50 μg/g silicon was not included in
this experiment, so that one can only speculate whether such a level would have
been as effective as the 250 μg/g.

The demonstration of the essentiality of silicon for higher animals is recent, so
that reliable data on the silicon content of human foods and dietaries are meager.
Furthermore, little is known of the extent of silicon absorption from various
sources; for example, some forms of silicon are very insoluble. Also, since
silicon is ubiquitous in the environment, the likelihood of a silicon deficiency
arising under natural conditions in humans or domestic animals might be questioned. Of possible significance here is the suggestion that silicon absorption
might be under hormonal regulation (30), and if so, a decline in hormonal
activity in senescence might result in decreased silicon absorption.

Foods of plant origin are normally much richer in silicon than those of animal
origin. In plants the amounts or proportions of silicon present as monosilicic acid
and solid silica vary with the species, stage of growth, and soil conditions under
which the plant has grown. Whole grasses and cereals may contain 3–4% of the
whole dry plants, as SiO_2, with levels up to 6% silica in some range grasses (10).
In leguminous plants total silicon concentrations are appreciably lower, the levels approximating those found in animal tissues, with a high proportion of the
relatively low amounts of silicon present as monosilicic acid. Solid silica is only
sparsely deposited in these species (6,10). Cereal grains high in fiber such as oats
are much richer in silicon than low-fiber grains such as wheat or maize (68). This
would suggest that the silicon content of patent white flour is significantly lower
than that of the whole wheat from which it is made. Substantial losses of silicon
occur in the refining of sugar. Thus Hamilton and Minski (41) reported the
following values (in micrograms of silicon per gram dry weight): Barbados
brown sugar, 735, Demerara sugar 60, refined sugar 2, and granulated sugar 4.

These workers also obtained a mean total silicon intake from human adult diets consumed in Great Britain of 1.2 ± 0.1 g/day. The authors were careful to point out that their data for silicon are subject to possible contamination because of contact between the samples and glass surfaces prior to receipt by the laboratory, although they state that "previous studies suggest that the degree of contamination is slight."

In mature gramineous plants most of the silicon present is in the form of solid mineral particles, known as opal phytoliths ($SiO_2 \cdot H_2O$) (6). The marked species differences among plants in the total amount of silicon absorbed and subsequently secreted as phytoliths is illustrated by one study in which ryegrass (*Lolium perenne*) was found to contain 23 times as much insoluble ash as alfalfa (*Medicago sativa*). This difference was reflected in the silicon and opal phytolith contents of the two species (6). In another study, prairie grass hay (mainly *Festuca scabrella*) averaged 2.92% total silicon dry basis compared with only 0.18% silicon in alfalfa hay (3).

Since the opal phytoliths of pasture plants are harder than the dental tissues of sheep, and the amounts ingested by grazing animals are so large and continuous, it has been suggested that these minerals may be a major cause of wear in sheep's teeth (5). At a level of 4% SiO_2 and a dry-matter intake of 1 kg/day, a sheep would ingest 40 g SiO_2/day, or 14 kg over a period of a year. This represents a very large amount of abrasive material. Moreover, it excludes the large quantities of silica that can occur as quartz particles on the surface of plants from soil contamination. This source of adventitious silica has been implicated as a significant source of wear of teeth in grazing sheep (43).

V. TOXICITY

A. Silicosis in Humans

Investigations in the area of silicon toxicity are almost invariably associated with the silicosis problem. Detailed consideration of the disease silicosis, which occurs in certain classes of miners as a result of the continued inhalation of silica particles into the lungs, lies outside the scope of this text. Particles of silica and asbestos (fibrous silicates of complex composition) have long been known to stimulate a severe fibrogenic reaction in the lungs and elsewhere in the body. This reaction arises initially from phagocytosis of silica particles by alveolar macrophages. Collagen synthesis by neighboring fibroblasts is stimulated by the death of these macrophages. The particular toxicity of silica to macrophages derives from the fact that the particles are taken up into lysosomes and readily damage lysosomal membranes through hydrogen-bonding reactions (1,65). Heppleston and Styles (44) have provided evidence that the macrophage–silica interaction results in the release of a factor of unknown nature that stimulates collagen formation, a finding of great interest in light of the involvement of silicon in collagen synthesis disclosed by research. In a second step that occurs at the same time, silica evidently stimulates the production of lipids. It appears that lipid may produce a systemic and even a local role designed to maintain a supply of macrophages to replace those destroyed under the local action of silica. This second aspect requires further investigation.

Malignant tumors of the pleura and peritoneum constitute a further manifestation of the toxicity of silicon in the form of asbestos (fibrous silicates) to humans and experimental animals (87). This is now identified as an important public health problem. Silicon (quartz) does not reproduce these carcinogenic effects. Physical shape and size of the carcinogenic particles appear to be more important than their chemical composition in the induction of asbestosis and cancer; for example, the particles must be an accurate size to produce the disease, and needlelike particles of alumina or nonsiliceous glass can cause similiar effects, at least in test animals where this type of cancer develops more rapidly (74).

There is no clear understanding of the cellular mechanisms that underlie the development of cancer associated with silicon particulate matter; however, there is good reason to believe that such an effect is not related simply to the presence of silicon as an element.

B.   Silica Urolithiasis

In contrast to the well-known occurrence of calculi composed chiefly of silica in some animals is a sparcity of information on urological problems in humans associated with silica.

Normally, urinary silica is readily excreted, but under some conditions part of the silicon of the urine is deposited in the kidneys, bladder, or urethra to form calculi or uroliths. Small calculi may be excreted without harm, while large calculi can block the passage of urine and cause death. Urinary calculi can be composed of various predominant minerals, particularly calcium, magnesium, phosphorus, and silicon. Silica urolithiasis is a serious problem in grazing wethers in Western Australia (69) and in grazing steers in the western regions of Canada (32,88) and the northwestern parts of the United States (71,83). This is in contrast to urolithiasis in humans, where oxalates, urates, or phosphates play the predominant role.

The silica of ovine and bovine calculi has been identified as amorphous opal (4,36), most of which is derived from the absorbed monosilicic acid. A small proportion of the opal occurs as phytoliths from plants, with occasional fragments of sponge spicules and diatoms embedded in the calculi (4). In addition to hydrated silica, siliceous uroliths contain small amounts of accessory elements, notably magnesium, calcium, and organic material. The exact nature of the organic material, or matrix, in siliceous calculi has not been determined, but the chemical studies of Keeler (49,53) indicate the presence of a glycoprotein containing a neutral carbohydrate moiety.

The factors responsible for the formation of siliceous calculi are poorly defined. Attempts to produce them in sheep and cattle by adding silicates to the diet (8,88) or by restricting water consumption (84) have not been successful, even when the urinary excretion level achieved was two- to threefold greater than the 70–80 μg /ml level at which silicon normally precipitates in bovine urine (52). The concentration in the urine of sheep (67) and cattle (3) usually exceeds that of a saturated solution of amorphous silica and may reach 467 ppm silicon in sheep. It is clear that high dietary intakes and high silica outputs in the urine, associated with supersaturation of the urine, are insufficient to explain the polymerization of the monosilicic acid, deposition of silica, and formation of calculi.

It has been suggested that among the accessory elements, magnesium might be involved in the formation of calculi (4). However, it is not possible to state whether these elements are simply occluded in the silica or whether they play a role in the aggregation of urolithic material.

The glycoprotein of the organic matrix has been assigned a critical role in the formation of urinary calculi in humans, through acting as a primary matrix that becomes secondarily mineralized. A similar theory has been adopted to explain the formation of siliceous calculi in cattle (50) and phosphatic calculi in sheep, cattle, and dogs (33). Jones and Handreck (48) have disputed this theory on the grounds that the main mechanism involved is "precipitation of the inorganic components which, in turn, depends upon both supersaturation and nucleation." They tend to favor the theory that foreign particles of silica and other foreign particles, such as have been found in calculi from sheep, act as nuclei for the deposition of silica. Solid (amorphous) silica is known to accelerate the polymerization and deposition of silica from supersaturated solution, but whether solid particles of silica are consistently implicated in the formation of siliceous calculi is unknown. In view of the marked dependence of urinary silica concentration on the rate of urine excretion in cattle, it has been suggested that any method that increases urine output could also be used to prevent urolith formation (3).

C. Silica Forage Digestibility

The digestibility of forage dry matter *in vivo* has been shown to be significantly depressed by increasing levels of silica (SiO_2 in ash from plant tissues) (81,85). These observations were investigated further by Smith and co-workers (82), who found that aqueous sodium silicate added to rumen cultures significantly depressed the organic matter digestibility *in vitro* of siliceous forages, already known to exhibit depressed organic matter digestibility because of silica accumulated in the plant tissues. The depression generally amounted to ~1 percentage unit of organic matter digestibility for each increase of 100 mg/liter in "soluble silica" concentration but was greater when a highly siliceous grass was used as the substrate. The effect of silicate was modified by adding glucose, urea, and/or a mixture of minerals (magnesium, manganese, zinc, cobalt, and copper), suggesting that availability of minerals to sustain cellulolytic microbial activity may be a major factor influencing the effect of soluble silica on forage digestion by ruminants.

REFERENCES

1. Allison, A. C., Harrington, J. S., and Birbeck, M. (1966). *J. Exp. Med.* **124,** 141.
2. Archibald, J. G., and Fenner, H. (1957). *J. Dairy Sci.* **40,** 703.
3. Bailey, C. H. (1967). *Am. J. Vet. Res.* **28,** 1743.
4. Baker, G., Jones, L. H. P., and Milne, A. A., (1961). *Aust. J. Agric. Res.* **12,** 473.
5. Baker, G., Jones, L. H. P., and Wardrop, I. D. (1959). *Nature (London)* **184,** 1583.
6. Baker, G., Jones, L. H. P., and Wardrop, I. D. (1961). *Aust. J. Agric. Res.* **12,** 426.
7. Baumann, H. (1960). *Hoppe-Seyler's Z. Physiol. Chem.* **319,** 38; **320,** 11 (1960).
8. Beeson, W. M., Pence, J. W., and Holan, G. C. (1943). *Am. J. Vet. Res.* **4,** 120.
9. Benke, G. M., and Osborn, T. W. (1978). *Food Cosmet. Toxicol.* **17,** 123.
10. Bezeau, L. M., Johnston, A., and Smoliak, S. (1966). *Can. J. Plant Sci.* **46,** 625.
11. Carlisle, E. M. (1969). *Fed. Proc., Fed. Am. Soc. Exp. Biol.* **28,** 374; *Science* **167,** 279 (1970).
12. Carlisle, E. M. (1970). *Fed. Proc., Fed. Am. Soc. Exp. Biol.* **29,** 565.
13. Carlisle, E. M. (1971). *Fed. Proc., Fed. Am. Soc. Exp. Biol.* **30,** 462.
14. Carlisle, E. M. (1972). *Fed. Proc., Fed. Am. Soc. Exp. Biol.* **31,** 700; *Science* **178,** 619 (1972).
15. Carlisle, E. M. (1973). *Fed. Proc., Fed. Am. Soc. Exp. Biol.* **32,** 930.
16. Carlisle, E. M. (1974). *Fed. Proc., Fed. Am. Soc. Exp. Biol.* **33,** 1758.
17. Carlisle, E. M. (1975). *Fed. Proc., Fed. Am. Soc. Exp. Biol.* **34,** 927.
18. Carlisle, E. M. (1976). *J. Nutr.* **106,** 478.
19. Carlisle, E. M., and Alpenfels, W. F. (1978). *Fed. Proc., Fed. Am. Soc. Exp. Biol.* **37,** 404.
20. Carlisle, E. M. (1979). *Fed. Proc., Fed. Am. Soc. Exp. Biol.* **38,** 553.
21. Carlisle, E. M. (1980). *J. Nutr.* **10,** 352.
22. Carlisle, E. M. (1980). *J. Nutr.* **10,** 1046.
23. Carlisle, E. M., and Alpenfels, W. F. (1980). *Fed. Proc., Fed. Am. Soc. Exp. Biol.* **39,** 787.
24. Carlisle, E. M. (1981). *Calcif. Tissue Int.* **33,** 37.
25. Carlisle, E. M., Berger, J. W., and Alpenfels, W. F. (1981). *Fed. Proc., Fed. Am. Soc. Exp. Biol.* **40,** 866.
26. Carlisle, E. M. (1982). *Nutr. Rev.* **40,** 193.
27. Carlisle, E. M., and Garvey, D. L. (1982). *Fed. Proc., Fed. Am. Soc. Exp. Biol.* **41,** 461.
28. Carlisle, E. M., and Suchil, C. (1983). *Fed. Proc., Fed. Am. Soc. Exp. Biol.* **42,** 398.
29. Carlisle, E. M. (1984). Unpublished data.
30. Charnot, Y., and Peres, G. (1971). *Annee Endocrinol.* **32,** 397.
31. Charnot, Y., and Peres, G. (1978). *In* "Biochemistry of Silicon and Related Problems" (G. Bendz and I. Lindquist, eds.), p. 269. Plenum, New York.

32. Connell, R., Whiting, F., and Forman, S. A. (1959). *Can. J. Comp. Med. Vet. Sci.* **23**, 41.
33. Cornelius, C. E., and Bishop, J. A. (1961). *J. Urol.* **85**, 842.
34. Dobbie, J. W., and Smith, M. J. B. (1982). *Scott. Med. J.* **27**, 17.
35. Emerich, R. J., Embay, L. B., and Olson, O. E. (1959). *J. Anim. Sci.* **18**, 1025.
36. Forman, S. A., Whiting, F., and Connell, R. (1959). *Can. J. Comp. Med. Vet. Sci.* **23**, 157.
37. Fregert, S. (1959). *Acta Derm./-/Venereol., Suppl.* **39**, 1.
38. Gallup, W. D., and Norris, L. C. (1939). *Poult. Sci.* **18**, 76.
39. Geissler, U., and Gerloff, J. (1965). *Nova Hedwigia* **10**, 565.
40. Goldwater, L. J. (1936). *J. Ind. Hyg. Toxicol.* **18**, 163.
41. Hamilton, E. I., and Minski, M. J. (1972–1973). *Sci. Total Environ.* **1**, 375.
42. Hamilton, E. I., Minski, M. J., and Cleary, J. J. (1972–1973). *Sci. Total Environ.* **1**, 341.
43. Healy, W. B., and Ludwig, T. (1965). *N. Z. J. Agric. Res.* **8**, 737.
44. Heppleston, A. W., and Styles, J. A. (1967). *Nature (London)* **214**, 521.
45. Holt, P. F. (1950). *Br. J. Ind. Med.* **7**, 12.
47. Jones, L. H. P., and Handreck, K. A. (1969). *J. Agric. Sci.* **65**, 129.
48. Jones, L. H. P., and Handreck, K. A. (1967). *Adv. Agron.* **19**, 107.
49. Keeler, R. F. (1960). *Am. J. Vet. Res.* **21**, 428.
50. Keeler, R. F. (1963). *Ann. N.Y. Acad. Sci.* **104**, 592.
51. Keeler, R. F., and Lovelace, S. A. (1959). *J. Exp. Med.* **109**, 601.
52. Keeler, R. F., and Lovelance, S. A. (1961). *Am. J. Vet. Res.* **22**, 617.
53. Keeler, R. F., and Swingle, K. F. (1959). *Am. J. Vet. Res.* **20**, 249.
54. Kelsay, J. L., Behall, K. M., and Prather, E. S. (1979). *Am. J. Clin. Nutr.* **32**, 1876.
56. King, E. J., and McGeorge, M. (1938). *Biochem. J.* **32**, 417.
57. King, E. J., Stacy, B. D., Holt, P. F., Yates, D. M., and Pickles, D. (1955). *Analyst* **80**, 441.
58. Kirchgessner, M. (1959). *Z. Tierphysiol., Tierernaehr. Futtermittelkd.* **14**, 270 and 278.
59. Leslie, J. G., Kung-Ying, T. K., and McGavack, T. H. (1962). *Proc. Soc. Exp. Biol. Med.* **110**, 218.
60. LeVier, R. R. (1975). *Bioinorg. Chem.* **4**, 109.
61. Loeper, J., Loeper, J., and Lemaire, A. (1966). *Presse Med.* **74**, 865.
62. Losee, F., Cutress, T. W., and Brown, R. (1974). *In* "Trace Substances Environmental Health-7" (D. D. Hemphill, ed.), p. 19. Univ. of Missouri Press, Columbia.
63. McGavack, T. H., Leslie, J. G., and Tang Kao, K. (1962). *Proc. Soc. Exp. Biol. Med.* **110**, 215.
64. Muller, S. A., Posner, H. S., and Firschein, H. E. (1966). *Proc. Soc. Exp. Biol. Med.* **121**, 844.
65. Nash, T., Allison, A. C., and Harrington, J. S. (1966). *Nature (London)* **210**, 259.
66. Needham, A. E. (1965). "The Uniqueness of Biological Materials." Pergamon, Oxford.
67. Nottle, M. C. (1966). *Aust. J. Agric. Res.* **17**, 175.
68. Nottle, M. C. (1962). Private communication.
69. Nottle, M. C., and Armstrong, J. M. (1966). *Aust. J. Agric.* **17**, 165.
70. Okuda, A., and Takahashi, E. (1964). *Symp. Int. Rice. Res. Inst.,* p. 123.
71. Parker, K. G. (1957). *J. Range Manage.* **10**, 105.
72. Policard A., Collet, A., Moussard, D. H., and Pregermain, S. (1961). *J. Biophys. Biochem. Cytol.* **9**, 236.
73. Sauer, F., Laughland, D. H., and Davidson, W. M. (1959). *Can. J. Biochem. Physiol.* **37**, 183 and 1173.
74. Selikoff, I. J. (1978). *In* "Biochemistry of Silicon and Related Problems" (G. Bendz and I. Lindquist, eds.), p. 311. Plenum, New York.
76. Schaible, P. J., and Bandemer, S. L. (1942). *Poult. Sci.* **21**, 8.
77. Schwarz, K. (1973). *Proc. Natl. Acad. Sci., U.S.A.* **70**, 1608.

78. Schwarz, K. (1978). *In* "Biochemistry of Silicon and Related Problems" (G. Bendz and I. Lindquist, eds.), p. 207. Plenum, New York.
80. Schwarz, K., and Milne, D. B. (1972). *Nature (London)* **239,** 333.
81. Smith, G. S., Nelson, A. B., and Boggino, E. J. A. (1971). *J. Anim. Sci.* **33,** 466.
82. Smith, G. S., and Urquhart, N. S. (1975). *J. Anim. Sci.* **41,** 882; Smith, G. S., and Nelson, A. B., *ibid.*, p. 891.
83. Swingle, K. F. (1953). *Am. J. Vet. Res.* **14,** 493.
84. Swingle, K. F., and Marsh, H. (1953). *Am. J. Vet. Res.* **14,** 16.
85. Van Soest, P. J., and Jones, L. H. P. (1965). *J. Diary Sci.* **51.**
86. Volkenheimer, G. Z. (1964). *Gastroenterology,* **2,** 57.
87. Wagner, C. (1966). *Perugia Quadrenn. Int. Conf. Cancer, 3rd,* p. 589.
88. Whiting, F., Connell, R., and Forman, S. A. (1958). *Can. J. Comp. Med. Vet. Sci.* **22,** 332.
89. Worth, G. (1952). *Klin. Wochenschr.* **30,** 82.

8

Lithium

WALTER MERTZ

U.S. Department of Agriculture
Agricultural Research Service
Beltsville Human Nutrition Research Center
Beltsville, Maryland

I. LITHIUM IN SOILS, PLANTS, AND ANIMAL TISSUES

Lithium (Li) belongs to the small number of trace elements that exhibit three distinct zones of activity in the intact organism depending on exposure levels: biological, pharmacodynamic, and toxic (25). Lithium's pharmacodynamic action, discovered by Cade in 1949 (8), has since found wide application in the treatment of manic–depressive psychoses (1). That action is associated with intakes that maintain plasma concentrations of ~1 mmol/liter, corresponding to 7000 ng/ml. Signs of toxicity have been reported at only twice that concentration. In contrast, the zone of biological activity encompasses plasma concentrations between <10 ng/ml and fractions of a nanogram per milliliter (18).

Lithium, a light metal with an atomic number of 3 and an atomic weight of 6.94, is widely distributed throughout the geosphere and biosphere, ranking twenty-seventh in abundance among the elements. It is one of the few elements for which experimental evidence has shown differences in the metabolism of its two isotopes lithium-6 (^{6}Li) and lithium-7 (^{7}Li) in a living system (6).

Before discussing the reported concentrations of lithium in the environment, it is necessary to assess the validity of the analytical methods used. The sensitive methods of neutron activation analysis are not suitable for lithium, but mass spectrometry, certain chemical detection methods, and highly sophisticated flame emission spectrometric methods can be expected to give reliable results in the concentration range of nanograms per gram. At these levels, contamination during sample preparation becomes a problem imposing stringent demands on the purity of reagents, containers, and the atmosphere of the work space. Furthermore, Standard Reference Materials (SRM), certified for lithium, are not yet available (only one, SRM No. 1571, orchard leaves, has been issued with an "information value" of 0.6 μg/g), and only very few reported lithium data are validated by simultaneous analysis of the SRM (18). In view of the rapid developments in analytical techniques, values reported in the literature for lithium in biological materials at concentrations in the nanograms per gram range should be carefully evaluated and cautiously interpreted, whereas the much higher concentrations of lithium in rocks and soil (and those in patients under lithium treatment) can be accepted as correct.

Older studies reporting concentrations of lithium in soils ranging from <10 to >100 μg/g are in agreement with later measurements (24) in the United States and in Europe (3). The latter study reported a mean lithium content of all soils examined of 28 μg/g, with much lower concentrations (2.9–4.7 μg/g) in swamp and muck soils of high organic content.

The lithium content of plants is correlated with the soil lithium concentration and is strongly dependent on plant species. It is therefore not surprising that the range of concentrations (per dry matter) reported is very wide—from 0.4 μg/g in Cyperaceae grown in France (5) to >1000 μg/g in pickleweed growing in lithium-rich basin soils (10). Fourfold differences in lithium concentrations have been reported in one species of plants grown on soils of different lithium content. A study in Eastern Europe concluded that leafy vegetables, roots, and fruits are relatively rich in lithium, in contrast to the seeds of most cereals and leguminous species (Table I). Most animal feeds of vegetable origin were reported within or somewhat above that range of concentrations. The values reported by Robinson were mostly lower by an order of magnitude (20).

Two of the latest studies of lithium in human and rat tissue arrived at very low values. Hamilton (12) reported the following lithium concentrations (in nanograms per gram wet weight): human ovaries 2, testes 3, blood 6, liver 7, lungs 60, and lymph nodes 200. Patt et al. (16) found similar low values in the tissue of rats fed experimental diets of 5 to 10 ng/g lithium content. Burt reported average levels of 0.3 ng/g in the blood of rats kept on very low-lithium diets (3–5 ng/g), as compared to ~10 times that concentration in lithium-supplemented controls (7).

A study using inductively coupled plasma emission spectrometry reported, in wet weight, lithium levels in fish muscle from the Pacific of 20 and 108 ng/g for Pacific cod and Pacific whiting, respectively (23).

Total dietary lithium intakes by human subjects have been variously estimated as 2000 (14), 107 (12), and 100 μg/day (22). Wittrig et al. (27) estimated the total lithium absorbed by the intestines from 1 day's diet at 20 μg, but Pickett and Hawkins (17) detected only 2.63 ng/ml (with 1% relative SD), which would

Table I
Lithium Content of Several Foods and Feeds Grown on
Naturally Lithium-Poor Soil of Eastern Europe[a,b]

Item	n	$\bar{X}$
Leguminous seeds		
Beans	32	5.7
Peas (animal feed)	5	0.74
Peas (for human consumption)	34	0.28
Cereal grains		
Corn	24	1.82
Winter rye	28	0.97
Winter wheat	35	0.67
Barley	66	0.67
Oats	37	0.48
Forages		
Corn stalks	23	3.2
Alfalfas	22	3.0
Red clover	16	2.6

[a]From Ref. (19).
[b]Lithium content given in micrograms per gram dry weight.

translate into an absorbed amount of only 2 to 5 µg/day, because it is known that urine is the major excretory route for lithium.

These data in their entirety do not allow an estimate of the human lithium intake, and validated analytical data are urgently needed.

II. LITHIUM METABOLISM

Little is known of the metabolic behavior of lithium from foods and dietaries at physiological levels. Medicinal lithium salts in drinking water are believed to be mostly absorbed and largely excreted in the urine (15,21). It can be postulated, in analogy to the behavior of sodium and potassium, that most lithium in foods may also be available for absorption and that the daily urinary lithium excretion would provide a reasonable index of dietary intake.

One series of studies demonstrated a highly specific tissue distribution of lithium in rats fed diets low or adequate in lithium, with 5 to 10 and >500 ng/g, respectively (16) (Table II).

The high concentrations of lithium in the two endocrine organs and the preservation of these high levels despite extreme low-lithium diets may indicate these organs as specific sites of action for lithium.

Table II
Lithium Concentrations in Tissues from Rats
on Low- and Adequate-Lithium Diets[a,b]

Rat tissue	Dietary lithium (ng/g)	
	<10	>500
Plasma	3.6	5.8
Whole blood	5.9	11.0
Liver	2.7	14.0
Kidneys	4.6	63.0
Thymus	16.0	18.0
Brain	20.0	25.0
Adrenals	57.0	61.0
Pituitary	140.0	130.0

[a]From Ref. (16).
[b]Tissue concentrations given in nanograms per gram wet weight.

Bone can serve as a store for lithium, and concentrations of 100 to 200 μg/g in lithium-supplemented rats have been reported as compared to 5 to 20 μg/g in those fed the deficient diets.

On the other hand, an independent study of lithium concentrations in goats fed experimental rations of different lithium content used a different analytical method and reported concentrations one to two orders of magnitude higher (3,4).

III. EFFECTS OF LITHIUM

Following the discovery by Cade (8) of a strong therapeutic effect of lithium in manic–depressive disorders, lithium salts are widely used with good success in patients with this disorder. The dosages required to raise serum lithium to effective levels of approximately 0.5–1.5 meq/liter are close to 2.5 g of a lithium salt, or ~500 mg of the element per day. Toxic symptoms have been reported from serum concentrations as low as 1.6 meq/liter (2). In a proportion of patients, particularly women, the lithium treatment depresses the thyroid function, probably by blocking thyroxine release (9). In rats lithium has been shown to lower the uptake and turnover rate of ^{125}I by the thyroid (13).

In his reviews of the pharmacology and toxicology of lithium, Schou (21) has discussed the various potential modes of action of lithium therapy, for example, the effects on amine and electrolyte metabolism, thyroid and kidney function, hormone-stimulated adenyl cyclase activity, and carbohydrate metabolism. He pointed out the need for great caution in the interpretation of experimental data obtained with lithium concentrations *in vitro* that would be very toxic *in vivo*. He concluded his 1976 review with the statement: "The mode of action of lithium in manic–depressive disorder is as yet unknown" (21).

This statement is equally applicable to the *biological* action of lithium, independently reported by two groups of investigators.

Anke *et al.* (3) in 1981 reported a series of five consecutive experiments using a total of 27 goats on lithium-low control ration (3.3 µg/g) and 28 goats whose diet was supplemented to 24 µg/g lithium. Twenty-six controls and 25 experimental animals were evaluated: Live-weight gain per day during the first 180 days was 77 and 107 g for low-lithium goats and controls, respectively ($p < .05$). A crossover experiment in which 7 goats were first fed the high-lithium diet for 84 days, followed by depletion, demonstrated that the effect of low-lithium rations on weight gain is almost immediate and that under the conditions of the experiment, few body reserves were available. The low-lithium ration also affected reproductive performance of female goats: 74% of control animals but only 50% of the low-lithium goats became pregnant after the first insemination ($p < .05$). In addition, the low-lithium status had a highly significant influence on intrauterine weight gain of the lambs: the newborn of low-lithium animals weighed 2.6 kg, versus 3.2 kg for the newborn of controls ($p < .001$). This difference in birth weight, although somewhat attenuated, persisted throughout the first 3 months of life.

An extension of these studies, reported in 1983 (4), confirmed the previous results and furnished additional data: although no effect of the low-lithium status on mortality of the lambs was seen, the mortality of lithium-deficient mother goats during the first year was significantly greater than that of controls (29 versus 12%, $p < .05$). Deficiency significantly reduced the lithium content of milk to ~30% of control values, but did not depress milk production after the sixth week of lactation.

Two studies of lithium deficiency in rats have been reported; a third one is still under way at the time of this writing (18). Patt *et al.* (16) fed a diet of 5 to 15 ng/g lithium to groups of five animals each and found no growth depression or other abnormalities in the deficient group and their offspring into the third generation, except for a reduction in the number of litters produced. All lithium-supplemented rats (500 ng/g) had borne litters at day 42 as compared to only 60% of the deficient animals.

The second study used a more deficient diet of 2 ng/g lithium and a larger number of animals. Again, there was no effect on growth, but reproduction and survival of the litters was significantly depressed in lithium deficiency (7). The size of live litters at birth in deficient and supplemented animals was 6 versus 8 (F_1), 6.4 versus 9.2 (F_2), and 6.9 versus 9.2 (F_3), respectively (all $p < .05$). The litter weight at birth was 30.3 versus 46.1 g (F_1), 36.3 versus 52.8 g (F_2), and 39.9 versus 49.6 g, respectively (all $p < .05$), and the survival rate of the newborn to day 21 was 58 versus 97% ($p < .05$). Both studies reported a strong dependence of lithium concentrations in blood, brain, liver, kidneys, spleen, heart, lungs, thymus, and bone on dietary lithium supply. The important observation of the first study that the lithium concentrations in pituitary and adrenals were not reduced by lithium deficiency, was confirmed and extended in the second study, which also identified thyroid, ovarian, and mammary tissue as well as hippocampus as lithium-conserving organs. Although these findings suggest a possible involvement of lithium in neuroendocrine function, no differences were found in the concentration of four neurotransmitters in the hypothalamus of deficient versus supplemented rats.

An epidemiological study detected a negative correlation between the concentration of lithium in the drinking water of ~100 cities in the United States and age-adjusted mortality from atherosclerotic heart disease, as well as mortality in general (26). There is a close positive correlation of lithium and water hardness, and it was postulated that the well-known correlation of water hardness and heart disease can be explained by the high lithium content of hard waters. This postulate is consistent with the higher correlation coefficients for lithium and atherosclerotic heart disease, significant for all four subgroups of race and sex, than those for water hardness of which only one correlation, that for white males, was significant.

In another study the lithium concentration in drinking water in Texas correlated closely with the lithium excretion in the urine of randomly selected subjects from 26 counties. Either one of these parameters was inversely associated with the admission rates to mental hospitals in that state. Also the rate of homicides in the counties examined declined with increasing lithium concentrations in drinking water (11).

Although none of these correlations prove a cause–effect relationship, they are not inconsistent with the known effects of higher, therapeutic lithium intakes.

IV. LITHIUM REQUIREMENTS AND SOURCES

The evidence from animal experiments discussed above strongly suggests that there is a lithium requirement, at least for the two species studied. Because of the substantial discrepancies among the analytical data no quantitative estimate can be made, not even in form of a range of adequate intakes. Similarly, the data on human lithium intake vary so widely that an estimate of the adequacy of our intake cannot be made. It is obvious that progress in the field of lithium metabolism and nutrition depends on greatly increased analytical activities, on validation of methods, and on their application to food analysis and metabolic studies.

REFERENCES

1. AMA Division of Drugs (1983). "AMA Drug Evaluation," 5th ed., A.M.A., Chicago, Illinois.
2. Amdisen, A. (1967). *Scand. J. Clin. Lab. Invest.* **20**, 104.
3. Anke, M., Grün, M., Groppel, B., and Kronemann, H. (1981) *In* "Mengen-und Spurenele-mente" (M. Anke and H. J. Schneider, eds.), p. 217. Karl-Marx-Univ., Leipzig, G.D.R.
4. Anke, M., Groppel, B., Kronemann, H., and Grün, M. (1983). *In* "Lithium, 4. Spurenele-ment-Symposium" (M. Anke, W. Baumann, H. Bräunlich, and C. Brückner, eds.), p. 58. Wiss. Publ., Friedrich-Schiller-Univ., Jena, G.D.R.
5. Bertrand, D. (1959). *C. R. Hebd. Seances Acad. Sci.* **249**, 787.
6. Bowen, H. J. M. (1956). *J. Nucl. Energy* **2**, 255.
7. Burt, J. (1982). Ph.D. dissertation, University of Missouri, Columbia.
8. Cade, J. F. J. (1949). *Med. J. Aust.* **2**, 349.
9. Cade, J. F. J. (1975). *Med. J. Aust.* **1**, 684.
10. Cannon, H. L., Harms, T. F., and Hamilton, J. C. (1975). *Geol. Surv., Prof. Pap. (U.S.)* **918.**
11. Dawson, E. B., Moore, T. D., and McGanity, W. J. (1972). *Dis. Nerv. Syst.* **33**, 546.
12. Hamilton, E. I. (1979). "Chemical Elements and Man," p. 490. Thomas. Springfield, Illinois.
13. Hullin, R. P., and Johnson, A. W. (1970). *Life Sci.* **9**, 9.
14. International Committee of Radiological Protection (1975). "Report of the Fact Group on Reference Man." Pergamon, Oxford.
15. Ljumberg, S., and Paalzow, L. (1969). *Acta Psychiatr. Scand., Suppl.* **207**, 68.
16. Patt, E. L., Pickett, E. E., and O'Dell, B. L. (1978). *Bioinorg. Chem.* **9**, 299.
17. Pickett, E. E., and Hawkins, J. L. (1981). *Anal. Biochem.* **112**, 213.
18. Pickett, E. E. (1983). *In* "Lithium, 4. Spurenelement-Symposium" (M. Anke, W. Baumann, H. Bräunlich, and C. Brückner, eds.), p. 66. Wiss. Publ., Friedrich-Schiller-Univ., Jena, G.D.R.
19. Regius, A., Pavel, J., Anke, M., and Szentimihalyi, S. (1983). *In* "Lithium, 4. Spurenele-ment-Symposium" (M. Anke, W. Baumann, H. Bräunlich, and C. Brückner, eds.), p. 45, Wiss. Publ. Friedrich-Schiller-Univ., Jena, G.D.R.
20. Robinson, W. O., Steinkoenig, L. A., and Mitter, C. R. (1971). *U.S., Dep. Agric., Tech. Bull.* **600.**
21. Schou, M. (1976). *Annu. Rev. Pharmacol. Toxicol.* **16**, 231.
22. Soman, S. D., Panday, V. K., Joseph, K. T., and Raut, S. R. (1969). *Health Phys.* **17**, 35.
23. Teeny, F. M., Gauglitz, E. J., Hall, A. S., and Houle, C. R. (1984). *J. Agric. Food Chem.* **32**, 852.
24. U.S. Geological Survey (1972). Geochemical Survey of Missouri: Flows and Progress for Fourth Six-Month Period, January–June (1971).
25. Venchikov, A. I. (1974). *In* "Trace Element Metabolism in Animals-2" (W. G. Hoekstra, J. W. Suttie, H. E. Ganther, and W. Mertz, eds.), p. 295. University Park Press, Baltimore, Maryland.
26. Voors, A. W. (1970). *Am. J. Epidemiol.* **92**, 164.
27. Wittrig, J., Anthony, E. J., and Lucarno, H. E. (1970). *Dis. Nerv. Syst.* **31**, 408.

9

Aluminum

ALLEN C. ALFREY

Department of Medicine
University of Colorado
Denver, Colorado

I. INTRODUCTION

In 1977, the fourth edition of Underwood's textbook (112) devoted a mere three pages to the discussion of aluminum (Al). It is not surprising, for in those years there was little evidence that aluminum was an essential element, nor had its systemic toxicity for animals been clearly defined. Neither was the metabolism of aluminum well understood, since the available analytical techniques lacked the precision, sensitivity, and specificity to measure aluminum in biological samples. Since the mid-1970s, however, there has been increasing interest in the biological importance of aluminum, as a result of evidence for the systemic toxicity of the element and the development of accurate analytical techniques. We can understand this more readily when we compare the 80 citings on the adverse effects of aluminum in the *Index Medicus* from the years 1966 to 1976 to the more than 268 listings found in this reference book between the years 1977 and 1984.

It has been suggested that because of its position in the periodic table and its abundance in the solar system, aluminum is an essential element (113). Still, there is no conclusive evidence that aluminum has any essential function in animals. Although aluminum promotes the reaction between cytochrome c and succinic dehydrogenase (45) and has been shown to be a necessary cofactor for the activation of guanine nucleotide-binding regulating protein by fluoride for the stimulation of anenylate cyclase activity (106), it is not at this time known to be involved in these reactions in the human body.

II. ALUMINUM CONCENTRATIONS

Aluminum is the second most plentiful element in the earth's crust. It does not normally exist in the metallic state but rather, because of its reactive nature, occurs in combination with oxygen, silicon, fluoride, and other elements and compounds. In many of its compounds aluminum is insoluble, and its concentrations in seawater are usually <1 μg/liter and normally quite low in most surface and ground freshwater supplies (77). It does, however, become much more soluble in an acid environment (41). As a result of "acid rain," aluminum levels in some northern lakes are increasing. A major source of aluminum in water is aluminum sulfate, which is commonly added to municipal water supplies as a coagulant to decrease water turbidity (77).

In the plant kingdom, aluminum is present virtually everywhere, at least in small quantities. Aluminum content is especially high in a number of families of subtropical plants (Teaceae, Euphorbiaceae and Caryophyllaceae), which are considered to be aluminum accumulators (22). One such species, Lycopodiaceae, contains as much as 70% aluminum oxide in its ash (50).

As analytical techniques have improved, normal concentrations of aluminum, especially in blood, have been found to be considerably lower than first estimated. It is generally accepted that the normal level of aluminum in plasma is ~5 μg /liter. Normal cerebral–spinal fluid aluminum content has been reported as 6 μg/liter, and biliary aluminum content as 3 μg/liter (3,114). Aluminum is bound in plasma to some large moiety with only ~5% present in the free or filterable state. Several studies would suggest that the major plasma-binding protein is transferrin (111). Tissue aluminum content is <4 mg/kg dry weight (Table I), and whole-body aluminum content has been estimated at <45 mg (5,6). With the exception of brain, which is slightly higher in aluminum content in advanced-age adults than in infants (70), the lungs are the only tissue store of aluminum found actually to increase with age (5). Since lung aluminum levels are higher but do not correlate with other tissue stores in the body, it is assumed that some inhaled aluminum is trapped in the pulmonary tissue and is not allowed to cross into the blood or have any appreciable effect on other tissue stores of aluminum (5).

Early studies suggested that the daily ingestion of aluminum from food sources varied between 24 and 36 mg (19). However, later evidence points to ingestion of fewer than 3 to 5 mg (36,37). It should be noted that aluminum is used as a filler in pickles and cheese and is a major component of baking powder. So, depending on one's eating habits, individual aluminum intake could be considerably higher. An additional source of aluminum intake that is frequently not taken into consideration is water. A U.S. Environmental Protection Agency survey would show that some municipal waters contain as much as 2 to 4 mg/liter (77). Another possible source of ingested aluminum is the leaching of this element from cooking utensils during the preparation of food. However, despite all the

Table I
Normal Tissue Aluminum Levels[a]

	Liver	Spleen	Bone	Heart	Muscle	Lungs	Brain gray matter
Mean	4.1	2.6	3.3	1.0	1.2	43	2.2
SD	1 7	2.1	2.9	0.8	1.2	43	1.3

[a]Given in milligrams per kilogram dry weight.

exposure, it would appear that little ingested aluminum is actually absorbed. This is based on the fact that total-body aluminum content is extremely low and does not increase with aging. Aluminum in urine, which appears to be the major avenue for elimination of the substance, is normally low, usually ≤20 μg/liter.

III. ALUMINUM METABOLISM

Since aluminum is abundant in our environment, we are continuously exposed to it through our lungs, skin, and gastrointestinal tract. However, despite this consistent exposure little aluminum is absorbed or retained.

Factors that modulate aluminum absorption from the gastrointestinal tract are poorly understood. One reason for the paucity of knowledge about aluminum absorption is the lack of an available isotope to measure absorption directly. Instead, indirect methods such as change in urinary aluminum excretion and alterations in tissue content have been utilized to estimate absorption. Since aluminum hydroxide and oxides are 100–1000 times more soluble at pH 4.2 than at pH 6.2–8.1, it has been suggested that aluminum might be more readily absorbed in the stomach or proximal duodenum, and therefore gastric pH might have a major effect on aluminum absorption (53).

On the basis of the finding that parathyroid hormone (PTH) administration enhances the body burden of aluminum in animals receiving large oral loads of this element, it has been suggested that PTH may augment absorption (68,69). However, this has not been directly tested. Evidence to the contrary shows that neither plasma nor bone aluminum levels correlate with PTH levels in uremic patients (7).

Since vitamin D has been shown to increase the absorption of a number of other elements including calcium, lead, strontium, beryllium, cesium, zinc, and cobalt, it has also been suggested that it might enhance aluminum absorption. Studies have not yet been carried out to test this possibility.

Aluminum forms tight complexes with fluoride, greatly increasing its solubility. This might suggest that an increased fluoride content of the diet would enhance aluminum absorption. However, the converse has been noted regarding aluminum's effect on fluoride absorption, showing that it reduces fluoride absorption (105). In fact, oral aluminum has been used to treat fluorosis (83). These findings would suggest that fluoride, in turn, might decrease aluminum absorption. So far, there are no conclusive studies to support that fluoride either enhances or diminishes aluminum absorption.

There is good evidence that when individuals are given large oral loads of aluminum, some of this excess aluminum is absorbed. This was initially reported by Kaehny et al. (52), who found that urinary output of aluminum increased from 16 to 275 µg/day in patients receiving 2.2 g/day oral aluminum supplementation. Subsequently, Gorsky et al. (36) found urine aluminum output to increase from 65 to 280 µg/day in subjects given 1.08 to 2.9 g/day of aluminum. Recker and associates (92) reported urine aluminum output to rise from 86 to 495 µg/day when individuals were fed 3.8 g of aluminum. Using a smaller oral aluminum load of only 125 mg/day as $AlCl_3$, Greger and Baier (37) found urine aluminum output to increase from 36 to 129 µg/day. In view of the small oral loads given, the Greger and Baier studies might suggest that aluminum is somewhat better absorbed when given as the chloride salt than when administered as aluminum hydroxide, aluminum carbonate, or dihydroxyaluminum acetate as was the case in the above studies.

Although it is clear that some aluminum is absorbed when large oral loads of aluminum are given, the major question is whether all of the absorbed aluminum is excreted by the kidneys. Balance studies carried out in normal subjects who were ingesting 1–3 g/day of aluminum show that these individuals were in a positive aluminum balance of 23 to 313 mg/day (18,36). In contrast, balance studies performed in individuals receiving 125 mg/day aluminum showed that all subjects excreted >96% of the ingested aluminum (19). The latter study would more closely correlate with the fact that tissue aluminum levels have uniformly been found to be low in normal individuals. However, at this time it has not been firmly established whether or not some aluminum is retained in normal subjects when they are given large oral loads of aluminum over a protracted period of time.

It would appear that the major pathway for the elimination of any absorbed or systemically administered aluminum is the kidneys. Using emission spectroscopy, it was initially reported that normal urinary aluminum excretion was 700–1000 µg/day (86,108). But later studies using flameless atomic absorption have found normal urinary aluminum excretion to be somewhere between 15 and 55 µg/day (36,53,92). Because of the binding of aluminum in plasma, aluminum clearance is only ~5% of the glomerular filtration rate (42). Under ordinary conditions the kidney appears to be able to eliminate all the absorbed aluminum. Still, even patients with normal renal function who are given large parenteral loads of aluminum retain some of the administered aluminum, because the kidneys' capacity to excrete the element is exceeded (56). Again, whether this also pertains to individuals ingesting large oral loads of aluminum is unknown.

Table II
Effect of Parathyroidectomy on Bone
Aluminum Uptake

Experimental group[a]	Aluminum levels (mg/kg)
Control	55 ± 9
Uremic	113 ± 16
Uremic (PTX)	80 ± 7

[a]All animals had received 1.5 mg/kg aluminum intraperitoneally for 79 days prior to study. PTX, Parathyroidectomized.

Although bile seems to be a major avenue for the elimination of many other trace elements, it does not seem to affect the elimination of aluminum, at least in acute situations. Kovalchick *et al.* (59) reported that only 0.1% of a parenterally administered aluminum load was eliminated via bile over an 8-hr period as opposed to 48% by the kidneys.

During aluminum loading the tissues most affected are liver, spleen, and bone, where concentrations may increase to 300 to 400 mg/kg dry weight (5,6). Other tissues such as brain and heart have less of an increase, although aluminum levels as high as 40 to 50 mg/kg have been documented. In contrast, skeletal muscle aluminum rarely increases to more than 4 to 6 mg/kg dry weight. Hair and skin aluminum level changes are inconsistent and do not correlate with other tissue stores, possibly as a result of environmental contamination of these two exposed tissues (5). Also, plasma aluminum levels appear to be a better indicator of ongoing aluminum loading than the status of the total-body aluminum (69).

Other tissue stores of aluminum are also not well correlated with each other, although high bone aluminum levels in general denote that other tissue levels of aluminum are also increased. This lack of proximate correlation between the various body burdens of aluminum may be because compartmentalization of aluminum is determined to some extent by factors other than the size of the aluminum load. It has been shown in uremic animals that aluminum is preferentially taken up in bone, whereas in nonuremic animals it appears to go initially to liver. This enhanced bone aluminum uptake in uremic animals appears to be partly a result of the animals' hyperparathyroid state, which can be blunted by parathyroidectomy (Table II) (10).

IV. ALUMINUM TOXICITY

Aluminum may be toxic to plants under certain conditions. Aluminum has been found to inhibit cell extension and division and interfere with plant mineral nutrition (67). Since aluminum binds to DNA in plant cell nuclei, it may limit template activity, which would explain part of the deleterious effects of aluminum on plant metabolism (66). Since aluminum has primarily existed in an insoluble state, plants have largely been protected from its toxicity. However, with the advent of acid rain, aluminum has been rendered more soluble in surface waters and has a greater potential for causing plant toxicity (51).

Aluminum is also toxic to fish and other aquatic biota. Labile inorganic aluminum in acidified lake water may affect trout survival (29). It has been suggested that this toxicity can be reduced by complexing aluminum with fluoride (29).

In humans, aluminum has three potential toxic effects: (1) a local effect in the gastrointestinal tract, (2) a potential to cause pulmonary damage if inhaled, and (3) a capacity to exert systemic toxicity if absorbed or parenterally administered.

Orally administered aluminum compounds reduce the absorption of a number of other elements and compounds including strontium, iron, fluoride, phosphorus, and, to a lesser extent, calcium (7,38,63,104). Because of this property, aluminum has been used therapeutically to treat fluorosis and to reduce phosphorus absorption in uremic patients. However, when given in large dosages over an extended period of time aluminum can cause severe phosphate depletion resulting in osteomalacia, as well as cause interference with phosphorylation processes in tissues (63,86). Aluminum compounds have also been shown to decrease cholesterol absorption (82). It has been suggested that aluminum combines with pectin to bind fats in nondigestible vegetable fibers, thus preventing the absorption of fats.

An additional local effect that aluminum has on the gastrointestinal tract is to decrease motility. Aluminum ions inhibit acetylcholine-induced contractions of rodent and human gastric smooth muscle (40). Free aluminum in the stomach has also been shown to reduce gastric emptying (49). A well-recognized complication of aluminum administration is constipation, which results at least in part from the inhibition of gastrointestinal motility.

Pulmonary disease resulting from the inhalation of aluminum powders was first described in 1934 (31). It was more fully characterized 14 years later when Shaver reported 11 patients with an occupational pneumoconiosis that resulted from inhaling bauxite ($Al_2O_3 \cdot 3H_2O$) (95). Another equally toxic aluminum compound is "pyro," which is a fine powder (80). These inhaled aluminum compounds produce pulmonary fibrosis resulting in a restrictive and obstructive airway disease. It has been shown that animals exposed to high concentrations of aluminum chlorhydrate, as found in aerosol antiperspirant, develop granulomatous lesions in the respiratory bronchioles (28).

Some of the first evidence for the systemic toxicity of aluminum was reported by Scherp and Church in 1937 (96) and continued by Kopeloff (58) in 1942 and again by Klatzo et al. (55) in 1965. These investigators noted that 9–14 days following an intracerebral injection of aluminum, rabbits would develop an unsteady gait, motor inhibition, ataxia, and seizures. Klatzo *et al.* (55) were the first to point out that histological examination of the brains of these animals showed neurofibrillary tangles. DeBoni and associates (27) described similar symptoms and lesions, as well as elevated brain aluminum levels, in animals that had received aluminum lactate or tartrate injections subcutaneously, and concluded that aluminum can cross the blood–brain barrier. Wisniewski *et al.* (117) have developed a chronic model of aluminum-induced neurofibrillary changes in the rabbit. These neurofilaments in the rabbit appear to accumulate as a result of slowed or blocked axonal transport caused by the aluminum. It is of interest, however, that whereas aluminum is neurotoxic in rabbits and cats, other animals such as rats and mice are relatively resistant to this toxicity.

The finding of neurofilaments in experimental aluminum intoxication has prompted the suggestion that Alzheimer's disease is caused by aluminum (24,25,73). However, the dissimilarities between the biochemical and histological alterations between Alzheimer's disease and aluminum encephalopathy have been reviewed by Wisniewski and Iqbal (116), casting some doubt on any association between the two. Furthermore, although brain aluminum levels have been discovered to be elevated in Alzheimer's by some investigators (24,25, 65,73), the findings have not been uniform (65,116). Similarly, aluminum has been implicated in the pathology of other diseases characterized by neurofibrillary changes such as amyotrophic lateral sclerosis and guamanian and Kii Peninsula amyotrophic lateral sclerosis (34). At the present moment, though, evidence is inconclusive that aluminum is important in the pathogenesis of any of the aforementioned diseases. However, there is additional evidence that aluminum can exert neurotoxicity in humans. Two cases of a progressive dementia-producing encephalopathy have been reported, one in an individual who had worked in a ball mill room in an aluminum powder factory (61,74). Both patients were found to have high brain, as well as other tissue, aluminum levels. Subsequently, a somewhat similar disease, dialysis dementia or dialysis encephalopathy, was seen in a large number of patients with renal failure who were undergoing chronic dialysis (4,5,90). The disease is characterized by dyspraxia of speech, parietal lobe signs such as directional disorientation, myoclonus, seizures, hallucinations, dementia, and death in 6 to 8 months after the onset of symptoms. Histological alterations in the brain are minimal and neurofibrillary changes are not present (17). Brain aluminum levels, as well as all other tissue aluminum levels, have consistently been found to be elevated in patients dying of this condition (90). Additional evidence that this disease was caused by aluminum came from the finding that this disease occurred in epidemic proportions in some dialysis units, whereas it was rarely seen in others (4,6,8,9,12,14,20,30, 32,33,71,90,94,97). It soon became apparent that dialysis centers having a high incidence of dialysis encephalopathy had high aluminum levels in the water used to prepare the dialysate, and centers without this disease had low aluminum levels in their water source. Additional evidence incriminating aluminum in the pathogenesis of this disease was the finding that the disease could virtually be eradicated by removing aluminum from the dialysate (5,8,14,71). Final proof implicating aluminum as the etiological agent for this disease is the fact that it can be cured by chelation—presumably aluminum chelation (2,79).

Besides its neurotoxicity, there is increasing evidence that aluminum has a toxic effect on the skeleton. This was first recognized in the late 1970s when large epidemiological studies showed that dialysis centers that had a high incidence of dialysis encephalopathy and large amounts of aluminum in their dialysate also had a high incidence of fracturing osteomalacia (88,91). This disease is characterized by low bone formation rates manifested by classical osteomalacia or so-called aplastic bone, a condition remarkable for absence of bone formation as determined by tetracycline labeling. Serum calcium levels are either normal, or actually elevated, and PTH levels, in contrast to other osteomalacic states, as well as uremia, may be subnormal (43,44,47,48,87). Clinical symptoms are pathological fracture, bone pain, and proximal myopathy. The disease is unresponsive to vitamin D therapy but has been reported to heal following removal of the aluminum exposure and after chelation (1,16,84). Further supporting aluminum's skeletal toxicity is the finding that parenterally administered aluminum can induce a bone disease in rats that is similar to that in humans (21,35,93).

Aluminum also appears to exert toxicity on the hematopoietic system. A microcytic hypochromic anemia not associated with iron deficiency has been described in patients with aluminum-induced osteomalacia and encephalopathy (85,101,109). This condition has been reported to improve with chelation therapy. Furthermore, both microcytic and macrocytic anemia has been produced in animals by parenterally administering aluminum (54,100). In animal studies it would appear that the type of anemia induced is at least partially dependent on the rate and amount of aluminum administered. It has been suggested that aluminum exerts systemic toxicity other than that found in neurological, skeletal, and hematopoietic functions, but so far the data on other types of toxicity are not convincing.

V. MECHANISM OF ALUMINUM TOXICITY

The mechanisms by which aluminum exerts its toxicity are not well understood, although aluminum has been shown to interfere with a variety of biological and enzymatic processes. Regarding its neurotoxicity, aluminum has been shown to bind to brain calmodulin, changing its configuration and its interaction with other proteins (102). Aluminum has also been shown to accelerate the rate of oxygen uptake by the succinodehydrogenase cytochrome system (113) and to increase acetylcholine hydrolysis and acetylcholinesterase activity (89). Additional support for the neurotoxicity of aluminum comes from *in vitro* studies that have shown that cultured neuroblastoma cells exposed to aluminum had a proliferation of neurofilament, a reduction in cellular RNA, an increased protein content, an elevated rate of [^{3}H]leucine incorporation, and reduced acetylcholine activity (76). DeBoni *et al.* (26,27) have also shown that human neurons when exposed to a culture medium containing aluminum undergo neuronal death with an overproduction of neurofilaments. Aluminum has also been shown to exert local toxicity when injected into the hypoglossal nucleus (15).

Regarding its skeletal toxicity, aluminum has been noted to be deposited between mineralized and unmineralized bone at the calcification front preventing tetracycline labeling, suggesting that it might locally prevent further mineralization of osteoid (64). In addition, aluminum has been shown to be deposited in the mitochondria of osteoblasts (23) and to inhibit bone phosphatases (62). Aluminum, especially in association with citrate, has been shown to inhibit crystal growth, which might also interfere with mineralization (107).

With relation to its causing anemia, aluminum has been shown to inhibit ferroxidase (ceruloplasmin) activity (46). Since aluminum is bound to transferrin, it has been suggested that it might prevent unloading of transferrin iron onto a specific tissue or receptor or else be unloaded instead of iron(III).

It has been suggested that inhibition of glycolysis and phosphorylation is the most toxic reaction caused by aluminum (103). Three independent investigative groups have shown that hexokinase, a magnesium-dependent phosphorylating enzyme, is inhibited by aluminum (39,110,115). Aluminum has also been shown to displace magnesium from ATP with the resulting stabilization of ATP, preventing phosphate transfer by Na$^+$,K$^+$-ATPase (39,60). It has been suggested that possibly all phosphate-transferring systems involving ATP and magnesium may be biological targets for aluminum (103).

VI. SOURCE OF ALUMINUM EXPOSURE AND TOXICITY

As stated above, skin, lung, and the gastrointestinal tract are formidable barriers to aluminum absorption. However, when exposure is markedly increased, these barriers can be overcome. It has been shown that workers exposed to aluminum fumes have urinary aluminum values 10–20 times greater than control values (81). Similarly, as stated above, individuals given large oral aluminum loads have even greater increases in urinary aluminum excretion. It has been assumed, though not documented, that all of the absorbed aluminum is normally eliminated through the kidney, and thus the element poses little threat of systemic toxicity. Although this requires additional study, there is strong evidence that aluminum is retained and systemic toxicity manifested if there is impairment of renal function. Aluminum-associated encephalopathy and osteomalacia have been repeatedly described in uremic patients whose only aluminum exposure was orally administered aluminum compounds (11,13,72,75,99). However, the most common cause of systemic aluminum toxicity has occurred when an individual has had parenteral aluminum exposure. The first group of patients recognized to have systemic aluminum intoxication were receiving parenteral aluminum during hemodialysis (52). It was found that the dialysate was contaminated with aluminum. The estimation was that as much as 2 to 4 mg of aluminum was given to patients during each dialysis. Patients were unable to eliminate any of the administered aluminum because of lack of renal function. When large parenteral loads of aluminum are given to individuals with normal renal function, the kidney's capacity to excrete aluminum is exceeded and some is retained. A number of patients on chronic total parenteral nutrition received 2–3 mg aluminum daily intravenously as a result of aluminum-contaminated casein hydrolysate, which was used as their amino acid source (57). These patients were documented to have very high serum, urine, bone, and liver aluminum levels as well as osteomalacia. Aluminum has been shown to be a very common contaminant in many intravenous solutions and is especially concentrated in human albumin (98).

It has been shown that premature infants receiving these solutions have elevated tissue stores of aluminum. It is unknown at this time whether any toxicity can be related to this amount of aluminum exposure.

VII.   DIAGNOSIS AND TREATMENT OF ALUMINUM INTOXICATION

The diagnosis of aluminum intoxication is dependent on the establishment of a high level of aluminum exposure either parenterally or orally, in association with compromised renal function. The clinical picture may consist of the combination of encephalopathy, osteomalacia, and anemia, although this is not invariable, and the patient may have only encephalopathy or osteomalacia. Bone histological study is especially useful in the diagnosis of aluminum-associated osteomalacia. Classically, the bone shows wider osteoid seams, failure of separation of tetracycline labels (supporting low bone formation), and a positive aluminum stain (aurin tricarboxylic acid) at the junction between calcified and noncalcified bone (64). Clinically, the diagnosis of aluminum encephalopathy is more difficult to make. It largely depends on the exclusion of other neurological diseases and the documentation of the clinical features described above. The EEG is of some help, for in contrast to other metabolic encephalopathies, including uremia, when there is generalized slowing of the rhythms, in dialysis encephalopathy there are bursts of slow-wave activity with relatively normal background activity between bursts (4). Plasma aluminum levels are of some help in identifying a population who are either intoxicated with aluminum or at risk of becoming so. In general, these patients will have plasma aluminum levels >100 μg/liter (78). However, this is not invariably the case, and a lower plasma aluminum level does not exclude the diagnosis of aluminum intoxication. More recently, the desferroxamine infusion test has been used in uremic patients to document aluminum excess and possible toxicity. An increase in plasma aluminum level of greater than 200 to 300 μg/liter 24–48 hr after a single injection of desferroxamine has been used as presumptive evidence of an excessive body burden of this element (78). However since desferroxamine is rapidly excreted in patients with normal renal function, this test is only applicable in patients with markedly impaired renal function.

Treatment of aluminum-associated osteomalacia and encephalopathy with desferroxamine has been very encouraging; both of these conditions can be completely cured. However, improvement may be slow, and therapy may require 3–6 months of chelation before marked clinical improvement is noted (1,2,16,79, 84).

In summary, there continues to be a group of patients at risk of developing aluminum intoxication. Primarily, they are individuals with compromised renal function who are receiving large oral loads of aluminum or patients who are chronically exposed to parenteral aluminum as a result of prolonged intravenous infusions, especially with aluminum-contaminated albumin, or uremic patients undergoing dialysis with aluminum-contaminated dialysate. At this time it is unknown whether industrial exposures or orally administered aluminum compounds given to individuals with normal renal function pose any risk of systemic aluminum intoxication. Although it is well documented that aluminum has multiple systemic effects, the encouraging aspect is that aluminum intoxication is treatable. In the future, it should also be possible to reduce the risk of aluminum intoxication by eliminating aluminum contamination in intravenous solutions, avoiding aluminum contamination of dialysate, and finding alternate phosphate-binding agents to replace aluminum compounds that are now commonly used in uremic patients for this purpose.

REFERENCES

1. Ackrill, P., Day, J. P., Garstang, F. M., Hodge, K. C., Metcalfe, P. J., Benzo, Z., Hill, K., Ralston, A. J., and Denton, J. (1982). *Proc. Eur. Dial. Transplant Assoc.* **19,** 203–207.
2. Ackrill, P., Ralston, A. J., Day, J. P., and Hodge, K. C. (1980). *Lancet* **2,** 692–693.
3. Alfrey, A. C. (1983). *Adv. Clin. Chem.* **23,** 69–91.
4. Alfrey, A. C. (1978). *Annu. Rev. Med.* **29,** 93–96.
5. Alfrey, A. C. (1980). *Neurotoxicology* **1,** 43–53.

6. Alfrey, A. C., Hegg, A., and Craswell, P. (1980). *Am. J. Clin. Nutr.* **33**, 1509–1516.
7. Alfrey, A. C., Hegg, A., Miller, N., Berl, T., and Berns, A. (1979). *Miner. Electrolyte Metab.* **2**, 81–87.
8. Alfrey, A. C., LeGendre, G. R., and Kaehny, W. D. (1976). *N. Engl. J. Med.* **294**, 184–188.
9. Alfrey, A. C., Mishell, M. M., Burks, J., Contiguglia, S. R., Rudolph, H., Lewin, E., and Holmes, J. H. (1972). *Trans.—Am. Soc. Artif. Intern. Organs* **18**, 257–261.
10. Alfrey, A. C., Sedman, A., and Chan, Y.-L. (1985). *J. Lab. Clin. Med.* **105**, 227–233.
11. Andreoli, S. P., Bergstein, J. M., and Sherrard, D. J. (1984). *N. Engl. J. Med.* **310**, 1079–1084.
12. Arieff, A. I., Cooper, J. D., Armstrong, D., and Lazarowitz, V. C. (1979). *Ann. Intern. Med.* **90**, 741–747.
13. Baluarte, J. H., Gruskin, A. B., Kiner, L. B., Foley, C. M., and Grover, W. D. (1977). *Proc.—Clin. Dial. Transplant Forum* **7**, 95–98.
14. Berkseth, R., Mahowald, M., Anderson, D., and Shapiro, F. (1978). *Am. Soc. Nephrol.* **11**, 36A.
15. Bizzi, A., Clark, R., Crane, L., Autilio-Gambetti, L., and Gambetti, P. (1984). *J. Neurosci.* **4**, 722–731.
16. Brown, D. J., Ham, K. N., Dawborn, J. K., and Xippel, J. M. (1982). *Lancet* **2**, 343–345.
17. Burks, J. S., Alfrey, A. C., Huddlestone, J., Norenberg, M. D., and Lewin, E. (1976). *Lancet* **1**, 764.
18. Cam, J. M., Luch, V. A., Eastwood, J. B., and deWardner, H. E. (1976). *Clin. Sci. Mol. Med.* **51**, 407–414.
19. Campbell, I. R., Cass, J. S., Cholak, J., and Kehoe, R. A. (1957). *AMA Arch. Ind. Health* **15**, 359–448.
20. Cartier, F., Allain, P., Gary, J., Chatel, M., Menault, F., and Pecker, S. (1978). *Nouv. Presse Med.* **7**, 97–102.
21. Chan, Y., Alfrey, A. C., Posen, S., Lissner, D., Hills, E., Dunstan, C. R., and Evans, R. A. (1983). *Calcif. Tissue Int.* **35**, 344–351.
22. Chenery, E. M. (1948). *Kew Bull.* **1**, 1973.
23. Cournot-Witmer, G., Plachot, J.-J., Bourdeau, A., Lieberherr, M., Jorgetti, V., Mendes, V., Halpern, S., Hemmerle, J., Drueke, T., and Balsan, S. (1985). *Kidney Int.* (in press).
24. Crapper, D. R., Krishnan, S. S., and Dalton, A. J. (1973). *Science* **180**, 511–513.
25. Crapper, D. R., Quittkat, S., Krishnan, S. S., Dalton, A. J., and DeBoni, U. (1980). *Acta Neuropathol.* **50**, 19–24.
26. DeBoni, U., Otvos, A., Scott, J. W., and Crapper, D. R. (1976). *Acta Neuropathol.* **35**, 285–294.
27. DeBoni, U., Scott, J. W., and Crapper, D. R. (1974). *Histochemistry* **40**, 31–37.
28. Drew, R. T., Gupta, R. N., Bend, J. R., and Hook, G. E. R. (1974). *Arch. Environ. Health* **28**, 321–326.
29. Driscoll, C. T., Jr., Gaker, J. P., Bisogni, J. J., Jr., and Schofield, C. L. (1980). *Nature (London)* **284**, 162–164.
30. Dunea, G., Mahuraker, S. D., Mamdani, B., and Smith, E. C. (1978). *Ann. Intern. Med.* **88**, 502–504.
31. Filip, D. (1934). *Rass. Med. Appl. Lav. Ind.* **5**, 128–144.
32. Flendrig, J. A., Kruis, H., and Das, H. A. (1976). *Prox. Eur. Dial. Transplant Assoc.* **13**, 355–361.
33. Galle, P., Chatel, M., Berry, J. P., and Menault, F. (1979). *Nouv. Presse Med.* **8**, 4091–4094.

34. Garruto, R. M., Fukatsu, R., Yanagihara, R., Gajdusek, D. C., Hook, G., and Fiori, C. E. (1984). *Proc. Natl. Acad. Sci. U.S.A.* **81,** 1875–1879.
35. Goodman, W. G., Gilligan, J., and Horst, R. (1984). *J. Clin. Invest.* **73,** 171–181.
36. Gorsky, J. E., Dietz, A. A., Spencer, H., and Osis, D. (1979). *Clin. Chem. (Winston-Salem, N.C.)* **25,** 1739–1743.
37. Greger, J. L., and Baier, M. J. (1983). *Food Chem. Toxicol.* **21,** 473–476.
38. Hall, G. J. L., and Davis, A. D. (1969). *Med. J. Aust.* **2,** 95.
39. Harrison, W. H., Codd, E., and Gray, R. M. (1972). *Lancet* **2,** 277.
40. Hava, M., and Hurwitz, A. (1973). *Eur. J. Pharmacol.* **22,** 156–161.
41. Hem, J. D. (1985). *Kidney Int.* (in press).
42. Henry, D. A., Goodman, W. G., Nudelman, R. K., DiDomenico, N. C., Alfrey, A. C., Slatopolsky, E., Stanley, T. M., and Coburn, J. W. (1984). *Kidney Int.* **25,** 362–369.
43. Hodsman, A. B., Sherrard, D. J., Alfrey, A. C., Brickman, A. S., Miller, A., Maloney, N. A., and Coburn, J. W. (1982). *J. Clin. Endocrinol. Metab.* **54,** 539–546.
44. Hodsman, A. B., Sherrard, D. J., Wong, E. G. C., Brickman, A. S., Lee, D. B. N., Alfrey, A. C., Singer, F. R., Norman, A. W., and Coburn, J. W. (1981). *Ann. Intern. Med.* **94,** 629–637.
45. Horecker, B. L., Statz, E., and Hogness, T. (1939). *J. Biol. Chem.* **128,** 251–256.
46. Huber, C. T., and Frieden, E. (1970). *J. Biol. Chem.* **245,** 3979–3984.
47. Hudson, G. A., Milne, F. J., Meyers, A. M., and Reis, P. (1980). *Kidney Int.* **17,** 532.
48. Hudson, G. A., Milne, F. J., Oliver, N. J., Reis, P., Murray, J., and Meyers, A. M. (1979). *S. Afr. Med. J.* **56,** 439–443.
49. Hurwitz, A., Robinson, R. G., Vats, T. S., Whittier, F. C., and Herrin, W. F. (1976). *Gastroenterology* **71,** 268–273.
50. Hutchinson, G. E., and Wollack, A. (1943). *Trans. Conn. Acad. Arts Sci.* **35,** 73–128.
51. Johnson, N. M., Likens, G. E., Feller, M. C., and Driscoll, C. T. (1984). *Science* **224,** 1424–1425.
52. Kaehny, W. D., Alfrey, A. C., Holman, R. E., and Schorr, W. J. (1977). *Kidney Int.* **12,** 361–365.
53. Kaehny, W. D., Hegg, A. P., and Alfrey, A. C. (1977). *N. Engl. J. Med.* **296,** 1389–1390.
54. Kaiser, L., Burnatowska-Hledin, M. A., Schwartz, K. A., and Mayor, G. H. (1984). *Kidney Int.* **25,** 194.
55. Klatzo, I., Wisniewski, H., and Streicher, E. (1965). *J. Neuropathol. Exp. Neurol.* **24,** 187–199.
56. Klein, G. L., Alfrey, A. C., Miller, N. L., Sherrard, D. J., Hazlet, T. K., Ament, M. E., and Coburn, J. W. (1982). *Am. J. Clin. Nutr.* **35,** 1425–1429.
57. Klein, G. L., Ott, S. M., Alfrey, A. C., Sherrard, D. J., Hazlet, A., Miller, N. L., Maloney, N. A., Berquist, W. E., Ament, M. E., and Coburn, J. W. (1982). *Trans. Assoc. Am. Physicians* **95,** 155–164.
58. Kopeloff, L. M., Barrern, S. W., and Kopelott, N. (1942). *Am. J. Psychiatry* **98,** 881–902.
59. Kovalchik, M. T., Kaehny, W. D., Jackson, T., and Alfrey, A. C. (1978). *J. Lab. Clin. Med.* **92,** 712–716.
60. Lai, J. C. K., Guest, J. F., Leung, T. K. C., Lim, L., and Davison, A. N. (1980). *Biochem. Pharmacol.* **29,** 141–146.
61. Lapresle, J., Duckett, S., Galle, P., and Cartier, L. (1975). *C. R. Seances Soc. Biol. Ses. Fil.* **169,** 282–285.
62. Lieberherr, M., Grosse, B., Cournot-Witmer, G., Thil, C. L., and Balsan, S. (1982). *Calcif. Tissue Int.* **34,** 280–284.
63. Lotz, M., Zisman, E., and Bartter, F. C. (1968). *N. Engl. J. Med.* **278,** 409–415.

64. Maloney, N. A., Ott, S., Alfrey, A. C., Coburn, J. W., and Sherrard, D. J. (1982). *J. Lab. Clin. Med.* **99,** 206–216.

65. Markesbery, W. R., Ehmann, W. D., Hossain, T. I. M., Alauddin, M., and Goodin, D. T. (1981). *Ann. Neurol.* **10,** 511–516.

66. Matsumoto, H., Morimura, S., and Takahashi, E. (1977). *Plant Cell Physiol.* **18,** 987–993.

67. Matsumoto, H., Morimura, S., and Takahashi, E. (1977). *Plant Cell Physiol.* **18,** 325–335.

68. Mayor, G. H., Keiser, J. A., Makdoni, D., and Ku, P. (1977). *Science* **197,** 1187–1188.

69. Mayor, G. H., Sprague, S. M., Hourani, M. R., and Sanchez, T. V. (1980). *Kidney Int.* **17,** 40–44.

70. McDermott, J. R., Smith, A. I., Iqbal, K., and Wisniewski, H. M. (1979). *Neurology* **29,** 809–814.

71. McDermott, J. R., Smith, A. I., Ward, M. K., Fawcett, R. W. P., and Kerr, D. N. S. (1978). *Lancet* **1,** 901–903.

72. McKinney, T. D., Dewberry, F. L., Stone, W. J., and Alfrey, A. C. (1978). *Abstr. Am. Soc. Nehphrol.* **11,** 46A.

73. McLachlan, D. R. C., and DeBoni, U. (1980). *Neurotoxicology* **1,** 3–16.

74. McLaughlin, A. I. G., Kazantzis, G., King, E., Teare, D., Porter, R. J., and Owen, R. (1962). *Br. J. Ind. Med.* **19,** 253–263.

75. Mehta, R. P. (1979). *Can. Med. Assoc. J.* **120,** 1112–1114.

76. Miller, C. A., and Levine, E. M. (1974). *Neurochemistry* **22,** 751–758.

77. Miller, R. G., Kopfler, F. C., Kelty, K. C., Stober, J. A., and Ulmer, N. S. (1984). *J.—Am. Water Works Assoc.* **77,** 84–91.

78. Milliner, D. S., Ott, S. M., Nebeker, H. G., Andress, D. L., Sherrard, D. J., Alfrey, A. C., Slatopolsky, E. A., and Coburn, J. W. (1985). Submitted for publication.

79. Milne, F. J., Sharf, B., Bell, P., and Meyers, A. M. (1983). *Clin. Nephrol.* **20,** 202–207.

80. Mitchell, J. (1961). *Br. J. Ind. Med.* **18,** 10–20.

81. Mussi, I., Calzaferri, G., Buratti, M., and Alessio, I.. (1984). *Int. Arch. Occup. Environ. Health* **54,** 155–161.

82. Nagyvary, J., and Bradbury, E. L. (1977). *Biochem. Biophys. Res. Commun.* **2,** 592–598.

83. Navia, J. M. (1970). *Adv. Chem. Ser.* **94,** 123–160.

84. Nebeker, H. G., Milliner, D. S., Ott, S. M., Sherrard, D. J., Alfrey, A. C., Abuelo, J. G., Wasserstein, A., and Coburn, J. W. (1984). *Kidney Int.* **25,** 173.

85. O'Hare, J. A., and Murnaghan, D. J. (1982). *N. Engl. J. Med.* **306,** 654–656.

86. Ondreicka, R., Kortus, J., and Ginter, E. (1971). *In* "Intestinal Absorption of Metal Ions, Trace Elements and Radionuclides" (S. C. Skoryna and D. Waldron, eds.), p. 293. Pergamon, Oxford.

87. Ott, S. M., Maloney, N. A., Coburn, J. W., Alfrey, A. C., and Sherrard, D. J. (1982). *N. Engl. J. Med.* **307,** 709–713.

88. Parkinson, I. S., Ward, M. K., Feest, T. G., Fawcett, R. W. P., and Kerr, D. N. S. (1979). *Lancet* **1,** 406–409.

89. Patocka, J. (1971). *Acta Biol. Med. Ger.* **26,** 845–846.

90. Pierides, A. M., Edwards, W. G., Jr., Cullum, U. X., Jr., McCall, J. R., and Ellis, H. (1980). *Kidney Int.* **18,** 115–124.

91. Platts, M. M., Goode, G. C., and Hislop, J. S. (1977). *Br. Med. J.* **2,** 657–660.

92. Recker, R. R., Blotcky, A. J., Leffler, J. A., and Rack, E. P. (1977). *J. Lab. Clin. Med.* **90,** 810–815.

93. Robertson, J. A., Felsenfeld, A. J., Haygood, C. C., Wilson, P., Clarke, C., and Llach, F. (1983). *Kidney Int.* **23,** 327–335.

94. Rozas, V. V., Port, K. F., and Rutt, W. M. (1978). *Arch. Intern. Med.* **138,** 1375–1377.

95. Schaver, C. G. (1948). *Occup. Med.* **5,** 718–728.

96. Scherp, H. W., and Church, C. F. (1937). *Proc. Soc. Exp. Biol. Med.* **36,** 851–853.
97. Schreeder, M. T. (1979). *Arch. Intern. Med.* **139,** 510–511.
98. Sedman, A. B., Klein, G. L., Merritt, R. J., Miller, N. L., Weber, K. O., Gill, W. L., Anand, H., and Alfrey, A. C. (1985). Submitted for publication.
99. Sedman, A. B., Miller, N. L., Warady, B. A., Lum, G. A., and Alfrey, A. C. (1984). *Kidney Int.* **26,** 201–204.
100. Seibert, F. B., and Wells, H. G. (1929). *Arch. Pathol.* **8,** 230–261.
101. Short, A. I. K., Winney, R. J., and Robson, J. S. (1980). *Proc. Eur. Dial. Transplant Assoc.* **17,** 226–233.
102. Siegel, N., and Hang, A. (1983). *Biochim. Biophys. Acta* **744,** 36–45.
103. Sorenson, J. R. J., Campbell, I. R., Tepper, L. B., and Lingg, R. D. (1974). *Environ. Health Perspect.* **8,** 3–95.
104. Spencer, H., Lewin, I., Belcher, M. J., and Scamachson, J. (1969). *Int. J. Appl. Radiat. Isot.* **20,** 507–716.
105. Spencer, H., Wiatrowski, E., Osis, D., and Norris, C. (1977). *J. Dent. Res.* **56,** B131.
106. Sternweis, P. C., and Gilman, A. G. (1982). *Proc. Natl. Acad. Sci. U.S.A.* **79,** 4888–4891.
107. Thomas, W. C. (1982). *Proc. Soc. Exp. Biol. Med.* **170,** 321–327.
108. Tipton, I. H., Stewart, P. L., and Martin, P. G. (1966). *Health Phys.* **12,** 1683–1689.
109. Touam, M., Martinez, F., Lacour, B., Bourdon, R., Zingraff, J., DiGiulio, S., and Drueke, T. (1983). *Clin. Nephrol.* **19,** 295–298.
110. Trapp, G. A. (1980). *Neurotoxicology* **1,** 89–100.
111. Trapp, G. A. (1983). *Life Sci.* **33,** 311–316.
112. Underwood, E. J. (1977). "Trace Elements in Human and Animal Nutrition," 4th ed., pp. 430–433. Academic Press, New York.
113. Valkovic, V. (1980). *Origins Life* **10,** 301–305.
114. Versieck, J., and Cornelis, R. (1980). *N. Engl. J. Med.* **302,** 468.
115. Viola, R. E., Morrison, J. F., and Cleland, W. W. (1980). *Biochemistry* **19,** 3131–3137.
116. Wisniewski, H. M., and Iqbal, K. (1980). *Neurotoxicology,* **1,** 121–124.
117. Wisniewski, H. M., Sturman, J. H., and Shek, J. W. (1980). *Ann. Neurol.* **8,** 479–490.

10

Other Elements: Sb, Ba, B, Br, Cs, Ge, Rb, Ag, Sr, Sn, Ti, Zr, Be, Bi, Ga, Au, In, Nb, Sc, Te, Tl, W

FORREST H. NIELSEN

U.S. Department of Agriculture
Agricultural Research Service
Grand Forks Human Nutrition Research Center
Grand Forks, North Dakota

I. INTRODUCTION

The advent of multielement analyses of biological samples using methodology such as neutron activation, inductively coupled plasma emission atomic absorption spectroscopy, and spark-source mass spectrometry has stimulated interest, from a biological viewpoint, in a large number of trace elements currently not known to be nutritionally essential. This chapter is a brief overview of some of those elements.

II. ANTIMONY (Sb)

Antimony has no known function in living organisms and is not among the more toxic elements. Data on the metabolism of this element are therefore meager. Smith (254) was one of the first to use neutron activation to examine a range of human organs and found antimony present in all of them. Most of the median values fell between 0.05 and 0.15 µg/g dry tissue, with the highest levels in the lungs (0.28 µg/g) and hair (0.34 µg/g). In a more detailed study of human lungs and some other organs, similar antimony levels were obtained, with the highest concentrations in the apex and the lowest in the base of the lungs (170). A feature of this investigation was the high antimony levels found in the lymph glands (0.34 and 0.43 μg/g fresh tissue). Means of organs examined by Yukawa *et al.* (305) ranged from 0.01 to 0.03 μg/g fresh tissue, except lung, which contained a mean of 0.23 μg/g. All the tissues, including lung, analyzed by Sumino *et al.* (266) contained between 0.01 and 0.11 μg antimony per gram of fresh tissue. The highest concentration (0.11 μg/g) was found both in skin and adrenal gland. Neither Yukawa *et al.* (305) nor Sumino *et al.* (266) analyzed hair or lymph gland. Hamilton *et al.* (103) also reported the presence of antimony in all human tissues examined, with mean concentrations generally lower (0.005–0.02 μg/g fresh tissue) than those reported by Smith (254), Yukawa *et al.* (305), or Sumino *et al.* (266), but similar to those obtained by Molokhia and Smith (171) for mouse organs and tissues. Again, the lungs (0.06 μg/g) and lymph nodes (0.20 μg/g) were two tissues with the highest concentrations of antimony. Rib was also high in antimony, with samples from 22 hard-water areas in England averaging 1.3 ± 0.2 μg antimony per gram of ash, and from 22 soft-water areas 1.7 ± 0.3 μg/g ash (103). Several other reports indicated tissue antimony concentrations similar to those found by Hamilton *et al.* (103). For example, reported concentrations include 0.0007–0.005 μg/g fresh human or beef heart (203,294), and 0.02 μg/g dry human fetal liver (246).

Versieck and Cornelis (282) reviewed the reported levels of antimony in plasma and serum. The means, all obtained by neutron activation analysis, varied from 0.52 to 5.2 ng/ml. Versieck and Cornelis (282) implied that some of the variation in findings might have been the result of contamination of the sample by the containers used during irradiation. Later analyses done when awareness of possible contamination existed showed the lower values. These analyses gave concentrations of 0.75 (295) and 0.52 ng/ml (131). Both Hamilton *et al.* (103) and Kjellin (137) found a mean of 5 ng antimony per gram of fresh blood, or close to 20 ng/g dry blood. Kosta *et al.* (141) reported that the mean antimony concentration in dry human colostrum was 0.5 ng/g, in transitional milk 0.55 ng/g, and in mature milk 1.2 ng/g.

In three separate studies of human dental enamel, similar antimony levels were obtained. Nixon *et al.* (188) reported a range of 0.005 to 0.67 μg/g, with a mean of 0.034 μg/g, for Scottish subjects and 0.070 for Egyptian subjects who had received antimony treatment for bilharziasis (schistosomiasis). Losee *et al.* (158) obtained a range of 0.02 to 0.34 μg antimony per gram of dental enamel from 28 U.S. subjects, with a higher mean concentration of 0.13 ± 0.01 μg/g. Curzon and Crocker (55) obtained a range of 0.0 to 3.0 μg/g dental enamel from 225 U.S. and New Zealand subjects, with a mean concentration of 0.20 μg/g.

Environmental exposure and illness affects the antimony concentration of some tissues. Bencko *et al.* (23) reported that hair from adult controls contained a mean of 0.12 ± 0.1 $\mu g/g$ antimony, but hair from occupationally exposed smelter workers contained a mean of 318 ± 130 $\mu g/g$. Tomza et al. (273) reported similar but less marked findings. The hair from their controls contained a mean antimony concentration of 0.26 ± 0.31 $\mu g/g$, whereas hair from lead–zinc smelter workers contained a mean of 1.65 ± 1.90 μg antimony per gram. Vanoeteren et al. (276) presented evidence that indicated a major portion of lung antimony could be present as a component of inhaled particles deposited in the lung. Examples of illness affecting tissue antimony are the reports of elevated levels in heart tissue from uremic patients (203) and in injured heart tissue from patients with myocardial infarction (294).

Published antimony levels in individual foods or animal feedstuffs are not very extensive, although it has long been known that foods stored in enamel vessels and cans may contain appreciable antimony concentrations (174). Oakes et al. (192) analyzed 21 fruits and vegetables and found that most of them contained <1.0 ng antimony per gram fresh weight. The only exceptions were sweet potato and green peppers, which contained 4.2 and 3.0 ng/g, respectively. Furr and co-workers (83,84) found that nuts were relatively high in antimony (50–300 ng/g dry weight), and that kernels of apricots and other fruits that may be sold in health food stores were slightly lower in antimony than were nuts (20–150 ng/g dry weight). They also reported that the antimony concentration in maple syrup was normally ~ 300 ng/g dry weight, but was 800 ng/g dry weight if obtained from trees near highways (85). Cowgill (52) reported that fresh banana pulp contained 8.3 ng antimony per gram. Becker et al. (17) found that the antimony content for swine feeds, piglet starter rations, fish meals, and various mixed feeds was generally between 20 and 60 ng/g. An occasional feed sample contained between 100 and 180 ng/g.

Studies on the antimony content of total diets have given widely divergent results. Murthy et al. (181) analyzed the total 7-day diets, including between-meal snacks, of children in institutions in 28 localities in the United States. The total antimony intakes varied from 247 to 1275 $\mu g/day$, and the total dietary concentrations varied from 209 to 693 $\mu g/g$. Hamilton and Minski (102) analyzed English total diets and obtained the much smaller mean daily intake of 34 ± 27 μg antimony. An even lower intake for Italians (1.5 $\mu g/day$) was reported by Clemente et al. (44). On the basis of the limited analyses of foods to date (see above), the latter values probably are most likely correct. However, significant seasonal and geographical variations were observed by Murthy et al. (181). Furthermore, Hamilton and Minski (102) reported an antimony content of 0.08 $\mu g/g$ in brown sugar compared with <0.002 $\mu g/g$ in refined white sugar. This suggests that seasonal, geographical, and refinement variations in foods would influence dietary intakes of antimony.

Interest in the metabolism of antimony developed following the discovery (41,42) that the trivalent antimony compound tartar emetic is effective in the treatment of schistosomiasis, or bilharziasis. Soluble antimony compounds, such as antimonites and tartrates, are poorly absorbed. After finding that antimony in an oral dose of potassium antimonyl tartrate (tartar emetic) was excreted more in the feces than the urine, Waitz et al. (285) concluded that gastrointestinal absorption of antimony was poor in monkeys, rats, and mice. Felicetti et al. (74) found that very little trivalent or pentavalent antimony tartrate administered via gavage was absorbed by hamsters. Four days after administration, only ~2% of the dose was retained and almost two-thirds of that was in the gastrointestinal tract.

The metabolism of antimony is affected by its valence state. Several workers have demonstrated the affinity of trivalent antimony for red blood cells (172,199) both in vivo and in vitro. In contrast, human red cells are almost impermeable to pentavalent antimony compounds (197). In most rodents, injected trivalent antimony is excreted primarily in the feces and pentavalent primarily in the urine (198), but in humans, both valence states of antimony are excreted in the urine (199).

An excellent review (10) of antimony toxicosis revealed that antimony apparently has a low inherent toxicity. This review suggested, on the basis of limited evidence, that a maximum tolerable level of antimony for the rabbit is 70–150 $\mu g/g$ diet. Generally, trivalent antimony is more toxic than pentavalent antimony.

The following are examples showing the low inherent toxicity of antimony. Lifetime studies with mice fed drinking water supplemented with antimony potassium tartrate at the level of 5 $\mu g/ml$ antimony revealed no demonstrable toxic effects on males and only a slight decrease in life span and longevity and some suppression of growth of older females (235). No evidence of carcinogenesis or tumorigenesis was obtained. Whanger and Weswig (297) found that 50 μg antimony as Sb_2O_5 did not promote the development of liver necrosis in selenium–vitamin E deficient rats.

III. BARIUM (Ba)

There is no conclusive evidence that barium performs any essential function in living organisms. The experiments with rats and guinea pigs done by Rygh (218) in the late 1940s apparently have been neither confirmed nor invalidated. Those species showed satisfactory growth and development when their specially purified diets were supplemented with a "complete" mineral mixture. The omission of either barium or strontium from the mineral supplement resulted in depressed growth.

According to Schroeder et al. (237), the "standard reference man" contains 22 mg barium, of which most is present in the bones. The remainder is widely distributed throughout the soft tissues of the body in very low concentrations that do not increase with age, except in the lungs, presumably from atmospheric dust. Human tissues reported to be relatively high in barium are as follows: bone, 18.0 ± 2.8 μg/g ash from hard-water areas and 19.3 ± 2.0 μg/g ash from soft-water areas (103), or 7 μg/g ash from 35 normal men and women (258); lymph nodes, 0.8 ± 0.3 μg/g fresh weight (103); dental enamel, 18.83 μg/g (55); hair, 2.5 μg/g from the scalp and 2.2 μg/g from the pubic area; and possibly the choroid of the eye, because 3.35 μg barium per gram fresh weight was found in rabbit choroid (86). Tipton and Cook (270) reported the following mean barium concentrations for normal adult human tissues: adrenals 0.02, brain 0.04, heart 0.05, kidney 0.10, liver 0.03, lung 0.10, spleen 0.08, and muscle 0.05 μg/g dry tissue. Hamilton *et al.* (103) obtained generally similar low values for human tissues: brain 0.006, kidney 0.01, liver 0.01, lung 0.03, muscle 0.02, and testis 0.02 μg/g fresh tissue. They found that the mean barium level in 103 samples of whole human blood was 0.1 ± 0.06 μg/g. Subsequent analyses revealed similar values for whole blood, 0.07–0.1 μg/g, and serum, 0.06–0.15 μg/ml (54,190, 202,289). Myocardial infarction apparently causes a decrease in barium in blood (190) and serum (289) and an increase in barium in the injured heart tissue (294). Other pathological disorders that apparently decrease blood barium are duodenal ulcer, chronic cholecystitis, primary carcinoma of liver, and hepatocirrhosis (190).

Barium is poorly absorbed from ordinary diets, with little retention in the tissues or excretion in the urine. Thus, in rats fed barium, 24-hr urinary barium excretion was 7% of the amount ingested, while three human subjects excreted in the urine 1.8, 1.9, and 5.8% of their respective dietary barium intake (29). Barium obviously crosses the placental barrier, because the element has been found in the tissues of all newborn infants examined (237). A relatively high level of barium has been found in dry fetal liver, 0.99 μg/g (246).

The mean barium intake from English diets has been estimated as 603 ± 225 μg/day (102) and from a U.S. hospital diet as 750 μg/day (237). Higher intakes by five adult human subjects (0.65–1.77 mg/day) were reported by Tipton and co-workers (271,272), and lower amounts (<0.30 to <0.59 mg/day) were present in the hospital diets studied by Gormican (91). In 19 U.S. states, 20 school lunches from each of 300 schools analyzed by Murphy *et al.* (179) contained from 0.09 to 0.43 mg barium, with a mean of 0.17 mg. On the basis of the lunches supplying one-third of the daily food intake, the total barium intakes would range from 0.27 to 1.29 mg/day and would average 0.51 mg/day.

Barium is usually associated with calcium and strontium in the food chain from plants to animals. Nonetheless, reports of the barium levels in human foods are limited. Robinson and co-workers (213,214) reported that BaO in a variety of fruits and vegetables ranged from 3 μg/g dry weight in apple to 80 μg/g dry weight in lettuce. They found an unusually high level of barium in Brazil nuts (range 700–3200 μg/g) that was not accompanied by unusual concentrations of strontium. The high level of barium in Brazil nuts was confirmed by Furr *et al.* (83), who found most nuts contained 0.1–2.6 µg/g dry weight. However, slightly higher values were found for black walnut (8.7 µg/g) and pecan (14 µg/g). Furr *et al.* (84) reported that kernels of apricots and other fruits that may be sold in health food stores contained 2.4–33 µg/g dry weight of barium. Oakes *et al.* (192) found that a variety of fruits and vegetables contained between 0.02 and <1.1 µg/g fresh weight. Other foods analyzed for barium content include maple syrup, ~5 µg/g dry weight (85), fresh banana pulp, 0.14 µg/g (52), and orange juice, with a range of 0.018 to 0.776 µg/g in single-strength juice (165).

Wide variation in barium levels apparently is characteristic of plants both within and among species. Bowen and Dymond (27) reported a range of 0.5 to 400 µg barium (mean of 10) per gram of different plant species growing on different soils. An earlier study (114) found 10–90 µg barium per gram of dry Kentucky hay, and 1–20 µg/g of dry cereal grains. In a comparison of red clover and ryegrass grown together on different soils in Scotland, Mitchell (169) found that clover contained 12–134 µg (mean 42) barium per gram and the grass 8–35 (mean 18) µg/g.

Barium apparently has a low order of toxicity by the oral route. For example, Hutcheson *et al.* (120) fed barium sulfate to three generations of mice at various levels up to 8 µg/g diet with no significant effects on growth, mortality, morbidity, or reproductive or lactational performance. In a life-term study, Schroeder and Mitchener (234) fed mice water containing 5 ppm barium (as barium acetate) with no observable effects on longevity, mortality, body weight, or tumor incidence. Barium sulfate is used in taking X-ray photographs of the intestinal tract. Even though the barium ion Ba^{2+} is extremely toxic when absorbed, barium sulfate is relatively insoluble, making it essentially nontoxic. Nonetheless, extrapolating from toxicity data and *in vitro* rumen work, the level of soluble barium in a diet probably should not exceed 20 µg/g (10). Even this level may be too high, because studies have associated high barium levels (1.1– 10 ppm) in drinking water to high cardiovascular mortality, and showed that chronically exposing rats to 10 or 100 ppm barium in the drinking water induced increased blood pressure (204).

IV. BORON (B)

A. Distribution

1. Animals

Boron is distributed throughout the tissues and organs of animals at concentrations mostly between 0.05 and 0.6 µg/g fresh weight, and several times these levels in the bones (3,78,103,115). Hamilton *et al.* (103), in their study of human tissues, found the following mean boron concentrations: liver 0.2, kidney 0.6, muscle 0.1, brain 0.06, testis 0.09, lung 0.6, and lymph nodes 0.6 μg/g fresh weight. One analysis indicated the following mean boron concentrations for the following dried human organs: kidney 0.94, liver 2.31, and spleen 2.57 μg/g (121). The mean boron level in 22 rib samples from humans living in hardwater areas in England was 10.2 ± 5 μg/g ash, and in 22 rib samples from those living in soft-water areas 6.2 ± 2 μg/g ash (103). These values are lower than those reported by Alexander *et al.* (3). They reported that the boron content of 116 samples of ashed bones from 33 humans ranged from 16 to 138 μg/g, with a mean of 61 ± 1.7 μg/g. Two different studies showed that human dental enamel varies widely in boron content. In one study (158), the range of 56 samples was 0.5–69 μg boron per gram of dry enamel, with the median of 9.1 and a mean of 18.2 ± 2.65 μg/g. In the other study (55), the range of 337 samples was 0.0–190.0 μg/g of enamel, with a mean of 8.4 μg/g. The boron content of hair is similar to that of soft tissue. An analysis of 63 samples of scalp hair and 107 samples of pubic hair gave boron values of 1.1 and 0.7 μg/g respectively (54). The reported blood boron values have been very consistent since the mid-1950s. The reported values include 0.141 in 1954 (78), 0.4 in 1972 (103), 0.2 in 1975 (202), and 0.1 μg/g in 1976 (54). Panteliadis *et al.* (202) reported that the mean concentration of boron in serum was 0.18–0.21 μg/g. Apparently, plasma boron is relatively high at birth (0.5–0.6 μg/ml) and decreases to 0.2 to 0.3 μg/ml within 5 days (80). One study indicated that human milk contains less boron than blood or plasma, about 0.06 to 0.08 μg/ml (80).

The study of the boron content of animal tissues and fluids apparently has received little attention. Cow's milk normally contains 0.5–1.0 ppm boron, with little variation due to breed or stage of lactation (115,200), but responds rapidly to changes in dietary boron intake. Owen (200) raised the boron level in milk from 0.7 to >3 ppm by adding 20 g of borax daily to the normal ration of cows. When young cattle were given drinking water (0.8 mg boron per liter) supplemented with 0, 15, 30, 60, or 120 mg boron per liter for 10 days, plasma boron concentrations were 2.7, 4.4, 5.3, 8.3, and 13.4 mg/liter, respectively (292). Mathur and Roy (163) reported that 17 samples of buffalo milk contained 1.04 ± 0.22 ppm boron. Their findings also indicated that 44.8% of the boron was in a soluble form, 37.6% was associated with the fat phase, and 17.6% was attached to the proteins coagulated by rennet. The finding of 1.7 μg boron per gram of dry mule deer metacarpus (263) indicates that animal bone contains appreciable levels of boron. Ploquin (208) reported that a variety of organs of sheep contained 0.7–3.0 μg boron per gram of dry tissue. The one exception was thyroid, which contained 25–30 μg/g. When sheep were raised on soil with a high boron content, the boron contents of the organs were elevated to 1.0 to 20.4 μg/g dry tissue. The greatest increases occurred in spleen (1.3 to 1.8 increased to 7.9 to 10.7 μg/g), kidney (1.6 to 2.0 increased to 10.3 to 13.7 μg/g), and brain (2.5 to

3.0 increased to 16.8 to 20.4 μg/g). The thyroid boron concentration stayed at about the same level, 27.2–32.7 μg/g dry tissue. This study confirms earlier observations of a marked increase in the boron levels of tissues, particularly the brain, when large amounts of boric acid are ingested (308).

2. Foods and Feedstuffs

An extensive study of the mineral content of >200 Finnish foods included boron (191,279,280). The average boron content (in micrograms per gram dry weight) in different food groups was as follows: cereals 0.92, meat 0.16, fish 0.36, dairy products 1.1, vegetable foods 13, and others 2.6. Foods that contained the highest levels of boron (in micrograms per gram wet weight) included these: soy meal 28, prune 27, raisin 25, almond 23, rose hips 19, peanut 18, hazel nut 16, date 9.2, and honey 7.2. Wines contained up to 8.5 μg boron per gram. The high level of boron in nuts, fruit kernels, and honey was also reported elsewhere (83,84,177). Other reports of high levels of boron in fruits and vegetables are those of Ploquin (208) and Szabo (267), who found the following concentrations: apple 468, pear 709, cabbage 80, tomato 1258, and red pepper 440 μg/g dry weight. Thus, it is obvious that foods of plant origin are rich sources of boron. Meat or fish apparently are poor sources of boron.

It is apparent from the preceding that the daily intake of boron by humans can vary widely depending on the proportions of various food groups in the diet. Some years ago Kent and McCance (135) reported daily boron intakes by human adults of 10 to 20 mg, the higher amounts being associated with the consumption of large quantities of fruits and vegetables. Ploquin (208) stated that, without including beverages, the daily intake of boron should be near 7 mg. However, he thought that beverages commonly consumed in France (wine, cider, beer) would increase the boron intake to, or above, the intakes reported by Kent and McCance (135). These levels contrast greatly with the mean boron intakes of 0.42 and 0.35 mg/day found by Tipton et al. (272) for two adults consuming self-selected diets. These values are near the reported average daily intake of 1.7 mg boron by Finnish people (279). In their study of the minerals in total diets made up from various sources to supply 4200 kcal/day, Zook and Lehmann (308) obtained an overall average boron content of 3.1 mg, with individual composites varying relatively little from 2.1 to 4.3 mg/day. These levels conform well with the mean of 2.8 ± 1.5 mg/day reported for English total diets (102).

Boron intakes by grazing animals vary with the soil type and boron status and plant species consumed, because these factors influence the boron content of plants. Boron deficiency occurs in some plants when vegetative dry-matter boron is <15 μg/g. Toxicosis usually occurs when dry-matter boron exceeds 200 μg/g (10). Monocotyledons generally contain less boron than dicotyledons. Thus, it is not surprising that legumes usually are richer in boron than grasses (19). Boron levels of 4 to 7 µg/g dry matter have been reported for European pasture grass (136) and Wisconsin grass hay (43). Animals fed a high-grain concentrate diet probably would ingest relatively low amounts of boron, because most grains contain <2 µg boron per gram (280).

B. Physiological Aspects

1. Metabolism

Boron in food, sodium borate, or boric acid is rapidly absorbed and excreted largely in the urine (135,200,272). Because of variations in the boron dose and length of the study, reported recoveries of ingested boron from the urine ranged from 30 to 92%. Kent and McCance (135) did human balance studies in which 352 mg of boron as boric acid was ingested on day 1. At the end of 1 week, >90% of the boron was recovered from the urine. Pfeiffer et al. (206) recovered 40% of the boron ingested by dogs in a 24-hr urine collection. Owen (200) found that, in two dairy cows, 57% of the boron from a control ration (271 mg/day) was excreted in the urine, but 71% of the boron from boronated rations (2058 and 2602 mg boron per day) was recovered from urine. Akagi et al. (2) reported that after 48 hr >70% of an oral dose of boron as sodium borate was excreted in the urine by rabbits and guinea pigs; after 120 hr >80% of the dose was excreted in the urine. Heifers fed tap water and 215.3 mg boron daily excreted 30% of the boron in urine (93). When the drinking water was supplemented with 150 and 300 ppm boron (4224.2 and 7219.3 mg/day intake), heifers excreted 67 and 69% of the boron daily in urine. On the basis of the finding that the percentage of renally filtered boron that is excreted increases with increasing boron intake, Weeth et al. (292) developed an equation for predicting the boron status of cattle.

2. Deficiency Signs

Boron has been known to be essential for higher plants since the 1920s (287), but evidence has now been obtained indicating that boron is required by animals. Between 1939 and 1944, several attempts to induce a boron deficiency in rats were unsuccessful, although the diets fed apparently contained only 155–163 ng/g boron (196,252,268). In 1945, Skinner and McHargue (252) reported that supplemental dietary boron enhanced survival and maintenance of body fat, and elevated liver glycogen in potassium-deficient rats. Those findings were not confirmed by Follis (77), who fed a different diet with an unknown boron content and different levels of boron supplementation. After those reports the study of boron as a possible nutrient for animals was neglected until 1981, when Hunt and Nielsen (117) reported that boron deprivation depressed growth and elevated plasma alkaline phosphatase activity in chicks fed inadequate cholecalciferol. Further experiments indicated that cholecalciferol deficiency enhanced the need for boron and that boron might interact, in some manner other than through an effect on cholecalciferol metabolism, with the metabolism of calcium, phosphorus, or magnesium (118). The relationship seemed strongest between boron and magnesium, because boron tended to normalize the abnormalities associated with magnesium deficiency in chicks. Boron did not consistently alleviate signs of calcium and phosphorus deficiency. Nonetheless, because the boron–magnesium ratio was quite low in both plasma and diet, boron apparently indirectly affected magnesium metabolism.

Experiments with rats suggest that the indirect influence of boron was the result of altered parathormone activity (185,186). Dietary boron markedly affected the response of rats to treatments that supposedly cause changes in parathormone activity. For example, findings indicated that boron-supplemented (3 μg/g diet) rats did not respond as readily as boron-deprived (0.3–0.4 μg/g diet) rats to magnesium deprivation, which supposedly causes, in contrast with chicks, a hyperparathyroid state in rats. Furthermore, when magnesium was low, the response to other treatments that apparently alter parathromone activity was more marked in boron-supplemented than boron-deprived rats. On the other hand, when dietary magnesium was adequate, the response of rats to treatments of high dietary aluminum and fluoride, and to low dietary calcium, were often more marked and in an opposite direction in boron-deprived than in boron-supplemented rats.

3. Function

Although the data need clarification by further experimentation, the preceding findings support the hypothesis that boron has an essential function that somehow regulates parathormone action, and, therefore, indirectly influences the metabolism of calcium, phosphorus, magnesium, and cholecalciferol. Other evidence to support this hypothesis was reported by Elsair and co-workers (72,73). They found that high dietary boron partially alleviated the fluoride-induced secondary hyperparathyroidism signs of hypercalcemia, hypophosphatemia, and depressed renal absorption of phosphorus in rabbits. Seffner and co-workers (244,245) found that dietary sodium borate, and magnesium silicate to a lesser extent, reduced the fluoride-induced thickening of the cortices of the long bones in pigs, and that dietary boron altered the histological characteristics of the parathyroid. Baer et al. (11) reported that boron corrected radiographic and histological changes caused by fluoride toxicity in bone. It would not be surprising to find that boron has a regulatory role involving a hormone such as parathormone, because boron is suspected of having a regulatory role in the metabolism of such plant hormones as auxin, gibberellic acid, and cytokinin (122,150), perhaps through control of the production of a second messenger, such as cyclic AMP, at the cell membrane level (122,209).

There are numerous other reports that suggest that boron has a close relationship with calcium metabolism, most likely at the cell membrane level. Palytoxin, an extremely poisonous animal toxin from coral, raises the permeability of excitable and nonexcitable membranes of animals. The binding of palytoxin to membranes is potentiated both by Ca^{2+} and borate (100). Aplasmomycin, a novel ionophoric macrolide antibiotic, which was isolated from strain SS-20 of *Streptomyces griseus*, has been determined to be a symmetrical dimer built around a boron atom (37). Pollard *et al.* (209) presented evidence that supported the view that boron can influence the conformation and activity of specific membrane components of plant cells. They found that in *Zea mays* the depressed absorption of rubidium, phosphate, and chloride caused by boron deficiency could be partially reversed within 1 hr after boron was resupplied. This rapid reversal coincided with an increase in the KCl-stimulated ATPase activity of a membrane fraction prepared from roots resupplied with boron. Culturing the diatom *Cylindrotheca fusiformis* under boron-deficient conditions leads to changes in ^{86}Rb uptake and photosynthesis, thus supporting the idea that the regulation of cation fluxes through the cell membrane is tightly linked to boron status (255). The accentuating effect of excess potassium on both boron toxicity and deficiency in plants was thought to be due to the influence of potassium on boron-regulated cell permeability (71). If boron is involved in cell membrane metabolism, it should interact with calcium, another element involved in membrane function and integrity. Evidence of such an interaction includes the findings that calcium reversed boron repression of mitotic activity in pea root-tip meristem cells (139) and reversed the adverse effect of boron on the chlorophyll content of leaves (219).

Further indications of a relationship between boron and calcium are that boron or boron compounds can influence calcium metabolism and that tissue boron levels change in animals with abnormal calcium metabolism. Curzon and Crocker (55) statistically associated boron in human teeth with low caries incidence. However, other studies (119,153) indicated that high levels of orally administered boron increased dental caries or antagonized the cariostatic effect of fluoride. Parturient paresis, a disorder of cows characterized by depressed plasma calcium and phosphate and elevated plasma magnesium, is treated by the injection of calcium borogluconate (152). Of course, the treatment is given for its calcium component. However, it is possible that the boron component could be contributing to the effectiveness of the treatment, because Littledike *et al.* (152) concluded that calcium therapy does not appreciably alter the basic course of hypocalcemia but rather corrects some biochemical, neuromuscular, and membrane transport defects occurring secondarily to hypocalcemia. Amine cyanoboranes and amine carboxyboranes (boron analogs of α-amino acids) effectively block induced arthritis (101). Tablets containing magnesium carbonate and sodium borate are touted as a remedy for arthritis (183,184).

The possibility that boron has a role in some enzymatic reaction cannot be overlooked, because boron has been shown to affect the activity of numerous enzymes *in vitro* and in plants. Lewin and Chen (149) stated that boron might have a role as a cofactor for some enzymatic reaction, rather than a structural component of a molecular constituent, because they found that boron controlled the growth rate of diatom cells but did not affect the final yield of cells in culture. Wolny (300) briefly reviewed the *in vitro* enzyme work and found that borate competitively inhibits two classes of enzymes. One class is the pyridine or flavin nucleotide-requiring oxidoreductases such as yeast alcohol dehydrogenase, aldehyde dehydrogenase, xanthine dehydrogenase, and cytochrome b_5 reductase. Borate apparently competes with the enzyme for NAD or flavin, because borate has a great affinity for cis-hydroxyl groups. The other class of borate-inhibited enzymes are those in which borate and boronic acid derivatives bind to the active enzyme site. These enzymes include chymotrypsin, subtilisin, and glyceraldehyde 3-phosphate dehydrogenase. An example of the enzyme changes occurring in plants with boron deficiency was reported by Agarwala *et al.* (1). They found that pollen grains of such plants had low activities of catalase, acid phosphatase, starch phosphorylase, and invertase, and high activities of ribonuclease and amylase. Other plant enzymes affected by boron nutrition have been reviewed by Dugger (69) and Lewis (150).

4. Requirements

The minimum amounts of boron required by animals to maintain health, based on the addition of graded increments to a known deficient diet in the conventional manner, have not been determined. However, on the basis of the finding that rats and chicks sometimes have altered mineral metabolism when fed diets containing 0.3–0.4 µg boron per gram (117,118,185,186), diets probably should contain more than this level of boron.

C. Toxicity

Boron has a low order of toxicity when administered orally. A maximum tolerable level of 150 µg boron (as borax) per gram of dry diet has been suggested for cattle (10). Three excellent reviews of the toxicity of boron (10,151, 208) indicate that toxicity signs generally occur only after the dietary boron concentration exceeds 100 µg/g. For example, Weir and Fisher (293) fed diets containing 117, 350, and 1170 µg/g boron to dogs and rats for 2 years. The animals showed no apparent signs of toxicity when dietary boron was 117 or 350 µg/g. After 2 months, rats fed 1170 µg/g diet exhibited coarse hair coats, scaly tails, a hunched position, swelling and desquamation of the pads of the paws, abnormally long toenails, bloody discharge of the eyes, and depressed hemoglobin and hematocrit. After 38 weeks, dogs fed 1170 µg/g diet exhibited reversible testicular degeneration and cessation of spermatogenesis. Studies with gerbils injected with borax indicate that degenerative changes in testes may be characterized by increased activity of phosphatases (247).

Toxic boron concentrations in drinking water are lower than those in feed, probably because absolute daily boron intake is greater when drinking water instead of diet is supplemented with a specific concentration of boron. Cows consuming 150 or 300 mg boron per liter of water exhibited inflammation and edema in the legs and around the dew claws, and reduced feed intake, growth, hematocrit, hemoglobin, and plasma phosphorus (93). Urinary phosphate excretion was decreased; however, a subsequent study (292) indicated an increased urinary phosphate excretion by cows fed 60 or 120 mg/liter of boron in the drinking water. Green *et al.* (92) found that when boron exceeded 150 mg/liter in drinking water, rats exhibited depressed growth, continued prepubescent fur, lack of incisor pigmentation, aspermia, and impaired ovarian development. When the boron content of drinking water was 300 mg/liter, rats also exhibited depressed plasma triglycerides, protein and alkaline phosphatase, and depressed bone fat and calcium (243). Growth and reproduction were not affected when rats were fed 75 mg boron per liter of drinking water (92). However, Krasovskii *et al.* (142) claimed that gonadotoxic effects occurred in rats fed as little as 6 mg/liter drinking water. On the other hand, Schroeder and Mitchener (234) found that 5 ppm boron in drinking water for life had no effect on the life span, longevity, or tumor incidence of mice. Dixon *et al.* (62) fed drinking water containing 0.3, 1.0, or 6.0 mg boron (as borax) per liter to rats for 90 days. No boron supplement affected growth, weights of testis, prostate or seminal vesicles, or any clinical serum chemistry parameters.

In humans, the signs of acute toxicity are well known and include nausea, vomiting, diarrhea, dermatitis, and lethargy (151,207). In addition, high boron ingestion induces riboflavinuria (207). The association between riboflavin and boron is not unusual, because Landauer (145) found that newly hatched chicks treated with boric acid at 96 hr of incubation exhibited "curled-toe paralysis," an abnormality associated with riboflavin deficiency. Landauer (145,146) found that boron-induced teratogenic abnormalities, including several types of skeletal abnormalities, were reduced by the administration of riboflavin. Other polyhydroxy compounds (D-ribose, pyridoxine hydrochloride, D-sorbitol hydrate) also reduced or abolished the teratogenic effects of boric acid on chick embryos (146).

V. BROMINE (Br)

Bromine is one of the most abundant and ubiquitous of the recognized trace elements in the biosphere. It has not been conclusively shown to perform any essential function in plants, microorganisms, or animals. However, there are numerous findings that indicate the possible essentiality of bromine should be studied further. Bromide can completely replace chloride to support growth of several halophytic algal species (166) and can substitute for part of the chloride requirement of chicks (148). A small significant growth response to dietary trace additions of bromide has been reported for chicks (116) and mice (26) fed a semisynthetic diet containing iodinated casein to produce a hyperthyroid-induced growth retardation. The bromine content of the basal diet was not given, and these indications of a growth requirement for bromine apparently have not been investigated further. Earlier, Winnek and Smith (298) were unable to demonstrate any effect on growth, health, or reproductive performance of suboptimally growing rats fed a diet containing <0.5 µg/g bromine over a period of 11 weeks, nor any improvement from a supplement of 20 µg/g diet of bromine as potassium bromide.

In 1981, Oe *et al.* (193) found unusually low bromide concentrations in serum and brain of patients subjected to chronic hemodialysis; apparently the artificial kidney removed bromine (283). They associated the insomnia exhibited by many hemodialysis patients with the bromine deficit. Subsequently, they did a double-blind trial in which either bromide or chloride was added to the dialysate of four patients on maintenance hemodialysis. Quality of sleep improved markedly in the two patients who received bromide but not in those who received chloride.

The findings of Oe *et al.* (193) are not the first association between bromide and quality of sleep. Before barbiturates were used, doctors prescribed bromide for sleep. Many years ago, Zondek and Bier (307) reported the possible presence of a bromine-containing sleep hormone in dog pituitary gland. A bromine-containing compound has been isolated from human cerebrospinal fluid, with properties corresponding to 1-methylheptyl-γ-bromoacetoacetate (synonym of 2-acetylbromoacetoacetate) (304). This organic bromine compound has been shown to "provoke paradoxical sleep" when administered intravenously to cats (274).

All animal tissues contain 50–100 times more bromine than iodine, except the thyroid, where the reverse is true. Species differences in tissue bromine concentrations are small, and the element does not accumulate to any marked degree in any particular organ or tissue (47,103,106). Claims that bromine is concentrated in the thyroid and pituitary glands (97) have not been substantiated (16,50,63). Reported mean bromine levels in adult human organs (in micrograms per gram fresh tissue) include the following: brain 1.7, hair 2.7 and 8.0, heart 2.0, liver 4.0, lung 5.2 and 7.5, lymph nodes 0.9, prostate 1.1 and 3.3, teeth 4.5, testis 5.1, and ovary 3.3 (55,103,143,175,273,276,294). On a dry basis, the reported bromine content of fetal liver, 8.93 µg/g (246), is similar to that of adult liver, 15.0 µg/g (134). Versieck and Cornelis (282) have summarized the reported values for plasma and serum bromine. Except for one that was usually high, the means were between 2.13 and 7.52 µg/ml. The bromine concentrations found in whole blood are in the same range (30,50,103). Kjellin (137) reported that the bromine concentration in cerebrospinal fluid was near that of blood, serum, and plasma, ~3.8 μg/ml.

The bromine levels in all tissues and fluids can be substantially raised by increasing dietary bromine intakes (158,298). According to Versieck and Cornelis (282), a number of people have been found with relatively high levels of plasma or serum bromine. Most of these people apparently were using bromine-containing drugs. Bromides are available to the public without prescription in popular nostrums, nerve tonics, headache remedies, cough cures, and other proprietary preparations. Before brominated oils as food additives were banned in the United Kingdom, they were blamed for the high bromine levels found in the fat of tissues from United Kingdom children (53). Lynn et al. (159) increased the bromine concentration of milk from a pretreatment level of 10 ppm to as high as 60 ppm by feeding 12.5 g NaBr to cows for 5 days.

The bromine levels in fluids and tissues might be affected by some pathological disorders. Bromine apparently is elevated in heart tissue injured by myocardial infarction (294) and in heart tissue from patients who suffered uremic heart failure (203).

Little information is available on bromide metabolism, although the ion apparently is well absorbed (see earlier) and is mostly excreted in the urine, whether ingested or injected (47,106). Kinetic experiments on bromine uptake by brush border membrane vesicles isolated from rabbit jejunum showed a purely diffusive mechanism with no apparent saturable component (98). The transport of bromide from mother to fetus apparently is not impeded, as Alexiou et al. (4) found that the bromine concentrations in sera of controls, mothers, and umbilical cords were 3.15, 3.78, and 4.12 μg/ml, respectively. A significant positive correlation existed between values of maternal and umbilical cord blood. After oral intake of 1 g of KBr/kg body weight, bromine was found to be slowly removed from the bloodstream, the biological half-life varying from 10 to 12 days (253). Bromide and chloride readily exchange to some degree in body tissues, so that administration of bromide results in some displacement of body chloride, and vice versa (162). This exchange occurs as a consequence of feeding large amounts (298) or injecting physiological quantities of bromide (106). Rabbit thyroid glands rendered hyperplastic by lack of iodine are richer in bromine than the blood (16). This suggests that the thyroid distinguishes imperfectly between bromine and iodine and seizes some bromine in the absence of sufficient iodine. The bromine accumulated in the thyroid is quickly lost when iodine is supplied and cannot be used for hormone synthesis (211). There is some evidence that injected bromide reduces ^{131}I uptake by the rat thyroid and that goiter can occur in rats fed bromide during their first year of life (45).

Human dietary intakes of bromine are large and variable. Duggan and Lipscomb (67) obtained an average bromine intake of 24 mg/day over 2 years from U.S. diets, while Hamilton and Minski (102) reported a mean intake of 8.4 ± 0.9 mg/day from English total diets, and Varo and Koivistoinen (279) calculated an average intake of 4.2 mg/day from Finnish diets. Actual intakes will be much higher where organic bromine compounds are used as fumigants for soils and stored grains (68,109). Bromine concentrations as high as 53 to 220 μg/g in grain (oats and corn) fumigated with methyl bromide, and marked increases in the bromine content of milk from cows fed such grain have been reported (159). Even in untreated grain samples, bromine concentrations ranged widely, from 0.5 to 25 μg/g (109). Becker *et al.* (17) found that some pig feed concentrates and fish meals had high levels of bromine (15–69 μg/g). Varo and co-workers (191,279,280) included bromine in their extensive study of the mineral content of Finnish foods. This survey showed that no food group was especially high in bromine. Most foods contained no more than a few micrograms of bromine per gram. An occasional grain or flour would contain a high level of bromine; however, bakery products made from those grains and flours contained no more bromine than other foods, and no occasional high value was seen. Furr *et al.* (83) found four kinds of nuts relatively rich in bromine. In their analyses, almond, Brazil nut, English walnut, and pistachio contained 20, 87, 76, and 16 μg bromine per gram dry weight, respectively.

As an ion in diets, bromine apparently has a low order of toxicity. Growing pigs tolerated 200 μg/g, growing chickens tolerated 5000 μg/g, and growing rats tolerated 4800 μg/g without adverse effects (10).

VI. GERMANIUM (Ge)

The biology of germanium has excited little interest, despite its relative abundance in the lithosphere, its chemical properties, and its position in the periodic table within the range of the biologically active trace elements. There is no evidence that germanium is essential in mammalian nutrition, and germanates have a low order of toxicity in mice and rats (216,228,232). Schroeder and Balassa (228) found that almost all of 125 samples of foods and beverages they analyzed contained detectable germanium. However, only 4 of these samples contained >2 μg/g, and 15 others >1 μg/g. Germanium was not detected in refined white flour, although it was present in whole wheat and was concentrated in bran. Cowgill (52) found only 5 ng germanium per gram of fresh banana pulp. The mean concentrations in vegetables and leguminous seeds reported by Schroeder and Balassa (228) were 0.15–0.45 μg/g fresh weight. Similar concentrations were observed in meat and dairy products. Subsequent analyses confirming these findings apparently have not been done.

Few data on the germanium content of normal human organs have appeared. Preliminary data of Hamilton *et al.* (103) showed the following mean concentrations: blood 0.2, liver 0.04, kidney 9.0, muscle 0.03, lung 0.09, lymph nodes 0.009, brain 0.1, and testis 0.5 μg/g fresh tissue. Shand *et al.* (246) found 0.32 μg/g germanium in dry human fetal liver. The liver, kidneys, heart, lungs, and spleen of laboratory mice and rats fed normal diets contained germanium in concentrations ranging from 0.10 to 2.79 μg/g fresh tissue (228,232). When mice were given 5 ppm germanium in the drinking water for their lifetime, higher concentrations were found in several organs, especially in the spleen, but rat tissues accumulated very little germanium under these conditions (228,232). The germanium apparently was slightly toxic to both species, resulting in reduced life span and increased incidence of fatty degeneration of the liver. The element was neither tumorigenic nor carcinogenic. A subsequent study showed that germanium affected the level of some trace elements in some organs (236). For example, germanium depressed chromium in liver, lungs, heart, kidneys, and spleen, and elevated copper in liver. Plant studies suggest that germanium interacts with boron (28). Germanium supplementation can delay boron deficiency signs in tomatoes and sunflowers, apparently by serving as a substitute, in a sparing role, allowing greater mobility of boron to sites that require the element.

Little is known of the metabolism of germanium ingested from ordinary diets. Rosenfeld and Wallace (216) found that oral doses of sodium germanate were absorbed rapidly and almost completely from the gastrointestinal tract within a few hours, and were excreted largely in the urine during the next 4–7 days. The data of Schroeder and Balassa (228) concerning the germanium content of urine (mean of 1.26 μg/ml) from four individuals suggest that in humans dietary germanium is also well absorbed and excreted largely via the kidneys. Schroeder and Balassa (228) calculated that adults ingest ~1.5 mg germanium per day, of which 1.4 mg appears in the urine and 0.1 mg in the feces. A lower mean level of 367 ± 159 μg/day was obtained for total English adult diets (102).

VII. RUBIDIUM (Rb) AND CESIUM (Cs)

Biological interest in rubidium and cesium has been stimulated by their close physicochemical relationship to potassium and their presence in living tissues in higher concentrations, relative to those of potassium, than in the terrestrial environment. Over a century ago Ringer (212) observed that rubidium was similar to potassium in its effect on the contractions of isolated frog heart. Relationships between potassium and rubidium, and between cesium and potassium, have been found in a variety of physiological processes. These relationships exist in such diverse actions as their ability to neutralize the toxic action of lithium on fish larvae, or to affect the motility of spermatozoa, the fermentative capacity of yeast, and the utilization of Krebs cycle intermediates by isolated mitochondria. Their extracellular ionic concentrations also influence the resting potential in nerve and muscle preparations and the configuration of electrocardiograms (210).

The described metabolic interchangeability suggests that rubidium or cesium might have the ability to act as a nutritional substitute for potassium. Rubidium, and to a lesser extent cesium, can replace potassium as a nutrient for the growth of yeast (147) and of sea urchin eggs (155). This nutritional replaceability can be extended to bacteria (161), but higher animals are more discriminating. Additions of rubidium or cesium to potassium-deficient diets prevent the occurrence of characteristic lesions in the kidneys and muscles in rats (76) and, for a short period, permit almost normal growth until death inevitably supervenes (108). Glendening et al. (90) obtained no evidence that rubidium is an essential element for rats fed purified diets with variable supplements of rubidium, sodium, and potassium. However, there were indications that rubidium partially substituted for potassium and was more toxic in low- than in high-potassium diets. Purified diets containing up to 200 $\mu g/g$ rubidium were nontoxic, but levels of 1000 $\mu g/g$ or more depressed growth, reproductive performance, and survival time in rats.

In addition to its possible interchangeability role with potassium, there is some evidence suggesting that rubidium has a role involving some neurophysiological mechanism. In heart, rubidium was found to be lower in conductive tissue than in adjacent muscle tissue (294). The rubidium content of brain differs significantly between defined functional regions and also decreases with age (107,112). Rubidium apparently can enhance the turnover of brain norepinephrine (265). Vis et al. (283) suggested that because rubidium causes electroencephalogram activation in monkeys (167) and rats (264), perhaps the depletion of rubidium prevents the normalization of low-wave power values in dialysis patients. They found that rubidium rapidly diffused through the membranes of the artificial kidney. However, electroencephalogram activation by rubidium has not been demonstrated in humans (75).

Rubidium is rapidly and highly absorbed and excreted by the digestive tracts of mammals (222). Rubidium resembles potassium in its pattern of absorption, distribution, and excretion in animals. On the basis of studies with brush border membrane vesicles isolated from rabbit jejunum, potassium and rubidium apparently share a transport system (98). All plant and animal cells are apparently permeable to rubidium and cesium ions at rates comparable with those of potassium (210). All soft tissues of the body have rubidium concentrations that are high compared with many trace elements, with a total-body content of ~360 mg in the adult man. Rubidium does not accumulate in any particular organ or tissue and is normally relatively low in bones (90,250,261). The level of rubidium in human dental enamel varies widely. Curzon and Crocker (55) found a range of 0.0 to 30.0 $\mu g/g$ rubidium, with a mean of 4.61 $\mu g/g$, for 256 samples. Rubidium was associated with high caries incidence. Later analyses (103,305) have revealed the following mean ~ubidium concentrations in human organs, expressed as micrograms per gram of fresh tissue: brain 4.0 ± 1.1, cerebrum 2.8 ± 1.1, cerebellum 2.4 ± 0.9, muscle 5.0 ± 0.5 and 4.0 ± 1.0, ovary 5.0 ± 0.9, kidney 3.3 ± 0.4, kidney cortex 5.2 ± 0.5, kidney medulla 5.0 ± 0.3, lung 3.5 ± 0.4 and 2.2 ± 0.9, lymph nodes 5.5 ± 1.1, liver 7.0 ± 1.0 and 6.4 ± 2.8, testis 19.6 ± 6.2, heart 3.0 ± 1.9, pancreas 4.0 ± 1.7, and spleen 4.0 + 1.6. The rubidium content of plasma is lower than that of erythrocytes (51,90). Thus, while reported rubidium values for plasma or serum generally range between 0.08 and 0.31 µg/ml (282), mean values reported for whole blood include 2.7 and 3.0 µg/ml (44,103). The levels of rubidium in organs and tissues can be influenced by species differences and diet. Rat tissue contains relatively high levels of rubidium. For example, Kemp et al. (134) found the following mean concentrations for rat, pig, and human liver, respectively: 72, 28, and 20 µg/g dry weight. An example of the influence of diet is the study of Lombeck et al. (156), which showed that dietetically treated patients with phenylketonuria and maple syrup-urine disease had depressed levels of rubidium in whole blood. In another study (90), rubidium concentrations ranging from 100 to 200 µg/g dry tissue were found in the muscles, liver, lungs, kidneys, heart, and brain of rats fed a standard diet, and much higher concentrations, up to 8000 and 12,000 µg/g, when the animals were fed toxic levels of rubidium. The rubidium retained in tissues, as a result of such high dietary intakes, or from injection, is slowly lost from the body. Following intravenous administration of rubidium-86 (^{86}Rb), even though probably <5% remained in blood (168), 39–134 days were required for one-half of the dose to be excreted in the urine and feces (32). Although the major route of rubidium excretion is apparently the urine, with a clearance rate slightly less than that of potassium (144), the intestine probably is also involved in excretion also. Schäfer and Forth (222) found that rubidium was excreted against a concentration gradient from blood into the lumen of both the small and large intestines of rats.

The occurrence of cesium in biological samples has not been extensively studied. However, it has been found that, like rubidium, the cesium content is much higher in erythrocytes than plasma. Cornelis et al. (51) found a mean cesium concentration in serum of 0.74 ng/ml, and in red blood cells of 4.82 ng/g. Several other reports (see 282) have also indicated that the cesium level in serum or plasma is in the range of 0.74 to 1.33 ng/ml. Both Clemente et al. (44) and Hamilton et al. (103) reported that whole blood contains ~5 ng/ml cesium. Like blood, other tissues contain cesium at a level of about one order of magnitude less than rubidium. Vanoeteren et al. (276) found 9.1 ng cesium per gram of homogenized human lung. Wester (294) found a median of 11.4 ng/g in human heart, and the cesium content was reduced in injured tissue of myocardial infarcted heart. These values are similar to those reported by Hamilton et al. (103).

They found the following mean cesium concentrations, in nanograms per gram fresh tissue, for human organs: kidneys 9.0, lymph nodes 20, testis <1.0, and ovaries 9.0.

Varo and co-workers (191,279,280) included rubidium in their mineral analyses of Finnish foods. Their results indicated that most foods contained 0.5–5 μg/g rubidium and that the daily intake of rubidium was 4.2 mg. This compares well with the reported rubidium intakes of 4.35 ± 1.54 mg/day with English total diets (102), 1.28–4.98 mg/day with U.S. diets (182), and 2.5 mg/day with Italian diets (44). Diets with the highest rubidium levels were probably rich in meats and dairy products. Brazil nuts are very rich in rubidium (83). Becker *et al.* (17) found that the rubidium content for swine feeds, piglet starter rations, and various mixed feeds ranged from 2.6 to 26.1 μg/g dry weight. Based on the analyses by Varo and associates (191,279,280), many animal feeds probably contain rubidium at these levels.

The cesium content of foodstuffs and feeds has not been examined extensively. Some isolated values have appeared, including 0.1–0.3 μg/g dry fruit kernels (84), 0.06–0.07 μg/g dry maple syrup (85), 0.1–0.3 μg/g dry nuts except Brazil nuts, which contained 1.3 μg/g (83), 9 ng/g fresh orange juice (166), and 12.1 ng/g fresh banana pulp (52). Most fruits and vegetables apparently are quite low in cesium. Duke (70) found 3–11 ng cesium per gram dry weight, and Oakes *et al.* (192) found <1–3.3 ng/g fresh weight, in a limited variety of examined fruits and vegetables. With such low values reported for individual foods, it is not surprising that daily cesium intakes have been reported to be only 13 ± 7 μg with English total diets (102), and 15 μg with Italian diets (44).

VIII. SILVER (Ag)

There is no evidence that silver is essential for any living organism, nor is it ranked among the more toxic trace elements. It occurs naturally in very low concentrations in soils, plants, and animal tissues, and can gain access to foods from silver-plated vessels, silver–lead solders, and silver foil used in decorating cakes and confectionary.

The fact that foods contain very little silver is supported by reports of nondetectable to less than a few nanograms of silver per gram of a variety of fruits and vegetables (192), and orange juice (165), the finding of only 10 ng silver per gram of fresh banana pulp (52), and the finding of 0.027 to 0.054 mg silver per liter (weighted average of 0.047 ± 0.007 mg/liter) of cow's milk (180). Further support comes from the reported human dietary silver intakes of 27 ± 17 μg/day (102) and <1.0 μg/day (44).

The level of silver in normal human tissues is also very low. Hamilton *et al.*

(103) found the following mean values, expressed as micrograms of silver per gram fresh tissue: brain 0.004, kidneys 0.002, liver 0.006, lungs 0.002, lymph nodes 0.001, muscle 0.002, testis 0.002, and ovaries 0.002. The mean silver concentration found for 93 samples of whole human blood was 0.008 ± 0.0002 µg/g. A median value of 0.0025 µg/g was found for fresh heart tissue (203,294). Mean values reported for hair range from 0.13 to 0.60 µg/g (44,54,175,273). Curzon and Crocker (55) found a range of 0.0 to 36.6 µg/g human dental enamel (mean 3.44 µg/g). In another study, the silver content of human dental enamel ranged from 0.01 to 0.77 µg/g, with a mean of 0.14 µg/g (158). An analysis of human rib bone ash revealed only 1.1 µg/g silver (103). In a review of the reported concentrations of silver in plasma or serum, Versieck and Cornelis (282) found the values varied widely from 0.68 to 113 ng/ml. Most likely, however, the silver content of plasma and serum is <1 ng/ml. The concentration of silver in some tissues might be affected by disease or silver exposure. For example, Tomza et al. (273) found a median concentration of 7.3 µg/g hair from exposed nonferrous smelter workers. Indraprasit et al. (121) found that chronic and acute renal failure markedly elevated liver silver.

In a study of silver metabolism, rats were killed at various intervals after being given radiosilver intramuscularly, intravenously, and by stomach tube (242). Even when silver was administered intravenously or intramuscularly, by 4 days postdosing most of the administered silver (~93%) was excreted via the feces, with relatively little (~0.3%) appearing in the urine. Oral administration of silver did not result in any appreciable retention of radiosilver in tissues. By the fourth day, ~99% of the dose was excreted in the feces. The silver probably appeared in the feces via the bile (138). In an earlier investigation, Kent and McCance (135) found negligible amounts of silver in human urine. Thus, silver apparently is poorly absorbed and is excreted mainly via the bile and feces.

Silver interacts metabolically with copper and selenium. In the rat, of the known dietary copper antagonists tested by Whanger and Weswig (296), silver was the strongest, with cadmium, molybdenum, zinc, and sulfate following in descending order. Hill et al. (111) first showed that silver accentuated signs of copper deficiency, and that copper could reverse the silver toxicity signs of depressed growth, hemoglobin, and aorta elastin content, and elevated mortality in chicks. Jensen and co-workers (124,205) also found that the depressed growth rate, reduced packed cell volume, and cardiac enlargement induced in turkey poults and chicks by adding 900 µg silver as acetate or nitrate per gram of practical diet could be prevented by 50 µg copper per gram of diet. The growth retardation in chicks was only partially corrected, probably because of inadequate levels of selenium and vitamin E relative to the large amounts of silver. The manner in which silver interferes with copper metabolism is not clear. In the chick experiment of Peterson and Jensen (205), the high silver intake markedly depressed the copper levels in the tissues, which would suggest depressed copper absorption, but total copper excretion was apparently unaffected by the treatment. Furthermore, Van Campen (275) observed very little effect of silver on ^{64}Cu uptake in the rat. Copper retention by liver was increased as dietary silver increased, but the copper retention of heart, kidneys, and spleen was not affected, and that of blood was decreased.

In 1951, Shaver and Mason (248) reported that administering either silver nitrate or silver lactate in the drinking water promoted necrotic liver degeneration in vitamin E-deficient rats. Similarly, Dam *et al.* (57) reported that 20 μg silver as silver acetate per gram of diet promoted exudative diathesis in vitamin E-deficient chicks. To date, silver has been shown to accentuate or induce vitamin E- and selenium-type deficiency signs in chicks, rats, turkeys, pigs, and ducklings (31,57,124,205,277,278,297). Also, silver has been shown to alleviate selenium toxicity (125). The reason for the antagonistic relationship between silver and selenium is not clear. Suggested reasons include the following:

1. Because silver is easily reduced, it could initiate peroxidation and thus elevate the requirement for selenium and/or vitamin E (297).
2. Silver complexes with selenium to prevent the formation or function of the biologically active selenoenzyme, glutathione peroxidase (278).

IX. STRONTIUM (Sr)

A. Distribution

1. Animals

Many years ago, Gerlach and Muller (89) reported that the strontium concentration of a wide variety of animal tissues ranged from 0.01 to 0.10 μg/g, with no evidence of accumulation in any particular species, soft organ, or tissue. Subsequent analyses showed similar concentrations in a variety of human organs. Values (in micrograms of strontium per gram fresh tissue) reported include: brain 0.08 ± 0.01, kidney 0.10 ± 0.02, liver 0.10 ± 0.03, muscle 0.05 ± 0.02, testis 0.09 ± 0.002, ovary 0.14 ± 0.06, lung 0.20 ± 0.02, and lymph nodes 0.30 ± 0.08 (103). On a dry basis, values reported include kidney 1.51, liver 0.18, fetal liver 0.38, and spleen 0.44 μg/g tissue (121,246). The strontium concentration in blood has been reported to be 0.02 ± 0.002 (103), 0.076 (190), and 0.174–0.196 (202) μg/g. Reported values for strontium in serum are similar, ranging from 0.012 (289) to 0.21 (202) μg/ml. A biology data book (6) has stated that the concentration of strontium in serum is 0.057 μg/ml, and that in blood is 0.033 μg/ml. Blood or serum strontium apparently is depressed in patients with cholecystitis or myocardial infarction, and elevated in patients with primary carcinoma of the liver (190,289).

In general, the metabolism and distribution of strontium mimics that of calcium. The major sites of retention of both elements are the skeleton and teeth, with aorta a distant third. Strontium is incorporated mainly in the mineral phase of bones and teeth (123,281). The strontium content of bone has attracted much interest because of the possible problem of strontium-90 (^{90}Sr) retention from radioactive fallout. The total strontium content of the standard reference man has been reported to be 323 mg, of which 99% is present in the bones (237). The bone ash of human fetuses and adults in an early study averaged 160 and 240 µg/g, respectively (113). These levels are similar to those subsequently reported for human rib bone—that is, 155.9 ± 14.6 and 138.7 ± 9.0 µg/g ash (103)— but are appreciably higher than the mean of 100 µg/g ash of human bones examined by others (129,258,270). There is some evidence that the strontium levels in human bones, as well as lungs and aorta, increase with age and vary among geographical regions (102,237,258,299). For example, significantly higher rib ash strontium concentrations were found in Far East adults (190 µg/g) and children (320 µg/g) than in U.S. adults (median value of 110 µg/g) and children (median value of 96 µg/g) (237). The different strontium levels in the bones and teeth of human fetuses from different parts of Israel have been attributed to differences in the strontium levels in the water supplies (87,299).

Strontium occurs in the enamel and dentin of teeth in concentrations that parallel the levels in the bones of these same individuals and of those from similar geographical locations (260,299). Strontium is deposited primarily before eruption, during tooth calcification, is mostly permanently retained, and is not affected by fluoride in the drinking water (261). Fifty-six samples of dental enamel from U.S. residents ranged from 21 to 280 µg strontium per gram dry weight, with a median of 96 and a mean of 121 ± 11 µg/g (158). Similar levels were found in dental enamel from Australian children (36).

2. Foods and Feedstuffs

The strontium levels of a wide range of foods have been reported (52,70,83–85,91,165,192,237). In general, foods of plant origin are appreciably richer sources of strontium than are animal products, except where the latter include bone. Strontium tends to be concentrated in the bran rather than in the endosperm of grains and in the peel of root vegetables (66). Brazil nuts are an especially rich source of strontium (83). Like calcium, milk and milk products can contribute a major percentage of strontium intake; values between 11 and 32% of total dietary intake have been reported (see 182). The daily strontium intake from food is usually no more than a few milligrams. The food strontium intakes of adults in seven regions of India ranged from 3.1 to 4.7 mg/day (257), a level much higher than the mean of 0.858 ± 0.144 mg/day obtained for total English diets (102). In three studies of U.S. institutional diets, average strontium intakes of 1.2 to 2.9

(91), 2.1 to 2.4 (237), and 0.792 to 2.43 (182) mg/day were found. In some areas, the drinking water can contribute a substantial proportion of total strontium intake. Wolf *et al.* (299) reported that in three areas of Israel the drinking water contained between 1.0 and 1.6 mg/liter strontium. At a daily water consumption of 1500 to 2500 ml, these levels would supply no less than 1.5–4.0 mg/day strontium, which is as much or more than the indicated (see above) amount ingested in food. Wasserman *et al.* (288) reviewed the literature on the strontium content of water and indicated that most drinking water contains <1 mg strontium per liter.

Little information is available on the normal strontium intakes of farm animals. Mitchell (169) found 15 samples of red clover, growing on different soils, to vary from 53 to 115 (mean 74) µg strontium per gram dry weight, and 15 comparable samples of ryegrass to range from 5 to 18 (mean 10) µg/g dry tissue. A later analysis of grass hay revealed a strontium content of 22 µg/g dry weight (43). On the basis of these analyses, strontium intakes would be much higher from leguminous than from gramineous forages, and intakes would be greatly influenced by soil types on which the plants are grown.

B. Physiological Aspects

1. Metabolism

Early studies showed that the metabolism of strontium was similar to but not identical with that of calcium. However, wherever there is a metabolically controlled passage of ions across a membrane (e.g., gastrointestinal absorption, renal excretion, lactation, and placental transfer), calcium apparently is transported more effectively than strontium (288). Nonetheless, physiological and nutritional variables that affect strontium metabolism are similar to those that affect calcium metabolism and usually operate in the same direction (288). These variables include the following:

1. Calcium and strontium are better absorbed by young than old animals.
2. The stresses of pregnancy and lactation increase the efficiency of absorption of calcium and strontium.
3. The intestinal site of greatest absorption of both calcium and strontium is the duodenum, but the site of most effective (efficiency × residence time) absorption is the ileum.
4. Factors that enhance calcium and strontium absorption are vitamin D, lactose, and specific amino acids such as lysine and arginine.
5. The inclusion of alginates (286), or fiber such as cellulose (173), in the diet depresses both strontium and calcium absorption.
6. Parathyroid hormone accelerates the resorption of bone strontium, as it does calcium (34).

7. Magnesium deficiency depresses both strontium and calcium absorption (189).

In addition, the absorption of strontium is increased under fasting conditions (13) and is decreased in the presence of food (260). Raising dietary calcium intakes from low to normal reduces strontium retention (201), and supplementation with calcium plus phosphorus is more effective in reducing strontium retention than calcium alone (104). Increasing the dietary levels of the alkaline earth elements also depresses radiostrontium retention, with strontium being the least effective (178). Intestinal strontium absorption by adults of various mammalian species ranges from 5 to 25%, with age changes in the same species varying from >90% in very young to <10% in old individuals (49,284).

Strontium is poorly retained by humans. In the adult individual, net retention is essentially zero, or a steady state exists (288). Excretion occurs via the kidneys and apparently the bile. Schäfer and Forth (222) found that strontium is excreted against a considerable concentration gradient from blood into bile. Harrison *et al.* (105) found that one adult who ingested 1.99 mg strontium per day over an 8-day period excreted daily 1.58 mg strontium in the feces and 0.39 mg in the urine. Tipton and co-workers (272) obtained the following mean strontium values for two adults over a 30-day period: ingestion 1.37 and 1.2, fecal excretion 0.81 and 0.97, and urinary excretion 0.24 and 0.42 mg/day. In a study of children of different ages, the strontium intakes ranged from 0.67 to 3.57 mg/day, most of which appeared in the feces (18). The normal strontium value for human urine has been suggested to be 0.4 mg/liter (202).

Absorbed strontium is carried in the blood to the tissues, where the little that is retained is preferentially deposited by two distinct processes in the bones and teeth. Hodges *et al.* (113) have described these as (1) a rapid incorporation phase attributed to the blood strontium deposited by ionic exchange, surface absorption, and/or preosseous protein binding, and (2) a slow incorporation of strontium into the lattice structure of the bone crystals during their formation. Both processes are believed to depend mainly on the strontium concentration of the blood, including the cord blood in view of the early appearance of strontium in the bones and teeth of human fetuses (299).

2. Essentiality, Function, and Interactions

There is no conclusive evidence that strontium is essential for living organisms, although in 1949 Rygh (218) reported that the omission of this element from the mineral supplement fed to rats and guinea pigs consuming a purified diet resulted in growth depression, an impairment of the calcification of bones and teeth, and a higher incidence of carious teeth. This report has been neither confirmed nor invalidated. However, findings of Colvin and co-workers (48) indicated that some of the calcium requirement of growing chicks could be spared by strontium. The possibility that strontium can partially fulfill the calcium requirement is supported by findings such as that of MacDonald (160), who found that the inhibition of beef brain cortex adenylate cyclase activity *in vitro* by a chelating agent EGTA, though reversed completely by Ca^{2+}, was partially reversed by Sr^{2+}. This suggests that strontium can substitute for calcium in some enzyme systems.

The suggestion that strontium deprivation decreases the incidence of carious teeth (218) has not been supported well by subsequent research. Only epidemiological data suggest strontium is cariostatic. Curzon and co-workers (56,259) described a curvilinear association between caries incidence and the strontium concentration in drinking water that was related to the enamel concentration of strontium. Schamschula *et al.* (223,224) found that plaque strontium concentration was inversely related to caries incidence. However, when strontium was specifically administered orally to rats, no evidence was obtained that indicated strontium is a cariostatic agent (119). Instead, relatively high oral ingestion of strontium during the pre-eruptive period caused an increase in dental caries, and during the period of amelogenesis eliminated the cariostatic action of fluoride, in rats. Feeding strontium to female rats during gestation and lactation resulted in increased caries incidence in offspring (249). Chaudhri (36) found that the strontium content was higher in decayed than in healthy teeth of Australian children. Furthermore, Beighton and McDougall (21) found that strontium had no significant effect on plaque bacterial composition.

C. Toxicity

A wide margin of safety exists between dietary levels of stable strontium likely to be ingested from ordinary foods and water supplies, and those that induce toxic effects. For example, 3.0, 6.0, or 12.0 mg strontium as carbonate per gram of diet fed to laying hens for 4 weeks on a diet containing 3.6% calcium did not affect body weights, feed conversion, egg weight, or egg production, despite marked increases in eggshell strontium contents (290). However, in 1922 it was shown that if dietary strontium was high enough, the syndrome known as "strontium rickets" appeared in experimental animals (251). The skeleton fails to mineralize even in the presence of adequate vitamin D. Later evidence suggests that strontium interferes with vitamin D metabolism, perhaps by reducing or inhibiting the ability of kidney enzymes to convert 25-hydroxycholecalciferol to 1,25-dihydroxycholecalciferol (194,195).

The bones of chicks exhibit a remarkable capacity to retain strontium, >5% in some cases, mainly at the expense of calcium, when high dietary strontium is fed (48,291). The higher incorporation of strontium in bone is accompanied by a growth depression and a reduction in bone ash, both of which are more severe at lower calcium intakes. Also, when strontium levels up to 12.0 mg/g diet were fed to laying hens, the strontium content of eggshell was found to increase stepwise from 0.1 to 6.4% (290). An inverse relationship between calcium and strontium in eggshell formation is apparent from this study, because calcium decreased as strontium increased, with the total of the two elements remaining approximately constant at 31% for all treatments.

The importance of dietary calcium to the toxicity of strontium is evident from studies with growing chicks and pigs. Feeding 6.0 mg strontium as carbonate per gram of diet to chicks depressed growth slightly at 1.0% dietary calcium, and severely depressed growth at 0.72% dietary calcium (291). In another strontium toxicity study, factorially arranged experiments were done with piglets fed rations with variable levels of strontium (0, 0.47, and 0.67%) and calcium (0.16, 0.55, and 0.89%) (15). The pigs fed strontium and 0.16% calcium were most severely affected by incoordination and weakness followed by posterior paralysis. Mild toxic effects were only occasionally observed in pigs fed strontium and 0.55 or 0.89% calcium. The bone deformities were also more marked in the pigs fed strontium and 0.16% calcium. These findings conform well with those of Forbes and Mitchell (79), who found little effect on the daily weight gain or feed efficiency when up to 0.1% strontium was fed to rats receiving adequate calcium.

X. TIN (Sn)

A. Distribution

1. Animals

Most of the data reported before 1970 on the distribution of tin in animal tissues, foods, and feedstuffs apparently are questionable because of analytical procedures that were inadequate, both from the point of view of recoveries and of the sensitivity of the method (20,241). In 1972, Hamilton *et al.* (103) used spark-source mass spectrometry following ashing by a low-temperature ($\sim$100°C) nascent-oxygen technique to obtain the following mean values (micrograms of tin per gram fresh tissue) for human organs: muscle 0.07 $\pm$ 0.01, kidney 0.2 $\pm$ 0.04, brain 0.06 $\pm$ 0.01, testis 0.3 $\pm$ 0.1, ovary 0.32 $\pm$ 0.19, liver 0.4 $\pm$ 0.08, lung 0.8 $\pm$ 0.2, and lymph nodes 1.5 $\pm$ 0.6. The mean tin concentration found for 102 samples of whole human blood was very low (0.009 $\pm$ 0.002 μg/g). The mean tin concentration of 22 samples of rib bones from humans living in hard-water areas was 4.1 $\pm$ 0.6 μg/g ash and for 22 samples of rib bone from those living in soft-water areas was 3.7 $\pm$ 0.6 μg/g ash. Subsequent studies of isolated human organs indicate that these values are in the appropriate range. Examples of the findings from these studies include these: 0.06–0.18 μg/g fresh heart tissue (203), 1.03, 1.83, and 1.19 μg/g dry kidney, liver, and spleen, respectively (121), and 0.13 μg/g dry fetal liver (246). The limited number of other analyses of whole blood, serum, or plasma done since 1970 (54,202,282,289) gave values (30–186 ng/ml) slightly higher than that reported by Hamilton *et al.* (103). A biology data book (6) has given a serum value of 33 ng/ml tin. The tin content of human milk is very low, apparently <3 ng/g on a dry-weight basis (141). Two groups have used spark-source mass spectrometry to study the mineral content of human dental enamel. Losee *et al.* (158) found the wide range of 0.03 to 7.10 μg tin per gram of dry enamel, with a median of 0.23 μg/g and a mean of 0.53 ± 0.4 μ/g. Curzon and Crocker (55) also found a wide range of 0.0 to 44.0 μg tin per gram of enamel, with a mean of 1.60 μg/g. Creason *et al.* (54) found that the mean tin concentrations were 1.1 and 0.7 μg/g in scalp and pubic hair, respectively.

2.   Foods and Feedstuffs

Data on the tin content of foods obtained by reliable methods of treatment and analysis are exceedingly meager. Mean daily intakes by U.S. adults have been reported to be 1.5 (272) to 3.5 mg (229), while the average English total diet was reported to supply only 187 ± 42 μg tin per day (102). If people are not ingesting relatively large amounts of acidic canned foods, their daily tin ingestion most likely is near the value reported for English diets. The few noncanned foods apparently accurately analyzed for tin indicate a tin content of markedly <1 μg/g. For example, Greger and Baier (94) reported that bottled apple juice, apple sauce, baked beans, sauerkraut, and tomato juice contained <10 ng/g. Cowgill (52) found only 3.6 ng tin per gram of fresh banana pulp. Nuts are apparently relatively rich in tin, because Furr *et al.* (83) found a variety of nuts contained from 0.3 to 3.5 μg/g dry weight. However, the richest sources of dietary tin are foods that have been contaminated by packaging and processing material containing tin. For example, in the study in which Greger and Baier (94) found <10 ng tin per gram in bottled foods, tin levels as high as 669 μg/g were found in some canned foods. Highest levels were found in foods stored in a refrigerator for 1 week after their containers had been opened. Examples of tin concentrations (micrograms per gram) in foods from freshly opened cans were as follows: cheddar cheese soup 0.5 ± 0.2, cranberry sauce 1.4 ± 0.2, grapefruit juice 40.9 ± 15.5, grapefruit sections 96.4 ± 16.1, orange juice 53.2 ± 2.7, pineapple 89.4 ± 12.5, and tomato sauce 149.8 ± 8.5. Lopez and Williams (157) could not detect tin in fresh tomatoes, but found 11 μg/g in canned tomatoes. Such findings show that large amounts of tin can accumulate in foods that are in contact with tin plate unless it is lacquered or coated with resin.

B. Physiological Aspects

1. Metabolism

Tin, when fed at relatively high levels, is poorly absorbed and retained by humans and is excreted mainly in the feces. Apparently the lower the dietary level of tin, the higher the percentage that is absorbed. Johnson and Greger (127) found that when the dietary intake of tin was 49.7 mg/day, fecal and urinary losses of tin were 48.2 ± 1.5 mg/day and 122 ± 52 μg/day, respectively. This gives an apparent retention of 1.3 ± 1.5 mg/day and an apparent absorption of 3%. However, when the dietary intake of tin was 0.11 mg/day, fecal and urinary loss of tin were 0.06 ± 0.02 mg/day and 29 ± 13 μg/day, respectively. This gives an apparent retention of 0.03 ± 0.03 mg/day and an apparent absorption of 50%. Another earlier study (33) also showed that fecal excretion of tin was high and approximated dietary intakes when the subjects were fed foods that supplied 9.4–190 mg tin per day. Rat studies gave findings similar to those from human studies. Hiles (110) showed that >97% of a single oral dose of 20 mg tin per kilogram body weight was excreted in the feces of rats. Rats fed 200 μg tin per gram of diet for 21 days excreted >90% of the ingested tin via the feces (95). Fritsch et al. (81) reported that when an oral dose of 50 mg tin labeled with tin-113 (^{113}Sn) per kilogram body weight together with various food components such as sucrose and ascorbic acid was administered by gastric intubation into rats, 90–99% of the ^{113}Sn was excreted in the feces within 48 hr, at which time fecal excretion and retention in the alimentary tract accounted for 98.7 to 99.9% of the dose. Only traces of ^{113}Sn were detected in a wide range of organs and tissues examined.

Once tin is absorbed, both the bile and urine are routes of excretion; this is suggested by the preceding findings and demonstrated by studies in which tin was administered intravenously. Hiles (110) found that from a single intravenous dose of 2 mg of Sn^{2+} per kilogram body weight, 30% was excreted in the urine and 11% was eliminated in the bile.

Low tissue accumulation and rapid tissue turnover are also characteristic of tin. Rats fed a single large dose of radiolabeled tin (110), or high levels of tin for several weeks (95,110,302), accumulated some tin in bone. The accumulation apparently was related to the level of tin fed (110,302). The biological half-life of the tin in bones was calculated to be 20–40 days (110). Also, tin tended to accumulate in liver and kidneys when doses >2 mg/day were fed (95,110,302). However, this tin apparently is quickly lost when dietary tin is reduced to normal low levels (127).

Hiles (110) found that when pregnant rats were dosed orally with radiolabeled tin, as tin fluoride, at a rate of 20 mg tin per day, starting at day 1 until conception, their fetuses contained insignificant amounts of ^{113}Sn. Schwarz

(240), using sensitive analytical techniques, also found tin levels in rat embryos and newborns hardly measurable. However, placental tin transfer probably occurs, because Theuer *et al.* (269) found tin in fetuses and placentas (0.1–1.3 μg/g) of rats. The higher values were found when the maternal diet contained tin salts.

2. Essentiality and Function

The only evidence that supports essentiality of tin is that a dietary supplement of tin, both in the inorganic or organic form, improved the growth of suboptimally growing, apparently riboflavin-deficient (176) rats but did not result in optimal growth (241). The use of riboflavin-deficient rats in tin essentiality studies would be of particular concern, because the oxidation–reduction potential of $Sn^{2+} \rightleftarrows Sn^{4+}$ is 0.13 V, which is near the oxidation–reduction potential of flavin enzymes. Attempts in another laboratory to show that tin deprivation depressed growth in riboflavin-adequate rats were unsuccessful, even though animals and dietary materials used were from the same sources as those used in the study with apparent riboflavin-deficient rats (187). Further attempts to obtain findings similar to those of Schwarz *et al.* (241) probably will be difficult because of apparent dietary and/or metabolic interactions between tin and riboflavin (187). Probably only certain special conditions will produce tin growth responses in rats. Thus, it cannot be stated unequivocally that tin deprivation reproducibly impairs a function from optimal to suboptimal. Without such evidence, tin should not be considered an essential trace element at this time. Moreover, the description of any possible biological function or nutritional requirement seems inappropriate.

C. Toxicity

Ingested tin has a low order of toxicity, no doubt because of the poor absorption, low tissue accumulation, and rapid tissue turnover of the element. Convincing evidence of human poisoning with tin from canned or other foods is difficult to find. Soluble salts of tin are gastric irritants. Canned foods and drinks containing from 250 to 700 μg/g tin have been cited as the cause for occurrences of nausea, vomiting, and diarrhea in large numbers of people (24). In subacute toxicity studies, mice and rats fed 5 μg Sn^{2+} per milliliter of drinking water from weaning to natural death grew normally, and the life span of mice of both sexes and of male rats was unaffected (228,232). Female rats displayed a reduced longevity and an increased incidence of fatty degeneration of the liver. Vacuolar changes in the renal tubules occurred in both female and male rats. Kidney and liver changes were also seen in other studies. When fed 265 μg $SnCl_2 \cdot 2H_2O$ per gram diet for 9 weeks, rats showed a marked depression in liver pyruvate dehydrogenase activity, pyruvate concentration, and $NAD:NADH_c$ ratios (5). After being injected with a single intraperitoneal dose of $SnCl_2 \cdot 2H_2O$ (44.4 mg/kg), rats exhibited necrosis of the renal proximal tubules (306). De Groot and co-workers (58,59) found that inorganic tin fed to rats at subacute toxicity levels (250, 500, 1500, and 5300 μg/g diet) depressed growth, feed efficiency, and hemoglobin levels at low iron and copper dietary levels. When fed 6.3 mg tin as $SnCl_2$/g diet for 13 weeks, rats exhibited pancreatic atrophy, testicular degeneration, renal calcification, and status spongiosis of the brain. Increasing the iron and copper levels prevented the hematological effect but not the growth inhibition.

In addition to copper and iron, toxic levels of tin can influence the metabolism of several other minerals. High amounts of $SnCl_2$, orally administered to rats, depressed the calcium concentration and acid and alkaline phosphatase activities in liver and femur (302). However, when a diet that supplied ~50 mg/day tin for 20 days was fed to men, no effect on serum calcium or on fecal and urinary losses of calcium was seen (127). On the other hand, this level of dietary tin resulted in an elevated loss of selenium and zinc in the feces, and a depressed loss in the urine (96,126). Overall zinc retention was lower in the tin-fed subjects. Tin supplementation had no effect on fecal or urinary losses of copper, iron, manganese, or magnesium (126). Rats fed 200 μg tin per gram of diet lost more zinc in their feces and retained lower levels of zinc in their tibias and kidneys than rats fed a diet containing 1 μg/g (95). Another study (40) showed that the subcutaneous injection of tin (2 mg/kg every other day for 14 days) caused a redistribution of zinc. Zinc was elevated in the liver and depressed in spleen, heart, brain, lungs, and muscle. In contrast to the preceding studies, Solomons *et al.* (256) found that in humans a high dose of tin (25–100 mg) had no overall effect on zinc uptake from a 12.5-mg zinc dose as zinc sulfate in 100 ml of soda pop.

The toxic action of tin on hematopoiesis (58,82,303) apparently can involve something more than the direct depression of iron absorption (58,59) or the copper content of blood and tissues such as brain, kidneys, and liver (40,128). Tin can interfere with porphyrin biosynthesis and enhance heme breakdown. Tin injected into rabbits (38) or high dietary tin fed to rats (128) inhibited the activity of 5-aminolevulinate dehydratase in blood, but not in liver. However, no effect of tin on 5-aminolevulinic acid concentrations was observed (38). Stannous chloride injected into rabbits also increased the concentration of coproporphyrin in blood and urine. Kappas and Maines (130) reported that tin (Sn^{2+}) is a potent inducer of heme oxygenase and thus enhances heme breakdown in the kidney. On the other hand, tin (Sn^{4+}) protoporphyrin IX is a potent competitive inhibitor of heme oxygenase, and thus of heme oxidation in liver, spleen, and kidneys (64). This latter finding has led to the suggestion that tin protoporphyrin IX may be useful in the chemoprevention of neonatal jaundice, or hyperbilirubinemia

(65,220). Cells of the ciliate *Tetrahymena pyriformis* GL overproduce and accumulate massive quantities of protoporphyrin IX (217). The presence of Sn^{4+} stimulates protoporphyrin IX accumulation in the culture of these cells. Because tin affects heme metabolism, it also affects heme-dependent functions such as cytochrome P_{450}-mediated drug metabolism (130).

XI. TITANIUM (Ti)

Titanium resembles aluminum in being abundant in the lithosphere and soils, and in being poorly absorbed and retained by plants and animals. The poor absorption and retention of titanium results in levels in the tissues of plants and animals being generally much lower than those in the environment to which the organisms are exposed. Bertrand and Voronca-Spirt (25) examined a variety of plants for titanium and recorded levels ranging from 0.1 to 5 µg/g, with a high proportion being close to 1 µg/g. Similar concentrations were reported by Mitchell (169) in his subsequent study of the mineral composition of red clover and ryegrass grown on different soils. A mean of 1.8 µg/g dry weight (range 0.7–3.8 µg/g) was obtained for the former species and 2.0 µg/g dry weight (range 0.9–4.6 µg/g) for the latter.

Very little is known of the titanium content of human foods. Furr and co-workers (83–85) found that on a dry-weight basis, a variety of tree nuts, fruit kernels, and maple syrup contained 0.5–6.1 µg titanium per gram; most values ranged between 2 and 3 µg/g. Cowgill (52) found 0.05 µg titanium per gram fresh banana pulp, while Oakes *et al.* (192) found a variety of fruits and vegetables contained <0.5 µg/g fresh weight. Duke (70) reported the following titanium concentrations: breadfruit 6.0, coconut 0.3, taro 80.0, yam 15.0, cassava 6.0, banana 0.2, rice 2.0, avocado 1.0, dry beans 2.0, and corn 2.0 µg/g dry weight. English total diets reportedly supply ~800 µg titanium per day (102). Tipton *et al.* (272) reported the 30-day mean total dietary titanium intakes of two individuals to be 320 and 410 µg/day. This study indicated urinary excretion of titanium was high, thus suggesting either considerable absorption from the diet or loss from previously retained tissue titanium. Both individuals were in negative balance, with approximately equal excretion via the feces and urine. No evidence has appeared that suggests absorbed titanium performs any vital function in animals or that it is a dietary essential for any living organism.

The reported titanium concentrations of human and animal tissues are extremely variable, with high levels commonly appearing in the lungs, probably as a result of dust inhalation (25,103,270). Tipton and Cook (270) found that most of the soft tissues of the adult human body contained from 0.1 to 0.2 µg/g titanium, but the lungs averaged >4 µg/g, with some samples containing >50 µg/g fresh tissue. Hamilton *et al.* (103) reported the following mean levels in human organs: muscle 0.2 ± 0.01, brain 0.8 ± 0.05, kidney cortex 1.3 ± 0.2, kidney medulla 1.2 ± 0.2, liver 1.3 ± 0.2, and lung 3.7 ± 0.9 μg/g fresh tissue. Yukawa *et al.* (305) found similar concentrations in human organs; however, they thought some of the tissue titanium was the result of sample contamination. On a dry-weight basis, Indraprasit *et al.* (121) found 1.44, 3.09, and 1.84 μg titanium per gram of kidney, liver, and spleen, respectively. Shand *et al.* (246) found a range of <0.15 to 5.21 μg/g dry fetal liver. High variability in titanium concentration was also found for human dental enamel and hair. In one study of 29 samples, the levels reported for enamel ranged from 0.1 to 4.8 μg/g dry weight, with a median of 0.12 and a mean of 0.46 ± 0.13 μg/g (158). In another study of 243 samples the titanium levels ranged from 0.0 to 31.4 μg/g, with a mean of 1.93 μg/g enamel (55). Moo and Pillay (175) gave mean titanium levels in hair as 11.9 ± 8.3 μg/g (without samples given a value of 0.0 because titanium could not be detected) or 7.9 ± 8.8 μg/g (all values including those given a value of 0.0) for cancer patients. They found one exceptional value of 78.3 μg/g hair. They also reported a value of 10.6 ± 19.3 μg/g hair from the general population. Hamilton *et al.* (103) found 0.07 μg titanium per gram of whole blood. A subsequent analyses revealed a slightly lower level, 0.03 μg/g blood (202). This study also found 0.03 μg titanium per milliliter of serum. Some of the variation in tissue titanium content may be caused by variation in dietary intake of titanium. Schroeder *et al.* (230) found that the titanium levels in a variety of mouse organs was increased from 0.13–1.10 to 1.66–8.80 μg/g fresh tissue when dietary (water and food) titanium was increased from 0.03 to 5.03 μg/g.

Titanium is essentially nontoxic in the amounts and forms that are usually ingested. The findings of Schroeder and co-workers (230,234,238) indicated that 5 μg/ml titanium in the drinking water fed to mice for their lifetime did not consistently affect growth, longevity, and tumor incidence. A review of titanium toxicity (10) questioned whether evidence of a specific oral toxicity had ever been found.

XII. ZIRCONIUM (Zr)

Zirconium is relatively abundant in the lithosphere but has no known biological significance. Reported values for the zirconium levels in plant and animal tissues are highly discordant. Schroeder and Balassa (227) found zirconium in all the organs examined from four male accident victims, and found especially high concentrations in fat (18.7 μg/g fresh tissue), liver (6.3 μg/g fresh tissue), and red blood cells (6.2 μg/g). In other tissues and blood serum, the levels were reported to lie between 1 and 3 μg/g fresh weight. These levels are questionable, because Hamilton *et al.* (103) found much lower zirconium levels for human tissues than those just cited. Brain, muscle, testis, ovary, liver, kidney, and lung tissues all averaged between 0.01 and 0.06 μg/g fresh tissue, with only the lymph nodes exhibiting a significantly higher mean concentration (0.3 ± 0.6 µg/g fresh tissue). A series of 98 samples of whole human blood revealed a mean of 0.02 ± 0.008 µg/g zirconium. A biology data book (6) has stated that serum contains about 0.04 µg/ml zirconium. In one study of human dental enamel (158), the range of values for 53 samples was <0.02–2.6 µg/g dry weight, with a median of 0.2 and a mean of 0.53 ± 0.08 µg/g zirconium. Another study (55) showed a range of 0.0 to 0.8 µg/g enamel, with a mean of 0.08 µg/g.

In addition to human and animal tissues, the food zirconium values given by Schroeder and Balassa (227) seem high. They found that meat, dairy products, vegetables, cereal grains, and nuts generally contained 1–3 µg zirconium per gram fresh weight. Appreciably lower levels of zirconium were found in fruits and seafoods. They estimated a daily oral zirconium intake by humans of ~3.5 mg. In contrast, Hamilton and Minski (102) reported that English total diets supplied 53 ± 34 µg zirconium per day. This latter value seems to be more reasonable than the former, because Duke (70) found that a limited variety of fruits, vegetables, and grains contained between 0.005 and 0.20 µg zirconium per gram dry weight. Oakes *et al.* (192) could not detect zirconium in a variety of fruits and vegetables, and found <0.6 µg/g fresh weight of a limited number of vegetables. Cowgill (52) found only 0.0092 µg/g of fresh banana pulp.

The metabolic movements of zirconium in the animal body apparently have not been studied directly. Its presence in the blood and tissues indicates that it is absorbed from ordinary diets, while its virtual absence from the urine (227) suggests that it is excreted by the intestine. Bone apparently accumulates the greatest amount of intraperitoneally injected zirconium (221). More detailed and direct evidence of the sites of absorption, retention, and excretion of zirconium is clearly desirable.

Zirconium compounds have a low order of toxicity for rats and mice, whether injected (239) or orally ingested (235,239). Schroeder *et al.* (235) studied the lifetime effects of adding 5 ppm zirconium as zirconium sulfate to the drinking water of mice. No effects on growth were observed, but there was a small reduction in survival time. The element was neither carcinogenic nor tumorigenic, and apparently did not accumulate in the tissues of mice. However, 5 µg zirconium per milliliter of drinking water elevated the level of copper in the liver and kidney of rats (236).

XIII. MISCELLANEOUS ELEMENTS

There are occasional reports indicating that elements other than those given separate sections in this chapter might be of nutritional importance. These elements include the following.

A. Beryllium (Be)

Certain beryllium compounds are carcinogenic when inhaled, and osteosarcomas have been produced by the intravenous injection of several beryllium salts (29,225). However, ingested beryllium is not especially toxic. Beryllium, fed to rats at the level of 5 μg/ml drinking water for a lifetime, was virtually innocuous as measured by median life span, longevity, incidence of tumors, and serum cholesterol and uric acid (234). Apparently beryllium is poorly absorbed through the gut, and thus ingestion is not a hazard, especially considering the low levels apparently present in food (52,102,165). However, like other trace elements given in high doses, ingested beryllium can be toxic, causing severe rickets not cured by cholecalciferol (99) and inhibiting alkaline phosphatase of several tissues (46,215). Although beryllium apparently occurs in very low levels in human tissues (78), it was found to occur in blood and serum at concentrations of 0.019 and 0.017 μg/g, respectively (202).

B. Bismuth (Bi)

Bismuth is used as a coloring agent in decorative cosmetics, in ointments for burns, to delineate viscous surfaces in X-ray analyses, as a fungicide, in the treatment of warts, and to regulate stool odor and consistency in colostomy patients (10). Despite rather extensive use of bismuth as a therapeutic agent for gastrointestinal disturbances, no evidence exists to indicate that it is an essential nutrient. However, bismuth is a relatively nontoxic element (10) and is found in low quantities in human tissues. Hamilton *et al.* (103) found the following concentrations for human organs: kidney 0.4 ± 0.1, lung 0.01 ± 0.001, and lymph nodes 0.02 ± 0.001 μg/g fresh tissue. Sumino *et al.* (266) found 0.030 ± 0.020 and 0.038 ± 0.031 μg bismuth per gram of fresh liver and kidney, respectively. The normal dietary intake of bismuth is low, probably <5 μg/day (102). Nonetheless, a transport mechanism apparently exists for bismuth, because the element is excreted against a considerable concentration gradient from blood into bile (222).

C. Gallium (Ga)

Gallium can bind to the iron transport protein transferrin (88). Furthermore, differences in serum iron-binding capacity have been suggested as the reason for sex differences in gallium-67 (^{67}Ga) uptake by blood and bone in mice (39). Radiogallium scanning is a clinically useful procedure for such things as the identification of suspected occult inflammatory foci. Gallium is normally found, albeit in low quantities, in human tissues. Hamilton *et al.* (103) reported the following concentrations (in micrograms per gram fresh tissue): brain 0.0006 ±

0.00003, kidney 0.00009 ± 0.0003, liver 0.0007 ± 0.0001, lung, 0.005 ± 0.002, lymph nodes 0.007 ± 0.002, muscle 0.0003 ± 0.00004, testis 0.0009 ± 0.0001, and ovary 0.002 ± 0.0005. Shand *et al.* (246) found 0.16 μg gallium per gram of dry fetal liver. Gallium fed at the level of 5 μg/ml of drinking water for lifetime was slightly toxic to mice (233). Gallium significantly but not markedly suppressed growth and reduced the life span of females.

D. Gold (Au)

Gold compounds have been used for the treatment of rheumatoid arthritis since the early 1940s. Although the manner in which gold compounds produce their beneficial effects is not known, and although other agents are known to suppress rheumatoid disease, gold continues to be used in the treatment of rheumatoid arthritis. Although most foods apparently contain only a few nanograms or less of gold per gram (83–85,192), and the dietary gold intake is probably <7 μg/day (102), gold is found, albeit in low quantities, in human tissues and blood. Wester (294) found a median value of 0.0338 ng/g fresh heart tissue. Vanoeteren *et al.* (276) found a range of 0.72 to 1.6 ng gold per gram of fresh lung tissue from six individuals. Kjellin (137) indicated that brain gray matter, brain white matter, cerebrospinal fluid, whole blood, and serum contained 0.024, 0.040, 0.0062, 0.055, and 0.080 ng/g or ml fresh weight, respectively. Normal values for hair have been reported to be 0.15 ± 0.13 (175) and 0.036 ± 0.055 (273) μg/g. Cancer patients were found to exhibit elevated hair gold concentrations of 1.5 ± 1.4 μg/g (175). Pregnancy apparently increases gold in blood, because maternal (13.1 ng/ml) and cord blood (12.7 ng/ml) gold concentrations were higher than gold concentrations in the blood of nonpregnant women (3.4 ng/ml) (4). Therapeutically injected soluble gold is excreted mainly in the urine, but some does appear in the feces (135).

E. Indium (In)

Yukawa *et al.* (305) consistently found indium in human tissues at concentrations between 0.01 and 0.07 μg/g fresh tissue. Tomza *et al.* (273) found a mean of 0.0045 μg indium per gram of hair. Some indium is absorbed from the gastrointestinal tract of animals, because it is excreted in small amounts in the urine of animals fed indium salts. Indium is relatively nontoxic orally. When fed 20–30 mg indium per day for 27 days, rats exhibited a rough coat and some weight loss (164). On the other hand, when mice were fed drinking water containing 5 μg/ml indium, they exhibited a lower incidence of tumors than mice fed the same level of scandium, gallium, rhodium, palladium, chromium, or yttrium (233). Mice fed 5 μg/ml indium exhibited elevated levels of copper and chromium concentrations in several organs including kidney and heart (236).

F. Niobium (Nb)

Schroeder and Balassa (226) considered niobium a much neglected element from a biological viewpoint. They based this statement on their finding that levels of niobium in tissues of humans and wild, domestic, and laboratory animals were comparable to those of copper and were exceeded among the trace elements only by iron, zinc, and rubidium. They found niobium in nearly all animal and vegetable foods they analyzed, with most in the range of 0.5 to 3.0 μg/g fresh weight. Thus, they suggested a daily human intake of $\geq$600 μg of niobium, of which approximately one-half was absorbed and excreted in the urine.

Subsequent reports by Hamilton and co-workers (102,103) gave values very divergent to those of Schroeder and Balassa (226). They found the following concentrations (micrograms per gram fresh weight) in human tissues: blood 0.004 $\pm$ 0.0005, kidney 0.01 $\pm$ 0.004, liver 0.004 $\pm$ 0.009, lung 0.02 $\pm$ 0.0001, lymph nodes 0.06 $\pm$ 0.007, muscle 0.03 $\pm$ 0.008, and testis 0.009 $\pm$ 0.04. Bone contained <0.07 μg niobium per gram of ash. They reported that English total diets supplied only 20 $\pm$ 4 μg niobium per day. Other research suggests that the latter lower values are most correct. For example, Cowgill (52) found 10.5 ng/g of fresh banana pulp; this contrasts with the finding of Schroeder and Balassa (226) of 320 ng/g of fresh banana. Like Hamilton *et al.* (103), who found low levels of niobium in bone, Curzon and Crocker (55) found that the range of niobium in 245 samples of human dental enamel was 0.0–1.0 μg/g, with a mean 0.17 μg/g. Finally, Schroeder and co-workers (235,236) found that 5 μg niobium per milliliter of drinking water, a level not much higher than their analyses showed would be normally present in food, was toxic to mice and rats. In mice, the niobium supplement increased the incidence of hepatic fatty degeneration, decreased median life span and longevity, and suppressed growth. In rats, the niobium supplement elevated copper, manganese, and especially zinc in a variety of organs, with the most marked changes occurring in heart and liver. Regardless of the true levels, the findings indicate niobium is present in living things and can affect biological mechanisms. Furthermore, when zirconium-95 (^{95}Zr) and niobium-95 (^{95}Nb) were injected into rats, bone accumulated the parent isotope (Zr) and blood the daughter isotope (Nb) (221). Gonads, kidneys, lungs, liver, and spleen also preferentially accumulated niobium over zirconium. Thus, further study of the importance of niobium in nutrition probably should be done.

G. Scandium (Sc)

Scandium has been described as essential for the growth of two organisms: the mold *Aspergillus niger* (262) and the fungi *Cercospora granati* Rawla (35).

Pehrsson and Lins (203) found that the concentration of scandium was higher in heart tissue from patients with uremic heart failure (range 0.02–0.9 ng/g fresh weight) than from nonuremic controls (range 0.003–0.1 ng/g fresh weight). The control values were obtained by Wester (294), who also stated that the scandium concentration in uninjured infarcted heart tissue was lower in patients with than without a history of arterial hypertension, and that scandium is higher in conductive tissue than in adjacent heart tissue. Scandium also was found to be lower in hair of cancer patients (0.006 ± 0.004 µg/g) than of the general population (0.07 ± 0.07 µg/g) (175). Vanoeteren et al. (276) found that the scandium concentration in lung differed markedly from one person to another, with a range of 0.49 to 3.0 ng/g fresh tissue. They suggested that inhaled dust particles may have caused the differences. Although the preceding indicates that tissue concentrations are relatively low, the apparent requirement of scandium by some lower forms of life and the redistribution of tissue scandium in pathological conditions suggest that this element might have physiological action in higher forms of life. If there is a requirement for scandium, it must be low because limited analyses show relatively low amounts of scandium in food. Duke (70) found a small variety of fruits, vegetables, and grains contained 0.005–0.1 µg scandium per gram dry weight. Oakes et al. (192) found generally <1 ng/g fresh weight in a limited variety of fruits and vegetables. Furr et al. (83,84) reported that maple syrup and a variety of nuts contained 2–4 ng scandium per gram dry weight; only Brazil nuts were higher, containing 20 ng/g. Clemente et al. (44) reported that the Italian dietary intake of scandium was only 170 ng/day and urinary excretion was <10 ng/day.

H. Tellurium (Te)

For over a century, the ingestion of tellurium compounds has been known to be associated with a garliclike odor of the breath, thus indicating that tellurium is absorbed by the gut, metabolized by tissues, and excreted through routes other than the feces. Other research has confirmed that homeostatic mechanisms for tellurium exist in animals and humans. Using sodium tellurite labeled with tellurium-127m (^{127m}Te), Weight and Bell (301) obtained findings indicating that 24% of the dose was absorbed from the colon in sheep and swine. They found urinary excretion was about three times fecal excretion when sodium tellurite was administered intravenously. The ratio was reversed following oral administration. Barnes et al. (14) found the ratio of fecal to urinary excretion was 15 and 12 upon the oral administration of tellurium to guinea pigs and rats, respectively. De Meio and Henriques (60) found that tellurium, as [^{121}Te]sodium tellurite administered intravenously, appeared in the bile in addition to the urine. Schroeder et al. (231) reported that the tellurium concentration in normal human urine was 0.63 µg/ml.

In his review of tellurium, Klevay (140) presented evidence that animals can synthesize dimethyltelluride.

The toxicity of orally administered tellurium is relatively low. When fed 375–1500 µg of elemental tellurium per gram of diet for 21 days, rats developed a garlic odor in their breath, viscera, and urine but showed no pathological changes; in addition, those fed 375 and 750 µg tellurium as TeO_2 per gram of diet developed redness and edema of the digits, temporary paralysis of the hind legs, and a loss of hair (61). Soluble tellurite or tellurate salts were toxic at concentrations of 25 to 50 µg/g diet. Van Vleet (277) observed a marked decrease in blood glutathione peroxidase over the last 6 weeks of a 10-week trial in pigs fed 500 µg tellurium as the tetrachloride per gram of diet. This level of dietary tellurium also increased the incidence of vitamin E–selenium deficiency-type signs of necrosis of cardiac and skeletal muscle. On the other hand, Whanger and Weswig (297) reported that 10 µg tellurium as TeO_2 per gram of diet did not enhance liver necrosis in selenium–vitamin E deficient rats. Instead, tellurium apparently increased the life span of the rats. Whanger and Weswig (297) suggested that decreased food consumption might have been partially responsible for the increased life span.

I. Thallium (Tl)

Thallium is a relatively toxic element. The absorption of soluble thallium salts is rapid and complete, which probably contributes to its high toxicity. Furthermore, thallium apparently accumulates in the human body with age. Accumulation of thallium can lead to autonomic dysfunction, with tachycardia and hypertension (12). On the other hand, thallium apparently has the ability to serve as the required monovalent cation in the activation of certain enzyme-catalyzed reactions, including pyruvate kinase (133).

Thallium apparently can be present in high concentrations in plants. Zyka (309) found such plants in a region in Yugoslavia and suggested that thallium was responsible for the toxic affect of the plants on cattle.

J. Tungsten (W)

Tungsten apparently is a component of formate dehydrogenase from *Clostridium thermoaceticum* (154). However, although an interaction between tungsten and molybdenum metabolism has been established (see Chapter 6), no known essential role for tungsten has been found for animals. In a life-term study with mice and rats, Schroeder and Mitchener (234) found a slight enhancement in growth of rats when the drinking water was supplemented with tungsten at 5 µg/ml. On the other hand, tungsten slightly shortened the longevity of mice and rats. Tungsten was virtually innocuous as judged by incidence of tumors and levels of serum cholesterol, glucose, and uric acid. In six experiments with growing, gravid, and lactating goats fed either 0.06 or 1.0 μg tungsten per gram of semisynthetic diet, dietary tungsten did not affect growth, hematocrits, hemoglobin, insemination, conception, or abortion rate, sex or number of kids per goat, or mortality of kids (7–9). However, the tungsten-low mothers tended to show higher mortality and an elevated reticulocyte count. Nonetheless, Anke and co-workers (8) concluded from their studies that either tungsten is not essential, or that tungsten at a dietary level of 60 ng/g meets the requirement for growth and reproduction in goats. Whanger and Weswig (297) found that 500 μg tungsten per gram of diet increased the life span of selenium–vitamin E deficient rats. However, they attributed this finding to decreased food intake caused by unpalatability of the tungsten-containing diet.

The suggestion that any requirement for tungsten would be small is supported by the apparent relatively low dietary intake of tungsten by humans and animals. Hamilton and Minski (102) reported that English total diets supplied <1 μg tungsten per day. Furr *et al.* (83,85) found that maple syrup and a variety of nuts contained only 0.02–0.10 μg tungsten per gram dry weight. Anke and co-workers (8) stated that normal ruminant feedstuffs of Middle Europe contained only between 0.15 and 0.50 μg/g dry weight. Nonetheless, any tungsten ingested probably is well absorbed. Bell and Sneed (22) found that an oral dose of a tracer level of $(NH_4)_2{}^{185}WO_4$ was eliminated principally in the urine. Kaye (132) reported that 40% of a single oral tracer dose of $K_2{}^{185}WO_4$, or $K_2{}^{187}WO_4$, was excreted within 24 hr, with approximately equal amounts appearing in the urine and feces. By 72 hr, 97% of the tungsten dose was excreted. Bone apparently was a retention site for tungsten.

REFERENCES

1. Agarwala, S. C., Sharma, P. N., Chatterjee, C., and Sharma, C. P. (1981). *J. Plant Nutr.* **3,** 329–336.
2. Akagi, M., Misawa, T., and Kaneshima, H. (1962). *Yakugaku Zasshi* **82,** 934–936.
3. Alexander, G. V., Nusbaum, R. E., and MacDonald, N. S. (1951). *J. Biol. Chem.* **192,** 489–496.
4. Alexiou, D., Grimanis, A. P., Grimani, M., Papaevangelou, G., and Papadatos, C. (1976). *Biol. Neonate* **28,** 191–195.
5. Allmann, D. W., Mapes, J. P., and Benac, M. (1975). *J. Dent. Res.* **54,** 189.
6. Altman, P. L., and Dittmer, D. S., eds. (1973). "Biology Data Book," Vol. 3, pp. 1751–1752. Fed. Am. Soc. Exp. Biol., Bethesda, Maryland.
7. Anke, M., Groppel, B., Grün, M., and Kronemann, H. (1983). *In* "Mengen-und Spurenelemente" (M. Anke, C. Brückner, H. Gürtler, and M. Grün, eds.), pp. 37–44. Karl-Marx-Univ., Leipzig, G.D.R.

8. Anke, M., Groppel, B., Kronemann, H., and Grün, M. (1984). *In* "Trace Elements in Man and Animals—5 Abstracts," p. 134. Aberdeen, Scotland.
9. Anke, M., Kronemann, H., Groppel, B., and Partschefeld, M. (1983). *In* "Mengen-und Spurenelemente" (M. Anke, C. Brückner, H. Gürtler, and M. Grün, eds.), pp. 45–51. Karl-Marx-Univ., Leipzig, G.D.R.
10. Anonymous (1980). *In* "Mineral Tolerance of Domestic Animals," pp. 24–39, 54–57, 60–70, 71–83, 84–92, and 510–513. Natl. Acad. Sci., Washington, D.C.
11. Baer, H. P., Bech, R., Franke, J., Grunewald, A., Kochmann, W., Melson, F., Runge, H., and Wiedner, W. (1977). *Z. Gesamte Hyg. Ihre Grenzgeb.* **23,** 14–20.
12. Bank, W. J., Pleasure, D. E., Suzuki, K., Nigro, M., and Katz, R. (1972). *Arch. Neurol. (Chicago)* **26,** 456–464.
13. Barnes, D. W. H., Bishop, M., Harrison, G. E., and Sutton, A. (1961). *Int. J. Radiat. Biol.* **3,** 637–646.
14. Barnes, D. W. H., Cook, G. B., Harrison, G. E., Loutit, J. F., and Raymond, W. H. A. (1954). *J. Nucl. Energy* **1,** 218–230.
15. Bartley, J. C., and Reber, E. F. (1961). *J. Nutr.* **75,** 21–28.
16. Baumann, E. J., Sprinson, D. B., and Marine, D. (1941). *Endocrinology* **28,** 793–796.
17. Becker, R. R., Veglia, A., and Schmid, E. R. (1975). *Wien. Tieraerztl. Monatsschr.* **65,** 47–51.
18. Bedford, J., Harrison, G. E., Raymond, W. H. A., and Sutton, A. (1960). *Br. Med. J.* **I,** 589–592.
19. Beeson, K. C. (1941). *Misc. Publ.—U.S., Dep. Agric.* **369,** 1–164.
20. Beeson, K. C., Griffiths, W. R., and Milne, D. B. (1977). *In* "Geochemistry and the Environment," Vol. 2, pp. 88–92. Natl. Acad. Sci., Washington, D.C.
21. Beighton, D., and McDougall, W. A. (1981). *Arch. Oral Biol.* **26,** 419–425.
22. Bell, M. C., and Sneed, N. N. (1970). *In* "Trace Element Metabolism in Animals" (C. F. Mills, ed.), pp. 70–72. Livingstone, Edinburgh, and London.
23. Bencko, V., Geist, T., Arbetová, D., and Dharmadikari, D. M. (1983). *In* "Lithium, 4. Spurenelement-Symposium" (M. Anke, W. Baumann, H. Bräunlich, and C. Brückner, eds.), pp. 355–361. Wiss. Publ., Friedrich-Schiller-Univ., Jena, G.D.R.
24. Benoy, C. J., Hooper, P. A., and Schneider, R. (1971). *Food Cosmet. Toxicol.* **9,** 645–656.
25. Bertrand, G., and Voronca-Spirt, C. (1929). *C.R. Hebd. Seances Acad. Sci.* **188,** 1199–1202; **189,** 73–74 and 221–223.
26. Bosshardt, D. K., Huff, J. W., and Barnes, R. H. (1956). *Proc. Soc. Exp. Biol. Med.* **92,** 219–221.
27. Bowen, H. J. M., and Dymond, J. A. (1955). *Proc. R. Soc. London, Ser. B* **144,** 355–368.
28. Brown, J. C., and Jones, W. E. (1972). *Plant Physiol.* **49,** 651–653.
29. Browning, E. (1969). "Toxicity of Industrial Metals," 2nd ed. Butterworth, London.
30. Brune, D., Samsahl, K., and Wester, P. O. (1966). *Clin. Chim. Acta* **13,** 285–291.
31. Bunyan, J., Diplock, A. T., Cawthorne, M. A., and Green, J. (1968). *Br. J. Nutr.* **22,** 165–182.
32. Burch, G. E., Threefoot, S. A., and Ray, C. T. (1955). *J. Lab. Clin. Med.* **45,** 371–394.
33. Calloway, D. H., and McMullen, J. J. (1966). *Am. J. Clin. Nutr.* **18,** 1–6.
34. Catsch, A. (1967). *In* "Strontium Metabolism" (J. M. A. Lenihan, J. F. Loutit, and J. H. Martin, eds.), pp. 265–281. Academic Press, New York.
35. Chahal, S. S., and Rawla, G. S. (1977). *Indian Phytopathol.* **30,** 47–50.
36. Chaudhri, M. A. (1981). *Nuklearmedizin, Suppl. (Stuttgart)* **18,** 1058–1065.
37. Chen, T. S. S., Chang, C.-J., and Floss, H. G. (1980). *J. Antibiot.* **33,** 1316–1322.
38. Chiba, M., Ogihara, K., and Kikuchi, M. (1980). *Arch. Toxicol.* **45,** 189–195.

39. Chilton, H. M., Witcofski, R. L., and Heise, C. M. (1981). *J. Nucl. Med.* **22,** 478–479.
40. Chmielnicka, J., Szymańska, J. A., and Snieć, J. (1981). *Arch. Toxicol.* **47,** 263–268.
41. Christopher, J. B., and Newlove, J. R. (1919). *J. Trop. Med. Hyg.* **22,** 113–114 and 129–143.
42. Christopherson, J. B. (1918). *Lancet* **2,** 325–327.
43. Church, D. C., and Martinez, A. (1978). *Feedstuffs* **50,** 22.
44. Clemente, G. F., Cigna Rossi, L., and Santaroni, G. P. (1979). *In* "Trace Substances in Environmental Health-12" (D. D. Hemphill, ed.), pp. 23–30. Univ. of Missouri Press, Columbia.
45. Clode, W., Sobral, T. M., and Baptista, A. M. (1961). *Adv. Thyroid Res., Trans. Int. Goitre Conf., 4th, 1960,* pp. 65–73.
46. Cochran, K. W., Zerwic, M. M., and DuBois, K. P. (1951). *J. Pharmacol. Exp. Ther.* **102,** 165–178.
47. Cole, B. T., and Patrick, H. (1958). *Arch. Biochem. Biophys.* **74,** 357–361.
48. Colvin, L. B., Creger, C. R., Ferguson, T. M., and Crookshank, H. R. (1972). *Poul. Sci.* **51,** 576–581.
49. Comar, C. L., and Wasserman, R. H. (1964). *In* "Mineral Metabolism: An Advanced Treatise" (C. L. Comar and F. Bronner, eds.), Vol. 2, Part A, pp. 523–572. Academic Press, New York.
50. Conway, E. J., and Flood, J. C. (1936). *Biochem. J.* **30,** 716–727.
51. Cornelis, R., Ringoir, S., Mees, L., Lameire, N., Wallaeys, B., and Hoste, J. (1981). *In* "Trace Element Metabolism in Man and Animals-4" (J. McC. Howell, J. M. Gawthorne, and C. L. White, eds.), pp. 530–533. Aust. Acad. Sci., Canberra.
52. Cowgill, U. M. (1981). *Biol. Trace Elem. Res.* **3,** 33–54.
53. Crampton, R. F., Elias, P. S., and Gangolli, S. D. (1971). *Br. J. Nutr.* **25,** 317–322.
54. Creason, J. P., Svendsgaard, D., Bumgarner, J., Pinkerton, C., and Hinners, T. (1977). *In* "Trace Substances in Environmental Health-10" (D. D. Hemphill, ed.), pp. 53–62. Univ. of Missouri Press, Columbia.
55. Curzon, M. E. J., and Crocker, D. C. (1978). *Arch. Oral Biol.* **23,** 647–653.
56. Curzon, M. E. J., Spector, P. C., and Iker, H. P. (1978). *Arch. Oral Biol.* **23,** 317–321.
57. Dam, H., Nielsen, G. K., Prange, I., and Sondergaard, E. (1958). *Nature (London)* **182,** 802–803.
58. De Groot, A. P. (1973). *Food Cosmet. Toxicol.* **11,** 955–962.
59. De Groot, A. P., Feron, V. J., and Til, H. P. (1973). *Food Cosmet. Toxicol.* **11,** 19–30.
60. De Meio, R. H., and Henriques, F. C., Jr. (1947). *J. Biol. Chem.* **169,** 609–623.
61. De Meio, R. H., and Jetter, W. W. (1948). *J. Ind. Hyg. Toxicol.* **30,** 53–58.
62. Dixon, R. L., Lee, I. P., and Sherins, R. J. (1976). *Environ. Health Perspect.* **13,** 59–67.
63. Dixon, T. F. (1935). *Biochem. J.* **28,** 86–89.
64. Drummond, G. S., and Kappas, A. (1981). *Proc. Natl. Acad. Sci. U.S.A.* **78,** 6466–6470.
65. Drummond, G. S., and Kappas, A. (1982). *Science* **217,** 1250–1252.
66. Duckworth, R. B., and Hawthorn, J. (1960). *J. Sci. Food Agric.* **11,** 218–225.
67. Duggan, R. E., and Lipscomb, G. Q. (1971). *J. Dairy Sci.* **54,** 695–701.
68. Duggan, R. E., and Weatherwax, J. R. (1967). *Science* **157,** 1006–1010.
69. Dugger, W. M. (1973). *Adv. Chem. Ser.* **123,** 112–129.
70. Duke, J. A. (1970). *Econ. Bot.* **24,** 344–366.
71. El-Kholi, A. F., and Hamdy, A. A. (1977). *Egypt. J. Soil Sci.* **17,** 87–97.
72. Elsair, J., Merad, R., Denine, R., Reggabi, M., Benali, S., Alamir, B., Hamrour, M., Azzouz, M., Khalfat, K., Tabet Aoul, M., and Nauer, J. (1982) *Fluoride* **15,** 75–78.
73. Elsair, J., Merad, R., Denine, R., Reggabi, M., Benali, S., Azzouz, M., Khelfat, K., and Tabet Aoul, M. (1980). *Flouride* **13,** 30–38.

74. Felicetti, S. A., Thomas, R. G., and McClellan, R. O. (1974). *Am. Ind. Hyg. Assoc. J.* **35**, 292–300.

75. Fieve, R. R., Meltzer, H., Dunner, D. L., Levitt, M., Mendlewicz, J., and Thomas, A. (1973). *Am. J. Psychiatry* **130**, 55–61.

76. Follis, R. H., Jr. (1943). *Am. J. Physiol.* **138**, 246–250.

77. Follis, R. H., Jr. (1947). *Am. J. Physiol.* **150**, 520–522.

78. Forbes, R. M., Cooper, A. R., and Mitchell, H. H. (1954). *J. Biol. Chem.* **209**, 857–865.

79. Forbes, R. M., and Mitchell, H. H. (1957). *AMA Arch. Ind. Health* **16**, 489–492.

80. Friis-Hansen, B., Aggerbeck, B., and Aas Jansen, J. (1982). *Food Chem. Toxicol.* **20**, 451–454.

81. Fritsch, P., De Saint Blanquat, G., and Derache, R. (1977). *Food Cosmet. Toxicol.* **15**, 147–149.

82. Fritsch, P., De Saint Blanquat, G., and Derache, R. (1977). *Toxocology* **8**, 165–175.

83. Furr, A. K., MacDaniels, L. H., St. John, L. E., Jr., Gutenmann, W. H., Pakkala, I. S., and Lisk, D. J. (1979). *Bull. Environ. Contam. Toxicol.* **21**, 392–396.

84. Furr, A. K., Parkinson, T. F., Bache, C. A., Gutenmann, W. H., Pakkala, I. S., Stoewsand, G. S., and Lisk, D. J. (1979). *Nutr. Rep. Int.* **20**, 841–844.

85. Furr, A. K., Parkinson, T. F., Winch, F. E., Jr., Bache, C. A., Gutenmann, W. H., Pakkala, I. S., and Lisk, D. J. (1979). *Nutr. Rep. Int.* **20**, 765–769.

86. Garner, R. J. (1959). *Nature (London)* **184**, 733–734.

87. Gedalia, I., Yariv, S., Nayot, H., and Eidelman, E. (1974). *In* "Trace Element Metabolism in Animals-2" (W. G. Hoekstra, J. W. Suttie, H. E. Ganther, and W. Mertz, eds.), pp. 461–464. University Park Press, Baltimore, Maryland.

88. Gelb, M. H., and Harris, D. C. (1980). *Arch. Biochem. Biophys.* **200**, 93–98.

89. Gerlach, W., and Muller, R. (1934). *Virchows Arch. Pathol. Anat. Physiol.* **294**, 210–212.

90. Glendening, B. L., Schrenk, W. G., and Parrish, D. B. (1956). *J. Nutr.* **60**, 563–579.

91. Gormican, A. (1970). *J. Am. Diet. Assoc.* **56**, 397–403.

92. Green, G. H., Lott, M. D., and Weeth, H. J. (1973). *Proc., Annu. Meet.—Am. Soc. Anim. Sci., West. Sect.* **24**, 254–258.

93. Green, G. H., and Weeth, H. J. (1977). *J. Anim. Sci.* **46**, 812–818.

94. Greger, J. L., and Baier, M. (1981). *J. Food Sci.* **46**, 1751–1753 and 1765.

95. Greger, J. L., and Johnson, M. A. (1981). *Food Cosmet. Toxicol.* **19**, 163–166.

96. Greger, J. L., Smith, S. A., Johnson, M. A., and Baier, M. J. (1982). *Biol. Trace Elem. Res.* **4**, 269–278.

97. Gruner, J., Sung, S. S., Tubiania, M., and Segarra, J. (1951). *C.R. Seances Soc. Biol. Ses Fil.* **145**, 203–206.

98. Gunther, R. D., and Wright, E. M. (1983). *J. Membr. Biol.* **74**, 85–94.

99. Guyatt, B. L., Kay, H. D., and Branion, H. D. (1933). *J. Nutr.* **6**, 313–324.

100. Habermann, E. (1983). *Naunyn-Schmiedeberg's Arch. Pharmacol.* **323**, 269–275.

101. Hall, I. H., Starnes, C. O., McPhail, A. T., Wisian-Neilson, P., Das, M. K., Harchelroad, F., Jr., and Spielvogel, B. F. (1980). *J. Pharm. Sci.* **69**, 1025–1029.

102. Hamilton, E. I., and Minski, M. J. (1972-1973). *Sci. Total Environ.* **1**, 375–394.

103. Hamilton, E. I., Minski, M. J., and Cleary, J. J. (1972-1973). *Sci. Total Environ.* **1**, 341–374.

104. Harrison, G. E., Howells, G. R., Pollard, J., Kostial, K., and Manitasevic, R. (1966). *Br. J. Nutr.* **20**, 561–569.

105. Harrison, G. E., Raymond, W. H. A., and Tretheway, H. C. (1955). *Clin. Sci.* **14**, 681–695.

106. Hellerstein, S., Kaiser, C., Darrow, D. D., and Darrow, D. C. (1960). *J. Clin. Invest.* **39**, 282–287.

107. Henke, G., Möllmann, H., and Alfes, H. (1971). *Z. Neurol.* **199**, 283–294.

458   Forrest H. Nielsen

108. Heppel, L. A., and Schmidt, C. L. A. (1938). *Univ. Calif., Berkeley, Publ. Physiol.* **8,** 189–205.
109. Heywood, B. J. (1966). *Science* **152,** 1408.
110. Hiles, R. A. (1974). *Toxicol. Appl. Pharmacol.* **27,** 366–379.
111. Hill, C. H., Starcher, B., and Matrone, G. (1964). *J. Nutr.* **83,** 107–110.
112. Höck, A., Demmel, U., Schicha, H., Kasperek, K., and Feinendegen, L. E. (1975). *Brain* **98,** 49–64.
113. Hodges, R. M., MacDonald, N. S., Nusbaum, R., Stearns, R., Ezmirlian, F., Spain, P., and McArthur, C. (1950). *J. Biol. Chem.* **185,** 519–524.
114. Hodgkiss, W. S., and Errington, B. J. (1940). *Trans. Ky. Acad. Sci.* **9,** 17–20.
115. Hove, E., Elvehjem, C. A., and Hart, E. B. (1939). *Am. J. Physiol.* **127,** 689–701.
116. Huff, J. W., Bosshardt, D. K., Miller, O. P., and Barnes, R. H. (1956). *Proc. Soc. Exp. Biol. Med.* **92,** 216–219.
117. Hunt, C. D., and Nielsen, F. H. (1981). *In* "Trace Element Metabolism in Man and Animals-4" (J.McC. Howell, J. M. Gawthorne, and C. L. White, eds.), pp. 597–600. Aust. Acad. Sci., Canberra.
118. Hunt, C. D., Shuler, T. R., and Nielsen, F. H. (1983). *In* "Lithium-4. Spurenelement-Symposium" (M. Anke, W. Bauman, H. Bräunlich, and C. Brückner, eds.), pp. 149–155. Wiss. Publ., Friedrich-Schiller-Univ., Jena, G.D.R.
119. Hunt, C. E., and Navia, J. M. (1975). *Arch. Oral Biol.* **20,** 497–501.
120. Hutcheson, D. P., Gray, D. H., Venugopal, B., and Luckey, T. D. (1975). *J. Nutr.* **105,** 670–675.
121. Indraprasit, S., Alexander, G. V., and Gonvick, H. C. (1974). *J. Chronic Dis.* **27,** 135–161.
122. Jackson, J. F., and Chapman, K. S. R. (1975). *In* "Trace Elements in Soil-Plant-Animal Systems" (D. J. D. Nicholas and A. R. Egan, eds.), pp. 213–225. Academic Press, New York.
123. Jacobsen, N., and Jonsen, J. (1975). *Pathol. Eur.* **10,** 115–121.
124. Jensen, L., Peterson, R. P., and Falen, L. (1974). *Poul. Sci.* **53,** 57–64.
125. Jensen, L. S. (1975). *J. Nutr.* **105,** 769–775.
126. Johnson, M. A., Baier, M. J., and Greger, J. L. (1982). *Am. J. Clin. Nutr.* **35,** 1332–1338.
127. Johnson, M. A., and Greger, J. L. (1982). *Am. J. Clin. Nutr.* **35,** 655–660.
128. Johnson, M. A., and Greger, J. L. (1984). *Fed. Proc., Fed. Am. Soc. Exp. Biol.* **43,** 680.
129. Jury, R. V., Webb, M. S. W., and Webb, R. J. (1960). *Anal. Chim. Acta* **22,** 145–152.
130. Kappas, A., and Maines, M. D. (1976). *Science* **192,** 60–62.
131. Kasperek, K., Iyengar, G. V., Kiem, J., Borberg, H., and Feinendegen, L. E. (1979). *Clin. Chem. (Winston-Salem, N.C.)* **25,** 711–715.
132. Kaye, S. V. (1968). *Health Phys.* **15,** 399–417.
133. Kayne, F. J. (1971). *Arch. Biochem. Biophys.* **143,** 232–239.
134. Kemp, K., Jensen, F. P., Møller, J. T., and Hansen, G. (1975). "Risø Report M-1822." Danish Atomic Energy Commission, Roskilde, Denmark.
135. Kent, N. L., and McCance, R. A. (1941). *Biochem. J.* **35,** 837–844 and 877–883.
136. Kirchgessner, M., Merz, G., and Oelschläger, W. (1960). *Arch. Tierernaehr.* **10,** 414–427.
137. Kjellin, K. G. (1981). *Clin. Toxicol.* **18,** 1237–1245.
138. Klaassen, C. D. (1979). *Toxicol. Appl. Pharmacol.* **50,** 49–55.
139. Klein, R. M., and Brown, S. J. (1982). *Environ. Exp. Bot.* **22,** 199–202.
140. Klevay, L. M. (1976). *Pharmacol. Ther., Part A* **1,** 223–229.
141. Kosta, L., Byrne, A. R., and Dermelj, M. (1983). *Sci. Total Environ.* **29,** 261–268.
142. Krasovskii, G. N., Varshavskaya, S. P., and Borisov, A. I. (1976). *Environ. Health Perspect.* **13,** 69–75.

143. Kubo, H., Hashimoto, S., and Ishibashi, A. (1976). *In* "Trace Substances in Environmental Health-9" (D. D. Hemphill, ed.), pp. 317–322. Univ. of Missouri Press, Columbia.

144. Kunin, A. S., Dearborn, E. H., Burrows, B. A., Relman, A. S., and Roy, A. M. (1959). *Am. J. Physiol.* **197**, 1297–1307.

145. Landauer, W. (1952). *J. Exp. Zool.* **120**, 469–508.

146. Landauer, W. (1953). *Proc. Soc. Exp. Biol. Med.* **82**, 633–636.

147. Lasnitski, A., and Szörényi, E. (1934). *Biochem. J.* **28**, 1678–1683.

148. Leach, R. M., Jr., and Nesheim, M. C. (1963). *J. Nutr.* **81**, 193–199.

149. Lewin, J., and Chen, C.-H. (1976). *J. Exp. Bot.* **27**, 916–921.

150. Lewis, D. H. (1980). *New Phytol.* **84**, 209–229.

151. Life Sciences Research Office (1980). "Evaluation of the Health Aspects of Borax and Boric Acid as Food Packaging Ingredients." Fed. Am. Soc. Exp. Biol., Bethesda, Maryland.

152. Littledike, E. T., Whipp, S. C., and Schroeder, L. (1969). *J. Am. Vet. Med. Assoc.* **155**, 1955–1962.

153. Liu, F. T. Y. (1975). *J. Dent. Res.* **54**, 97–103.

154. Ljungdahl, L. G., and Andreesen, J. R. (1975). *FEBS Lett.* **54**, 279–282.

155. Loeb, R. F. (1920). *J. Gen. Physiol.* **3**, 229–236.

156. Lombeck, I., Kasperek, K., Feinendegen, L. E., and Bremer, H. J. (1980). *Biol. Trace Elem. Res.* **2**, 193–198.

157. Lopez, A., and Williams, H. L. (1981). *J. Food Sci.* **46**, 432–434 and 444.

158. Losee, F., Cutress, T. W., and Brown, R. (1974). *In* "Trace Substances in Environmental Health-7" (D. D. Hemphill, ed.), pp. 19–24. Univ. of Missouri Press, Columbia.

159. Lynn, G. E., Shrader, S. A., Hammer, O. H., and Lassiter, C. A. (1963). *J. Agric. Food Chem.* **11**, 87–91.

160. MacDonald, I. A. (1975). *Biochim. Biophys. Acta* **397**, 244–253.

161. MacLeod, R. A., and Snell, E. E. (1950). *J. Bacteriol.* **59**, 783–792.

162. Mason, M. F. (1936). *J. Biol. Chem.* **113**, 61–74

163. Mathur, O. N., and Roy, N. K. (1981). *Indian J. Dairy Sci.* **34**, 321–326.

164. McCord, C. P., Meek, S. F., Harrold, G. C., and Heussner, C. E. (1942). *J. Ind. Hyg. Toxicol.* **24**, 243–254.

165. McHard, J. A., Foulk, S. J., Jorgensen, J. L., Bayer, S., and Winefordner, J. D. (1980). *ACS Symp. Ser.* **143**, 363–392.

166. McLachlan, J., and Craigie, J. S. (1967). *Nature (London)* **214**, 604–605.

167. Meltzer, H. L., Taylor, R. M., Platman, S. R., and Fieve, R. R. (1971). *Nature (London)* **223**, 321–322.

168. Mitchell, P. H., Wilson, J. W., and Stanton, R. E. (1921). *J. Gen. Physiol.* **4**, 141–148.

169. Mitchell, R. L. (1957). *Research (London)* **10**, 357–362.

170. Molokhia, M. M., and Smith, H. (1967). *Arch. Environ. Health* **15**, 745–750.

171. Molokhia, M. M., and Smith, H. (1969). *Bull. W. H. O.* **40**, 123–128.

172. Molokhia, M. M., and Smith, H. (1969). *J. Trop. Med. Hyg.* **72**, 222–225.

173. Momčilović, B., and Gruden, N. (1981). *Experientia* **37**, 498–499.

174. Monier-Williams, G. W. (1949). "Trace Elements in Foods." Wiley, New York.

175. Moo, S. P., and Pillay, K. K. S. (1983). *J. Radioanal. Chem.* **77**, 141–147.

176. Moran, J. K., and Schwarz, K. (1978). *Fed. Proc., Fed. Am. Soc. Exp. Biol.* **37**, 671.

177. Morse, R. A., and Lisk, D. J. (1980). *Am. Bee J.* **120**, 522–523.

178. Mraz, F. R. (1962). *Proc. Soc. Exp. Biol. Med.* **110**, 273–275.

179. Murphy, E. W., Page, L., and Watt, B. K. (1971). *J. Am. Diet. Assoc.* **58**, 115–122.

180. Murthy, G. K., and Rhea, U. (1968). *J. Dairy Sci.* **51**, 610–613.

181. Murthy, G. K., Rhea, U., and Peeler, J. T. (1971). *Environ. Sci. Technol.* **5**, 436–442.

182. Murthy, G. K., Rhea, U.S., and Peeler, J. T. (1973). *Environ. Sci. Technol.* **7**, 1042–1045.
183. Newnham, R. E. (1981). *In* "Trace Element Metabolism in Man and Animals-4" (J. McC. Howell, J. M. Gawthorne, and C. L. White, eds.), pp. 400–402. Aust. Acad. Sci., Canberra.
184. Newnham, R. E. (1984). *In* "Trace Element Metabolism in Man and Animals-5 Abstracts," p. 234. Aberdeen, Scotland.
185. Nielsen, F. H. (1984). *In* "Trace Elements in Man and Animals-5 Abstracts," p. 26. Aberdeen, Scotland.
186. Nielsen, F. H. (1984). *In* "Trace Substances in Environmental Health-18" (D. D. Hemphill, ed.), pp. 47–52. Univ. of Missouri Press, Columbia.
187. Nielsen, F. H., Milne, D. B., and Zimmerman, T. J. (1982). *Proc. N. D. Acad. Sci.* **36**, 62.
188. Nixon, G. S., Livingston, H. D., and Smith, H. (1967). *Caries Res.* **1**, 327–332.
189. Nordio, S., Donath, A., Macagno, F., and Gatti, R. (1971). *Acta Paediatr. Scand.* **60**, 449–455.
190. Nozdryukhina, L. R. (1978). *In* "Trace Element Metabolism in Man and Animals-3" (M. Kirchgessner, ed.), pp. 336–339. Tech. Univ. Munich, F.R.G.
191. Nuurtamo, M., Varo, P., Saari, E., and Koivistoinen, P. (1980). *Acta Agric. Scand., Suppl.* **22**, 57–76 and 77–87.
192. Oakes, T. W., Shank, K. E., Easterly, C. E., and Quintana, L. R. (1978). *In* "Trace Substances in Environmental Health-11" (D. D. Hemphill, ed.), pp. 123–132. Univ. of Missouri Press, Columbia.
193. Oe, P. L., Vis, R. D., Meijer, J. H., van Langevelde, F., Allon, W., Meer, C. V. D., and Verheul, H. (1981). *In* "Trace Element Metabolism in Man and Animals-4" (J. McC. Howell, J. M. Gawthorne, and C. L. White, eds.), pp. 526–529. Aust. Acad. Sci., Canberra.
194. Omdahl, J. L., and DeLuca, H. F. (1971). *Science* **174**, 949–951.
195. Omdahl, J. L., and DeLuca, H. F. (1972). *J. Biol. Chem.* **247**, 5520–5526.
196. Orent-Keiles, E. (1941). *Proc. Soc. Exp. Biol. Med.* **44**, 199–202.
197. Otto, G. F., Jachrowski, L. A., and Wharton, J. D. (1953). *Am. J. Trop. Hyg.* **2**, 495–516.
198. Otto, G. F., and Maren, T. H. (1950). *Am. J. Hyg.* **51**, 370–385.
199. Otto, G. F., Maren, T. H., and Brown, H. W. (1947). *Am. J. Hyg.* **46**, 193–211.
200. Owen, E. C. (1944). *J. Dairy Res.* **13**, 243–248.
201. Palmer, R. F., and Thompson, R. C. (1964). *Am. J. Physiol.* **207**, 561–566.
202. Panteliadis, C., Boenigk, H.-E., and Janke, W. (1975). *Infusionstherapie* **2**, 377–382.
203. Pehrsson, S. K., and Lins, L.-E. (1983). *Nephron* **34**, 93–98.
204. Perry, H. M., Jr., Erlanger, M. W., and Perry, E. F. (1983). *Fed. Proc., Fed. Am. Soc. Exp. Biol.* **42**, 1172.
205. Peterson, R. P., and Jensen, L. S. (1975). *Poult. Sci.* **54**, 771–775 and 795–798.
206. Pfeiffer, C. C., Hallman, L. F., and Gersh, I. (1945). *JAMA, J. Am. Med. Assoc.* **128**, 266–274.
207. Pinto, J., Huang, Y. P., McConnell, R. J., and Rivlin, R. S. (1978). *J. Lab. Clin. Med.* **92**, 126–134.
208. Ploquin, J. (1967). *Aliment. Vie* **55**, 70–113.
209. Pollard, A. S., Parr, A. J., and Loughman, B. C. (1977). *J. Exp. Bot.* **28**, 831–841.
210. Relman, A. S. (1956-1957). *Yale J. Biol. Med.* **29**, 248–262.
211. Richards, C. E., Brady, R. O., and Riggs, D. S. (1949). *J. Clin. Endocrinol.* **9**, 1107–1121.
212. Ringer, S. (1882-1883). *J. Physiol. (London)* **4**, 370–379.
213. Robinson, W. O., and Edgington, G. (1945). *Soil Sci.* **60**, 15–28.
214. Robinson, W. O., Whetstone, R. R., and Edgington, G. (1950) *U.S., Dep. Agric., Tech. Bull.* **1013**, 1–36.
215. Roche, J., Thoai, N.-V., and Loewy, J. (1950). *C. R. Seances Soc. Biol. Ses Fil.* **144**, 638–640.

216. Rosenfeld, G., and Wallace, E. J. (1953). *Arch. Ind. Hyg. Occup. Med.* **8,** 466–479.
217. Ruben, L., Lageson, J., Hyzy, B., and Hooper, A. B. (1982). *J. Protozool.* **29,** 233–238.
218. Rygh, O. (1949). *Bull. Soc. Chim. Biol.* **31,** 1052–1061, 1403–1407, and 1408–1412.
219. Sankaran, N., Morachan, Y. B., and Sennaian, P. (1973). *Madras Agric. J.* **60,** 1022–1023.
220. Sassa, S., Drummond, G. S., Bernstein, S. E., and Kappas, A. (1983). *Blood* **61,** 1011–1013.
221. Sastry, B. V. R., Owens, L. K., and Ball, C. O. T. (1964). *Nature (London)* **20,** 410–411.
222. Schäfer, S. G., and Forth, W. (1983). *Biol. Trace Elem. Res.* **5,** 205–217.
223. Schamschula, R. G., Adkins, B. L., Barmes, D. E., Charlton, G., and Davey, B. G. (1977). *J. Dent. Res.* **56,** C62–C70.
224. Schamschula, R. G., Bunzel, M., Agus, H. M., Adkins, B. L., Barmes, D. E., and Charlton, G. (1978). *J. Dent. Res.* **57,** 427–432.
225. Schepers, G. W. H. (1961). *Prog. Exp. Tumor Res. 1961,* 203–244.
226. Schroeder, H. A., and Balassa, J. J. (1965). *J. Chronic Dis.* **18,** 229–241.
227. Schroeder, H. A., and Balassa, J. J. (1966). *J. Chronic Dis.* **19,** 573–586.
228. Schroeder, H. A., and Balassa, J. J. (1967). *J. Nutr.* **92,** 245–252; *J. Chronic Dis.* **20,** 211–224.
229. Schroeder, H. A., Balassa, J. J., and Tipton, I. H. (1964). *J. Chronic Dis.* **17,** 483–502.
230. Schroeder, H. A., Balassa, J. J., and Vinton, V. H., Jr. (1964). *J. Nutr.* **83,** 239–250.
231. Schroeder, H. A., Buckman, J., and Balassa, J. J. (1967). *J. Chronic Dis.* **20,** 147–161.
232. Schroeder, H. A., Kanisawa, M., Frost, D. V., and Mitchener, M. (1968). *J. Nutr.* **96,** 37–45.
233. Schroeder, H. A., and Mitchener, M. (1971). *J. Nutr.* **101,** 1431–1438.
234. Schroeder, H. A., and Mitchener, M. (1975). *J. Nutr.* **105,** 421–427 and 452–458.
235. Schroeder, H. A., Mitchener, M., Balassa, J. J., Kanisawa, M., and Nason, A. P. (1968). *J. Nutr.* **95,** 95–101.
236. Schroeder, H. A., and Nason, A. P. (1976). *J. Nutr.* **106,** 198–203.
237. Schroeder, H. A., Tipton, I. H., and Nason, A. P. (1972). *J. Chronic Dis.* **25,** 491–517.
238. Schroeder, H. A., Vinton, W. H., Jr., and Balassa, J. J. (1963). *J. Nutr.* **80,** 39–47.
239. Schubert, J. (1947). *Science* **105,** 389–390.
240. Schwarz, K. (1974). *In* ''Trace Element Metabolism in Animals-2'' (W. G. Hoekstra, J. W. Suttie, H. E. Ganther, and W. Mertz, eds.), pp. 355–380. University Park Press, Baltimore, Maryland.
241. Schwarz, K., Milne, D. B., and Vinyard, E. (1970). *Biochem. Biophys. Res. Commun.* **40,** 22–29.
242. Scott, K. G., and Hamilton, J. G. (1950). *Univ. Calif., Berkeley, Publ. Pharmacol.* **2,** 241–262.
243. Seal, B. S., and Weeth, H. J. (1980). *Bull. Environ. Contam. Toxicol.* **25,** 782–789.
244. Seffner, W., and Teubener, W. (1983). *Fluoride* **16,** 33–37.
245. Seffner, W., Teubener, W., and Geinitz, D. (1983). *In* ''Mengen-und Spurenelemente'' (M. Anke, C. Brückner, H. Gürtler, and M. Grün, eds.), pp. 200–203. Karl-Marx-Univ., Leipzig, G.D.R.
246. Shand, C. A., Aggett, P. J., and Ure, A. M. (1984). *In* ''Trace Elements in Man and Animals. 5. Abstracts,'' p. 205. Aberdeen, Scotland.
247. Sharma, M. P., Mathur, R. S., and Mehta, K. (1978). *Experientia* **34,** 1374–1375.
248. Shaver, S. L., and Mason, K. E. (1951). *Anat. Rec.* **109,** 382.
249. Shaw, J. H., and Griffiths, D. (1961). *Arch. Oral Biol.* **5,** 301–322.
250. Sheldon, J. H., and Ramage, H. (1931). *Biochem. J.* **25,** 1608–1627.
251. Shipley, P. G., Park, E. A., McCollum, E. V., Simmonds, N., and Kinney, E. M. (1922). *Bull. Johns Hopkins Hosp.* **33,** 216–220.

252. Skinner, J. T., and McHargue, J. S. (1945). *Am. J. Physiol.* **143**, 385–390.
253. Sklavenitis, H., and Comar, D. (1967). *Nucl. Act. Tech. Life Sci. Proc. Symp., 1967*, pp. 435–444.
254. Smith, H. (1967). *J., Forensic Sci. Soc.* **7**, 97–102.
255. Smyth, D. A., and Dugger, W. M. (1980). *Plant Physiol.* **66**, 692–695.
256. Solomons, N. W., Marchini, J. S., Duarte-Favaro, R.-M., Vannuchi, H., and Dutra de Oliveira, J. E. (1983). *Am. J. Clin. Nutr.* **37**, 566–571.
257. Soman, S. D., Panday, V. K., Joseph, K. T., and Raut, S. J. (1969). *Health Phys.* **17**, 35–40.
258. Sowden, E. M., and Stitch, S. R. (1957). *Biochem. J.* **67**, 104–109.
259. Spector, P. C., and Curzon, M. E. J. (1978). *J. Dent. Res.* **57**, 55–58.
260. Spencer, H., Li, M., Samachson, J., and Laszlo, D. (1960). *Metab., Clin. Exp.* **9**, 916–925.
261. Steadman, L. T., Brudevold, F., and Smith, F. A. (1958). *J. Am. Dent. Assoc.* **57**, 340–344.
262. Steinberg, R. A. (1939). *J. Agric. Res.* **59**, 749–763.
263. Stelter, L. H. (1980). *J. Wildl. Dis.* **16**, 175–182.
264. Stolk, J. M., Conner, R. L., and Barchas, J. D. (1971). *Psychopharmacologia* **22**, 250–260.
265. Stolk, J. M., Nowack, W. J., Barchas, J. D., and Platman, S. R. (1970). *Science* **168**, 501–503.
266. Sumino, K., Hayakawa, K., Shibata, T., and Kitamura, S. (1975). *Arch. Environ. Health* **30**, 487–494.
267. Szabo, A. S. (1979). *Lebensmittelindustrie* **26**, 549–550.
268. Teresi, J. D., Hove, E., Elvehjem, C. A., and Hart, E. B. (1944). *Am. J. Physiol.* **140**, 513–518.
269. Theuer, R. C., Mahoney, A. W., and Sarett, H. P. (1971). *J. Nutr.* **101**, 525–532.
270. Tipton, I. H., and Cook, M. J. (1963). *Health Phys.* **9**, 103–145.
271. Tipton, I. H., Stewart, P. L., and Dickson, J. (1969). *Health Phys.* **16**, 455–462.
272. Tipton, I. H., Stewart, P. L., and Martin, P. G. (1966). *Health Phys.* **12**, 1683–1689.
273. Tomza, U., Janicki, T., and Kosman, J. (1983). *In* "Lithium-4. Spurenelement-Symposium" (M. Anke, W. Baumann, H. Bräunlich, and C. Brückner, eds.), pp. 362–368. Wiss. Publ., Friedrich-Schiller-Univ., Jena, G.D.R.
274. Torii, S., Mitsumori, K., Inubushi, S., and Yanagisawa, I. (1973). *Psychopharmacologia* **29**, 65–75.
275. Van Campen, D. R. (1966). *J. Nutr.* **88**, 125–130.
276. Vanoeteren, C., Cornelis, R., Versieck, J., Hoste, J., and De Roose, J. (1982). *J. Radioanal. Chem.* **70**, 219–238.
277. Van Vleet, J. F. (1976–1977). *Am. J. Vet. Res.* **37**, 1415–1420; **38**, 1393–1398.
278. Van Vleet, J. F., Boon, G. D., and Ferrans, V. J. (1981). *Am. J. Vet. Res.* **42**, 789–799.
279. Varo, P., and Koivistoinen, P. (1980). *Acta Agric. Scand., Suppl.* **22**, 165–171.
280. Varo, P., Nuurtamo, M., Saari, E., and Koivistoinen, P. (1980). *Acta Agric. Scand., Suppl.* **22**, 27–35, 37–55, 89–113, 115–126, 127–139, and 141–160.
281. Vaughan, J. M. (1970). "The Physiology of Bone." Oxford Univ. Press (Clarendon), London and New York.
282. Versieck, J., and Cornelis, R. (1980). *Anal. Chim. Acta* **116**, 217–254.
283. Vis, R. D., Oe, P. L., and Verheul, H. (1981). *Turun Yliopiston Julk., Sar. D* **13**, 166–167.
284. Volf, V. (1971). *In* "Intestinal Absorption of Metal Ions, Trace Elements and Radionuclides" (S. C. Skoryna and D. Waldron-Edward, eds.), pp. 277–292. Pergamon, Oxford.
285. Waitz, J. A., Ober, R. E., Meisenhelder, J. E., and Thompson, P. E. (1965). *Bull. W.H.O.* **33**, 537–546.
286. Waldron-Edward, D., Paul, T. M., and Skoryna, S. C. (1964). *Can. Med. Assoc. J.* **91**, 1006–1010.

287. Warrington, K. (1923-1926). *Ann. Bot. (London)* **37**, 630–670; **40**, 27–42.
288. Wasserman, R. H., Romney, E. M., Skougstad, M. W., and Siever, R. (1977). *In* "Geochemistry and the Environment," Vol. 2, pp. 73–87. Natl. Acad. Sci., Washington, D.C.
289. Webb, J., Kirk, K. A., Jackson, D. H., Niedermeier, W., Turner, M. E., Rackley, C. E., and Russell, R. O. (1976). *Exp. Mol. Pathol.* **25**, 322–331.
290. Weber, C. W., Doberenz, A. R., and Reid, B. L. (1968). *Poult. Sci.* **47**, 1731–1732.
291. Weber, C. W., Doberanz, A. R., Wyckoff, R. W. G., and Reid, B. L. (1968). *Poult. Sci.* **47**, 1318–1323.
292. Weeth, H. J., Speth, C. F., and Hanks, D. R. (1981). *Am. J. Vet. Res.* **42**, 474–477.
293. Weir, R. J., Jr., and Fisher, R. S. (1972). *Toxicol. Appl. Pharmacol.* **23**, 351–364.
294. Wester, P. O. (1965). *Acta Med. Scand., Suppl.* **439**, 1–48.
295. Wester, P. O. (1973). *Acta Med. Scand.* **194**, 505–512.
296. Whanger, P. D., and Weswig, P. H. (1970). *J. Nutr.* **100**, 341–348.
297. Whanger, P. D., and Weswig, P. H. (1978). *Nutr. Rep. Int.* **18**, 421–428.
298. Winnek, P. S., and Smith, A. H. (1937). *J. Biol. Chem.* **121**, 345–352.
299. Wolf, N., Gedalia, I., Yariv, S., and Zuckermann, H. (1973). *Arch. Oral Biol.* **18**, 233–238.
300. Wolny, M. (1977). *Eur. J. Biochem.* **80**, 551–556.
301. Wright, P. L., and Bell, M. C. (1966). *Am. J. Physiol.* **211**, 6–10.
302. Yamaguchi, M., Saito, R., and Okada, S. (1980). *Toxicology* **16**, 267–273.
303. Yamaguchi, M., Sugii, K., and Okada, S. (1981). *J. Pharm. Dyn.* **4**, 874–878.
304. Yanagisawa, I., and Yoshikawa, H. (1973). *Biochim. Biophys. Acta* **329**, 283–294.
305. Yukawa, M., Suzuki-Yasumoto, M., Amano, K., and Terai, M. (1980). *Arch. Environ. Health* **35**, 36–44.
306. Yum, M. N., Conine, D. L., Martz, R. C., Forney, R. B., and Stookey, G. K. (1976). *Toxicol. Appl. Pharmacol.* **37**, 363–370.
307. Zondek, H., and Bier, A. (1932). *Klin. Wochenschr.* **11**, 759–760.
308. Zook, E. G., and Lehmann, J. (1965). *J. Assoc. Off. Agric. Chem.* **48**, 850–855.
309. Zyka, V. (1970). *Sb. Geol. Ved, Technol., Geochem.* **10**, 91–96.

11

Soil–Plant–Animal and Human Interrelationships in Trace Element Nutrition

W. H. ALLAWAY

Department of Agronomy
Cornell University
Ithaca, New York

I. INTRODUCTION

The fact that animals did not thrive, or suffered various disorders, when restricted to some areas and remained healthy in other areas, was recognized in Europe as early as the eighteenth century. These observations focused attention on the soil factors involved, because areas considered satisfactory for stock and those classified as unthrifty or unhealthy were often adjacent, which would minimize climatic differences as causal factors. Furthermore, animals transferred from unhealthy to healthy areas usually recovered, which suggested the existence of nutritional rather than infectious disorders. Since about the mid-1930s it has been discovered that many of the maladies of this type resulted from the inability of the soils of the affected areas to supply, through plants that grow on them, the mineral nutrients required by humans and animals in appropriate, safe, nontoxic amounts and in the proper balance with other factors in the environment.

The investigations leading to an understanding of some of these regional nutritional problems have at times had striking effects on the health of people or the success of agriculture in particular areas. The studies of Chatin (22), leading to an understanding of the relationship between iodine in soils, plants, and waters, and the incidence of goiter in humans, have formed the basis for a virtual elimination of this nutritional problem in areas where, at one time, it was very prevalent. Similarly, an understanding of the relationship between the selenium concentration in locally produced foods and the occurrence of Keshan disease in China has led to a dramatic improvement in human health in that area (96). The discovery that traces of molybdenum were essential for the normal growth of plants was the basis for establishing a productive pastoral agriculture in large areas of Australia that were previously of very low productivity for both plants and, consequently, animals (9). In the case of cobalt deficiency in parts of Australia, the field investigations leading to a solution to this problem also resulted in the establishment of a new essential element for animal nutrition (53,83). A very interesting account of the history of the development of research on the relation of soils to the nutrition of human and animals in the United States is given in the book by Beeson and Matrone (12).

Nutritional abnormalities involving interrelationships among soils, plants, and animals may arise as simple deficiencies or excesses of single elements, or they may occur as more complex interrelationships between the concentrations of trace elements in soils, plants, and animals and the concentration of accessory factors that may affect the need for, or the efficiency of, the dietary trace elements. An example of a simple relationship among soils, plants, and animals is found just east of the Cascade Mountains in central Oregon. Here, the soils are developed from volcanic deposits that were very low in selenium when they were deposited. All of the soils in the area contain very little selenium. All of the plants that grow in the area are low in selenium, and white muscle disease (WMD) in lambs and calves was endemic to the area prior to the discovery of selenium as an essential element and the correction of this deficiency by selenium supplementation of the livestock. The relation between iodine in soils and plants and the occurrence of endemic goiter in different areas of the world is another fairly simple relationship between a trace element and a health problem. In many areas of the world all of the soils, the potable waters, and all of the plants are low in iodine. These are the areas where endemic goiter is prevalent; however, even in these areas the occurrence of goiter may be conditioned by the presence of goitrogens in certain plants.

The selenium toxicity found in the Great Plains and intermountain area of North America is an example of a somewhat more complex relationship among soils, plants, and animal health. In this area, localized areas of alkaline, unweathered soils developed from certain Cretaceous age geological deposits, are very high in available selenium. Where selenium accumulator plants dominate the vegetation of the area, cases of acute selenium poisoning in cattle, horses, and sheep were detected. Where the vegetation of the area did not include these accumulator plant species, the problem was present but much less acute.

The incidence of the disease *Phalaris* staggers in sheep and cattle provides a further example of a complex soil–plant–animal–trace element interrelationship.

This condition is unknown in most areas where the grass *P. tuberosa* is grown. On cobalt-deficient or marginally cobalt-deficient soils, the neurotoxic substance present in this plant species induces demyelination in animals consuming the plant unless they are treated with cobalt salts or pellets, or the soils are fertilized with cobalt salts. It seems that normal soils produce herbage with cobalt levels adequate to meet the normal needs of ruminants, plus sufficient cobalt to enable them to detoxicate the neurotoxic substance in *Phalaris*. Marginally cobalt-deficient soils are capable of fulfilling the former needs, but not the latter. Severely cobalt-deficient soils result in the growth of plants carrying insufficient cobalt to fulfill either of these needs satisfactorily. The incidence of *Phalaris* staggers thus depends on the interaction between a soil factor and a plant factor, each of which can vary independently in a particular environment.

Other complex interactions involving soils, plants, and animals may involve interactions between trace elements themselves or between trace elements and major elements in the diet. Molybdenosis and hypercuprosis in animals are due to unbalanced ratios of molybdenum and copper in the animal diet. The molybdenum–copper interaction may also be conditioned by the levels of dietary sulfate. Molybdenum–copper problems are further complicated by the differences in the availability to plants of molybdenum in different kinds of soils. In much of the western United States molybdenosis due to high concentrations of molybdenum in the plants is found only on the poorly drained soils where soil molybdenum is present in forms highly available to plants. Well-drained areas, even though the soils may be formed from parent materials equally high in molybdenum, rarely produce plants containing sufficient molybdenum to cause molybdenosis in livestock (47).

II. SOIL, WATER, AND AIR AS SOURCES OF TRACE ELEMENTS

Food, and not the water or the atmosphere, normally supplies a major proportion of the total daily trace element intakes by animals and humans. This generalization does not apply in the endemic fluorosis regions where the water supplies constitute the principal source of the high fluoride intakes. In these regions the high levels of fluoride present in the water seldom bear any relationship to the fluoride status of the soils and herbage of the affected areas, because the water usually comes from deep wells drawing from other soil or rock formations. A high inverse correlation between the iodine content of the drinking water and the incidence of goiter has long been known, but only some 10% of the total intake of this element by humans comes from the water supply in goitrous and nongoitrous regions alike.

The major source of iodine to land plants is an indirect cycling of iodine from the oceans to the atmosphere and then to the soil from rainfall (29). Areas distant from the oceans in the direction of the prevailing winds and areas occupied by young soils or geological deposits, where there has been insufficient time for airborne iodine to accumulate, are likely to produce plants having low iodine concentrations.

Public water supplies are normally low in trace elements and generally supply <10% of the total human intakes. Wells and springs in some areas have been found to contain unusually high concentrations of arsenic, lithium, strontium, boron, and selenium. These may contribute substantial quantities of these elements to total intake.

Since the late 1950s, concerns over the introduction of trace elements and many other compounds into the environment as a result of human activity have greatly increased. These environmental concerns have generated increasing interest on the part of trace element scientists in determining the entire cycle of trace elements from geochemical reserves through mining and manufacturing and into the soil, air, water, and food supply, and their potential recycling through waste disposal systems and back into rivers or spread on soils used for production of food crops. Investigations of trace element cycles are now being reported at a number of annual and special conferences, and an extensive literature has developed from these. Only a few examples of this literature can be cited here (21,34,38,62,63,79).

Studies of the cycling of trace elements from manufactured products into the food chain have shown that the atmosphere is a minor pathway of trace element distribution except in the case of lead. In the case of lead, areas close to sources of lead emission either from manufacturing or auto exhaust have been found to have high lead concentrations in forage crops and on the leaves of food crops. Seed crops grown in the same area are rarely high in lead. Large quantities of selenium are emitted from fossil fuel combustion (31). This selenium may be carried down in rain, but it appears to be a minor source of selenium for crops.

Cadmium deposited into rivers from industrial sources in Japan has been carried into irrigation waters used to grow rice. High concentrations of cadmium in the rice occurred, and the *itai-itai* disease syndrome was observed in people eating this rice (44). It is not established whether the cadmium was taken up by the roots of the plants or absorbed through the leaves and translocated to the seed.

Direct ingestion of soil by animals and people may at times contribute important amounts of trace elements to the body burden of some trace elements. Direct ingestion of street and house dusts is thought to be a major source of lead in the bodies of children residing in some cities. Grazing animals may ingest up to 2 kg of soil per day (33). This would constitute a superabundant source of silicon dioxide and a very important source of iron. Young pigs raised on concrete floors are subject to iron deficiency, while this problem is infrequent on animals raised outside. Direct ingestion of soil may also be important in the cobalt, zinc, and manganese nutrition of grazing animals.

Direct ingestion of soil may supply compounds that react with the copper in the pasture plants to decrease its availability to the grazing animal. Cases of copper deficiency have been recorded from animals grazing when these same forages did not create copper deficiency when they were harvested and used as hay. This difference has been attributed to the fact that iron compounds from the ingested soil reacted with the copper to make it unavailable to the grazing animal (78).

The process involved in movement of trace elements into the roots of plants and then translocated into the edible part is the major process controlling the trace element intake of animals and humans.

III.   FACTORS AFFECTING TRACE ELEMENT CONCENTRATIONS IN PLANTS

A.   Rock Formation, Weathering, and Transport

Trace elements are brought to the surface of the earth by upwelling and volcanic emissions of igneous rocks and by hydrothermal solutions. The trace element content of these materials varies, depending on their source in the deeper layers of the earth, the volatility of the different elements, the amount of oxygen present, and the rate at which the molten magma cools to form solids. On the surface of the earth these rocks are subjected to weathering processes that break them into smaller fragments, dissolve certain constituents, and transport materials loosened from the surface to other places. Some trace elements may be segregated during these weathering and transport processes, and their concentrations in the weathered and transported materials may vary markedly from their concentrations in freshly solidified igneous rock.

Materials fragmented or dissolved from igneous rocks may be washed or blown into bodies of water where they settle out or are precipitated. After long periods of time and pressure from overlying sediments they form sedimentary rocks. The formation of sedimentary rocks may involve the precipitation of a cementing agent such as Fe_2O_3 or $CaCO_3$ along with the settling out of solid particles of insoluble material. Sedimentary rocks may incorporate the remains of living organisms present in the environment where they are formed. Sedimentary rocks may be uplifted and folded or further compressed by tectonic processes and become exposed to weathering processes followed by additional cycles of deposition and the formation of younger sedimentary rocks. Alternately, soluble minerals such as $CaCO_3$ may leach away, leaving a deposit made up of the less soluble minerals coprecipitated or settled out concurrent with the formation of limestone. All of these processes can result in fractionation of the trace elements and their concentration in some places and their depletion in others.

Weathered-rock materials on the earth's surface are moved, often over very long distances, by wind, flowing water, and glaciers. During these movements there is frequent mixing of rock material from various sources and also segregation of materials varying in particle size, density, and solubility. The distribution of trace elements in rocks and their redistribution during rock weathering has been described by Hodgson (35), West (92), Beeson and Matrone (12), and Thornton (79).

It is evident that the distribution of trace elements on the surface of the earth is usually the product of several cycles of rock weathering, transport, and deposition. Even so, there are examples of a relationship of trace elements in specific rocks or materials derived from these rocks to trace element nutritional problems in livestock. The Cretaceous geological age, which ended about 63 million years ago, was evidently a period when large quantities of selenium were brought to the surface of the earth by volcanic action (16). A very high percentage of the recognized field cases of selenium toxicity in animals have been in areas where cretaceous sedimentary rocks, or materials weathered from these rocks, are the dominant soil-forming material. Plants growing on the floodplain of the lower Mississippi River in the United States usually contain about 10 times as much selenium as plants growing on nearby uplands (46,87). The selenium in the plants on the alluvial floodplains of the Mississippi is most likely due to transport of selenium by the Mississippi and Missouri rivers from cretaceous rocks in the Missouri watersheds of the northern Great Plains—a distance of over 1000 miles.

B.   Relationship of Trace Element Nutrition Problems to Different Kinds of Soil

The material weathered and transported from rocks is the parent material from which soils form. In wet areas the soil parent material may be dominated by organic residues of aquatic plants with just a few mineral particles, but in dry, warm sites the soil parent material is almost completely dominated by mineral particles. The kind of soil found in any location depends on the nature of the soil parent material, the climate of the site, the amount of time the material has been exposed to weathering, the vegetation that has dominated the site, the topography as it influences drainage or the amount of oxygen present, the movement of water and the rate at which material is removed from the surface by erosion, and the burrowing of insects and animals and the activities of humans. In view of the large number of combinations of these factors that can occur on the surface of the earth, it is not surprising that very many kinds of soil have been recognized and described by pedologists. Various systems of taxonomic classification of the many kinds of soil have been developed (76). In some countries, maps showing their distribution on farms or broader areas have been prepared and are available to investigators of trace element problems.

The weathering of rock and mineral particles during soil formation in sites where soluble weathering products are removed by downward leaching proceeds under a general pattern that has been described by Jackson and Sherman (39) and by Barshad (10). In this process the least stable and most soluble minerals disappear first, and after long periods of weathering only the most stable and insoluble materials remain. Many of these are secondary minerals formed from decomposition products of easily weathered primary materials. In this weathering process, bases such as sodium, potassium, calcium, and magnesium are normally removed in earlier stages with less basic materials next. Old, highly weathered soils of warm, humid areas tend to be primarily composed of oxides of silicon, aluminum, and iron, plus stable secondary minerals such as kaolinite. Where removal of products of decomposition is restricted by poor drainage, the soil formed may develop secondary minerals and characteristics that are very different from those found in soils developed from the same parent material on nearby well-drained sites.

During the weathering of rock and mineral particles to form soils, a variety of organic compounds derived from the decomposing remains of plants and microorganisms react with some of the trace elements present to increase their solubility and facilitate their relocation to other parts of the soil profile or removal by deep percolating waters. Elements taken up by plant roots are continuously recycled to the soil surface when plant residues are left, and the concentration of these elements in surface layers may be maintained while other elements not so recycled may be depleted.

As a result of these soil development processes, the trace elements in soils are predominantly converted to insoluble forms or stable complexes, or they may be leached out of the soil profile. In weathered soils on well-drained sites only a very small fraction of the trace elements present in the original soil parent material remains in soluble forms that are potentially available to plants. The chemistry of the trace elements in soils in relation to their availability to plants has been reviewed by Hodgson (35), Cataldo and Wildung (20), and in a monograph edited by Mortvedt et al. (60). A substantial effort has been directed toward identification of the chemical species of trace elements that are most important in plant nutrition. These chemical forms differ for the different elements, but for nearly all trace elements required by plants, their ionic forms in solution are generally present at extremely low concentrations. Labile adsorption complexes on organic or inorganic solid phases constitute the primary available sink for supplying plants with many of the trace elements. Soluble organic complexes or chelates may be of great importance in supplying plants with metals such as zinc, but formation of very stable complexes with organic matter is an important source of copper unavailability to plants (8).

Firm relationships between certain kinds of soil and trace element nutrition of grazing animals have been established. These relationships are usually for a specific kind of soil developed from a specific parent material, and are described for the British Isles by Thornton (79), and for the United States by Kubota and Allaway (45).

C. Plant Factors Affecting Trace Element Concentrations in Human and Animal Diets

1. Trace Element Requirements of Plants

Many of the soil-related trace mineral nutrition problems of humans and animals arise because the mineral requirements and tolerances of normally growing plants differ both quantitatively and qualitatively from those of humans and animals. Animals may not require boron but this is a very important trace element in plants and is widely used in fertilizers. High levels of available boron, in soils may be toxic to some plants, but boron toxicity to animals other than experimental ones has not been reported. Plants do not require iodine or selenium, and they may grow normally and produce optimum yields even though they contain insufficient levels of these elements to meet human and animal requirements. At the same time, certain plants grow normally even though the concentration of selenium in the plant is high enough to cause selenosis in animals. Sodium and chlorine are "major elements" in animal nutrition but are found only in trace amounts in plants other than halophytes. Sodium is essential only for certain plant species, and plant deficiencies of sodium and chlorine are uncommon. Although some plants have shown growth response to silicon, it is not currently considered as essential for plants. Ruminant animal nutritionists are concerned with the effect of high silicon concentrations in certain forage grasses contributing to low digestibility of those grasses (85).

Of the trace elements listed by Mertz (56) as essential to animals, only manganese, iron, copper, zinc, molybdenum, and possibly nickel are considered essential to all plants. Cobalt is essential for the nitrogen-fixing microorganisms in the nodules on legume roots. Among the trace elements essential for plants, certain plants grow normally and produce optimum yields even though they contain less zinc, iron, manganese, cobalt, and copper than required by domestic animals and poultry, and normally growing plants may contain excessive levels of copper and molybdenum for grazing animals.

In studies of the critical levels of trace elements required by plants for optimum growth and yield, agronomists and plant nutritionists usually analyze dry leaf tissue, because reactions taking place in the leaf frequently determine ultimate yield of seeds and fruits as well as of leaves. Human and animal nutritionists are often concerned with the trace element concentrations in seeds and fruits, and analyze these on the basis of water content as consumed or that of "grocery store" weight. This difference, in part of the plant studied and moisture content as analyzed, must be reconciled in attempts to relate data of plant nutrition to studies of human and animal nutrition.

2. Trace Element Uptake by Plant Roots

Trace elements move to the plant root by diffusion in response to concentration gradients brought about by depletion of the soluble forms of these elements at the root surface and by mass flow of the soil solution to the root in the transpiration stream. As these processes proceed, the root is normally lengthening and contacting new soil and shortening the pathways for diffusion and mass flow. The process of movement of trace elements to plant roots has been reviewed by Wilkinson (93). The processes of diffusion and mass flow vary in importance in different kinds of soil, with varying levels of soil water, and for different trace elements. As the concentration of certain trace elements in the soil solution is depleted by plant uptake, labile adsorbed or crystalline forms of these elements may dissolve into the soil solution.

The trace element concentration in plants is not an accurate reflection of the concentration of soluble and labile solid forms of these elements in the soil. Trace element uptake by plants is usually a selective, active transport process with different mechanisms involved for different elements (25,58,81). In some cases where trace elements are present in plants at very low concentrations, the accumulation by plants of these elements can be accounted for by passive transport of the element in the transpiration stream.

Mineral elements enter the plant root as ions. Where these elements contact the exterior of the root in nonionic complexed forms, the complex is apparently broken down in the exterior of the root and the element passes a selective barrier into the xylem in ionic form. The most widely accepted models of ion accumulation by roots include specific "carriers" that transport the ions across critical membranes. Iron is reduced to the divalent form during accumulation by plant roots, even though it may contact the root surface as trivalent iron. Some elements compete with each other in the root uptake process, that is, a high concentration of one element may depress the uptake of another. In other cases, the presence of a certain element in the soil solution may enhance the uptake of one or more other elements. The pH and redox potential of the soil or culture solution may also influence the rate of uptake of trace elements. The impact of these factors may be different for different elements. In nearly all cases, increasing the concentration of an element present originally at very low concentrations in the culture solution results in an increase in the rate of uptake of that element until some limiting concentration is reached above which there are only slight increases in ion uptake for increasing concentrations of the element in the external medium.

The current concepts of the mechanisms of ion accumulation by plant roots have been derived partly from very short-duration experiments in which excised plant roots have been exposed to solutions containing the various elements in soluble ionic forms. Very frequently radioisotopes of the different elements have been utilized to facilitate measurements of the rate of uptake from solution. There are some problems in translating the results of this type of experiment to the composition of intact plants growing for long periods in soil. For example, molybdenum is taken up more rapidly from acid than from alkaline culture solutions, but the molybdenum concentration in field-grown plants is normally higher from alkaline than from acid soils. Apparently the effect of alkaline soil conditions in increasing the concentration of soluble molybdenum in the soil offsets the reduction in rate of uptake of the dissolved molybdenum.

3.   Genetic Controls over Trace Element Concentrations in Plants

A number of trace element nutrition problems of animals are associated with differences in the inherent capability of different plant species to accumulate different elements. The problem of selenosis in range livestock in the western United States was due primarily to their consumption of selenium accumulator plants (73). The black gum tree (*Nyssa sylvatica*) takes up ~100 times as much cobalt as broomsedge growing on the same soil (11). Lists of accumulator plants for various plant species have been prepared by Bowen (14).

Even among common forage, feed, and food crops, differences among species may be an important factor in the distribution of trace element nutritional problems. Over broad areas of the United States the forage legumes generally contain adequate concentrations of cobalt to meet the requirements of grazing ruminants, while forage grasses usually are deficient in cobalt for these animals (45). Manganese deficiency in poultry is common when their diet consists primarily of corn (*Zea mays*), sorghum grain, or barley. These grains contain substantially lower concentrations of manganese than do wheat or oats.

The genetic basis for differences in mineral uptake by plants has been reviewed by Epstein and Jeffries (26). Even within a given plant species there may be nutritionally important differences in trace element accumulation by different strains or varieties. Thus, different strains of ryegrass growing on the same soil have been found to accumulate markedly different concentrations of iodine (41). Differential zinc accumulation by different inbred strains of corn (maize) has been described (54). Soybeans growing on the same soils have shown 10-fold differences in selenium concentration among different varieties (87). Genetic differences in the uptake of trace elements have been reviewed by Brown *et al.* (15) and Gerloff and Gabelman (27). Opportunities for improving cereals as sources of essential trace elements are described by Graham (30).

Plant breeding approaches to trace element problems have been primarily directed toward improving crop yield under conditions of deficient supplies of available nutrient elements and toward development of plants that will tolerate unfavorable conditions of soil acidity or excess aluminum and manganese. Genetic controls over the ability to accumulate iron from soils having low levels of available or soluble iron are expressed through differences in secretion by roots of reducing agents capable of converting ferric to ferrous iron.

Genetic studies have pointed to important controls over the accumulation of strontium by plants (69). This may point to opportunities to control the concentration of potentially toxic trace elements such as cadmium through extension of this type of work. At present no new varieties of food or feed crops have been developed for the sole purpose of increasing human or animal intake of essential trace elements.

4. Translocation Chemical Forms and Seasonal Changes of Trace Elements in Plants

After trace elements are taken up by the root they are transferred to the upper part of the plant in the xylem. Mass flow in the transpiration stream is probably responsible for a large part of the xylem transport for some trace elements, but there is upward movement of some of these elements in plants growing in humid atmosphere where losses of water from leaves are very limited. General aspects of mineral transport in plants are described by Lauchli (49), and movements of trace elements have been reviewed by Tiffin (80) and by Tinker (81).

Movements of trace elements within plants differ for the different trace elements and with differences in the external supply of the element. These differences may result in important seasonal differences in the nutritional quality of the plant, or of different parts of the plant, for humans and animals. In their movement upward in the xylem and into leaves or into the phloem and to the seed, the trace elements must pass selective barriers each time they move from one tissue to another.

In moving from the roots through vascular tissues and into leaves or storage tissues such as seeds and tubers, the ionic forms of trace elements that are taken up by roots usually change to organic complexed forms. Most of the iron in plants moves as citrate in the xylem, but monoferric phytate has been found in wheat seeds (59). Zinc and copper move in vascular tissues as anionic complexes, possibly with amino acids. Chromium from bean leaf supernatant migrated as an anionic complex similar to chromium citrate or chromium aconitate (37). Manganese moves as the Mn^{2+} cation. In many selenium accumulator plants selenium is converted to the nonprotein amino acid selenium methylselenocysteine, while in nonaccumulator plants selenium moves through vascular tissues as SeO_4^- or selenoamino acids, and much of the selenium in wheat seed occurs as protein-bound selenomethionine (64).

A decline with time in the concentration of certain trace elements in the leaves of forages may be the result of dilution of a limited supply of the trace element with continued growth of the vegetative tissues or from transfer to other tissues such as seeds. In humid areas a decline in trace element concentrations may also be due to leaching losses of certain elements from the leaves. Murray *et al.* (61) reported declines in zinc concentration as the season progressed in seven different range grasses in southern Idaho. Beef calves grazing these ranges during the late summer months showed weight gain responses to zinc supplementation. In five of the seven grasses, copper concentrations also declined as the season progressed. Iron and manganese concentrations in these same grasses did not show consistent trends in concentration.

Jones *et al.* (42) found opaline silica to be concentrated in the glumes and nodes of oats with less in the leaves and very low amounts in the seed. Silica appears to move passively in the transpiration stream and accumulate with time in those plant tissues where transpiration is greatest. Cadmium (55) in cadmium-polluted areas in Wales, and lead (57) in Scotland are higher in pasture grasses during the winter than during the spring and summer. In the case of cadmium this increase may be due to the breakdown of selective mechanisms that exclude cadmium from the tops as the plant becomes senescent. In the case of lead the higher winter concentrations may be due to continuing deposition of airborne lead without the diluting effect of continuing growth of the plant.

Translocation of trace elements to seeds is very important in the trace element nutrition of humans and monogastric animals, since seeds of cereals and pulses constitute the major plant food in their diets. The processes involved in accumulation of trace elements by seeds have not recieved the degree of study given to accumulation of trace elements by roots. In movement of trace elements from the soil to the seed, at least three selective barriers—from soil to root xylem, from xylem to phloem, and from phloem to seed—must be passed. In addition, seeds, protected as they are by the husks or pods, do not transpire as much water as do leafy tissues and therefore are not as subject to accumulation of elements such as silicon that may move passively in the transpiration stream, nor are they as likely to accumulate elements from airborne deposition.

The concentration of some trace elements, especially the metals, in seeds is normally less affected than that of leaves and husks by the external supply of these elements. This has probably been an important factor in the protection of seed consumers from potentially toxic elements such as lead and cadmium. For some of the other elements, important differences in seed concentrations of trace elements are associated with differences in the available levels of these elements in the soil.

The selenium concentration in seeds of wheat (51), corn (66), rice (13), and soybeans (87) shows a relationship to the level of available selenium in the soil. The probable importance of selenium in the hard red wheats grown in the Great Plains states of the United States and the Prairie Provinces of Canada has been pointed out (5). Wheat in this part of North America is primarily grown on neutral to alkaline soils developed on Tertiary or Pliestocene deposits that contain some selenium from older underlying selenized Cretaceous formations. Wheat from this broad region of North America is the dominant source of wheat in international trade and may have been very important in the prevention of human selenium deficiency in the wheat-importing countries.

A striking example of the effect of external supply of copper on the movement of copper in wheat is provided by Loneragan et al. (50). Copper did not move from the older leaves to younger leaves, and no grain was produced on a copper-deficient sand. Supplementation with copper in amounts adequate for optimum yield of grain promoted movement of copper to younger leaves and into grain. The grain on the copper-adequate treatments contained 1.2 μg/g copper. At a luxury level of copper supplementation the grain contained 5.3 μg/g copper. In both copper-adequate and luxury plants, copper concentrations were higher in grain than in straw, and copper moved from older to younger leaves of these treatments coincident with movement of organic nitrogen compounds.

Addition of zinc to culture solutions increased the zinc concentration in seeds of peas (80). The level of zinc added was probably in excess of the level needed for optimum plant growth and yield. Zinc in the pea seeds grown in the high zinc cultures was similar in bioavailability to rats to zinc in seeds grown on the low-zinc cultures. These two studies indicate that seed crops containing high levels of zinc and copper will be produced only where the level of plant-available forms of these elements in the soil is higher than necessary for optimum yield.

D. Effect of Agronomic Practices on Trace Element Concentrations in Crops

1. Soil Selection

Many of the uses of soil selection to avoid trace element deficiencies and toxicities have been accomplished before the problem was identified, and frequently areas of trace element deficiency or toxicity have been abandoned from a combination of causes, among which the trace element problem was not the primary reason for the change. Thus, dairying in the United States moved westward from New England partly because dairy cattle were unthrifty as a result of cobalt deficiency and partly because large tracts of more productive land became available upon settlement of lands further west. Similarly, large areas of seleniferous lands in South Dakota were deliberately removed from agricultural use during the 1930s, partly because of selenium toxicity in livestock but also because these lands were not sufficiently productive for the type of farming being used (2).

2. Trace Element Fertilization

Trace element fertilization has often been used to raise the yield of crops, with only secondary considerations given to increasing the trace element concentration in plants to meet the nutritional requirements of animals. Where the trace element is required by both plants and animals, a double benefit of increased crop growth and improved nutritional quality may sometimes be obtained.

Cobalt fertilizers have been used to improve the pasture production and to supply cobalt for grazing animals. Applications in the range of 100 to 600 g cobalt sulfate per acre, depending on the kind of soil, may maintain cobalt levels in plants above animal requirements for several years (1,65,68,70). Copper fertilizers have been used primarily to increase the yield of crops on copper-deficient soils. The use of copper fertilizers may have concurrent value in increasing the copper content of the crops to prevent copper deficiency in animals. Copper fertilizers or copper fungicides may cause high copper concentration in plants, and where the molybdenum concentration in the plant is low, copper toxicity may result. Todd has described a situation in northern Ireland where the use of copper fungicides for many years has resulted in a number of cereal crops that induce copper toxicity when fed to housed sheep in the winter (82). It is possible that this situation could be corrected by the use of molybdenum fertilizers to increase the molybdenum concentration in the crop. Most uses of molybdenum fertilizers have been to increase the yield of crops.

Application of selenite to selenium-deficient soil has resulted in alfalfa containing sufficient selenium to protect lambs from WMD (7). On medium-textured soils, 2.24 kg of SeO_3/Se per hectare has provided protective but nontoxic levels of selenium for 3 years (17). Different forms of selenium have very different availability to the plant and recovery from the soil. Selenites are normally tightly absorbed by most acid soils, and only ~2% of the applied selenite–selenium will be removed in the top of the plants in the first 2–3 years after application. Selenates are not tightly absorbed by soils, and the high percentage of applied selenate may be recovered in the plant the first year. Annual topdressings of pastures with selenate using as little as 10 g/ha selenium have been used successfully in New Zealand (86). Similar applications are being considered in Finland and in the Keshan disease area of China for the protection of both animals and people from selenium deficiency. Annual foliar sprays of 5 to 10 g selenium per hectare have provided protective levels of selenium in the barley grain (28) and corn silage (19).

Application of zinc to the soil or culture solution can provide seed crops that contain enhanced levels of biologically available zinc (48,88). However, the levels of zinc fertilization required are apparently in excess of those needed to provide for maximum yield of the crop itself. Therefore, any attempt to use zinc fertilization for the purpose of improving the amount of zinc in human or animal diets would need to be monitored so that the producer could be compensated for plants of high zinc level, since there would be no compensation in terms of increased yield for these crops. Use of zinc fertilizers may reduce the plant uptake of cadmium from soil, and the bioavailability of cadmium in the high-zinc plants is decreased (90). On the other hand, there seems to be little opportunity to use iron or chromium fertilization for the purpose of increasing intake of people or animals. The iron concentration of a plant is more determined by genetics and by ion balances than by the level of available iron in the soil or culture media (88,91). The addition of either trivalent or hexavalent chromium to the soil has not resulted in plants of notably enhanced chromium concentration (18).

3. Effects of Major Element Fertilizers

The use of major element fertilizers (those containing nitrogen, phosphorus, potassium, and sulfur) may have varied effects on the trace element concentration in plants. Where the trace element concentration in the surface soil is marginal for plant growth, the use of major element fertilizers may enhance root growth so that the roots penetrate lower soil horizons where supplies of some of the trace elements may be more abundant and, consequently, increase trace element concentration in the fertilized crop. On the other hand, a long continued application of major element fertilizers to soil that is marginally deficient in one or more of the trace elements, such as cobalt, copper, or zinc, may lead to enhanced removal of these limiting trace elements in the harvested crop and ultimately lead to nutritionally important declines in the concentration of these elements. Some of the possible consequences of continued use of major element fertilizers in Australia have been discussed by Rooney et al. (72).

Another aspect of the use of major element fertilizers is involved with the interactions between the ions in the major element fertilizer and trace elements. These are most common with the phosphate ion that is present in many major element fertilizers. Peck et al. (67) have described a field experiment in which the zinc concentration of peas and field beans was depressed when phosphorus fertilization was increased unless concurrent additions of zinc fertilizers were made. On the zinc-fertilized treatments, increasing phosphorus application increased the zinc concentration in the crops. Heavy application of sulfur may depress selenium uptake by plants. Field experiments indicate that this effect is not of practical importance on soils that are low or marginal in both sulfur and selenium (77).

Some of the trace elements are present as impurities in fertilizers. The addition of phosphorus fertilizers is usually accompanied by an addition of iron, cadmium, and, possibly, selenium that are present as impurities in the phosphate rock used to prepared the fertilizer. Studies by Williams and David (94) have indicated that the cadmium concentration in some Australian crops is higher where phosphorus fertilizers have been used for long periods of time. Selenium is present in phosphate rocks, but the addition of phosphorus fertilizers has not been shown to be the cause of plants containing sufficient selenium to be toxic to animals. At the same time, there may be instances where selenium concentrations in crops have been increased to the level of preventing selenium deficiency in animals because of the selenium impurity in the phosphate rock applied (71). Fluoride is the major impurity in phosphorus fertilizers, particularly phosphates of the single superphosphate strength. The level of fluoride in plants that have been extensively fertilized with single superphosphate may be higher immediately after superphosphate application, but soon decline to very low levels since the fluoride is tied up in unavailable forms in the soil.

4. Effects of Liming on Trace Element Concentrations in Crops

Lime ($CaCO_3$) is very frequently added to agricultural soils to permit growth of legumes and to provide calcium to the plant. The primary purpose of the addition of lime has been to raise the pH of the soil. Additions of lime may depress the uptake of zinc, copper, and cobalt, and increase the plant uptake of molybdenum and selenium. On some molybdenum-deficient soils, the use of $CaCO_3$ will increase the availability of soil molybdenum to the plant to the point of correcting molybdenum deficiency in the plant (4). There have been no recorded instances where increases in molybdenum concentration in plants, to the point of causing molybdenosis in animals, have occurred. A high pH in the soil favors the oxidation of reduced forms of selenium such as Se^{2-} and SeO_3 to the more soluble and more plant-available SeO_4 form. While the selenium concentrations in plants grown on some low-selenium soils are increased by liming, these increases have not been sufficient to provide crops that will protect animals from selenium deficiency (6). Liming may be especially useful in depressing the uptake of cadmium by plants from soils that have been polluted with cadmium.

5. Effects of Drainage on Trace Element Concentration in Plants

Normally, the drainage of agricultural soil permits plant roots to explore deeper soil horizons and to have access to trace elements that are not available in poorly drained soils. However, drainage may reduce the availability of molybdenum to plants. Nearly all of the soils that produce plants containing sufficient molybdenum to cause molybdenosis in animals are poorly drained. The drainage of these soils may cause oxidation of the iron oxides in the soils from ferrous to ferric, and ferric molybdate complexes are much less soluble than are the ferrous molybdenum complexes. Drainage to correct molybdenosis in grazing animals has not been used extensively because of its high cost in relation to the cost of copper supplementation of the animals at risk (4).

6. Crop Selection and Management

The selection of crops to avoid trace element deficiencies in animals is an old practice. The basis for this practice is explained in the earlier sections concerning genetic effects on the trace element uptake by plants. In large areas of the United States the grasses contain too little cobalt to meet the requirements of grazing ruminants; however, most of the legumes such as alfalfa and the clovers contain adequate levels of cobalt for this purpose. The addition of legumes to pasture and hay mixtures used by ruminant animals may, in these areas, be an adequate way of meeting the cobalt deficiency of grazing animals. The time of harvest and the use of the crop has been shown to have a marked effect on the availability of copper in the crop to cattle. Animals may be copper deficient when pasturing some swards of grass, but free of this deficiency when consuming dry hay cut from these same swards (32).

IV. DIAGNOSIS AND CORRECTION OF SOIL-RELATED TRACE ELEMENT DEFICIENCIES AND TOXICITIES

A. Diagnosis of Trace Element Problems in Animals

Acute trace element deficiencies in animals can normally be diagnosed by reference to specific signs and symptoms of the deficiency or toxicity. Comparison of the symptoms with those of established deficiencies or toxicities in other places, plus analysis of tissues such as blood, plasma, hair, liver, or muscle is fairly rapid under the conditions available in many countries today. The diagnosis of marginal states is much more difficult. For example, zinc deficiency may lead to inappetence and a failure to gain weight, although no visible symptoms are apparent until the zinc deficiency reaches critical proportions. Similarly, copper excess may accumulate to the point of a sudden hemolytic jaundice with no previous indications of toxicity symptoms. Selenium deficiency, on the other hand, is most often evidenced early by visible symptoms such as WMD, mulberry heart in animals, or other fatal disorders.

When a trace element deficiency is suspected, analysis of the feed of the animals may be very helpful in cases of stall- or pen-fed animals where the entire diet may be sampled and analyzed for the element in question. In the case of grazing animals that are confined in distinct pastures, where only one or two plant species and one or two kinds of soil are present in the pasture, analysis of the forage may be equally helpful. For animals grazing over extensive areas, such as range lands where a number of plant species and a number of different soil conditions are present in the area grazed by the animals over a period of time, diagnosis on the basis of analysis of the forage becomes very difficult. The samples collected for analysis may not represent the actual intake of the animals.

If the deficiency is to be related to specific kinds of soil and plant species, it may be necessary to establish unequivocally that a regional or areal problem of poor performance does indeed exist. In some cases, a suspected poor performance of animals in a certain area may be due to factors of contagious disease, parasites, genetic incapability to grow or reproduce, or a total deficiency in intake of all required nutrients. With respect to some of the newly discovered trace elements such as chromium and silicon, it seems unlikely that deficiencies of these elements in livestock will occur as field cases. The establishment of these trace elements as being essential was dependent on elaborate procedures for the purification of the diet and environment. It seems unlikely that under natural conditions, diets that contained extremely low concentrations of elements like silicon, chromium, and vanadium would occur. However, it should be kept in mind that assumptions of this type have frequently proved to be erroneous in the past.

B. Correction of Soil-Related Trace Element Deficiencies in Animals

Where a deficiency of a particular element in animal production is confirmed, the appropriate methods for the correction of the problem are very frequently determined by the techniques used in animal production. Where the animals are fed in stalls or pens, addition of the deficient element to the entire mixed diet is probably the most effective way of correcting the deficiency. Where the animals are pastured in small and well-defined pastures with just a few plant species and kinds of soil present, the element may be added to the soil if it is one that is readily taken up by plants. However, if the element is not readily taken up by plants, addition of the deficient element to the soil will be ineffective. An example of the importance of animal management techniques in the correction of a soil-related trace element deficiency is evidenced by selenium. In places where many of the animals are grazed on small pastures, topdressing of the pastures with selenium in conjunction with fertilization with major elements has been quite effective (86). Where animals forage over larger, and usually more variable and less productive rangelands, addition of selenium to the soil would be ineffective and excessively expensive. In these cases, addition of the selenium to mineral supplements, injection of the animals with selenium compounds, or the use of selenium-bearing rumen bolus pellets may be the most effective way of counteracting the problem. Where the element is required by both plants and animals and an addition of the deficient trace element will result in an increase in growth of the plant, as well as correcting the mineral deficiency in the animal, fertilization is obviously an appropriate method of dealing with the problem.

C. Diagnosis of Trace Element Deficiencies and Excesses in Humans

When the fourth edition of this book was prepared, Dr. Underwood considered the relation between the distribution of iodine in soils and plants and the distribution of endemic goiter to be the only confirmed relationship between trace elements in soils and plants and human health. When reports of the occurrence of Keshan disease in parts of China in relation to the selenium concentration in local food products became more available, it became evident that selenium deficiency and Keshan disease is a second confirmed relationship between the levels of a trace element in soils and plants and human health.

Selenium toxicity in people in China and in Venezuela may be a similar relationship between soils and human health. In these two cases, however, the use of specific diets due to general food shortages in the area cannot be completely eliminated (43,95).

In New Zealand the adjacent cities of Napier and Hastings have significant different caries incidence despite similar fluoride contents of the drinking water. The higher molybdenum content of the Napier vegetables was suggested as a likely factor in the lower prevalence of caries among children of that city (52). Direct evidence in support of this suggestion is not available.

Numerous other links between human health and soil have been proposed, but the implications of specific soils or trace elements are still uncertain. Most of these tenuous relationships between soil and human health are based on associations revealed by correlation of epidemiological data on human health with some factor in the soil or food supply of the area. There have been a number of studies relating inverse associations of selenium to the incidence of various types of cancer in the United States (23,40,74,75). It has been pointed out that the residents of the United States have approximately three times the blood selenium levels of residents of New Zealand (5), and yet, a comparable increased incidence of cancer in New Zealand in comparison to the United States has not appeared (84). A possible explanation for this anomaly may be provided by the work of Hoekstra (36), who has pointed out that the inhibitory effect of dietary selenium on the action of carcinogens in animal experiments compared adequate levels versus barely subtoxic levels instead of deficient levels as compared to adequate. A slower rate of cell division and a longer cell life in the high-selenium treatments in these animal experiments may have affected the incidence of cancer.

The difficulty in conducting experiments on humans is very possibly responsible for the difficulty in developing firm associations between soils and human health. Epidemiological studies in which the level of some trace elements in the soils and plants of an area are compared with disease incidence are the primary research technique useful in uncovering these relationships. Epidemiological studies are normally possible only with fatal diseases. Despite the disadvantages of epidemiological investigations and the method of association for uncovering relationships between trace elements in soils and plants and human health, it should be pointed out that the relationship between iodine in soils and plants and the occurrence of endemic goiter was brought to light by essentially the method of association.

In the industrialized countries, regional and international movements of food may obscure any effects of the soil of local areas on the incidence of human disease. It seems most probable that any such associations to come to light in the future will involve underdeveloped countries having a substantially, if not completely, local food supply and in addition, will involve a trace element that is deficient in nearly all of the soils of a relatively wide area (3).

The presence of livestock in an area can be very useful in the diagnosis of trace element problems in humans. Some kinds of livestock are normally fed substantially on plants produced within the area and are thus more likely to show evidence of trace element deficiencies and toxicities that are related to soils. Selenium deficiency as the cause of Keshan disease in China was suspected because the disease occurred in areas where pigs were known to be affected by mulberry heart, a known selenium problem (24).

Multielement surveys in which soils or rocks are sampled on a grid pattern and analyzed for a number of elements and the results compared to epidemiological data on human health, have not been effective in developing firm relationships between soils and human or animal health.

D. Correction of Soil-Related Trace Element Problems in Humans

The use of direct supplementation in the form of iodized salt has been the most practical method of correcting iodine deficiency in people. This has been simple, safe, and fairly widely adopted. In developing countries where salt may be available from local deposits that are low in iodine, it may be less practical to utilize iodized salt for the prevention of goiter.

Keshan disease in China has been substantially prevented by a direct supplementation program in which all of the population at risk receives selenium supplements on a regular basis in a well-controlled program. It is possible that selenium fertilization of the soils of the low-selenium areas can also be used to prevent human selenium deficiency. Because of the possibility of selenium toxicity, the use of this method requires skilled applicators and institutional controls that may not be available in the developing countries.

In developing countries, direct supplementation of the population at risk or food fortification with the deficient elements may be prevented by the lack of a well-organized food processing industry and public health services. In the developed countries food fortification can probably be used for the protection of the population from deficiency of zinc and possibly other elements.

Looking to the future, the methods that will be most useful in preventing soil-related trace element deficiencies in people are likely to be interregional movement of foods, food fortification, and direct supplementation of the population at risk. The prevention of soil-related mineral toxicities of people will probably be accomplished by avoiding the problem soils for food production.

REFERENCES

1. Adams, S. N. *et al.* (1969). *Aust. J. Soil Res.* **7**, 29.
2. Allaway, W. H. (1968). *Adv. Agron.* **20**, 235–274.
3. Allaway, W. H. (1972). *Ann. N.Y. Acad. Sci.* **199**, 17–25.
4. Allaway, W. H. (1976). *Molybdenium Environ.* **1**, 317–339.
5. Allaway, W. H. (1979). *In* "Trace Substances in Environmental Health-12" (D. D. Hemphill, ed.), pp. 3–10. Univ. of Missouri Press, Columbia.
6. Allaway, W. H., Cary, E. E., and Ehlig, C. F. (1967). *In* "Selenium in Biomedicine: A Symposium" (O. H. Mute, ed.), pp. 273–296. AVI Publ., Westport, Connecticut.
7. Allaway, W. H., Moore, D. P., Oldfield, J. E., and Muth, O. H. (1966). *J. Nutr.* **88**, 411–418.
8. Alloway, B. J., and Tills, A. R. (1984). *Outlook Agric.* **13**, 32–42.
9. Anderson, A. J. (1956). *Adv. Agron.* **8**, 163–202.
10. Barshad, I. (1964). *In* "Chemistry of the Soil" (F. E. Bear, ed.), pp. 1–52. Van Nostrand-Reinhold, Princeton, New Jersey.
11. Beeson, K. C., Lazar, V. A., and Boyce, S. G. (1955). *Ecology* **36**, 155–156.
12. Beeson, K. C., and Matrone, G. (1976). "The Soil Factor in Nutrition." Dekker, Basel.
13. Bieri, J. G., and Akmad, K. (1976). *J. Agric. Food Chem.* **24**, 1073–1074.
14. Bowen, H. J. M. (1966). "Trace Elements in Biochemistry." Academic Press, New York.
15. Brown, J. C., Ambler, J. R., Chaney, R. J., and Roy, C. D. (1972). *In* "Micronutrients in Agriculture" (J. J. Mortvedt, P. M. Giordano, and W. L. Lindsay, eds.), pp. 389–418. Soil Sci. Soc. Am., Madison, Wisconsin.
16. Byers, H. G., Miller, J. T., Williams, K. T., and Lakin, H. W. (1938). *U.S., Dep. Agric., Tech. Bull.* **601.**
17. Cary, E. E., and Allaway, W. H. (1973). *Agron J.* **65**, 922–925.
18. Cary, E. E., Allaway, W. H., and Olson, O. E. (1977). *J. Agric. Food Chem.* **25**, 300–309.
19. Cary, E. E., and Rutzke, M. (1981). *Agron. J.* **73**, 1083–1085.
20. Cataldo, D. A., and Wildung, R. E. (1983). *Sci. Total Environ.* **28**, 159–168.

21. Chappell, W. R. *et al.*, eds. (1976). Dekker, New York.
22. Chatin, G. (1851). *C. R. Hebd. Seances Acad. Sci.* **33**, 529–531.
23. Clark, L. C., Graham, G. F., Crounse, R. G., Grimson, R., Hulba, B., and Shy, C. M. (1984). *Nutr. Cancer* **6**, 13–32.
24. Diplock, A. T. (1981). *Philos. Trans. R. Soc. London, Ser. B* **294**, 105–117.
25. Epstein, E. (1972). "Mineral Nutrition of Plants, Principles and Perspectives." Wiley, New York.
26. Epstein, E., and Jeffries, R. L. (1964). *Annu. Rev. Plant. Phys.* **15**, 169–184.
27. Gerloff, G. C., and Gabelman, W. H. (1983). *In* "Encyclopedia of Plant Physiology, New Series (A. Pirson and M. H. Zimmerman, eds.), Vol. 15B, pp. 453–480. Springer-Verlag, Berlin and New York.
28. Gissell-Nielson, G. (1981). *Commun. Soil Sci. Plant Anal.* **12**(6), 631–642.
29. Goldschmitt, V. M. (1954). "Geochemistry." Oxford Univ. Press (Clarendon), London and New York.
30. Graham, R. D. (1985). *Adv. Plant Nutr.* **1** (in press).
31. Gutenmann, W. H., Bach, C. A., Youngs, W. D., and Lisk, D. J. (1976). *Science* **191**, 966–967.
32. Hartmans, J., and Bosman, M. S. N. (1970). *In* "Trace Element Metabolism in Animals" (C. F. Mills, ed.), p. 362. Livingstone, Edinburgh and London.
33. Healy, W. B. (1973). *In* "Chemistry and Biochemistry of Herbage" (G. W. Butler and R. W. Bailey, eds.), Vol. 1, pp. 567–588. Academic Press, New York.
34. Hemphill, D. D., ed. (1964). "Trace Substances in Environmental Health," Vol. I. Univ. of Missouri Press, Columbia.
35. Hodgson, J. F. (1963). *Adv. Agron.* **15**, 119–159.
36. Hoekstra, W. G. (1984). *In* "Trace Element Metabolism in Man and Animals-5 Abstract," p. 57. Aberdeen, Scotland.
37. Huffman, E. W. D., and Allaway, W. H. (1973). *J. Agric. Food Chem.* **21**, 982–986.
38. Hutchinson, T. C., ed. (1977). "International Conference on Heavy Metals in the Environment." University of Toronto, Toronto, Ontario, Canada.
39. Jackson, M. L., and Sherman, D. (1953). *Adv. Agron.* **5**, 219–318.
40. Jansson, B., Seibert, B. G., and Speer, J. F. (1975). *Cancer* **36**, 2373–2384.
41. Johnson, J. M., and Butler, G. W. (1957). *Physiol. Plant.* **10**, 100.
42. Jones, L. N. P., Wilne, A. A., and Wadham, S. M. (1963). *Plant Soil* **18**, 358–371.
43. Kerdel-Vargas, V. (1966). *Econ. Bot.* **20**, 187–195.
44. Kobayachi, J. (1971). *Proc. Int. Water Pollut. Res. Conf., 5th, 1970*, pp. 1–7.
45. Kubota, J., and Allaway, W. H. (1972). *In* "Micronutrients in Agriculture" (J. J. Mortvedt, P. M. Giordano, and W. L. Lindsay, eds.), pp. 525–554. Soil Sci. Soc. Am., Madison, Wisconsin.
46. Kubota, J., Allaway, W. H., Carter, D. L., Cary E. E., and Lazar, V. A. (1967). *J. Agric. Food. Chem.* **15**, 448–453.
47. Kubota, J., Lazar, V. A., Langan L. N., and Beeson, K. C. (1961). *Soil Sci. Soc. Am. Proc.* **25**, 227–232.
48. Lantzsch, H. J., Scheuermann, S. E., and Marschner, H. (1981). *In* "Trace Element Metabolism in Man and Animals-4" (J. C. McHowell, J. M. Gawthorne, and C. L. White, eds.), pp. 107–110. Aust. Acad. Sci., Canberra.
49. Lauchli, A. (1979). *Prog. Bot.* **41**, 44–54.
50. Loneragan, J. F., Snowball, K., and Robson, A. D. (1980). *Ann. Bot. (London)* [N.S.] **45**, 621–632.
51. Lorenz, K. (1981). *In* "Selenium in Biomedicine" (J. E. Spallholz *et al.*, eds.), pp. 449–453. AVI Publ. Co., Westport, Connecticut.

52. Ludwig, T., and Healy, W. B. (1960). *Nature (London)* **186,** 195.
53. Marston, H. R. (1935). *J. Counc. Sci. Ind. Res. (Aust).* **8,** 111.
54. Massey, H. F., and Loeffel, F. A. (1967). *Agron. J.* **59,** 214–217.
55. Matthews, H., and Thornton, I. (1982). *Plant Soil* **66,** 181–193.
56. Mertz, W. (1981). *Science* **213,** 1332–1338.
57. Mitchell, R. L., and Reith, J. W. S. (1966). *J. Sci. Food Agric.* **17,** 437–440.
58. Moore, D. P. (1972). *In* "Micronutrients in Agriculture" (J. J. Mortvedt, P. M. Giordano, and W. L. Lindsay, eds.), pp. 171–198. Soil Sci. Soc. Am., Madison, Wisconsin.
59. Morris, E. R., and Ellis, R. (1976). *J. Nutr.* **106,** 753–760.
60. Mortvedt, J. J., Giordano, P. M., and Lindsay, W. L., eds. (1972). "Micronutrients in Agriculture." Soil Sci. Soc. Am., Madison, Wisconsin.
61. Murray, R. B., Wayland, H. J., and van Soest, P. J. (1978). *USDA For. Serv. Res. Pap.* **INT-199.**
62. National Academy of Sciences (1971). "Medical and Biological Effects of Environmental Pollutants (A Series). "Nat. Acad. Sci., Washington, D.C.
63. Nriagu, J. O., ed. (1984). "Changing Metal Cycles and Human Health." Springer-Verlag, Berlin and New York.
64. Olson, O. E., Novacek, E. I., Whitehead, E. I., and Palmer, I. S. (1970). *Phytochemistry* **9,** 1181–1188.
65. Ozanne, P. G., Greenwood, E. A., and Shaw, T. C. (1963). *Aust. J. Agric. Res.* **14,** 39.
66. Patrias, C. and Olson, O. E. (1969). *Feedstuffs* **41**(43), 32.
67. Peck, N. H., Grunes, D. L., Welch, R. M., and MacDonald, G. E. (1980). *Agron. J.* **72,** 528–534.
68. Powrie, J. K. (1960). *Aust. J. Agric. Sci.* **23,** 198.
69. Rasmusson, D. C., Smith, L. H., and Kleese, R. A. (1964). *Crop Sci.* **4,** 566–589.
70. Reisenhauer, H. M. (1960). *Nature (London)* **186,** 375.
71. Robbins, C. W., and Carter, D. L. (1970). *Soil Sci. Soc. Am. Proc.* **34,** 506–509.
72. Rooney, D. R., Wren, N. C., and Leaver, D. D. (1977). *Aust. Vet. J.* **53,** 9–16.
73. Rosenfeld, I., and Beath, O. A. (1964). "Selenium," 2nd ed. Academic Press, New York.
74. Schrauzer, G. N. (1978). *Adv. Exp. Med. Biol.* **91,** 323–344.
75. Shamberger, R. J., and Willis, C. E. (1971). *CRC Crit. Rev. Clin. Lab. Sci.* **2,** 211–221.
76. Soil Survey Staff (1975). "Soil Taxonomy," USDA-SCS Handb. No. 436. U.S. Dept. Agric., Washington, D.C.
77. Spencer, K. (1982). *Aust. J. Exp. Agric. Anim. Husb.* **22** (118–119), 420–427.
78. Suttle, N. F., Alloway, B. J., and Thornton, I. (1975). *J. Agric. Sci.* **84,** 249–254.
79. Thornton, I., ed. (1983). "Applied Environmental Geochemistry." Academic Press, London.
80. Tiffin, L. O. (1972). *In* "Micronutrients in Agriculture" (J. J. Mortvedt, P. M. Giordano, and W. L. Lindsay, eds.), pp. 199–230. Soil Sci. Soc. Am., Madison Wisconsin.
81. Tinker, P. B. (1981). *Philos. Trans. R. Soc. London, Ser. B* **294,** 41–55.
82. Todd, J. R. (1976). *Molybdenum Environ.* **1,** 33–49.
83. Underwood, E. J., and Filmer, J. F. (1935). *Aust. Vet. J.* **11,** 84.
84. Van Rij, A. M., Robinson, M. F., Godfrey, P. J., Thornson, C. D., and Rhea, H. M. (1979). *In* "Trace Substances in Environmental Health-12" (D. D. Hemphill, ed.), pp. 157–163. Univ. of Missouri Press, Columbia.
85. Van Soest, P. J. (1982). "Nutritional Ecology of the Ruminant," pp. 72–74. O&B Books, Inc., Corvallis, Oregon.
86. Watkinson, J. H. (1984). *Abst. Int. Symp. Selenium Biol. Med., 3rd, 1984,* p. 61.
87. Wauchope, R. D. (1978). *J. Agric. Food Chem.* **26,** 226–228.
88. Wein, E. M., van Campen, D. R., and Rivers, J. M. (1975). *J. Nutr.* **105,** 459–466.
89. Welch, R. M., House, W. A., and Allaway, W. H. (1974). *J. Nutr.* **104,** 733–740.

90. Welch, R. M., House, W. A., and van Campen, D. R. (1978). *Nutr. Rep. Int.* **17,** 35–42.
91. Welch, R. M., and van Campen, D. R. (1975). *J. Nutr.* **105,** 253–256.
92. West, T. S. (1981). *Philos. Trans. R. Soc. London, Ser. B* **294,** 19–39.
93. Wilkinson, H. F. (1972). *In* ''Micronutrients in Agriculture'' (J. J. Mortvedt, P. M. Giordano, and W. L. Lindsay, eds.), pp. 139–169. Soil Sci. Soc. Am., Madison, Wisconsin.
94. Williams, C. A., and David, D. J. (1976). *Soil Sci.* **121,** 86–93.
95. Yang, G., Wang, S., Zhou, R., and Sun, S. (1983). *Am. J. Clin. Nutr.* **37,** 872–881.
96. Zhu, L. (1981). *In*''Trace Element Metabolism in Man and Animals-4.'' (J. C. McHowell, J. M. Gawthorne, and C. L. White, eds.), pp. 514–517. Aust. Acad. Sci., Canberra.

Index